ENVIRONMENTAL FLUID MECHANICS

Theories and Applications

EDITED BY
Hayley H. Shen
Alexander H.-D. Cheng
Keh-Han Wang
Michelle H. Teng
Clark C.K. Liu

ASCE American Society of Civil Engineers
1801 ALEXANDER BELL DRIVE
RESTON, VIRGINIA 20191–4400

Abstract: Fluid mechanics has evolved for over two thousand years since Archimedes time. Utilization of fluid mechanical laws in hydraulic systems saw its greatest growth during the last three centuries. Applying fluid mechanics to treat environmental problems has become increasingly important since only three decades ago. To extract the most relevant materials from the entire field of fluid mechanics in training environmental engineers is a challenging task. Under the request of the American Society for Civil Engineers, the Engineering Mechanics Division – Fluids Committee undertook the responsibility to edit this book. The aim of this book is to introduce recent developments of fluid mechanics theories that have been applied to environmental problems. This book begins with chapters more inclined towards principles of fluid mechanics, followed by contemporary applications. Problems covering river, lake, coastal, and ground water areas are presented. The materials presented can serve as a basis to construct an advanced undergraduate or introductory graduate level course. This book is also intended to provide environmental engineering practitioners with some recent developments in fluid mechanics applicable to environmental problems. This book project was approved by the Executive Committee of the Engineering Mechanics Division, American Society for Civil Engineers on January 1999.

Library of Congress Cataloging-in-Publication Data

Environmental fluid mechanics : theories and applications / edited by Hayley H. Shen ...
[et al.].

p. cm.

Includes bibliographical references and index.

ISBN 0-7844-0629-4

1. Fluid mechanics. 2. Environmental engineering. I. Shen, Hayley H.

TA357 .E58 2002
628.1'68--dc21 2002074679

ISBN 0-7844-0629-4
Manufactured in the United States of America.

Acknowledgments

Chapter 1
The Editorial Board

Hayley Shen
Professor
Dept. Civil and Environmental Engineering
Clarkson University
Potsdam, NY 13699-5710
hhshen@clarkson.edu

Alex Cheng
Professor and Chair
Department of Civil Engineering
University of Mississippi
P.O. Box 1848
University, MS 38677-1848
acheng@olemiss.edu

Keh-Han Wang
Associate Professor
Dept. Civil and Environmental Engineering
University of Houston
Houston, TX 77204-4791
kwang@central.uh.edu

Michelle H. Teng
Associate Professor
Department of Civil Engineering
University of Hawaii at Manoa
2540 Dole Street, Holmes Hall 383
Honolulu, HI 96822
teng@wiliki.eng.hawaii.edu

Clark Liu
Professor
Department of Civil Engineering
University of Hawaii at Manoa
2540 Dole Street, Holmes Hall 383
Honolulu, HI 96822
liu@wiliki.eng.hawaii.edu

Chapter 2

Philip J.W. Roberts
Professor
School of Civil and Environmental Engineering
Georgia Institute of Technology
Atlanta, Georgia 30332-0355
proberts@ce.gatech.edu

Donald R. Webster
Assistant Professor
School of Civil and Environmental Engineering
Georgia Institute of Technology
Atlanta, Georgia 30332-0355
dwebster@ce.gatech.edu

Chapter 3

Joseph H.W. Lee
Professor
Dept. of Civil Engineering
The University of Hong Kong
Hong Kong, China
hreclhw@hkucc.hku.hk

G.Q. Chen
Professor
Centre for Environmental Sciences and Dept. of Mechanics
Peking University
Beijing, China
chen@ns.nlspku.ac.cn

C.P. Kuang
Senior Engineer
Department of River & Harbour Engineering
Nanjing Hydraulic Research
ckuang@hkucc.hku.hk

Chapter 4

Scott A. Socolofsky
Res. Assoc.
Inst. for Hydromech.
Univ. of Karlsruhe
76128 Karlsruhe, Germany
socolofs@alum.mit.edu

Brian C. Crounse
Engineer
Carlisle & Company
30 Monument Sq.
Concord MA 01742, USA
brian.crounse.94@alum.dartmouth.org

E. Eric Adams
Sr. Res. Engr. and Lecturer
Dept. of Civ. & Envirn. Engrg.
Mass. Inst. of Tech.
Rm. 48-325
Cambridge, MA 02139, USA
eeadams@mit.edu

Chapter 5

Vincent H. Chu
Professor
Dept. of Civil Engineering and Applied Mechanics
McGill University
Montreal, Canada H3A 2K6
vincent.chu@mcgill.ca

Chapter 6

Keh-Han Wang
Associate Professor
Department of Civil and Environmental Engineering
University of Houston
Houston, TX 77204-4791
khwang@uh.edu

Chapter 7

Gour-Tsyh Yeh
Professor
Department of Civil and Environmental Engineering
Penn State University
gty2@psu.edu

Presently:
Provost Professor
Department of Civil and Environmental Engineering
University of Central Florida
Orlando, FL 32816-2450
gyeh@mail.ucf.edu

Ming-Hsi Li
Post-doctor Researcher
Division of Flood Mitigation
National Science & Technology Program for Hazards Mitigation.

Presently:
Assistant Professor
Institute of Hydrology
National Central University
Chunli 32054
Taiwan
mli@ns.naphm,ntu.edu

Malcolm D. Siegel
Principal Member Technical Staff
P.O. Box 5800
Sandia National Laboratories
Albuquerque, NM 87185
msiegel@sandia.gov

Chapter 8

Chin-Tsau Hsu
Department of Mechanical Engineering
Hong Kong University of Sciences and Technology
Clear Water Bay
Kowloon, Hong Kong
mecthsu@ust.hk

Chapter 9

Ne-Zheng Sun
Professor
Department of Civil and Environmental Engineering
UCLA
Los Angeles, CA 90095
nezheng@ucla.edu

Alexander Yishan Sun
Environmental Engineer
R&D Division
Tetra Tech, Inc.
Lafayette, CA 94549
aysun45@yahoo.com

Chapter 10

Der-Liang Frank Young
Professor
Department of Civil Engineering and Hydrotech Research Institute
National Taiwan University
Taipei, Taiwan-10617
Republic of China
dlyoung@hy.ntu.edu.tw

Chapter 11

Jan-Tai Kuo
Professor, Department of Civil Engineering
Director of Hydrotech Research Institute
National Taiwan University
1 Sec. 4, Roosevelt Rd.
Taipei, Taiwan
kuoj@ccms.ntu.edu.tw

Ming-Der Yang
Assistant Professor
Department of Civil Engineering
National Chung-Hsing University
250 Kuo-Kuang Rd.
Taichung 402, Taiwan
mdyang@dragon.nchu.edu.tw

Chapter 12

Clark C.K. Liu
Professor
Department of Civil Engineering
University of Hawaii at Manoa
Honolulu, HI 96822 USA
liu@wiliki.eng.hawaii.edu.

Jenny Jing Neill
Engineer
M&E Pacific, Inc.
1001 Bishop Street
Honolulu, HI 96813 USA
jenny_li@air-water.com

Contents

Contents

Chapter 1

INTRODUCTION

The Editorial Board

By environmental fluid mechanics we refer to the study of principles of fluid mechanics and their application for the identification, investigation, and solution of environmental problems. Environmental fluid mechanics is as essential for environmental engineers as hydraulics is for sanitary engineers.

Sanitary engineering deals with municipal water supply and wastewater disposal. A city obtains its freshwater supply from surface or groundwater sources. After being used, freshwater becomes wastewater, which is collected by a sewerage system and conveyed to a treatment plant. The wastewater is treated and then discharged into the receiving water body, often a river. Hydraulics, a branch of fluid mechanics, was developed for the study of water transportation and distribution (Rouse and Ince, 1957). It focuses on flow in open channels and closed conduits. Groundwater hydraulics, on the other hand, was developed for the investigation of groundwater supply.

Environmental engineering, now an integral part of civil engineering, has evolved from traditional sanitary engineering. To face the modern-day environmental challenges, environmental engineering now includes air pollution control, solid waste management, and these water-related subjects: (1) waste assimilative capacity analysis of receiving waters, (2) remediation of contaminated soil and groundwater, and (3) evaluation of environmental problems of global scale, such as global warming and acid rain. Engineering design and analysis for these subjects requires a good understanding of the fate of pollutants in the environment.

The fate of pollutants in a natural or man-made environment is determined by the joint actions and interactions of process kinetics and transport mechanics. Over the years, the study of the fate of pollutant in environmental systems has focused on the process kinetics using simplified transport mechanics. Following this approach, an environmental system can be designed conceptually as an ideal reactor, either completely stirred tank reactor (CSTR) or plug flow reactor (PFR). For a completely stirred tank reactor, which is the simplest system that can be used to simulate transport mechanics of an environmental system, the chemical concentration of the inflow mixes with the liquid inside the tank immediately and completely. In an ideal plug flow reactor, the pollutant is instantaneously mixed in the lateral and vertical directions and the longitudinal movement of the pollutants is due only to advective transport. This approach has been successfully applied to the design and analysis of water and wastewater treatments, or man-made environmental systems.

Mathematical model with detailed process kinetics and simplified transport mechanics has also been applied to water quality analysis of natural systems. A noted example is the Streeter and Phelps (1925) model of dissolved oxygen in a receiving river. By making a number of simplifying assumptions in transport mechanics and reaction kinetics, Streeter and Phelps developed a utilitarian approach to river water quality analysis—they assumed that the environmental system of a receiving river could be represented by a steady-state ideal plug flow reactor (PFR).

Water quality analysis with simplified transport mechanics has become inadequate after the enactment of the 1972 Federal Water Pollution Control Act Amendment (now called the Clean Water Act). This act established minimum effluent limits – secondary treatment for municipal discharges and best available treatment for industrial discharges. Sometimes even more stringent effluent limits were necessary to meet special water quality standards of the receiving water. As the level of wastewater treatment gets higher so does the cost of treatment. An intelligent resources allocation for environmental protection demands a more accurate water quality analysis. And, an accurate water quality analysis requires a better understanding of the fate of pollutant, especially the effects of transport mechanics.

The Clean Water Act has been revised at the frequency of about every five years since 1972. As a part of its first revision in 1977, US Congress created a waiver of secondary treatment, through which coastal communities might be allowed to discharge effluent of less than secondary quality into ocean if it would not adversely affect the marine water environment. Evaluation of initial dilution of effluent from ocean outfall requires a detailed analysis of the turbulent mixing of buoyant jet, a subject beyond the traditional hydraulics.

Environmental fluid mechanics has evolved in response to the need for a more accurate analysis of environmental mixing in inland and coastal waters, groundwater, as well as the atmosphere. The study of turbulent mixing and longitudinal dispersion started with pioneering works of G. I. Taylor (Taylor, 1953). At present, most river and estuary water quality analysis are conducted following Taylor's original concept (Csanady, 1973; Fischer et al., 1979). The scope of environmental fluid mechanics has further expanded as other fluid mechanics principles are being applied to the study of the fate of pollutants in the environment. Density stratified flow was studied and found many applications in water and air (Yih, 1980). Buoyant jets and plumes were investigated for smoke emission and diffuser discharge of wastewater from ocean outfalls (Turner, 1973).

This book of Environmental Fluid Mechanics was prepared by following its historical development and by emphasizing emerging subjects. The first part of this book, from Chapter 2 to Chapter 5, gives the theories of environmental fluid mechanics. The second part, from Chapter 6 to Chapter 12, gives the application of environmental fluid mechanics in water quality analysis of rivers, estuaries, lakes and reservoirs, and groundwater.

The goal of this book is to provide instructors a reference for a course in Environmental Fluid Mechanics. With the background materials and examples given in this book, instructors can select topics of special interest, supplement them with more fundamental details and examples, to construct an advanced undergraduate or introductory graduate level course. This book is also intended to provide environmental engineering practitioners with some contemporary fluid mechanics applicable to environmental problems.

The book is organized as follows.

Chapter 2 reviews the essential parts of turbulence theory. After discussing various length and time scales, the diffusion mechanism is introduced. Relative magnitude of longitudinal, transverse and depthwise diffusion are used to simplify the species transport equation. Applications of the diffusion equation to rivers, estuaries and coastal waters are given. Interesting examples of possible plume tracking scheme by aquatic animals are discussed.

Chapter 3 describes mixing of turbulent jets and plumes in a current. This is a common type of flow encountered in near-field mixing of wastewater effluent from an ocean outfall effluent and in many water treatment situations. The problem is illustrated by a detailed study of a jet in crossflow. A 3D numerical solution of an advected line puff is presented and a practical application example is given.

Chapter 4 introduces new theoretical analysis and experimental results on multi-phase plumes in stratified and flowing environments. The theoretical analysis includes re-examination of the governing dimensionless parameters and presentation of a double plume model, incorporating the effects of plume peeling and bubble dissolution. These new analyses were verified through experiments. The results are applicable to studying behavior of multi-phase plumes in the deep ocean such as carbon sequestration and the fate if oil released from an oil-well blowout.

Chapter 5 investigates the transport process in domains with large horizontal extent and relatively small vertical dimensions. Systematic analysis of different length scales is presented. The respective roles of the horizontal and vertical turbulence on the transport process are considered. Large eddy simulation and Reynolds averaged Navier Stokes simulation models are developed. The results of a series of numerical simulations conducted using these models are verified with laboratory and field data.

Chapter 6 presents a comprehensive overview of the development of a three-dimensional hydrodynamic and salinity transport model for applications to the study of the dynamics of an estuary. Both the mathematical model and the numerical algorithm for solving these equations are described in detail. Case studies of Chesapeake Bay and Galveston Bay by applying the comprehensive model were carried out to evaluate the model performance.

Chapter 7 studies fluid flows and reactive chemical transport in variably saturated porous media. The fundamental physical, chemical, and biological processes that control the movement of fluids, the migration of chemicals, and their interactions with the media are formulated. A numerical model is developed and validated. Six examples are employed to illustrate the applications of the model.

Chapter 8 presents the mathematical model of flow and heat and mass transfer through porous media. The macroscopic transport equations of momentum, energy and mass are obtained by averaging the microscopic equations over a representative elementary volume, which lead to the closure problem requiring information about dispersion, interfacial tortuosity and interfacial transfer. The closure relations are constructed from earlier knowledge and new analyses based on experimental and numerical evidences.

Chapter 9 addresses the general problem of parameter identification in all environmental system models. A variety of methods for determining unknown parameter such as the hydraulic conductivity, dispersion coefficients, sorption and release coefficients, and chemical reaction coefficients are discussed.

Chapter 10 provides the formulation of laterally averaged two-dimensional stratified lake modeling with free surfaces. As an example of how to apply this model, a finite element analysis is adopted to study the Te-Chi reservoir located at central Taiwan.

Chapter 11 reviews the commonly used models for reservoir and lake water quality management. Those models range from simplified zero-dimensional to two- and three-dimensional models. Case studies for Feitsui Reservoir using simplified models and a two-dimensional model, CE-QUAL-W2, are demonstrated. In addition, model calibration is also shown in the examples using remote sensing data. A detailed example integrating GIS software and modeling is given.

Chapter 12 provides an alternative approach to river water quality modeling by applying the linear systems theory. Dissolved oxygen variations in receiving rivers under either steady or unsteady condition were simulated by using this alternative modeling approach. The results are useful for predicting and managing water quality in a river environment.

Recently, ASCE has put out a manuscript related to Environmental Fluid Mechanics: *Waterbody Hydrodynamic and Water Quality Modeling: An Introductory Workbook and CD-ROM on Three-Dimensional Waterbody Modeling*, edited by John Eric Edinger, ASCE Press, (2001), ISBN: 0-7844-0550-6. Textbooks designed to address environmental fluid mechanics have also begun to appear. For instance: *Environmental Fluid Mechanics*, edited by Hillel Rubin and Joseph Atkinson, Marcel Dekker, Inc. (2001), ISBN: 0-8247-8781-1.

References

Csanady, G.T., *Turbulent Diffusion in Environment*, Reidel Pub. Co., 1973.

Fischer, H.B., List, E.J., Koh, R.C.Y., Imberger, J. and Brooks, N.H., *Mixing in Inland and Coastal Waters*, Academic Press, 1979.

Rouse, H. and Ince, S., History of Hydraulics, Iowa Institute of Hydraulic Research, Iowa City, 1957.

Streeter, H.W. and Phelps, E.B., *A Study of the Pollution and Natural Purification of the Ohio River, III. Factors Concerned in the Phenomena of Oxidation and Reaeration*, U.S. Public Health Service Publication, No. 149, 1925.

Taylor, G.I., "Dispersion of soluble matter in solvent flowing slowly through a Tube," *Proc. Roy. Soc., Ser. A*, **219**, 186-203, 1953.

Turner, J.S., *Buoyancy Effects in Fluids*, Cambridge Univ. Press, 1973.

Yih, C. -S., *Stratified Flows*, Academic Press, 1980.

Chapter 2

TURBULENT DIFFUSION

Philip J. W. Roberts[1] and Donald R. Webster[2]

ABSTRACT

Almost all flows encountered by the engineer in the natural or built environment are turbulent, resulting in rapid mixing of contaminants introduced into them. Despite many years of intensive research into turbulent diffusion, however, our ability to predict mean contaminant distributions is often quite crude and to predict statistical variations of concentration fluctuations even cruder. This chapter reviews basic ideas of turbulence and the mechanisms whereby scalar quantities, such as contaminants, are mixed. The evolution equations for scalar quantities are derived, the engineering assumptions used to make them tractable are discussed, and typical solutions are presented. Applications to various situations of engineering interest are given, including diffusion in rivers, estuaries, and coastal waters. The complexities of the diffusion process are demonstrated by the use of new optical experimental techniques. New modeling techniques are discussed, and research questions are posed.

1. INTRODUCTION

Turbulent diffusion is very efficient in rapidly decreasing the concentrations of contaminants that are released into the natural environment. Despite intensive research over many years, however, only crude predictions of these concentrations can be made. Most mathematical models of turbulent diffusion, particularly engineering models, predict only time-averaged concentrations. While this may often be sufficient, for example, water quality standards are usually written in terms of time-averaged values, more information on the statistical variation of concentration fluctuations may sometimes be needed. This could include prediction of the peak exposures of humans to air pollutants, of aquatic organisms to water contaminants, of the probability of combustion of flammable gases accidentally released into the atmosphere, or of the information available to an organism attempting to navigate through a

[1] Professor, School of Civil and Environmental Engineering, Georgia Institute of Technology, Atlanta, Georgia 30332-0355, proberts@ce.gatech.edu

[2] Assistant Professor, *ibid*, dwebster@ce.gatech.edu

turbulent chemical odor plume to its source. These and similar topics are becoming increasingly important in turbulent diffusion research.

Consider the photograph of a passive tracer released into a turbulent flow shown in Figure 1. This photograph was obtained in a study of chemical odor plumes. Even a cursory inspection of this image shows that the time-average tracer concentration at any point would be a very poor measure of the contaminant signal there. This signal consists of a small mean value with intermittent fluctuations that range from zero to levels that are orders of magnitude higher than the mean. Present mathematical models usually do not predict higher order measures of these signals such as their intermittency, peak values, probability density functions, and spatial correlations.

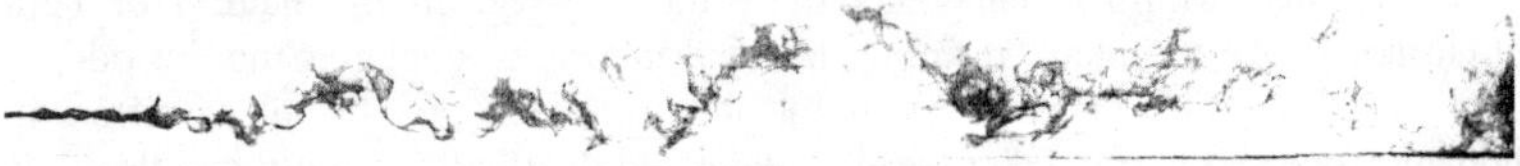

Figure 1. Chemical plume released iso-kinetically into fully developed turbulent open-channel flow. The release height is ¼ of the channel depth.

New experimental techniques are now beginning to provide fresh insights into these questions. These include non-intrusive optical techniques, particularly planar laser-induced fluorescence (PLIF) to measure tracer concentration levels and particle image velocimetry (PIV) to measure velocity. They enable simultaneous measurement of instantaneous whole fields of tracer concentration and velocity from which detailed statistical measures of their spatial variability and correlations can be obtained. In this chapter, we will use some of these techniques to illustrate the complexities of turbulent diffusion.

The purpose of this chapter is to provide an introductory overview of the essential mechanisms of turbulent diffusion and some methods of predicting mean concentration distributions for a few select applications of engineering interest. We consider only the case of turbulent diffusion, that is, the spreading of a scalar quantity due to irregular turbulent velocity fluctuations. This excludes mixing due to the combined effect of diffusion plus shear in the mean velocity, sometimes known as shear-flow dispersion. We also exclude the effects of buoyancy due to density differences between the discharged fluid and the receiving fluid, and the suppression of turbulence due to density stratification. We first provide a review of turbulent flows, particularly those features that are important to turbulent diffusion. We then discuss the mechanisms whereby turbulence induces rapid mixing; derive the time-averaged equations of species conservation, and present methods of estimating the resulting turbulent diffusion coefficients. We then give examples of applications of interest to engineers and biologists. We conclude with a discussion of some of the newer techniques for turbulent diffusion simulations, and pose some research questions.

2. BASIC CHARACTERISTICS OF TURBULENT FLOWS

The first step in understanding turbulent diffusion and the fate of tracer concentrations is understanding some basic characteristics of turbulence.

2.1 The Nature of Turbulence

Turbulence is difficult to define exactly; nevertheless, there are several important characteristics that all turbulent flows possess. These characteristics include unpredictability, rapid diffusivity, high levels of fluctuating vorticity, and dissipation of kinetic energy.

The velocity at a point in a turbulent flow will appear to an observer to be "random" or "chaotic." The velocity is unpredictable in the sense that knowing the instantaneous velocity at some instant of time is insufficient to predict the velocity a short time later. A typical velocity record is shown in Figure 2.

The unpredictable nature of turbulence requires that we describe the motion through statistical measures. The velocity typically will be described as a time-averaged value plus some fluctuation. Time-average quantities are denoted with an over bar:

$$\overline{u} = \frac{1}{T}\int_0^T u\,dt \tag{1}$$

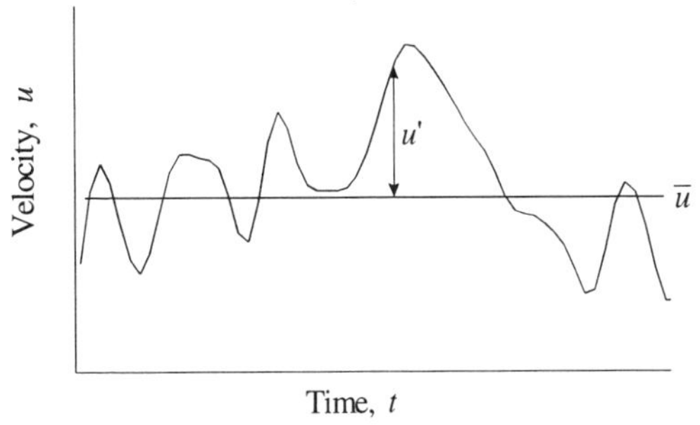

Figure 2. Sample turbulent velocity record.

where T is a time much longer than the longest turbulent fluctuations in the flow. A time record such as shown in Figure 2 is called statistically stationary if the mean quantities remain constant over the period of interest.

For a stationary velocity record, the instantaneous velocity can be decomposed into the sum of time-averaged and fluctuating contributions (called the Reynolds decomposition):

$$u = \overline{u} + u' \tag{2}$$

where u' is the fluctuating component (i.e. the deviation from the mean value) as shown in Figure 2. By definition, the time-average fluctuation is zero. Higher order statistical quantities, such as the variance, are used to describe the magnitude of the fluctuations:

$$\tilde{u}^2 = \overline{u'^2} = \int_0^T \left(u - \overline{u}\right)^2 dt \tag{3}$$

The square root of the variance of the velocity fluctuations (the standard deviation, $\sqrt{u'^2}$) is denoted by $\tilde{u}$ and is defined as the turbulence intensity.

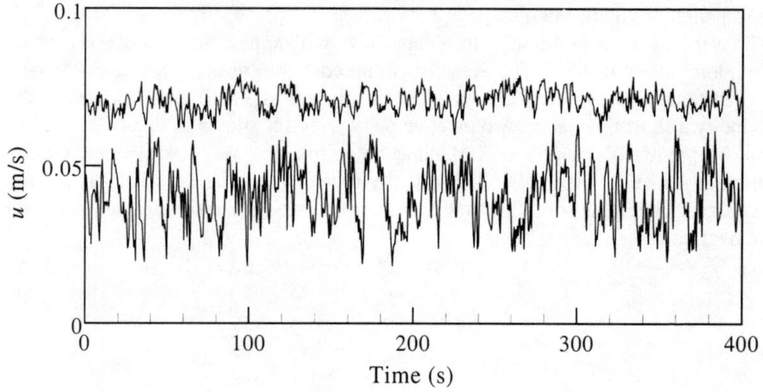

Figure 3. Velocity time series obtained in a turbulent open- channel flow at z/d=0.03 (bottom) and z/d=0.72 (top).

Actual velocity records obtained at two depths in the open channel flow photographed in Figure 1 are shown in Figure 3. The distance from the wall is z and water depth is d. The time-averaged velocity is greater farther from the wall, as would be expected in a boundary layer. The turbulence intensity also varies with distance from the wall, being significantly larger near the wall. The variation of the time-averaged velocity and turbulence intensity with distance from the lower bed are shown in Figure 4. The average velocity increases monotonically from zero at the wall; the turbulence intensity increases rapidly from zero at the wall to a local maximum near the wall and then monotonically decreases.

The velocity fluctuations act to efficiently transport momentum, heat, and tracer concentration. This turbulent transport is significantly more effective than molecular diffusion. Thus, the second characteristic of turbulence is a high rate of *diffusivity*. In fact, it is common to model the transport due to the fluctuations by defining an effective diffusion coefficient called the eddy diffusivity.

While the velocity fluctuations are unpredictable, they do possess a spatial structure. A turbulent flow, on close examination, consists of *high levels of fluctuating vorticity*. At any instant vortical motion, called eddies, are present in the flow. These eddies range in size from the largest geometric scales of the flow down to small scales where molecular diffusion dominates. The eddies are

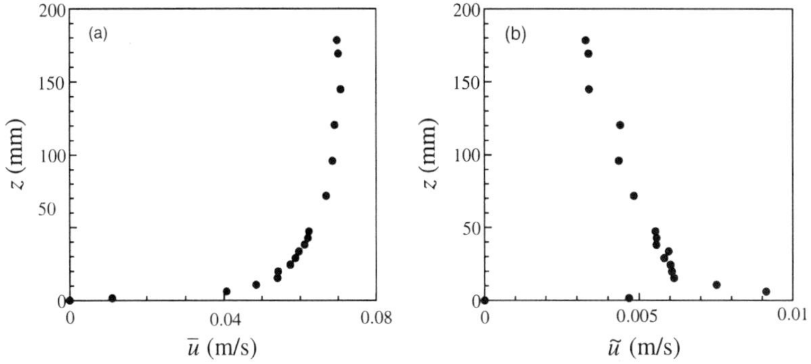

Figure 4. (a) Time-averaged, and (b) standard deviation velocity profiles for turbulent flow in an open-channel

continuously evolving in time, and the superposition of their induced motions leads to the fluctuating time records such as those shown in Figures 2 and 3. Turbulent kinetic energy is passed down from the largest eddies to the smallest though a process called the energy cascade. At the smallest scales, the energy is dissipated to heat by viscous effects. Thus, the fourth characteristic of turbulent flows is *dissipation of kinetic energy*. To maintain turbulence, a constant supply of energy must be fed to the turbulent fluctuations at the largest scales from the mean motion.

2.2 Length Sales in Turbulent Flows

Motions in a turbulent flow exist over a broad range of length and time scales. The length scales correspond to the fluctuating eddy motions that exist in turbulent flows. The largest scales are bounded by the geometric dimensions of the flow, for instance the diameter of a pipe or the depth of an open channel. These large scales are referred to as the *integral* length and time scales.

Observations indicate that eddies lose most of their energy after one or two overturns. Therefore, the rate of energy transferred from the largest eddies is proportional to their energy times their rotational frequency. The kinetic energy is proportional to the velocity squared, in this case the fluctuating velocity that is characterized by the standard deviation. The rotational frequency is proportional to the standard deviation of the velocity divided by the integral length scale. Thus, the rate of dissipation, ε , is of the order:

$$\varepsilon \propto \frac{\tilde{u}^3}{l} \tag{4}$$

where l is the integral length scale.

Interestingly, the rate of dissipation is independent of the viscosity of the fluid and only depends on the large-scale motions. In contrast, the scale at which the dissipation occurs is strongly dependent on the fluid viscosity. These arguments allow an estimate of this dissipation scale, known as the Kolmogorov microscale, η, by combining the dissipation rate and kinematic viscosity in an expression with dimensions of length:

$$\eta \propto \left(\frac{\nu^3}{\varepsilon}\right)^{1/4} \tag{5}$$

Similarly, time and velocity scales of the smallest eddies can be formed:

$$\tau \propto \left(\frac{\nu}{\varepsilon}\right)^{1/2} \quad v \propto (\nu\varepsilon)^{1/4} \tag{6}$$

An analogous length scale can be estimated for the range at which molecular diffusion acts on a scalar quantity. This length scale is referred to as the Batchelor scale and it is proportional to the square root of the ratio of the molecular diffusivity, D, to the strain rate of the smallest velocity scales, γ :

$$L_B \propto \left(\frac{D}{\gamma}\right)^{1/2} \tag{7}$$

The strain rate of the smallest scales is proportional to the ratio of Kolmogorov velocity and length scales, $\gamma \propto v/\eta \propto (\varepsilon/\nu)^{1/2}$. Thus, the Batchelor length scale can be recast into a form that includes both the molecular diffusivity of the scalar and the kinematic viscosity:

$$L_B \propto \left(\frac{\nu D^2}{\varepsilon}\right)^{1/4} \tag{8}$$

The ratio of the Kolmogorov and Batchelor length scales equals the square root of the Schmidt number, Sc:

$$\frac{\eta}{L_B} \approx \left(\frac{\nu}{D}\right)^{1/2} \approx Sc^{1/2} \tag{9}$$

For the open channel flow example discussed in this chapter, the mean velocity is 50 mm/s and the integral length scale is roughly half the channel depth, i.e. 100 mm. The fluid was water at 20°C with a kinematic viscosity of $1\times10^{-6} m^2/s$. Therefore, the Kolmogorov length and time scales are 0.7 mm and 0.5 s, respectively. Assuming a diffusivity, $D = 1\times10^{-9} m^2/s$ for the chemical tracer photographed in Figure 1, the Batchelor scale is 0.02 mm, which is 35 times smaller than the Kolmogorov microscale. Thus, we would expect a much finer structure of the concentration field than the velocity field.

2.3 Energy Cascade

The energy spectrum characterizes the turbulent kinetic energy distribution as a function of length scale. (The power spectrum is usually described in terms of a wave number, k, which is the inverse of length, but we will focus our discussion in terms of physical length scales.) The spectrum indicates the amount of turbulent kinetic energy contained at a specific length scale. This section describes some universal features of the energy spectrum for turbulent flows.

As described in the previous section, the large turbulent length scales in the flow dictate the rate of dissipation. These large length scales draw energy from the mean flow, then transfer the energy to successively smaller scales until it is dissipated at the Kolmogorov microscale. This process is called the *energy cascade*.

The energy distribution at the largest length scales is generally dictated by the flow geometry and mean flow speed. In contrast, the smallest length scales are many orders of magnitude smaller than the largest scales and hence are isotropic in nature. In between, we can describe an inertial subrange bounded above by the integral scale and below by the Kolmogorov microscale, $\eta \leq L \leq l$. In this range, the spectrum will only be a function of the length scale and the dissipation rate. (The spectrum depends on the dissipation rate because the largest length scales set the rate and the energy is transferred through this range.) With this dependence, dimensional reasoning yields Kolmogorov's $k^{-5/3}$ law:

$$E = \alpha \varepsilon^{2/3} k^{-5/3} \tag{10}$$

where k is the wavenumber and α is a constant of order one.

A typical energy spectrum is shown in Figure 5. The large length scales have the most energy and the distribution in that range depends on the boundary conditions. The smaller scales have less energy by several orders of magnitude. In between, the turbulent kinetic energy varies in the inertial subrange in proportion to $k^{-5/3}$. Also indicated in the figure is the Batchelor scale, which is more than an order of magnitude smaller than the Kolmogorov scale.

2.4 Evolution Equations

Turbulent flows must instantaneously satisfy conservation of mass and momentum. Thus, in principle, the incompressible continuity and Navier-Stokes equations can be solved for the instantaneous turbulent flow field. The difficulty with this approach is that an enormous range of scales must be accounted for in the calculation. To accurately simulate the turbulent field, the calculation must span from the largest geometric scales down to the Kolmogorov and Batchelor length scales. Even with the fastest, largest modern supercomputers, such a calculation can be achieved only for simple geometries at low Reynolds numbers. In many situations, engineers and scientists are satisfied with an accurate assessment of the time-averaged flow quantities. For instance, the time-averaged

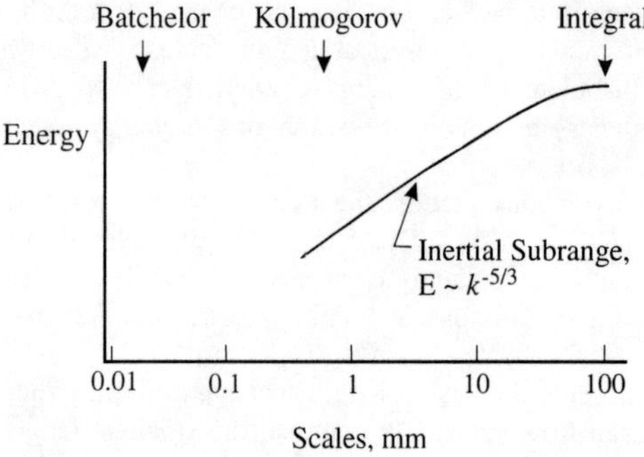

Figure 5. Typical turbulence energy spectrum, with length scales for the open channel flow of Figures 1 through 4 indicated.

velocity and pressure distribution is sufficient to calculate the wind load on a skyscraper. To derive the time-averaged flow equations we start with the instantaneous conservation equations, substitute the Reynolds decomposition (Eq. (2)), and time average the equations, yielding:

$$\frac{\partial \overline{u}_i}{\partial x_i} = 0 \tag{11}$$

$$\rho\left(\frac{\partial \overline{u}_i}{\partial t} + \overline{u}_j \frac{\partial \overline{u}_i}{\partial x_j}\right) = -\frac{\partial \overline{p}}{\partial x_i} + \mu \frac{\partial^2 \overline{u}_i}{\partial x_j^2} - \rho \frac{\partial}{\partial x_j} \overline{u_i' u_j'} \tag{12}$$

where we have employed indicial notation to indicate vectors and the standard Einstein summation convention. These Reynolds-averaged equations for the mean velocity and pressure are very similar to the instantaneous continuity and Navier-Stokes equations. The primary difference is the addition of the $\rho \frac{\partial}{\partial x_j}\left(\overline{u_i' u_j'}\right)$ term in the time-averaged momentum equation. This term is called the Reynolds stress tensor; it physically corresponds to the transport of momentum due to the turbulent fluctuations. This turbulent transport generally dominates that due to molecular diffusion. An equation for the time-averaged scalar transport equation can be derived in the same way, as will be shown in Section 3.3.

While the evolution equations for the time-averaged quantities are valid, they cannot be solved because several new unknown quantities have been introduced, specifically $\overline{u_i' u_j'}$. This dilemma is referred to as the "closure problem"; in other words, the mathematical problem is not closed because there are more unknowns than equations.

2.5 Turbulent Kinetic Energy Budget

An evolution equation for the kinetic energy can be derived for both the mean and turbulent components of the flow. The mean kinetic energy equation is:

$$\underbrace{\frac{D}{Dt}\left(\frac{1}{2}\overline{u}_i^2\right)}_{\substack{\text{total change in}\\ \text{mean kinetic energy}}} = \underbrace{\frac{\partial}{\partial x_j}\left(\frac{-\overline{p}\overline{u}_j}{\rho} + 2\nu\overline{u}_i S_{ij} - \overline{u_i'u_j'}\overline{u}_i\right)}_{\text{transport}} - \underbrace{2\nu S_{ij}S_{ij}}_{\substack{\text{viscous}\\ \text{dissipation}\\ \text{(typically small)}}} + \underbrace{\overline{u_i'u_j'}\frac{\partial \overline{u}_i}{\partial x_j}}_{\substack{\text{loss to}\\ \text{turbulence}}} \qquad (13)$$

where $S_{ij} = \frac{1}{2}\left(\frac{\partial \overline{u}_i}{\partial x_j} + \frac{\partial \overline{u}_j}{\partial x_i}\right)$ is the mean strain rate tensor.

The equation indicates that the total change in kinetic energy of the mean flow results from the combined effects of transport, viscous dissipation, and loss to turbulence. The loss to turbulence is the dominant term on the right hand side of the equation. As discussed in the previous section, the mean flow feeds energy to the large turbulent scales. The viscous dissipation is generally small for the mean flow because the gradients of mean velocity are mild. The transport terms represent the spatial movement of mean kinetic energy.

The budget of turbulent kinetic energy is:

$$\underbrace{\frac{D}{Dt}\left(\frac{1}{2}\overline{u_i'^2}\right)}_{\substack{\text{total change in}\\ \text{turbulent kinetic energy}}} = \underbrace{\frac{\partial}{\partial x_j}\left(\frac{-\overline{p'u_j'}}{\rho} + 2\nu\overline{u_i s_{ij}} - \frac{1}{2}\overline{u_i'^2 u_j'}\right)}_{\text{transport}} - \underbrace{2\nu\overline{s_{ij}s_{ij}}}_{\substack{\text{viscous}\\ \text{dissipation}}} - \underbrace{\overline{u_i'u_j'}\frac{\partial \overline{u}_i}{\partial x_j}}_{\substack{\text{shear}\\ \text{production}}} \qquad (14)$$

where $s_{ij} = \frac{1}{2}\left(\frac{\partial u_i'}{\partial x_j} + \frac{\partial u_j'}{\partial x_i}\right)$ is the strain rate tensor for the fluctuating field.

The shear production term is identical to the loss to turbulence term in the mean equation, although opposite in sign. These terms correspond to kinetic energy transfer from the mean scales to the turbulent scales. In this equation, the viscous dissipation is not small; in fact, the dissipation of turbulent kinetic energy is an important characteristic of every turbulent flow as discussed in the previous sections. Again, the transport terms correspond to the spatial movement of the turbulent kinetic energy.

3. MECHANISMS OF MIXING IN TURBULENT FLOWS

As the previous brief review has shown, turbulent flows contain irregular motions over a wide range of length and time scales. The major question in this chapter is: How do these motions contribute to mixing, resulting in the plume patterns seen in Figure 1 and the concomitant rapid decay of contaminant concentrations with distance from the source? In this section, we discuss these issues, and derive the equations of mass conservation in turbulent flows.

3.1 Molecular Diffusion

Consider first mass transport due to molecular diffusion. The rate of mass transport in the x-direction is given by Fick's law:

$$q = -D\frac{\partial c}{\partial x} \tag{15}$$

where q is the solute flux, i.e. the mass transport rate per unit area per unit time, c is the mass concentration, i.e. the mass of tracer per unit volume, and D is the molecular diffusion coefficient. Eq. (15) can be readily generalized to three dimensions:

$$\vec{q} = -D\vec{\nabla}c \tag{16}$$

where the arrows indicate vector quantities.

Eqs. (15) and (16) state that the rate of mass transport due to molecular diffusion in any direction is directly proportional to the concentration gradient in that direction. The equations are analogous to Fourier's law of heat conduction, which states that the heat (energy) flux due to conduction is proportional to the temperature gradient. In both cases, the negative sign indicates that the direction of transport is down the gradient (i.e. from hot to cold, or high to low concentration). For other transport processes, it is sometimes assumed that the flux is also proportional to the concentration gradient. These processes are then called Fickian processes, in analogy to Eq. (16), although the transport mechanism can be other than molecular diffusion.

In a flowing fluid, another major transport mechanism occurs due to the flow itself. The magnitude of this transport in the x-direction is uc, or more generally in three dimensions $\vec{u}c$. This is called advective transport. (It is often called convective transport, but we prefer to reserve the word convective for motions induced by buoyancy effects).

We can derive the equations for conservation of species by applying mass conservation to an arbitrarily-shaped control volume (see, for example, Fischer et al., 1979). The result, where the transport mechanisms are molecular diffusion and advective transport, is:

$$\frac{\partial c}{\partial t} + u\frac{\partial c}{\partial x} + v\frac{\partial c}{\partial y} + w\frac{\partial c}{\partial z} = D\left(\frac{\partial^2 c}{\partial x^2} + \frac{\partial^2 c}{\partial y^2} + \frac{\partial^2 c}{\partial z^2}\right) \tag{17}$$

This equation is known as the advective-diffusion equation and is closely analogous to the heat conduction equation. Because of this, solutions to similar heat conduction problems can sometimes be utilized in mass diffusion problems. Many solutions to the heat conduction equation are presented in the classic texts by Crank (1956), and Carslaw and Jaeger (1959). In addition, Fischer et al. (1979) discuss some fundamental properties of, and solutions to, the advective-diffusion equation.

The molecular diffusion coefficient, D, is a property of both the fluid and the diffusing solute. For low tracer concentrations (i.e. dilute solutions), D is constant, and values can be obtained from tables such as those in CRC Handbook of Chemistry and Physics (1999). For example, the diffusion coefficient for salt

(NaCl) diffusing into water is about 1.5×10^{-9} m^2/s. For gases diffusing into air the diffusion coefficient is much higher; for methane into air, it is about 1.8×10^{-5} m^2/s.

To illustrate the consequences of these values, consider the distance, *L*, diffused by some material in a time *t* given a diffusion coefficient *D*. Simple scaling indicates:

$$L \propto \sqrt{Dt} \quad \text{or} \quad t \sim \frac{L^2}{D} \tag{18}$$

Suppose this refers to sugar deposited at the bottom of a coffee cup. For a cup height of 5 cm, the time for sugar to become uniformly mixed through the cup by molecular diffusion is, from Eq. (18), of the order of 30 days! Clearly, molecular diffusion is a very slow process. The mechanism to produce full mixing quickly is known to all: Stir the coffee to produce advection and turbulence, which results in uniform mixing in a few seconds. If molecular diffusion were the only process acting to diffuse the plume shown in Figure 1 (i.e. if the channel flow were laminar), it would maintain it's identity as a thin streak with negligible mixing for very long distances from the source.

An important property of mixing is the relationship that exists between the variance of the spatial concentration distribution in a diffusing cloud in various situations to the diffusion coefficient. In Section 4, solutions to the advective-diffusion equation are presented for constant diffusion coefficients. In each case the concentration distribution is a Gaussian function proportional to $\exp\left(-r^2/4Dt\right)$, where *r* is the distance from the centerline. The standard deviation of the concentration distribution, σ (which is also a measure of the characteristic width of a plume) is therefore:

$$\sigma = \sqrt{2Dt}$$

which is consistent with the scaling of Eq. (18). It follows that

$$\frac{1}{2}\frac{d\sigma^2}{dt} = D \tag{19}$$

This result will be used later in the chapter when modeling the eddy diffusivity due to turbulent mixing.

3.2 Mixing In Turbulent Flows

So how does turbulence result in such rapid mixing? Consider a patch of material in a turbulent flow, as shown in Figure 6. Within the turbulent flow is a wide range of length scales, or eddy sizes, (Figure 5) ranging from the integral scale down to the Kolmogorov scale. Eddies that are smaller than the patch size down to the Kolmogorov scale. Eddies that are smaller than the patch size continually distort it resulting in steep concentration gradients, which are then smoothed by molecular diffusion. The role of eddies that are larger than the patch size is to translate the entire patch without contributing to its mixing. The mixing process is therefore due to distortion, stretching, and convolution of the original patch whereby the original volume is distributed irregularly over a larger volume, so that the concentration, averaged over some finite volume, decreases. In the absence of molecular diffusion, however, such a process would not reduce actual peak concentrations at a point; the reduction of these peaks is therefore very dependent on molecular diffusion.

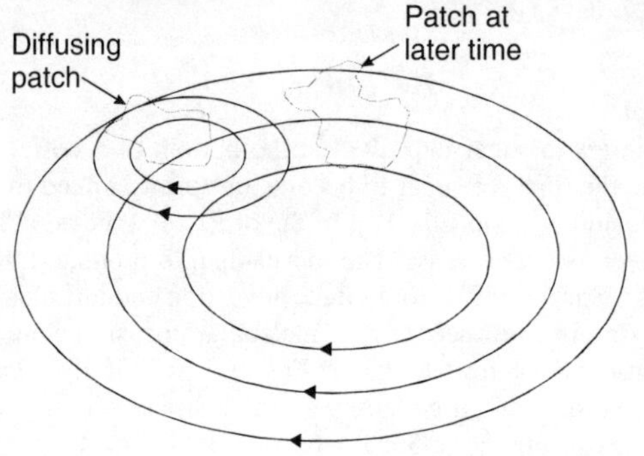

Figure 6. Schematic depiction of a patch diffusing in a turbulent flow.

3.3 Conservation Equations

We can derive a conservation equation for turbulent flows from the advective-diffusion equation by decomposing the velocity and concentration into the sum of their mean and fluctuating parts and then time-averaging the result. This is similar to the process whereby the evolution equations were derived in Section 2.4. Thus:

$$c = \overline{c} + c' \tag{20}$$

where $\overline{c}$ is the time-averaged concentration and c' the instantaneous fluctuation, or deviation, from the mean. On substituting this into Eq. (17), along with the velocity decompositions, (Eq. (2)), and time-averaging the result, we obtain:

$$\frac{\partial \overline{c}}{\partial t} + \overline{u}_j \frac{\partial \overline{c}}{\partial x_j} = D \frac{\partial^2 \overline{c}}{\partial x_j^2} - \frac{\partial}{\partial x_j} \overline{u_j' c'} \tag{21}$$

As before, the time-averaged transport equation is similar to the instantaneous equation (Eq. (17)) with the addition of the $\frac{\partial}{\partial x_j} \overline{u_j' c'}$ term, which physically corresponds to the transport of c by turbulent fluctuations. The x-component of the terms on the right-hand side of Eq. (21) can be written:

$$\frac{\partial}{\partial x}\left(D \frac{\partial \overline{c}}{\partial x} - \overline{u'c'} \right)$$

from which it can be seen that both terms in the parentheses represent mass transport. The first is the transport due to molecular diffusion (Fick's law, Eq. (15)), and the second is a turbulent flux that arises due to the correlation between

u' and c'. Because D is usually a very small quantity, $\overline{u'c'} >> -D\dfrac{\partial c}{\partial x}$ and the molecular transport term is neglected compared to the turbulent flux. Note, however, that molecular diffusion is still an important mechanism for mixing at the smallest scales, as discussed in Section 3.2.

It is usual to drop the bar terms at this point, so $c = \bar{c}$, and $u = \bar{u}$, etc., and Eq. (21) then becomes:

$$\frac{\partial c}{\partial t} + u\frac{\partial c}{\partial x} + v\frac{\partial c}{\partial y} + w\frac{\partial c}{\partial z} = -\frac{\partial}{\partial x}\overline{u'c'} - \frac{\partial}{\partial y}\overline{v'c'} - \frac{\partial}{\partial z}\overline{w'c'} \tag{22}$$

Again, this equation cannot be solved because three new unknown quantities have been introduced, specifically $\overline{u_i'c'}$. This is the "closure problem" again; in other words, the mathematical problem is not closed because there are more unknowns than equations. To circumvent this problem, the unknown quantities are often modeled, at least for common engineering problems, with eddy diffusivity coefficients, ε_i, defined as :

$$\overline{u'c'} = -\varepsilon_x\frac{\partial c}{\partial x} \qquad \overline{v'c'} = -\varepsilon_y\frac{\partial c}{\partial y} \qquad \overline{w'c'} = -\varepsilon_z\frac{\partial c}{\partial z} \tag{23}$$

Eq. (23) assumes that the diffusion process is Fickian, i.e. the turbulent mass transport is proportional to the mean concentration gradient. While this simple model can be effective, the coefficients are strongly flow dependent, vary within the flow field, and are not known a priori. As a result, estimation of eddy diffusivity coefficients often relies on empirical data.

As already stated, the turbulent transport is much greater than the molecular transport, i.e. $\varepsilon_i >> D$. With this assumption, Eq. (22) becomes:

$$\frac{\partial c}{\partial t} + u\frac{\partial c}{\partial x} + v\frac{\partial c}{\partial y} + w\frac{\partial c}{\partial z} = \frac{\partial}{\partial x}\left(\varepsilon_x\frac{\partial c}{\partial x}\right) + \frac{\partial}{\partial y}\left(\varepsilon_y\frac{\partial c}{\partial y}\right) + \frac{\partial}{\partial z}\left(\varepsilon_z\frac{\partial c}{\partial z}\right) \tag{24}$$

This equation is the most usual starting point for water and air quality models (with the possible addition of terms to account for creation and loss of species due to chemical or biological processes). The assumptions made in deriving it should be kept in mind, however.

When using Eq. (24), obvious questions are: what are the values of the eddy diffusion coefficients in any particular situation, and how do they depend on any reasonably obtained or measured mean properties of the flow? It should be reiterated that these coefficients are properties of the flow, and cannot therefore be found in any standard tables or handbooks of fluid properties.

3.4 Estimation of Eddy Diffusion Coefficients.

G. I. Taylor published one of the most important results in turbulent diffusion theory, which provides a link between the eddy diffusion coefficients and turbulent flow properties, in 1921. To illustrate his theory, consider two realizations of an experiment in which two particles are released into a turbulent flow, as sketched in Figure 7.

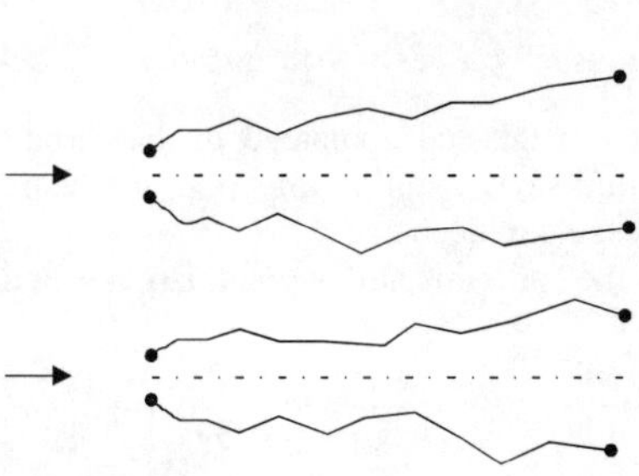

Figure 7. Trajectories of two particles released into a turbulent flow.

Because the turbulent velocity fluctuations are irregular, the results of each trial differ. On average, however, the particles wander apart from each other, and the rate at which they wander apart can be related to a diffusion coefficient. This is easier to imagine by considering individual particles, released from the coordinate origin at different times, as shown in Figure 8. The particle location after travel time T is:

$$\vec{X} = \int_0^T \vec{u}dt$$

where $\vec{u}$ is the velocity of the particle as it travels (i.e. its Lagrangian velocity).

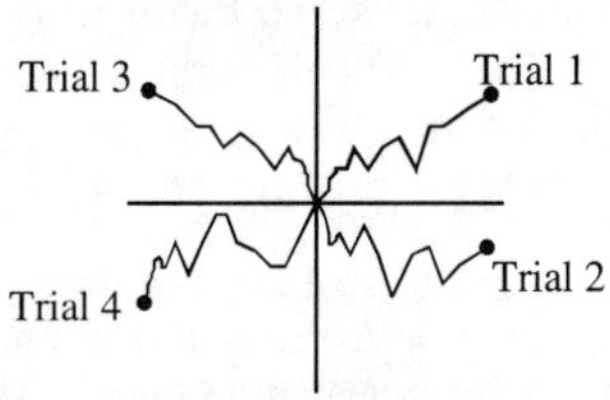

Figure 8. Possible trajectories of a particle released at the coordinate origin at various times into a turbulent flow.

The mean position of the particles, averaged over many releases in a stationary, homogeneous turbulent field, is clearly zero, i.e. the origin. The variance of their displacements is not zero, however, and is given by:

$$\overline{X^2(t)} = 2\tilde{u}^2 \int_o^t \int_o^{t'} R_L \, d\tau dt' \tag{25}$$

where R_L is the autocorrelation of the velocity:

$$R_L(\tau) = \frac{\overline{u(t)u(t+\tau)}}{\tilde{u}^2} \tag{26}$$

and $\tilde{u}^2 = \overline{u'(t)u'(t)}$ is the variance of the velocity fluctuations. The autocorrelation is a measure of the memory of the flow, in other words how well correlated future velocities are with the current value.

It would be expected that the shape of the autocorrelation function would have the form sketched in Figure 9. It should tend to zero for long times, in other words, the particle eventually "forgets" its original velocity. For short times, however, the velocity is strongly correlated with its original velocity.
We can define a time scale T_L for this process by:

$$T_L = \int_0^\infty R_L d\tau \tag{27}$$

from which it can be seen that the area under the rectangle of width T_L is the same as that under the curve of. R_L . T_L is known as the Lagrangian time scale of the flow, and it gives rise to a definition of a Lagrangian length scale:

$$L_L = \tilde{u}T_L \tag{28}$$

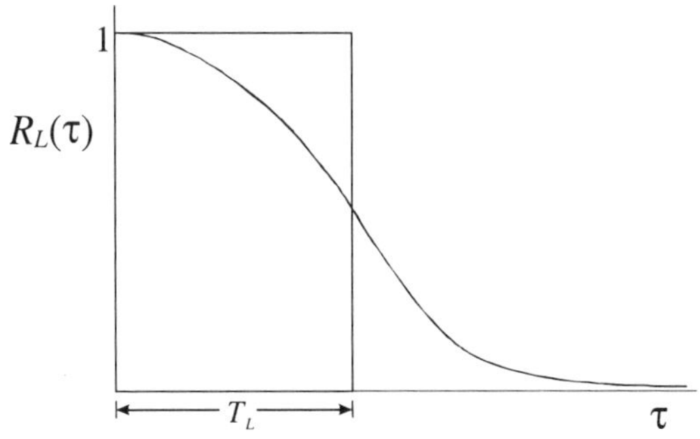

Figure 9. Autocorrelation function.

For times less than T_L and distances smaller than L_L, the velocities are generally well correlated and R_L approaches one. The Lagrangian length scale is closely related to the integral length scale, l, because the length over which the flow is well correlated corresponds to the size of the largest eddies.
We are interested in two limiting cases: very long times and very short times compared to T_L. For long times, (i.e. $t \propto T_L$) Eq. (25) becomes:

$$\overline{X^2(t)} = 2\tilde{u}^2 T_L t + \text{constant}$$

On differentiating this expression with respect to time, we obtain:

$$\frac{1}{2}\frac{d\overline{X^2(t)}}{dt} = \tilde{u}^2 T_L \tag{29}$$

The standard deviation of the displacement, $\sqrt{\overline{X^2(t)}}$, therefore increases in proportion to $t^{1/2}$ because the distance traveled is analogous to a random walk, that is, uncorrelated steps. By analogy to Eq. (19), the left hand side of Eq. (29) can be taken as a diffusion coefficient, therefore:

$$\varepsilon \propto \tilde{u}^2 T_L \propto \tilde{u} L_L \tag{30}$$

Taylor did not give a diffusion coefficient in his original analysis; also, Eq. (30) is only valid for travel times longer than T_L. An important consequence for this case is that the eddy diffusion coefficient is constant, as given by Eq. (30).

In the other limit of short travel times (i.e. $t \ll T_L$), the autocorrelation is very close to one ($R_L \approx 1$). Eq. (25) then becomes:

$$\overline{X^2(t)} = \tilde{u}^2 t^2$$

so the standard deviation of the displacements increases in proportion to t because of complete correlation between steps. On differentiation, this becomes:

$$\varepsilon = \frac{1}{2}\frac{d\overline{X^2(t)}}{dt} = \frac{1}{2}\tilde{u}^2 t \tag{31}$$

The diffusion coefficient is therefore not constant for short travel times; it increases in proportion to time because the particle displacements are highly correlated and the standard deviation of the displacement increases linearly.

For practical problems, it is more convenient to discuss the variation of the diffusion coefficient with patch or cloud size. Over a time T_L, a particle travels an rms distance $\tilde{u}T_L$. But $\tilde{u}T_L$ is the Lagrangian length scale L_L (Eq. (28)), so in other words the size of the diffusing cloud, L, should be much larger than the Lagrangian length scale, L_L for Eq. (30) to apply and for the diffusion coefficient to be constant. For smaller clouds, i.e. $L < L_L$, the diffusion coefficient increases with cloud size, and for very small clouds, Eq. (31) applies.

This result implies that the diffusion coefficient increases with cloud size while the cloud size is smaller than L_L. This phenomenon is known as relative diffusion – in other words, the magnitude of the diffusion coefficient is relative to the cloud size. When the patch size lies within the inertial subrange, Batchelor (1952) shows that the rate of increase of the mean square separation of the particles is:

$$\frac{d\overline{s^2}}{dt} \propto \varepsilon^{1/3}\left[\overline{s^2}\right]^{2/3} \tag{32}$$

where s is the separation between particles. This leads to the celebrated "4/3 power law" for diffusion:

$$\varepsilon = \alpha L^{4/3} \tag{33}$$

where α is a constant depending on the energy dissipation rate, and L is a measure of the cloud size. A similar result was first obtained by Richardson (1926) in conjunction with atmospheric diffusion. Equation 33 is frequently used for open water and atmospheric diffusion problems.

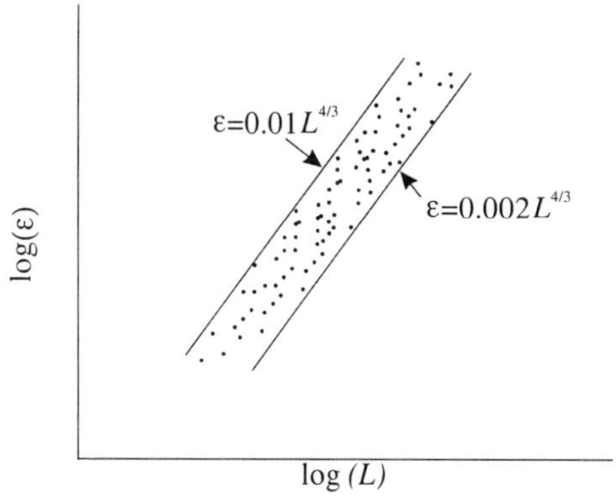

Figure 10. Relative diffusion in the ocean.

Observations of diffusing dye patches in the open ocean show considerable scatter when compared to Eq. (33). Some results are shown in Figure 10. The experiments can be approximately bracketed with 0.01 < α < 0.002 $cm^{2/3}/s$ (see Fischer et al., 1979, Figure 3.5), ε in cm^2/s, and L in cm.

4. SOLUTIONS TO THE ADVECTIVE-DIFFUSION EQUATION

4.1 Introduction

In this section, we present solutions to the advective-diffusion equation (Eq. (17)) for several simple boundary and initial conditions. These solutions, while idealized, provide insight into the basic transport mechanism and provide a means for understanding situations that are more complex. Solutions provided here are referenced in future sections regarding specific flow applications.

4.2 Continuous Line Source In Two-Dimensions, Constant Diffusion Coefficient

Consider a steady release of contaminated fluid from a line source into a steady uniform flow with $\vec{u} = (U,0,0)$ as shown in Figure 11. The objective is to predict the concentration distribution in the x-y plane. This configuration models, for example, a continuous release into a deep river from a long multiport diffuser. The mass flow rate from the source per unit length along the z-axis is $\dot{m}$ (e.g. the units of $\dot{m}$ are kg/m/s). As the transport due to diffusion is significantly greater in the y-direction than in the x-direction, due to the steeper concentration gradients in the y-direction, the governing equation reduces to:

$$u\frac{\partial c}{\partial x} = D\frac{\partial^2 c}{\partial y^2}$$

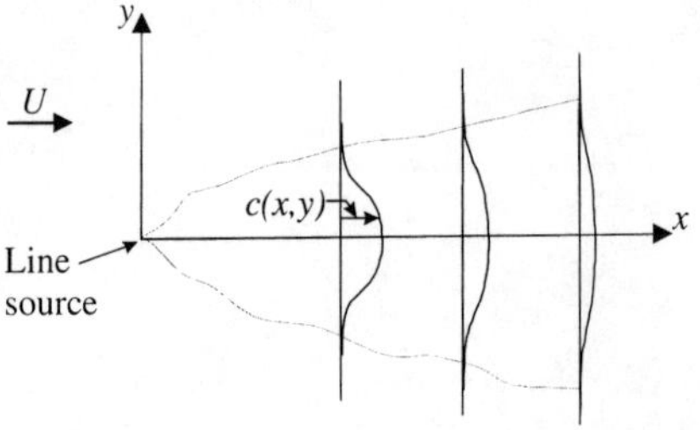

Figure 11. Diffusion from a continuous line source.

Applying the transformation $x = Ut$ yields the simple one-dimensional diffusion equation:

$$\frac{\partial c}{\partial t} = D\frac{\partial^2 c}{\partial y^2}$$

Let us first consider solutions to this equation by means of dimensional analysis. The downstream mean concentration is given by:

$$c = f\left(y, t, \dot{m}/U, D\right)$$

where $\dot{m}/U$ is the amount of mass of the contaminant "picked up" by the passing flow at the source. The concentration at any point in the field must be proportional to the contaminant mass flowrate divided by some characteristic length. Equation 18 defines a characteristic length, proportional to the distance that the contaminant diffuses in time *t*, $\sqrt{Dt}$. Thus,

$$c(y,t) = \frac{\dot{m}}{U\sqrt{4\pi Dt}} f\left(\frac{y}{\sqrt{4Dt}}\right)$$

where we have added arbitrary constants to make the solution mathematically more convenient. Inserting this functional form into the governing equation and defining a similarity variable, $\eta = y/\sqrt{4Dt}$, yields an ordinary differential equation whose solution is:

$$c(y,t) = \frac{\dot{m}}{U\sqrt{4\pi Dt}} \exp\left(-\frac{y^2}{4Dt}\right)$$

Finally, transforming back into spatial coordinates, $t = x/U$, we obtain:

$$c(x,y) = \frac{\dot{m}}{U\sqrt{4\pi Dx/U}} \exp\left(-\frac{y^2 U}{4Dx}\right) \tag{34}$$

The solution at several distances from the line source is sketched on Figure 11. As the contaminant advects downstream in the *x*-direction, diffusion acts to spread the contaminant in the y-direction and decrease the centerline value in proportion to $x^{-1/2}$.

4.3 Continuous Point Source, Constant Diffusion Coefficient

Consider now a variation of the previous example in which a point source is exchanged for the line source (Figure 12). Among other examples, this configuration corresponds approximately to release from a smokestack into a crossflow. Define $\dot{m}$ as the mass flow rate from the source (whose units are now kg/s). Again, the velocity field is uniform flow in the x-direction, $\vec{u} = (U, 0, 0)$. The solution is analogous to the previous example except the contaminant spreads in the z-direction as well as the y-direction. The solution is:

$$c(x, y, z) = \frac{\dot{m}}{4\pi D x} \exp\left(-\frac{(y^2 + z^2)U}{4Dx} \right) \tag{35}$$

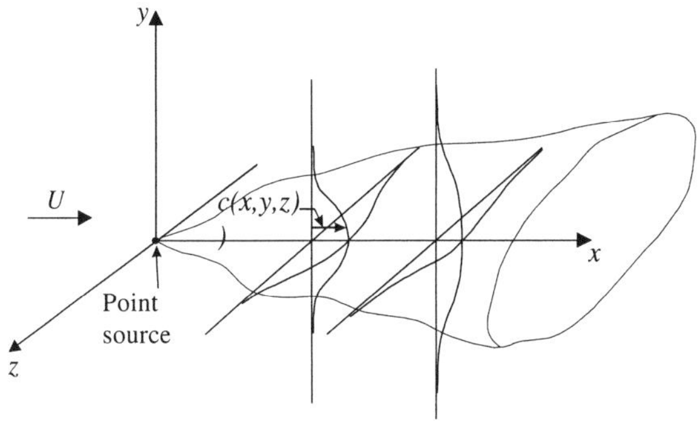

Figure 12. Diffusion from a continuous point source.

The concentration distributions at several distances downstream are sketched on Figure 12. The centerline concentration decreases more rapidly than for the line source as the distribution spreads in both the z- and y-directions; it decreases in proportion to x^{-1}.

In each of these examples it was assumed that the streamwise diffusion was negligible compared to the cross-stream diffusion. Close to the source, this assumption is not valid. Thus, the solutions discussed above are only valid for $x \gg 2D/U$.

4.4 Continuous Line Source of Finite Length - Variable Diffusion Coefficient

This situation arises when sewage or other wastewaters are discharged from outfalls with fairly long diffusers into essentially unbounded waters such as a wide estuary or coastal waters (Roberts, 1996) as sketched in Figure 13. For this case, the advective-diffusion equation, Eq. (24), can be formulated as:

$$u\frac{\partial c}{\partial x} = \frac{\partial}{\partial y}\left(\varepsilon_y \frac{\partial c}{\partial y} \right) - kc \tag{36}$$

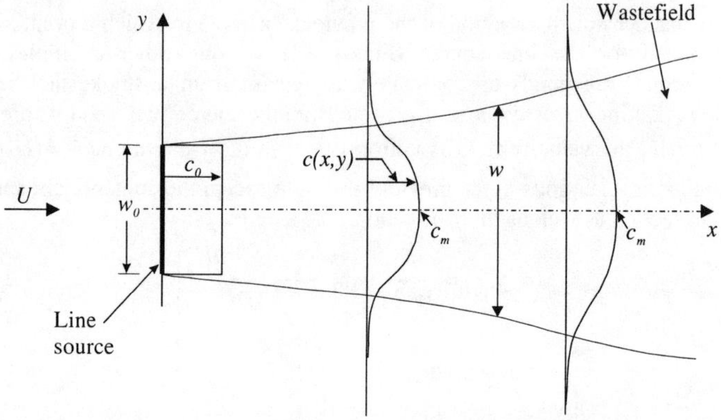

Figure 13. Diffusion from a continuous line source of finite length.

where we have assumed steady-state conditions and neglected diffusion in the *x*- and *z*-directions. Also, because bacterial decay is important for sewage discharges, we have included a decay term, -*kc*, which corresponds to a first-order decay process with *k* the decay constant. For zero decay, i.e. a conservative substance, the following solutions still apply with *k* set equal to zero.

Solutions to Eq. (36) for various assumptions about the variation of the diffusion coefficient ε_y were obtained by Brooks (1960). He defined the wastewater field width, *w*, in terms of the second moment of the concentration distribution, σ :

$$w = 12\sigma^2 = 12\frac{\int_{-\infty}^{\infty} y^2 c(x, y)dy}{\int_{-\infty}^{\infty} c(x, y)dy} \tag{37}$$

so that *w*, and therefore ε_y, are functions of *x* only. Assuming that ε_y follows the "4/3 law", Eq. (33), Brooks obtained the solution to Eq. (36) as:

$$c(x, y) = \frac{c_0 e^{-kt}}{2\sqrt{\pi \varepsilon_0 t'}} \int_{-w_o/2}^{w_o/2} \exp\left[-\frac{(y - y')^2}{4\varepsilon_0 t'} \right] dy' \tag{38}$$

where $t = x/U$ and $t' = x'/U$ and x' satisfies the equation:

$$\frac{dx'}{dx} = \frac{\varepsilon}{\varepsilon_0} = \left(\frac{w}{w_o}\right)^{4/3}$$

where w_o is the length of the diffuser. Of particular interest is the centerline (maximum), concentration, c_m. This is obtained by putting $y = 0$ into Eq. (38), yielding:

$$c_m(x) = c_0 e^{-kt} erf \sqrt{\frac{3/2}{\left(1 + \frac{2}{3}\beta \frac{x}{w_o}\right)^3 - 1}} \tag{39}$$

where β is $12\varepsilon_o/Uw_o$ and $erf(\eta) = \frac{2}{\sqrt{\pi}} \int_0^{\eta} e^{-y^2} dy$ is the standard error function. For large distances from the source, i.e. $\beta \frac{x}{w_o} >> 1$, Eq. (39) can be approximated by:

$$c_m(x) \approx c_0 \frac{9}{2\sqrt{\pi}} \left(\beta \frac{x}{w_o}\right)^{-3/2} e^{-kt} \tag{40}$$

The variation of the wastewater field width is given by:

$$\frac{w}{w_o} = \left(1 + \frac{2}{3}\beta \frac{x}{w_o}\right)^{3/2} \tag{41}$$

The implications of this solution will be discussed further in Section 6.2.

5. EXAMPLE: POINT SOURCE DIFFUSION

To illustrate the complexities and effects of turbulent diffusion, we consider a relatively well-defined situation: the diffusion from a small source in a turbulent shear flow. This is the case shown in Figure 1, which is an iso-kinetically released plume in a smooth bed, open-channel flow. The idealized mean concentration field for a constant diffusion coefficient is given by Eq. (35).

The instantaneous concentration field is much more complex, however. Using planar laser-induced fluorescence (PLIF), detailed spatial measurements of the instantaneous tracer concentration within the plume were obtained. (For details of the methodology and further results, see Webster et al., 1999, and Rahman et al., 2000). A typical instantaneous concentration distribution is shown in Figure 14.

It is clear that the concentration distribution is extremely patchy, with isolated pockets of high concentration that have steep gradients at their edges. In between the patches are large expanses with zero concentration. By averaging over many images similar to that shown in Figure 14, the time-averaged concentration field can be obtained, as shown in Figure 15. This distribution varies smoothly in space with moderate spatial concentration gradients, in contrast to the patchy instantaneous distribution.

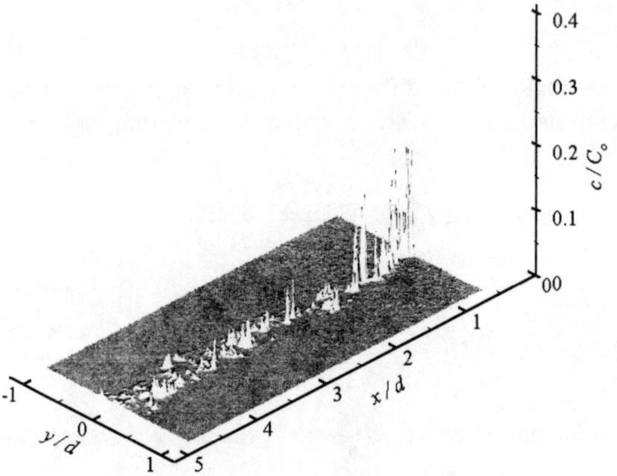

Figure 14. Instantaneous concentration distribution in a plane on the centerline of a plume in an open channel flow.

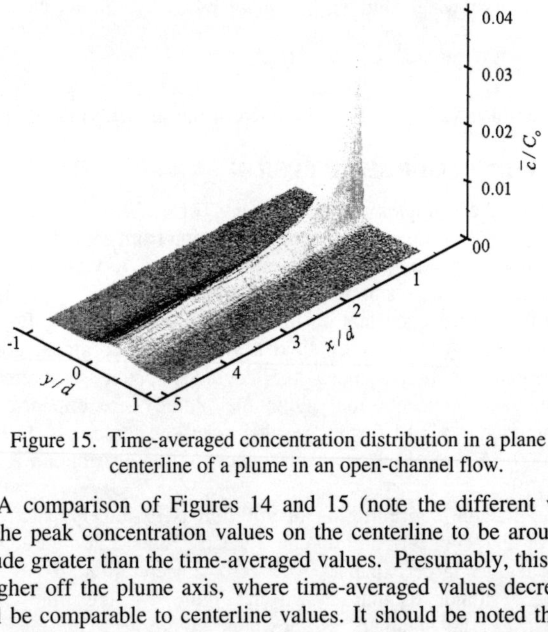

Figure 15. Time-averaged concentration distribution in a plane on the centerline of a plume in an open-channel flow.

A comparison of Figures 14 and 15 (note the different vertical scales) shows the peak concentration values on the centerline to be around an order of magnitude greater than the time-averaged values. Presumably, this ratio would be even higher off the plume axis, where time-averaged values decrease, but peaks can still be comparable to centerline values. It should be noted that peak values are very dependant on the sample size; peak values increase with decreasing sample size until the smallest concentration scale, the Batchelor scale, is reached.

For this case, the Batchelor scale is about 0.02 mm (Section 2.2) and the sample size about 1 mm, so the actual peaks could be much higher than those shown in Figure 14, and the peak to time-average ratios even higher.

The time-average and standard deviation of the concentration fluctuations along the plume centerline are shown in Figure 16. The time-averaged concentration decreases very rapidly with distance. This can be thought of in kinematic terms as the spreading of a fixed mass of tracer over an increasing volume by the action of the various eddy sizes. Initially, the time-average value decreases more rapidly than x^{-1} (see Eq. (35)), which indicates a relative diffusion regime. Between x/d = 2 and 5, the time-averaged concentration decreases approximately in proportion to x^{-1}, which agrees with Eq. (35) and

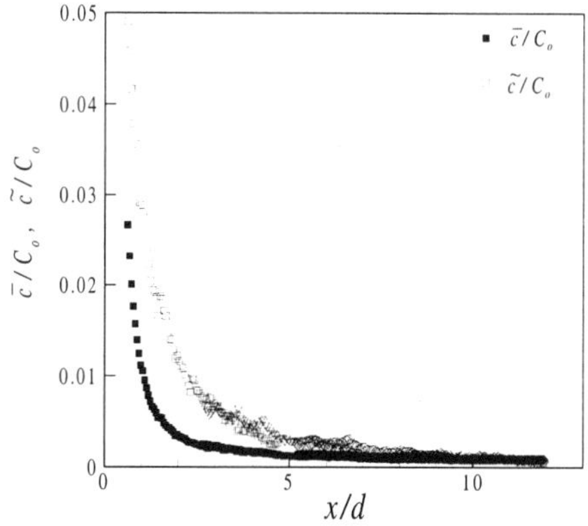

Figure 16. Variation of concentration properties along centerline of a plume in an open-channel flow.

implies a constant diffusion coefficient. Beyond x/d = 5, the rate of dilution slows, which suggests that the plume mixing may be influenced by the free surface and bed.

The standard deviation of the concentration fluctuations is greater than the time-average values along the centerline because, as seen in Figure 14, the concentration field consists of very large, but brief, spikes of concentration resulting in large fluctuations about the mean. The equations discussed previously (for example, Eq. (24)) only apply to time-averaged values and cannot predict the evolution of the concentration fluctuations. The behavior of these fluctuations is now receiving increasing attention, with additional measurements and further attempts to model them. A conservation equation for concentration fluctuations can be derived in a similar manner to that of turbulent kinetic energy (Eq. (14)). For the idealized case of a steady point source, the equation is (Pasquill and Smith, 1983):

$$\bar{u}\frac{\partial \overline{c'^2}}{\partial x} = -2\left[\overline{w'c'}\frac{\partial \bar{c}}{\partial z} + \overline{v'c'}\frac{\partial \bar{c}}{\partial y}\right] - \left[\frac{\partial}{\partial z}\overline{w'c'^2} + \frac{\partial}{\partial y}\overline{v'c'^2}\right] - S \qquad (42)$$

where S is the rate of reduction of the mean square fluctuations by molecular diffusion. Equation 42 makes the usual assumptions that gradients in the x-direction can be neglected compared to those in the y and z directions. Because of the usual closure problems, Eq. (42) cannot be solved directly. Some solutions, with certain assumptions, are given by Csanady (1967ab). Gifford (1959) proposed a model for atmospheric diffusion in which the plume is represented as discs in a plane normal to the mean wind speed.

Several studies have been reported in which concentration fluctuations were measured. The motivation for these studies includes pollution transport, boundary layer meteorology, and chemical plume tracking. Measurements of chemical plumes released into turbulent boundary layers in the laboratory (Fackrell and Robins, 1982; Nakamura, 1987; Bara et al., 1992; and Yee et al., 1993) and field (Gifford, 1960; Murlis and Jones, 1981; Jones, 1983; Murlis, 1986; Hanna and Insley, 1989; and Mylne et al., 1996) show highly intermittent concentration time-records. The intermittency is a result of the filamentous and unpredictable nature of the plume as illustrated in Figure 1. Measurements have typically consisted of time records of temperature or concentration at individual points in the flow, which have been analyzed to meet the specific focus of the study. For instance, Yee et al. (1993) attempted to match standard PDF shapes to their concentration record, while Murlis (1996) examined the importance of burst duration and intermittency to moth plume tracking. The new experimental techniques, for example PLIF, are now being applied to measure the spatial variation of instantaneous concentration distributions for the first time. These data should provide new insight into the instantaneous plume structure and enable more rigorous testing of models of the mixing and dilution processes.

6. APPLICATIONS

6.1 Rivers

A common civil and environmental application of turbulent diffusion theory is prediction of the mixing of pollutants in rivers and streams. Because of the importance of this topic, it has been extensively researched over many years.

Consider first the idealized case of mixing of a continuous discharge from a point source in a straight, rectangular channel of constant cross-section as shown in Figure 17. For steady-state conditions, longitudinal diffusion is small compared to longitudinal advection, i.e. $\frac{\partial}{\partial x}\left(\varepsilon_x \frac{\partial c}{\partial x}\right) << u\frac{\partial c}{\partial x}$. The mean transverse and vertical velocities, v and w, are zero so the advective-diffusion equation, Eq. (24), reduces to:

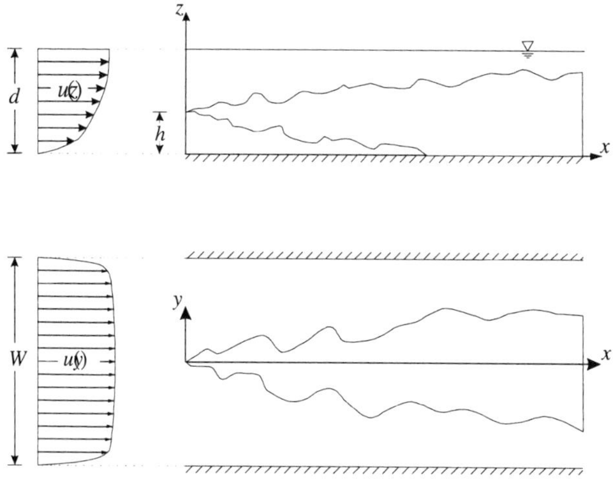

Figure 17. Diffusion from a point source into a straight, rectangular channel.

$$u\frac{\partial c}{\partial x}=\frac{\partial}{\partial y}\left(\varepsilon_y\frac{\partial c}{\partial y}\right)+\frac{\partial}{\partial z}\left(\varepsilon_z\frac{\partial c}{\partial z}\right) \tag{43}$$

in which ε_y and ε_z are the transverse and vertical diffusion coefficients. The problem is now reduced to determination of ε_y and ε_z and their dependency on the turbulence characteristics of the flow.

Turbulence in open-channel flows has been extensively studied. The measurements of turbulence intensity shown in Figure 18 are typical. This is the same data shown in Figure 4b replotted in normalized form as $\tilde{u}/u^*$ where u^* is the friction velocity, defined as $u^*=\sqrt{\tau_o/\rho}$, where τ_o is the wall shear stress, and ρ the water density. It can be seen that $\tilde{u}$ is of the same order as u^*.

The largest length scales, or eddy sizes, of the turbulence in an open channel flow are smaller than the channel depth, *d*. A typical value is around half the channel depth. Thus, when the plume size becomes comparable to the depth, the diffusion coefficients are constant and Eq. (30) applies. Applying $\tilde{u} \propto u^*$ and $L_L \propto d$ to Eq. (30) we obtain:

$$\varepsilon_y \propto du^* \text{ and } \varepsilon_z \propto du^* \tag{44}$$

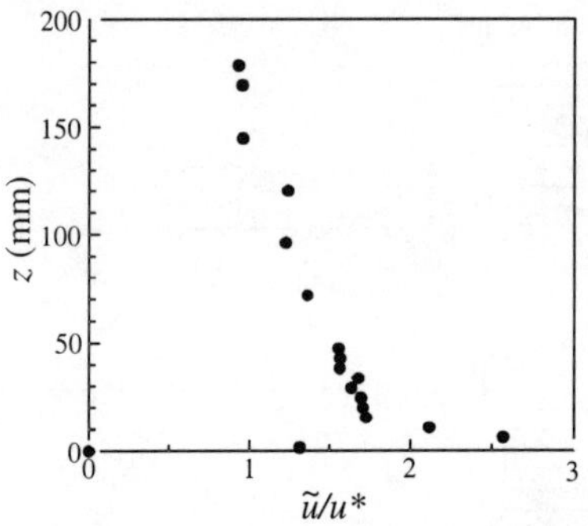

Figure 18. Normalized turbulence intensity for flow over a smooth bed open-channel flow.

Both the eddy length scales and the turbulence intensity vary over the depth, therefore, the vertical diffusion coefficient, ε_z, also varies with depth. The variation of ε_z with depth can be obtained by means of the "Reynolds analogy" whereby it is assumed that the mass diffusion coefficients are the same as the momentum diffusion coefficients. Using this assumption and a logarithmic velocity profile, Elder (1959) obtained the following expression for the depth-averaged value of ε_z for wide, open-channel flows:

$$\overline{\varepsilon_z} = 0.067du^* \tag{45}$$

and this result has been experimentally confirmed in a flume by Jobson and Sayre (1970).

If the channel is narrow relative to the depth, then the channel walls can affect the turbulence and transverse diffusion coefficient, ε_y. A more general statement of Eq. (44) for transverse mixing is then:

$$\frac{\varepsilon_y}{du^*} = f\left(\frac{W}{d}\right) \tag{46}$$

where W is the channel width. Many experiments have been reported to evaluate the effect of *W/d*. The experiments performed up to 1979 are summarized in Fischer et al. (1979), which show the value of ε_y/du^* to range between 0.1 and 0.2 with no systematic dependence on *W/d*. Based on these results, Fischer et al. (1979) recommended use of the formula $\varepsilon_y = 0.15du^* \pm 50\%$; in other words, there is the possibility of an error of ±50% when using this formula.

More recent experiments have looked more carefully at the role of channel width and also the friction factor on lateral turbulent diffusion. Webel and Schatzman (1984) reported that if the flow is fully rough and $W/d \geq 5$ the walls exert no influence. For this case, Eq. (46) becomes (Webel and Schatzman, 1984):

$$\frac{\varepsilon_y}{du^*} = 0.13 \tag{47}$$

As *W/d* decreases below about five, the wall effect causes the value of ε_y/du^* to increase. Discharges in the wall region, which extends for a distance of about 2.5*d* from the wall, experience a higher diffusion coefficient than releases in the center of the channel. They also find that diffusion dominates lateral mixing for straight laboratory channels. These findings were confirmed in Nokes and Wood (1988) in which it was suggested that $\varepsilon_y/du^* = 0.134$ is a lower bound for lateral diffusion when the chief mixing mechanism is turbulence generated at the channel floor. This value applies for wide channels, and becomes independent of the friction factor, f, for $f > 0.055$. Nokes and Wood (1988) found the diffusion coefficient to be independent of the channel width when $W/d > 8$.

Natural streams differ from these ideal channels in at least three major ways. First is that the cross-section may vary irregularly, second the channel will probably meander and not be straight, and third there may be large sidewall irregularities. The effect of these on vertical mixing is not known, and we know of no experiments in which vertical mixing has been measured in the field. It is usual to assume that Eq. (45) also applies to natural channels. As will be shown later, vertical mixing in rivers is usually quite rapid compared with transverse mixing and precise quantification of the rate of vertical mixing is not usually important.

Bends and sidewall irregularities generally increase the rate of transverse mixing. A previous summary of much field data on mixing in rivers (see Fischer et al. (1979), table 5.2), shows that ε_y/du^* ranges between about 0.3 and 0.8 for reasonably straight channels. Most rivers fall in the range 0.4 to 0.8, and Fischer et al. (1979) recommends using:

$$\frac{\varepsilon_y}{du^*} = 0.6 \pm 50\% \tag{48}$$

in the absence of any better information or field measurements. This equation is quite useful in that it contains only hydraulic parameters that can be fairly readily estimated for any particular river.

A bend in a river causes secondary circulations due to centrifugal forces, as shown in Figure 19. Such circulations would clearly considerably increase the rate of lateral mixing, which is now due to both advection and turbulent diffusion. If experimental results are parameterized as turbulent diffusion, the apparent diffusion coefficient increases markedly if the river is sharply curving. In a stretch of the Missouri River which included a 90° and a 180° bend, Yotsukura and Sayre (1976) found $\varepsilon_y/du^* \approx 3.4$. Similarly high values of ε_y/du^* between 3.0 and 4.9 were also reported in a very sinuous section of the Ogeechee River by Pernik (1985), and in field tests on the Mississippi River (Dematracopoulos and Stefan, 1983) values between 0.24 and 4.65 were observed. The large transverse coefficients reported by Holly and Nerat (1983) suggest that secondary currents can play an important role in transverse mixing, and that the secondary

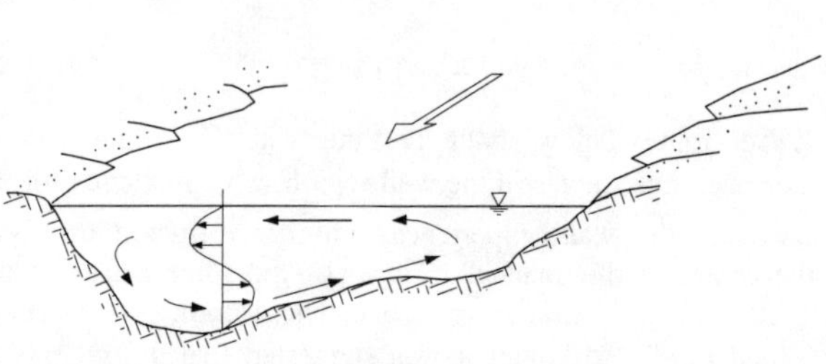

Figure 19. Secondary circulation in river bends.

circulations must be accounted for explicitly. A classification scheme, which can be used to indicate whether secondary currents will be strong enough to induce additional transverse transport, was presented by Almquist and Holley (1985).

It can be seen that our ability to predict diffusion coefficients is quite limited. If reliable knowledge of the value of ε_y is needed in a particular case, it will be necessary to perform a field experiment to measure it directly.
In order to apply Eqs. 45, 47, and 48, to rivers or any open channel flow, knowledge of u^* is necessary. Various empirical equations can be used to estimate u^*, including:

$$u^* = \left(gR_hS\right)^{1/2} \tag{49}$$

$$u^* = 3.1nUR_h^{-1/6} \quad \text{[SI units only]} \tag{50}$$

where g is the acceleration due to gravity, R_h is the hydraulic radius equal to the cross-sectional area divided by the wetted perimeter, S is the river slope, and n is the Manning coefficient. If the river slope is not known, Eq. (50) must be used. These equations pose yet another problem: what is the value of n?

Manning's coefficient, n, is a type of roughness parameter and its estimation for rivers is probably more of an art than a science. Some guidance is given, however, by books such as Chow (1950) in which values of n for various rivers are quoted. For natural streams the values range from about 0.025 for clean and straight sections up to 0.15 for very weedy and vegetated reaches. Streams wider than about 30 m have somewhat smaller n values. Flood plains can have very high n values, up to 0.2, especially if covered with heavy strands of trees. Other excellent sources of information on n values are Barnes (1967) and Arcement and Schneider (1984). Both these publications contain color photographs of rivers and their computed n values. It may be possible to find a river in these publications similar to the one of interest.

The implications of the results quoted above for mixing in typical rivers can be illustrated by means of an example. Let us consider first the rates with which materials mix vertically and transversely. The time, t, required for an effluent to mix a distance L with a diffusion coefficient ε is given by Eq. (18): $t \propto L^2/\varepsilon$. Thus, the ratio of the time required to mix transversely across the river, t_y, to that required to mix vertically, t_z, is:

$$\frac{t_y}{t_z} = \frac{(W/d)^2}{\left(\varepsilon_y/\overline{\varepsilon_z}\right)^2} \tag{51}$$

To make this definite, suppose we have a channel with dimensions 30 m wide by 1 m deep. According to Eqs. 45 and 48, $\varepsilon_y/\overline{\varepsilon_z} \approx 10$, so Eq. (51) becomes: $t_y/t_z \propto 90$. Thus, material mixes over the depth much quicker than it mixes across the width. This is a fairly typical result, and we usually go one step further to assume that vertical mixing is instantaneous compared to horizontal. Because of this relatively rapid vertical mixing, it will often be found that there is little variation in properties such as temperature over the depth of a river.

Consider, for example, the discharge from a point source in the middle of a river. Assuming the vertical mixing to be instantaneous is equivalent to replacing the source by a line source extending over the depth of the river. The problem is then two-dimensional, and when the plume size becomes comparable to the river depth the diffusion coefficient becomes constant, so Eq. (43) becomes:

$$U\frac{\partial c}{\partial x} = \varepsilon_y \frac{\partial^2 c}{\partial y^2} \tag{52}$$

Note that we have changed our notation slightly so that c is the time-averaged value and U is now the average river velocity equal to Q/A where Q is the river discharge and A the cross-sectional area. The solution to this equation for a steady release into a steady, uniform current is (Eq. (34)):

$$c = \frac{\dot{m}}{Ud\left(4\pi\varepsilon_y x/U\right)^{1/2}} \exp\left(-\frac{y^2}{4\varepsilon_y x/U}\right) \tag{53}$$

for an infinitely wide river, where $\dot{m}$ is the mass flow rate of pollutant. The maximum concentration occurs on the plume centerline (y = 0), and the concentration distribution about the maximum is a Gaussian, or normal, distribution. The plume width, w, is usually defined as four standard deviations of the Gaussian distribution, which is:

$$w = 4\left(2\varepsilon_y x/U\right)^{1/2} \tag{54}$$

So the plume grows in proportion to $x^{1/2}$ downstream and therefore reaches the banks, i.e. $w = W$, at a distance x_1 downstream given by:

$$x_1 = \frac{UW^2}{32\varepsilon_y} \tag{55}$$

Beyond this distance, Eq. (53) no longer applies as the effects of the banks must be considered. The rather lengthy analytical solution for this case is given in Fischer et al. (1979) (Eq. (5.9)). We rarely need the equation in this full form, however, as we are mostly concerned with the variation of maximum concentration, bank concentration, and the distance for uniform mixing. It is shown in Fischer et al. (1979) that these can be expressed in non-dimensional form as in Figure 20. The non-dimensional distance is $x' = x\varepsilon_y / UW^2$ and concentration is expressed as c/c_o where c_o is the far-field concentration when the effluent is well-mixed over the river cross-section; it is given by $c_o = \dot{m}/UdW$.

Theoretically, the distance at which the plume becomes uniformly mixed over the river cross-section is infinite. Figure 20 shows, however, that for a dimensionless downstream distance x' greater than about 0.1, the concentration varies by less than 5% of the mean over the cross-section. Taking this to define the length, L_c, required for "complete mixing" we obtain:

$$L_c = 0.1 \frac{UW^2}{\varepsilon_y} \tag{56}$$

for a centerline discharge. A comparison of Eqs. (55) and (56) shows that the distance for complete mixing is about three times the distance at which the plume first reaches the banks.

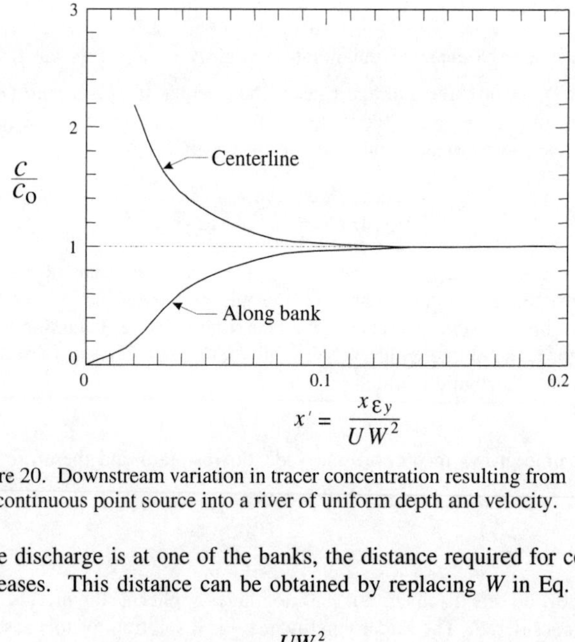

Figure 20. Downstream variation in tracer concentration resulting from a continuous point source into a river of uniform depth and velocity.

If the discharge is at one of the banks, the distance required for complete mixing increases. This distance can be obtained by replacing W in Eq. (56) by $2W$:

$$L_s = 0.4 \frac{UW^2}{\varepsilon_y} \tag{57}$$

In other words, a side-discharge requires four times the distance for complete mixing than does a centerline discharge. (Assuming the lateral diffusion coefficient is constant across the channel. In actuality, it is somewhat higher near the banks, so the ratio will be somewhat less than four times). Clearly, however, if it is necessary to discharge effluent from an open-ended pipe more rapid mixing will be accomplished by placing the discharge in the center of the river.

What are these distances for mixing in typical rivers? Suppose we have a discharge of effluent in the middle of a river flowing with an average speed U of 0.4 m/s that is 300 m wide and 3 m deep. The hydraulic radius, R_h, is 2.9 m. Assuming the Manning coefficient is 0.04, Eq. (50) gives the shear velocity, u^* = 0.042 m/s. Note that the shear velocity is roughly one tenth of the mean velocity, a fairly typical result. The diffusion coefficients are, from Eqs. 45 and 48, $\overline{\varepsilon_z} \approx 0.0084\ m^2/s$, and $\varepsilon_y \approx 0.076\ m^2/s$. For a bottom discharge, the distance required for the effluent to be mixed over the depth can be crudely estimated by replacing W with d and ε_y with ε_z in Eq. (57). The result implies that the effluent will be well-mixed vertically about 170 m downstream of the injection point.

The distance x_1 at which the plume reaches the river banks can be estimated from Eq. (55). Substituting the values above, we obtain $x_1 \approx 14{,}800$ m or about 15 km! The distance for uniform mixing from Eq. (56) is $L_c \approx 47{,}400$ m or 47 km. Clearly, these distances are very much greater than the distance for vertical mixing, justifying our assumpticn of two-dimensionality or very rapid vertical mixing.

These distances are probably surprisingly large to someone approaching the subject for the first time. The downstream distance where the plume reaches the banks is about fifty river widths, i.e. the plume remains slender for long distances downstream. A photograph of just such a case is given in Fischer et al. (1979), p. 115. Of course, in a real river, bends, cross-sectional changes, or other obstructions, could speed up the mixing. Nevertheless, this example shows that caution should be used in applying one-dimensional models right from the source, or in assuming very rapid cross-sectional mixing. It also shows the value of a multiport diffuser across the river that can cause this mixing to occur very rapidly.

The solution, Eq. (53), to the advective-diffusion equation, Eq. (24), is one of the simplest possibilities, and is given mainly for illustrative purposes. It is not possible in this review to give other solutions, but Holley and Jirka (1986), Ch. 5, provide solutions to other one-, two-, and three-dimensional problems.

Another method of analyzing diffusion in rivers is the ray method of Smith (1981). The depth topography can have a strong influence on lateral spreading, and use of the ray method shows that contaminant concentration is greatest in shallow water and towards the outside of bends.

6.2 Estuaries and Coastal Waters

Diffusion in estuaries and coastal waters has a number of applications, including prediction of mixing of wastewater discharges from sewage treatment plants, thermal effluent from power plants, and accidentally released oil spills.

For shallow waters, it is often assumed that the equation for open channels, Eq. (48) applies. This seems reasonable for cases where the turbulence is predominantly generated by bottom shear. For deeper waters other formulations are usually used. For a fairy small source in the initial stages of growth, Csanady (1973) recommends a constant diffusion coefficient $\varepsilon \approx 0.1$ m^2/s. For longer travel times or for larger source sizes, diffusion coefficients can be much larger, and the size of the source can significantly affect the rate at which it diffuses. Equation 39 is often used to model this situation. This solution assumes a line source of finite length and a variable diffusion coefficient that varies as the 4/3 power of plume size. Diffusion is two-dimensional, i.e. in the lateral direction only. Because the presence of density stratification inhibits vertical diffusion, this is a reasonable, and conservative, assumption. The dilution can be expressed as a far-field dilution $S_f = c_o / c_m$ where c_o is the initial concentration of some tracer (assumed uniform along the line source), and c_m is the maximum (centerline) concentration at some distance from the source. Rearranging Eq. (39) for a conservative constituent, i.e. $k = 0$:

$$S_f = \left[erf\left(\frac{3/2}{(1+8\alpha L^{-2/3} t)^3 - 1} \right)^{1/2} \right]^{-1} \qquad (58)$$

Equation (58) shows that the far field dilution depends only on the travel time, t. It is useful for examining the role of turbulent diffusion for diffusers of various lengths. Some computed values of the far-field dilution S_f assuming an upper value for α of 0.01 cm$^{2/3}$/s, are given in Table 1.

It can be seen that, whereas dilution by oceanic turbulence can be quite effective for short diffusers, it is relatively minor for long diffusers. The physical interpretation of this result is that the time needed for the centerline concentration to be reduced is the time required for eddies at the plume edges to "bite" into it. For a wide field produced by a long diffuser, the eddies have farther to go so it takes them longer to get to the centerline. The rate of decay is much smaller than for a small point source, for example, Figure 15.

	Far field dilution, S_f	
Travel time, t (hr)	Diffuser length, L (m)	
	35	700
1	2.4	1.0
3	7.4	1.4
10	35.5	3.2
20	95.9	6.9

Table 1. Far field dilutions for diffusers of various lengths.

6.3 Chemical Plume Tracking

Many aquatic and terrestrial animals rely on sensory cues to track turbulent odor plumes in order to locate food and mates. It is not practical for animals, such as blue crabs, to use the time-averaged concentration because they do not monitor the plume at a particular location long enough to obtain converged statistics (Elkinton et al., 1984, Moore and Atema, 1991). Thus, these animals must be using instantaneous observations of the odor plume to make tracking decisions. In this section, we discuss the usefulness of a sensory cue, namely bilateral comparison, available to animals such as blue crabs in a turbulent odor plume (Webster et al., 2001).

Animals, such as blue crabs and lobsters, have chemosensors on their appendages, which are separated horizontally. Several investigators have hypothesized that animals may be using bilateral comparison of these chemosensors to orient toward the source location (e.g. Reeder and Ache, 1980, Atema, 1996). To assess the usefulness of bilateral comparison, we evaluate the plume data presented in Section 5. Figure 21a shows the spatial correlation between the instantaneous centerline concentration, c_o, and the instantaneous concentration at distance y from the centerline, c_y. The correlation is identically one at the centerline and decreases rapidly with increased spacing because the dimensions of dye filaments are smaller than the sensor spacing. The area under the correlation curve increases dramatically as the plume grows downstream.

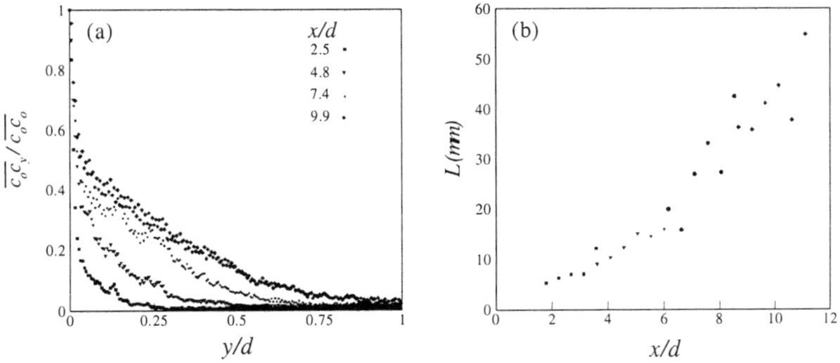

Figure 21. (a) Spanwise spatial correlation of absolute concentration at four locations. (b) Spanwise integral length scale.

This trend can be quantified by defining an integral spanwise length scale, L, calculated from the area under the correlation curve. As shown in figure 21b, this integral length scale increases with distance from the source. A searcher with sensors separated by a distance larger than the integral length scale, L, can better assess the instantaneous gradients and therefore identify the plume centerline more easily. In other words, with sensor spacing greater than the integral length scale, L, the searcher has sufficient spatial contrast to adjust toward the centerline of the plume from instantaneous measurements. With smaller sensor spacing, the contrast is insufficient to make useful decisions based on the instantaneous concentration field. Since the plume is growing downstream, there is an advantage to animals that can continually maintain sensor spacing greater than L by either moving an appendage or having a broad array of sensors. This conclusion is consistent with Weissburg (2000), who defined the spatial integration factor (SIF) as the dimensionless ratio of the sensor spacing to the plume width and suggested that a large SIF allows the detection of a plume edge.

7. OTHER MODELING TECHNIQUES

7.1 Random Walk

The analytical solutions presented in Sections 4 and 6, while being very useful in providing insight into turbulent diffusion processes, are quite limited in practice. They only apply for flows with uniform and steady velocity, uniform diffusion coefficient, and simple geometries. They can also only predict mean concentration distributions. For real water quality problems, numerical, computer-based solutions are usually used. These fall into two main types.

First is numerical solution of the advective-diffusion equation, Eq. (24). Decay or production of species due to chemical, biological, or other processes can be readily incorporated (for example, the biological decay term in Eq. (36)), and multiple species and their chemical interactions can also be added. A separate hydrodynamic model is needed which is solved first to obtain the velocity field; this velocity field is then provided as input to the water quality model. An example of this procedure for coastal waters is given by Connolly et al. (1999).
The second major type is particle tracking, or random-walk models. In these models, mass is represented by discrete particles; at each timestep, the displacement of the particle follows a "random-walk." Following the description of Feynman et al. (1963) we can imagine a particle moving such that there is no correlation between the directions of two consecutive steps. In each timestep the particle moves a distance that is random with a Gaussian distribution and an average value L. After N steps, the rms distance traveled is proportional to the square root of the number of steps:

$$R_N^{rms} = L\sqrt{N}$$

As N is synonymous with time, this result is consistent with the observation in Section 3.1 that the distance diffused is proportional to $t^{1/2}$. This result is useful in relation to the eddy diffusivity model for times longer than the Lagrangian time scale, Eq. (30). In a flowing fluid, the displacement consists of an advective, deterministic, component and an independent random component. Again, the deterministic component must be supplied by a separate hydrodynamic model.

Random walk models have been extensively used in groundwater problems, and are now being increasingly used in surface water problems. We cannot review them in detail here, but they have been applied to rivers (Jeng and

Holley, 1986, Pearce et al. 1990), to estuaries and coastal waters (Chin and Roberts, 1985, Dimou and Adams, 1994), and to the atmosphere (Luhar and Britter, 1992, Luhar and Sawford, 1995). Many more examples can be found in the literature.

8. FUTURE RESEARCH ISSUES

Probably the biggest research advance in recent years has been the rapid development of experimental instrumentation and computers. The rapid development of computers is familiar to all; their increasing power has led to their widespread use to solve diffusion problems of increasing complexity. The range of problems that can be solved continues to grow, but full solutions of the turbulent equations over the entire range of length scales in practical flows is still not in sight.

Perhaps less familiar is the rapid development of experimental techniques and instrumentation for both laboratory and field experiments. In the laboratory, the use of LIF is particularly useful, for example, the techniques that led up to Figures 14, 15, and 16. These optical techniques allow measurement with non-intrusive optical techniques of one million points or more simultaneously, at rates of 60 Hz or greater. Vast amounts of tracer concentration data can therefore be obtained. These can be combined with PIV techniques, which allow similar whole-field measurements of instantaneous velocity. The measurements of instantaneous concentration and velocity allow computation of mean values, but also many other properties, for example, their fluctuations, spatial correlations, and fluxes (Webster et al., 2001). These instrumentation advances result, in turn, from rapid advances in CCD sensors, image processing and acquisition, mass storage, and increasing computer power. These will undoubtedly continue in the future and should prove especially useful in conjunction with the development of mathematical turbulence models.

At the same time, field instrumentation has also rapidly developed. It is now possible to perform real-time monitoring of mixing in water environments with submersible fluorometers combined with packages that also measure currents, turbulence, and other properties (Petrenko et al., 1998, Roldao et al., 2000).

Demands on turbulence diffusion theory are now coming from new areas. For example, prediction of how animals or robots can seek the source of a turbulent chemical odor plume (Weissburg and Zimmer-Faust, 1994), and prediction of peak exposures of animals and organisms to contaminants. These require knowledge of characteristics that are not usually sought, for example, peak concentrations, probability density functions of concentration fluctuations, burst length and structure, and intermittency.

9. CONCLUSIONS

Turbulent diffusion is a complex process that is very efficient at mixing pollutants in the natural environment thereby reducing the concentrations of potentially harmful contaminants to safe levels. Despite many years of research, it is still poorly understood, and can only be rather crudely predicted in many cases. In this chapter, we have given an introductory overview of the most important features of turbulence relevant to turbulent diffusion, and the processes whereby it occurs. We presented the equations for species conservation, and gave their solutions and examples in simple cases. Mathematical modeling techniques were briefly introduced for more complex situations.

Demands for more reliable predictions, and predictions of quantities that have received little attention in the past are now increasing. These are driven by increasing environmental awareness, more stringent environmental standards, and application of diffusion theory in new areas. These lead to the need to quantify and predict, for example, instantaneous peak concentrations, intermittency of concentration fluctuations, the durations of concentration bursts, their onset slopes, and many other characteristics.

One of the most exciting areas of diffusion research is the rapid development of instrumentation techniques in the laboratory and field. These have improved our ability to measure concentration fields enormously over the past ten years or so. The challenge now is to incorporate these new data into improved understanding and improved mathematical models of turbulent diffusion.

REFERENCES

Almquist, C. W., and Holley, E. R. (1985). "Transverse Mixing in Meandering Laboratory Channels with Rectangular and Naturally Varying Cross Sections." Report 205, Center for Research in Water Resources, Univ. of Texas at Austin, Austin, Texas.

Arcement, G. J., and Schneider, V. R. (1984). "Guide for Selecting Manning's Roughness Coefficients for Natural Channels and Flood Plains." Report No. FHWA-TS-84-204, U.S. Dept. of Transportation, Federal Highway Administration.

Atema, J. (1996). "Eddy Chemotaxis and Odor Landscapes: Exploration of Nature with Animal Sensors." *Biol. Bull.*, 191, 129-138.

Bara, B.M., Wilson, D.J., and Zelt, B.W. (1992). "Concentration Fluctuation Profiles from a Water Channel Simulation of a Ground-level Release." *Atmospheric Environment*, 26A, 1053-1062.

Barnes, H. H. (1967). "Roughness Characteristics of Natural Channels." Water Supply Paper 1849, U.S. Geological Survey.

Batchelor, G. K. (1952). "Diffusion in a Field of Homogeneous Turbulence. II. The Relative Motion of Particles." *Proc. Camb. Phil. Soc.*, 48, 345-362.

Brooks, N. H. (1960). "Diffusion of Sewage Effluent in an Ocean Current." *First International Conference on Waste Disposal in the Marine Environment*, pp. 246-267, University of California, 1959.

CRC Handbook of Chemistry and Physics (1999).

Carslaw, H. S. and Jaeger, J. C. (1959). *Heat Conduction in Solids*, Oxford University Press, Second Edition.

Chin, D. A., and Roberts, P. J. W. (1985). "Model of Dispersion in Coastal Waters." *J. Hydr. Eng.*, ASCE, 111(1), 12-28.

Chow, V. T. (1950). *Open-Channel Hydraulics*, McGraw-Hill.

Connolly, J. P., Blumberg, A. F., and Quadrini, J. D. (1999). "Modeling Fate of Pathogenic Organisms in Coastal Waters of Oahu, Hawaii." *Journal of Environmental Engineering*, ASCE, 125(5), 398-406.

Crank, J. (1956). *The Mathematics of Diffusion*, Oxford University Press.

Csanady, G. T. (1967a). "Concentration Fluctuations in Turbulent Diffusion." *J. Atm. Sc.*, 24, 21-28.

Csanady, G. T. (1967b). "Variance of Local Concentration Fluctuations." *Boundary Layers and Turbulence, Physics of Fluids Supplement*, 576-578.

Csanady, G. T. (1973). *Turbulent Diffusion in the Environment*, D. Reidel Publ. Co., Boston.

Demetracopoulos, A. C., and Stefan, H. G. (1983). "Transverse Mixing in Wide and Shallow River: Case Study." *J. Hydr. Div.*, ASCE, 109(3), 685-699.

Dimou, K. N., and Adams, E. E. (1994). "A Random-Walk, Particle Tracking Model for Well-mixed Estuaries and Coastal Waters." *Est., Coastal and Shelf Sc.*, 37, 99-110.

Elder, J. W. (1959). "The Dispersion of Marked Fluid in Turbulent Shear Flow." *J. Fluid Mech.*, 5, 544.

Elkinton, J. S., Carde, R. T., Mason, C. J. (1984). "Evaluation of Time-average Dispersion Models for Estimating Pheromone Concentration in a Deciduous Forest." *J. Chem. Ecol.*, 10, 1081-1108.

Fackrell, J. E., and Robins, A. G. (1982). "Concentration Fluctuations and Fluxes in Plumes from Point Sources in a Turbulent Boundary Layer." *J. Fluid Mech.*, 117, 1-26.

Feynman, R. P., Leighton, R. B., and Sands, M. (1963). *The Feynman Lectures on Physics*, Addison Wesley, New York.

Fischer, H. B., List, E. J., Koh, R. C. Y., Imberger, J., and Brooks, N. H. (1979). *Mixing in Inland and Coastal Waters*, Academic Press, New York.

Gifford, F. A. (1959). "Statistical Properties of a Fluctuating Dispersion Model." in *Atmospheric Dispersion and Air Pollution*, F. N. Frankiel and P. A. Sheppard, Editors, Academic Press, 117.

Hanna, S. R. and Insley, E. M. (1989). "Time Series Analysis of Concentration and Wind Fluctuations." *Boundary-Layer Meteorology*, 47, 131-147.

Gifford, F. A. (1960). "Peak to average concentration ratios according to a fluctuating plume dispersion model." *Int. J. Air Poll.*, 3, 253.

Holley, E. R., and Jirka, G. H. (1986). "Mixing in Rivers." Tech Rept. No. E-86-11, U.S. Army Corps of Engineers, Vicksburg, Miss.

Holly, F. M., and Nerat, G. (1983). "Field Calibration of Stream-Tube Dispersion Model." *J. Hydr. Div.*, ASCE, 109, 1455.

Jeng, S. W., and Holley, E. R. (1986). "Two-Dimensional Random Walk Model for Pollutant Transport in Natural Rivers." Center for Research in Water Resources, Univ. of Texas, Austin, Texas.

Jobson, H. E., and Sayre, W. W. (1970). "Vertical Transfer in Open Channel Flow." *J. Hydr. Div.*, ASCE, 96, 703.

Jones, C. D. (1983). "On the Structure of Instantaneous Plumes in the Atmosphere." *J. Hazardous Materials*, 7, 87-112.

Luhar, A. K., and Sawford, B. L. (1995). "Lagrangian Stochastic Modeling of the Coastal Fumigation Phenomenon." *J. Appl. Meteor.*, 34, 2259-2277.

Luhar, A. K., and Britter, R. E. (1992). "Random-walk Modeling of Buoyant-plume Dispersion in the Convective Boundary Layer." *Atmospheric Environment*, 26A(7), 1283-1298.

Moore, P. A. and Atema, J. (1991). "Spatial Information in the Three-Dimensional Fine Structure of an Aquatic Odor Plume." *Biol. Bull.*, 181, 408-418.

Murlis, J. (1986). "The Structure of Odour Plumes." in *Mechanisms in Insect Olfaction*, T. L. Payne, M. C. Birch, and C. E. J. Kennedy, Editors, Oxford: Clarendon.

Murlis, J., and Jones, C. D. (1981). "Fine-scale Structure of Odour Plumes in Relation to Insect Orientation to Distant Pheromone and Other Attractant Sources." *Physiological Entomology*, 6, 71-86.

Mylne, K. R., Davidson, M. J., and Thomson, D. J. (1996). "Concentration Fluctuation Measurements in Tracer Plumes using High and Low Frequency Response Detectors." *Boundary-Layer Meteorology*, 79, 225-242.

Nakamura, I., Sakai, Y., and Miyata, M. (1987). "Diffusion of Matter by a Non-buoyant Plume in Grid-generated Turbulence." *J. Fluid Mech.*, 178, 379-403.

Nokes, R. I., and Wood, I. R. (1988). "Vertical and Lateral Turbulent Dispersion; Some Experimental Results." *J. Fluid Mech.*, 187, 373-394.

Pasquill, F., and Smith, F. B. (1983). "Atmospheric diffusion," E. Horwood, New York.

Pearce, B. R., et al. (1990). "Thermal Plume Study in the Delaware River: Prototype Measurements and Numerical Simulation." *IAHR International Conference on Physical Modeling of Transport and Dispersion*, 13B.7-13B.12, Cambridge, Mass.

Pernik, M. (1985). *Mixing Processes in a River-Floodplain System*, M. S. Thesis, School of Civil Engineering, Georgia Institute of Technology, Atlanta.

Petrenko, A. A., Jones, B. H., and Dickey, T. D. (1998). "Shape and Initial Dilution of the Sand Island, Hawaii Sewage Plume." *J. Hydr. Eng.*, ASCE, 124(6), 565-571.

Rahman, S., Dasi, L. P., Webster, D. R., and Roberts, P. J. W. (2000). "Characteristics of Turbulent Plumes using PLIF Technique." *2000 ASCE Joint Conference on Water Resources Engineering and Water Resources Planning & Management*, Minneapolis, MN.

Reeder, P. B., and Ache, B. W. (1980). "Chemotaxis in the Florida Spiny Lobster, *Panulirus argus*." *Anim. Behav.*, 28, 831-839.

Richardson, L. F. (1926). "Atmospheric Diffusion Shown on a Distance-neighbor Graph." *Proc. Roy. Soc. London*, A110, 709-739.

Roberts, P. J. W. (1996). "Sea Outfalls." in *Environmental Hydraulics*, V. P. Singh and W. Hager, Editors, Kluwer Academic Publishers, Dordrecht.

Roldao, J., Carvalho, J. L. B., and Roberts, P. J. W. (2000). "Field Observations of Dilution on the Ipanema Beach Outfall." *1st World Congress of the IWA*, Paris.

Smith, R. (1981). "Effect of Non-Uniform Currents and Depth Variations upon Steady Discharges in Shallow Water." *J. Fluid Mech.*, 110, 373.

Taylor, G. I. (1921). "Diffusion by Continuous Movements." *Proc. London Math. Soc.*, 2(20), 196.

Webel, G., and Schatzmann, M. (1984). "Transverse Mixing in Open Channel Flow." *J. Hydr. Eng.*, ASCE, 110(4), 423-435.

Webster, D. R., Rahman, S., and Dasi, L. P. (2001). "On the Usefulness of Bilateral Comparison to Tracking Turbulent Chemical Odor Plumes." To appear in *Limnology and Oceanography*.

Webster, D. R., Roberts, P. J. W, Rahman, S., and Dasi, L. P. (1999). "Simultaneous PIV/PLIF Measurements of a Turbulent Chemical Plume." *1999 International Water Resources Engineering Conference*, Seattle, WA.

Webster, D. R., Roberts, P. J. W., and Ra'ad, L. (2001). "Simultaneous DPTV/PLIF Measurements of a Turbulent Jet." *Experiments in Fluids*, 30(1), 65-72.

Weissburg, M. J. (2000). "The Fluid Dynamical Context of Chemosensory Behavior." *Biol. Bull.*, 198, 188-202.

Weissburg, M. J., and Zimmer-Faust, R. K. (1994). "Odor Plumes and How Blue Crabs Use Them in Finding Prey." *J. Experimental Biology*, 197, 349-375.

Yee, E., Wilson, D. J., and Zelt, B. W. (1993). "Probability Distributions of Concentration Fluctuations of a Weakly Diffusive Passive Plume in a Turbulent Boundary Layer." *Boundary-Layer Meteorology*, 64, 321-354.

Yotsukura, N., and Sayre, W. W. (1976). "Transverse Mixing in Natural Channels." *Water Resour. Res.*, 12(4), 695-704.

Chapter 3

MIXING OF A TURBULENT JET IN CROSSFLOW - THE ADVECTED LINE PUFF

Joseph H.W. Lee[1], G.Q. Chen[2], and C.P. Kuang[3]

ABSTRACT

The mixing of turbulent jets and plumes in a current is illustrated by a detailed study of a jet in crossflow. The practical importance of dilution prediction in environmental applications and simple equations based on self-similarity and the line puff analogy are first discussed. A 3D numerical solution of an advected line puff - a turbulent non-buoyant jet introduced at no excess horizontal momentum into a horizontal crossflow, is then presented. The numerical results of the standard two-equation k - ε model show that the advected line puff is characterized by a longitudinal transition, from an initially near-vertical jet through a bent-over phase during which the jet becomes almost parallel with the main freestream, to a sectional vortex-pair flow with double concentration maxima; the computed flow details and scalar mixing characteristics can be described by self-similar relations beyond a dimensionless distance of around 20 - 30. The aspect ratio and maximum concentration ratio for the kidney-shaped sectional puff are all found to be around 1.2. A loss in vertical momentum is observed and the added mass coefficient of the puff motion is found to be approximately 1. The predicted puff flow and mixing rate are well-supported by experimental and field data. The analogy between the bentover phase of a jet in crossflow (the advected line puff) and the corresponding line puff is examined. The coefficients in the self-similar relations are derived from the numerical solution, and a practical application example is given.

1. INTRODUCTION

Jets and plumes are turbulent flows produced by momentum and buoyancy forces. Examples include cooling tower and smokestack emissions, fires and

---

[1]Professor, Dept. Civil Engineering, The University of Hong Kong, Hong Kong, China, hreclhw@hkucc.hku.hk

[2]Professor, Centre for Environmental Sciences and Dept. of Mechanics, Peking University, Beijing, China, chen@ns.nlspku.ac.cn

[3] Senior Engineer, Dept. River & Harbour Engineering, Nanjing Hydraulic Research Institute, Nanjing, China, ckuang@hkucc.hku.hk

volcano eruptions, deep sea vents, atmospheric thermals, marine outfall sewage discharges, thermal effluents from power stations, and ocean dumping of sludge. The prediction of the path and mixing of jets and plumes are important for sound environmental control and impact assessment. The study of this important turbulent shear flow is a key to understanding a variety of environmental and industrial mixing problems.

A good grasp of jets and plumes greatly facilitates the understanding of water quality control. Environmental regulations and water quality objectives are mostly set in terms of pollutant concentrations in receiving waters. It is important to be able to predict the concentration distribution in the vicinity (the near field) of a discharge for a given discharge design and location, waste load, and environmental conditions. This is required for defining mixing zones and to minimize the impact of discharges on sensitive receivers (e.g. nearby beaches for amenities, wetland reserve, or fisheries). The proper design of a submarine outfall - a hydraulic structure, can have a profound effect on the water quality actually observed near the discharge.

Environmental discharges as turbulent jets and plumes form a class of 'active' designs for which the engineer can control the amount of mixing, and hence the degree of dilution, near the discharge point. Within this near field, the discharge itself generates the flow field and mixing. The velocity and concentration fields are hence coupled and have to be solved simultaneously. An important engineering parameter is the dilution. The dilution at a point is defined as the ratio of the discharge concentration (or concentration excess above the ambient) to the concentration (excess) at the point. This is a measure of the mixing capacity of the discharge.

The mixing of jets and plumes in an otherwise stagnant fluid has been well studied. The fluid motion for this case is self-similar, and simple analytical solutions for straight jets and plumes can be developed from either experiments or integral models (e.g. Fischer 1979; Chu and Lee 1996). The jet/plume characteristics (velocity, width, volume flux, dilution) at any height can be reliably predicted, and outfall design is often based on the calculation of a 'stillwater dilution' - i.e. the receiving water is assumed to be stationary. On the other hand, both experiments and field studies have shown the presence of even a weak current enhances dilution considerably (e.g. Wright 1977; Lee and Neville-Jones 1987). For effluent discharges in shallow waters, environmental guidelines often cannot be met if the realistic effect of a current (which is present for the greater part of the time) is simply ignored. Nevertheless, the mixing of a jet in crossflow is a considerably more difficult fluid mechanics problem. A reasonable understanding and reliable predictions have been made possible only by recent advances of numerical and experimental modelling techniques.

In this Chapter a detailed study of a turbulent jet in crossflow is presented. First the problem is stated together with some experimental observations and an overview of previous work. Second, an asymptotic analysis based on the use of length scales is given; simple relations for the straight jet and the bent-over jet are derived. Third, a fully three-dimensional turbulence model calculation is presented in Section 3 for an advected line puff - a special 3D flow which is particularly representative of the bent-over jet in crossflow. The numerical predictions of the flow and scalar field are discussed and compared with experimental data. Finally, an application example is provided and some concluding remarks on related work given.

2. THE JET IN CROSSFLOW PROBLEM

When a jet is directed into a crossflow, the interaction of the jet momentum with the current results in a complicated turbulent shear flow. Different kinds of vortices are generated close to the source, in particular a pair of twin-vortices in the bent-over phase of the jet. Fig.1 and Fig.2 show a side view as well as a section view of a flow visualization of a vertical momentum jet in crossflow using the laser-induced fluorescence (LIF) technique (Chu and Lee 1996). It is observed that close to the jet source there are ring vortices characteristic of shear entrainment (similar to jet in stagnant fluid), and wake vortices due to the blockage of the current by the jet itself. In the bent-over phase, however, the jet cross-section is characterised by a vortex pair; the scalar field is bifurcated as seen in the LIF image in Fig.2. The double vortex induces a large scale entrainment that is manifested in the form of tornado vortices drawn into the jet (Fric and Roshko 1994, Chu and Lee 1996). As expected, the mixing is greatly enhanced by the vortex pair - an environmentally friendly gesture of nature as effluents as a rule are discharged into moving fluids. We seek a prediction of the turbulent-mean velocity and scalar field of this flow.

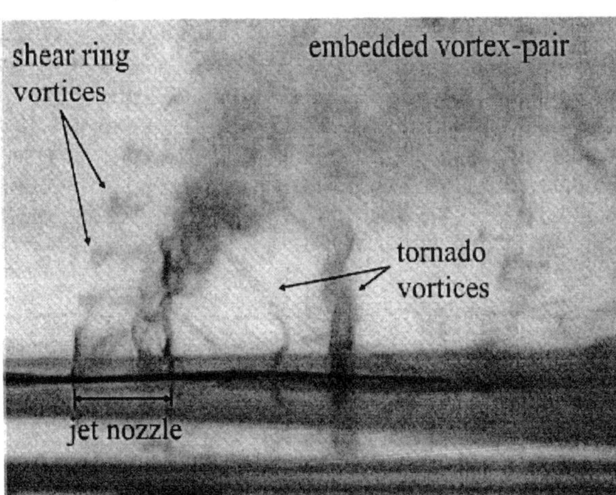

Fig.1 Momentum jet in crossflow: observed vortices (Chu and Lee 1996)

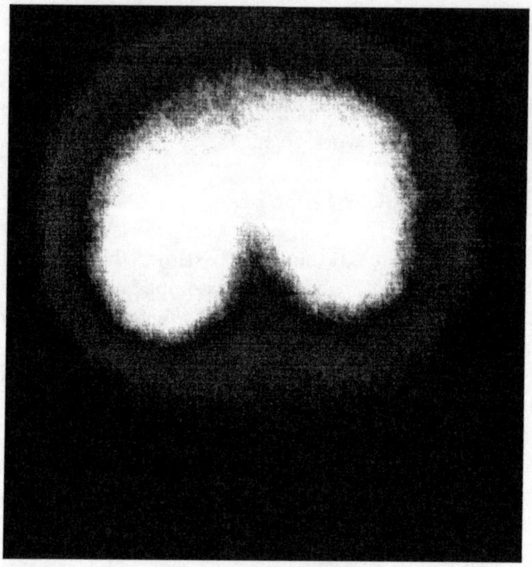

Fig.2 Scalar concentration field of a bent-over jet cross-section (Chu 1996)

2.1 Previous Work

There is an extensive literature on the jet in crossflow problem. Many experimental studies (e.g. Keffer & Baines 1963; Kamotani & Greber 1972; Rajaratnam 1976; Moussa,Trischka & Eskinazi 1977; Andreopoulos & Rodi 1984) have revealed some of the essential features of the fluid dynamics, among which the most striking is a transition from an initially vertical jet through a bent-over phase during which the jet becomes almost parallel with the main free stream, to a secondary vortex pair flow in the transverse section of the jet. However, most of these studies were concentrated in a region very close to the source, in the order of 5-10 diameters. On the other hand, for environmental mixing studies, the region of 'active mixing' is typically in the order of the water depth, which can extend to beyond 100 jet diameters downstream

To study the mixing of a passive scalar carried in the jet fluid, visual trajectory and limited tracer concentration measurements in the bent-over phase of a momentum-dominated buoyant jet in crossflow have been made (Ayoub 1971; Chu & Goldberg 1974; Wright 1977). However, most of the concentration measurements were limited to a vertical traverse along the jet centerline; the measured dilution rates also exhibited considerable scatter. On the other hand, numerical studies (Patankar, Basu & Alpay 1977; Chien & Schetz 1975; Demuren 1983; Chu 1985; Sykes, Lewellen & Parker 1986) failed to offer any detailed view of the scalar field, due to the relatively low-resolution and small computational domain used. In the systematic numerical study of Sykes, Lewellen

& Parker (1986), the computational domain extends only 15 jet diameters downstream from the jet source; the computed cross-sectional scalar concentration field fails to reveal any double peak structure. A basic knowledge of the scalar field is useful not only for illustration and gaining insight into the flow; it is needed for interpretation of experimental data or linking cross-section average jet properties predicted by integral models with local properties. For example, the aspect ratio of the jet cross-section, and the ratio of the maximum concentration in the cross-section to that along the jet centerline axis are two important parameters which are not resolved by previous studies.

As the only analytical endeavour, the bent-over jet in crossflow problem was treated by Yih (1981) via a two-term approximation of power expansion, assuming a constant x-velocity and negligible diffusion in the direction of the crossflow. While the flow field was illustrated for an unrealistic small (by a factor of 50) value of dimensionless eddy viscosity for which the approximation is valid, the mixing process was not studied. Using this free shear layer model, a self-similar solution for the bent-over jet has recently been obtained numerically (Lee, Li and Cheung 1999). By the nature of the computation, although the jet bifurcation was revealed, satisfactory comparison with data was obtained only for some of the parameters.

2.2 Length Scale Analysis and the Line Puff analogy

Consider a round turbulent jet, of diameter D, initial velocity U_{jet}, and tracer concentration C_o discharging into a steady uniform crossflow U_a in the horizontal +x direction (Fig.3). The jet is oriented at an angle of θo to the x-axis, with initial vertical velocity $Wo = U_{jet} sin\, \theta o$. The jet volume flux is $Q = U_{jet}\, \pi D^2 / 4$; the vertical jet momentum flux is $Mvo = QU_{jet} sin\, \theta o$. The behaviour of a jet is characterized by its momentum flux. The velocity induced by the vertical momentum varies as $w \bullet M_{vo}^{1/2} z^{-1}$ (see later discussion). As the jet-induced velocity decreases, the jet becomes bent over due to entrainment of ambient fluid with crossflow momentum (at velocity Ua). At a vertical distance when $w \sim Ua$, or $z \bullet M^{1/2}{}_{vo}/U_a$, heuristically we would expect the jet to be significantly bent over. Thus a crossflow momentum length scale $Lmv = M^{1/2}{}_{vo}/U_a$ can be defined. For $z/Lmv = zUa/M^{1/2}{}_{vo} << 1$, the jet-induced velocity is significantly larger than the ambient current, and the jet is only slightly affected by the current. For $z/Lmv = zUa/M^{1/2}{}_{vo} >> 1$, the jet-induced velocity is much less than the ambient current, and the jet is significantly bent over. These two asymptotic regimes are referred to as the *momentum-dominated near field* (MDNF, or the jet-dominated regime) and *momentum-dominated far field* (MDFF, or the current-dominated regime) respectively. Note that in the general case the horizontal exit velocity Uo need not be the same as the ambient velocity Ua.

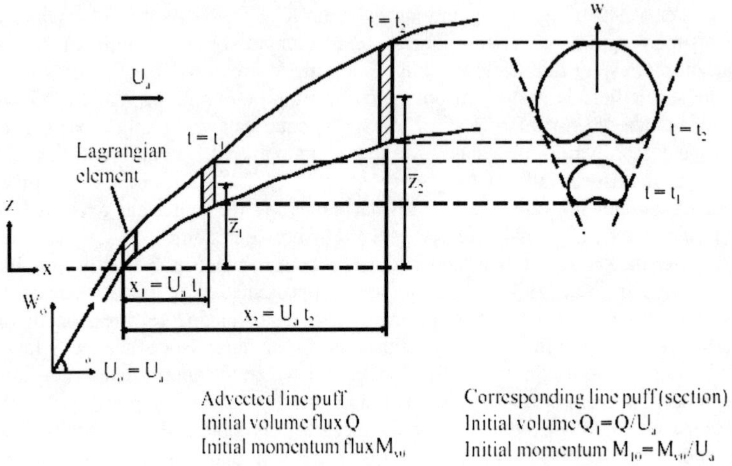

Fig.3 An advected line momentum puff in a crossflow

For $z/Lmv = zUa/M^{1/2}_{vo} << 1$ the jet is only slightly bent over. Thus the mixing is similar to that of a momentum jet in still fluid in this MDNF. In this region, the jet velocity and concentration are radially-symmetrical, self-similar, and can be well-approximated by Gaussian distributions. By regarding the jet as only advected by the current ($x = Uat$), the trajectory and dilution relations can be obtained from the solutions of a classical pure jet (Wright 1977; Fischer 1979; Chu and Lee 1996 - see later discussion).

For $z/Lmv >> 1$, the jet is significantly bent over; it is expected that the horizontal exchange of momentum between the jet and the ambient current is complete - i.e. the x-velocity of the jet is approximately equal to *Ua*. Consider a control volume with outer boundaries far removed from the source, along which the pressure is assumed to be hydrostatic (Fig.4). By momentum conservation, the vertical jet momentum flux across a vertical section, $M_v = \int_{x-section} uw\, dA$, where u,w are the streamwise (x) and vertical (z) velocities, is then equal to the initial jet discharge momentum flux *Mvo*. In the bent-over phase, since $u \approx Ua$, we have $Mv \approx Ua \int w\, dA$.. The flow in successive vertical x-sections can then be regarded as equivalent to that of a line puff with initial momentum $Mlo = Mvo/Ua$ imparted in the +z direction, and initial volume $Qlo = Q/Ua$, per unit x-length.

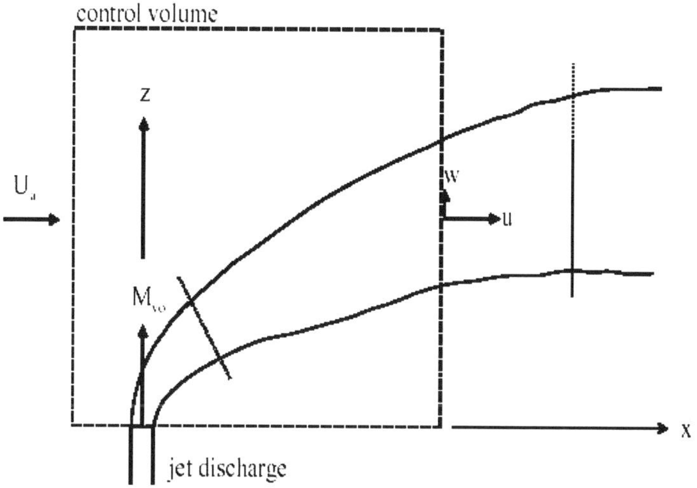

Fig.4 A turbulent jet in crossflow: vertical momentum balance

Intuitively, the jet in crossflow can be viewed as issuing line momentum puffs at a steady rate (Fig.3). As far as the mixing in the bent-over phase is concerned, the action of the jet is to impart a vertical (+z) kinematic momentum flux of $M_{vo} = QW_o$ to the flow streaming by at U_a, where $Q = \pi \frac{D^2}{4} U_{jet}$ is the initial jet volume flux. In a continuous manner, the jet distributes a vertical momentum of $M_{vo}\,\Delta t$ to an ambient fluid element of length $U_a\,\Delta t$. Each of these Lagrangian elements (advected momentum puffs) receives an impulse of $\frac{M_{vo}\,\Delta t}{U_a\,\Delta t}$ per unit x-length as they pass by the source. To an observer moving with the crossflow, the ambient environment is stagnant while the turbulent element rises as a line puff. The mixing of the 3D bent-over jet at successive *x*-sections ($x1$, $x2$) then correponds to that of an equivalent line puff at corresponding times ($t1$, $t2$) via a Galilean transformation. This analog, to be confirmed by the experiments, is expected to hold in the bent-over stage of the jet, $z \bullet L_{mv}$.

2.3 Similarity Relations

A two-dimensional line puff is a transient flow (in the y-z plane) resulting from injecting a patch of fluid with initial volume Q_{lo} $[L^2]$ (per unit x-length) and significant initial momentum M_{lo} $[L^3/T]$, into a water body at rest (Fig. 5). The initial tracer concentration is C_o. The unsteady fluid flow is characterized by a turbulent core (the 'colored patch').

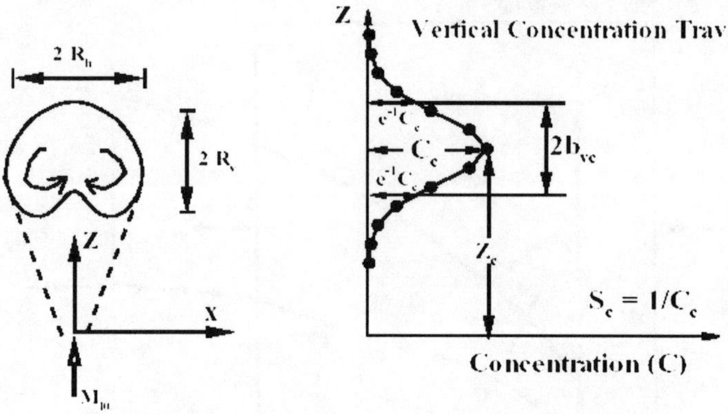

Fig.5 Definition of section puff parameters

Assuming self-similarity, the puff can be described by a characteristic vertical location *z*, half-width b, vertical velocity *W*, and concentration *C*. It can be shown by dimensional analysis that a characteristic time scale can be defined from *Qlo* and *Mlo* to express the relative importance of the initial volume and momentum flux:

$$t_c = \frac{Q_{lo}^{\frac{3}{2}}}{M_{lo}^{\frac{1}{2}}} = \frac{Q^{3/2}}{U_a^{1/2} M_{vo}} \tag{1}$$

or, using a Galilean transformation, a characteristic length scale.

$$L_x = U_a t_c = \frac{Q^{3/2} U_a^{1/2}}{M_{vo}} \tag{2}$$

As the flow is entirely driven by the initial impulse, for *t/tc* >> 1 the effect of the source volume will be negligible, and the variation of puff characteristics depends only on *Mlo* and the time t. We therefore have the following similarity relations for the commonly concerned quantities, including the maximum velocity decay, spreading rate, trajectory, and scalar dilution of the line puff:

$$W \sim M_{lo}^{1/3} t^{-2/3} \quad \text{or} \quad \frac{W_o}{W_m} \sim \left(t^*\right)^{2/3} \tag{3}$$

$$b \sim z \tag{4}$$

$$z \sim M_{lo}^{1/3}\, t^{1/3} \quad \text{or} \quad z \sim Q_{lo}^{1/2} \left(t^*\right)^{1/3} \tag{5}$$

where dimensionless time $t^* = t/tc$, with the time scale $t_c = Qlo^{3/2} / Mlo$. Conservation of tracer mass also implies:

$$\frac{C_o}{C} \sim \frac{b^2}{Q_{lo}} \sim \frac{M_{vo}^{2/3}\, t^{2/3}}{Q_{lo}} \sim \frac{z^2}{Q_{lo}} \tag{6}$$

These relations indicate that the puff Reynolds number Re and the circulation around one half of the puff, •, both • *Wb*, decrease slowly as $\sim t^{-1/3}$.

Given the equivalence between successive vertical x-sections of a bent-over jet and the line puff at corresponding times $t = x/Ua$, the following relations can be obtained for the spreading rate, trajectory, and dilution of the 3D bentover jet in terms of the crossflow momentum length scale *Lmv*.

$$b_{vc} \sim z_c \tag{7}$$

$$\frac{z_c}{L_{mv}} \sim \left(\frac{x}{L_{mv}}\right)^{1/3} \tag{8}$$

$$\frac{S_c Q}{U_a L_{mv}^2} \sim \left(\frac{z_c}{L_{mv}}\right)^2 \tag{9}$$

$$\frac{W_o}{W_m} \sim \left(x^*\right)^{2/3} \tag{10}$$

where *bvc* is the vertical centerline half-width defined by the $e^{-1}Cc$ points, *zc* is the vertical location defined by the maximum centerline concentration *Cc*, *Sc* = *Co/Cc* is referred to as centerline minimum dilution, as illustrated in Fig.5; and *Wo* and *Wm* stand for the initial and maximum vertical velocity of the jet, respectively, with the dimensionless downstream distance $x^* \equiv x/Lx$ (the jet exit is located at x = 0). For the sake of reference, Eq.(8) and (9) can be expressed alternatively as:

$$z_c = \left(Q/U_a\right)^{1/2} \left(x^*\right)^{1/3} \tag{11}$$

$$S_c \sim z_c^2 \Big/ \left(Q/U_a\right) \tag{12}$$

2.4 Some Remarks on the Line Puff

The above similarity relations furnish a useful means of correlating experimental data; the proportionality constants can be derived by best-fitting experimental data or numerical results with these relations. It should be noted, however, that these relations are strictly an outcome of dimensional analysis and the assumption of self-similarity. The coefficients do not provide any information on the structure of the flow. For example, the large discrepancies between the dilution constants derived from different data sets cannot be explained without reference to the scalar field. In addition, in view of the fast decay of the puff velocity ($W \bullet t^{-2/3}$), the duration or extent of the puff flow that can be studied is rather short. Previous investigations of a line puff or jet in crossflow are limited in several respects. First, it is difficult to generate two-dimensional puffs in the laboratory and to carry out detailed measurements in these puffs. In a classic study (Richards 1965), line puffs were simulated by short-pulsed water jets. Based on visual observations, it was concluded that the flow was self-similar and very much like that of line thermals. However, the observed puff spreading rate exhibited great scatter, and differed from that of line thermals by a factor of two (Scorer 1978). Little quantitative details on the flow was given, and the passive scalar field was not studied. Second, the considerable literature on the jet in crossflow problem dealt mainly with the flow details in the near field; there is also significant scatter and discrepancy in the correlation of the probe-based measurements of vertical centerline concentration profiles (Fan 1967; Ayoub 1971; Chu 1974; Wright 1977). Third, the role of the added mass effect in the momentum conservation has not been examined. This is possibly related to the scarcity of data on the scalar and flow fields in the bent-over jet; even the jet boundary could not be clearly defined, and the vortex flow has also not been measured.

More recently, the concept of advected line puff (ALP) - a special configuration of a near-vertical jet introduced at no excess horizontal momentum into a horizontal crossflow, was advanced and experimentally studied (Knudsen & Wood, 1986, 1988; Lee, Rodi & Wong, 1996; Gaskin, 1995; Chu 1996), as an indirect methodology to study the line puff. In the recent experimental studies, emphasis was given to measure the hitherto unavailable passive scalar field of a turbulent line puff using the analogy between a line puff and advected line puff. The measurements in the steady ALP flow are much more reproduceable than those in the unsteady 2D puff; empirical constants for the measured trajectory and mixing rates in the self-similar regime were satisfactorily derived. The advected puff width, trajectory, concentration dilution and maximum vertical velocity decay are found to follow self-similar laws similar to those of a line puff. In addition, detailed numerical studies on time dependent line puffs using the standard and renormalization (RNG) two-equation k - ε models (Lee, Rodi & Wong, 1996; Lee & Chen 1998) have been carried out. While some of the features of the bent-over jet are captured in these numerical calculations, the constants in the similarity laws characterizing scalar dilution and maximum vertical velocity decay differ substantially from the ALP measurements.

In the next section, a numerical solution of the flow and scalar field of the 3D advected line puff is presented. The computed scalar field in the bent-over jet will be studied using the line puff relations; the coefficients in the characteristic relations will be derived and compared with experiments on 3D advected line puffs (Wong 1991; Chu 1996).

3. NUMERICAL MODELLING OF ADVECTED LINE PUFFS

3.1 Advected Line Puffs

The **advected line puff** is produced by discharging the jet at an angle such that the x-component of the source velocity is the same in magnitude as the crossflow, with $Uo \equiv Ujet\cos\theta o = Ua$ (Fig.3). The jet imparts a vertical (+z) excess (kinetic) momentum flux of $Mv0 = QUjet\sin\theta o$ to the flow streaming by at velocity Ua, where Q is the initial jet volume flux. In the absence of body forces, the jet excess momentum flux is conserved. With no initial excess xmomentum, the averaged excess x-velocity of the jet is zero and the jet may be assumed to be globally advected with a horizontal velocity of Ua. The motion and characteristics of this advected jet are investigated.

We perform a 3D numerical experiment of an advected line puff in very much the same way as if one were doing a laboratory experiment. A near vertical non-buoyant jet is discharged at no excess horizontal momentum into a horizontal crossflow, and the fluid motion and scalar mixing of such a jet with the surrounding crossflow are investigated. As a first approximation, the standard two-equation k - ε. model is adopted for turbulence closure. The numerical results are then compared with laboratory experimental data and discussed.

3.2 Governing Equations

We take the Reynolds-averaged equations for constant density incompressible flow:

$$\frac{\partial U_i}{\partial x_i} = 0 \tag{13}$$

$$U_j \frac{\partial U_i}{\partial x_j} = -\frac{1}{\rho}\frac{\partial p}{\partial x_i} + \frac{\partial \tau_{ij}}{\partial x_j} \tag{14}$$

where Ui = fluid velocity in (x,y,z) direction, ρ= density, p = dynamic pressure, τij = stress tensor. τij is equal to the sum of the viscous and Reynolds-stresses. The eddy-viscosity model (Boussinesq hypothesis) provides the following expression for the stresses:

$$\tau_{ij} = \nu_{eff}\left(\frac{\partial U_j}{\partial x_i} + \frac{\partial U_i}{\partial x_j}\right) - \frac{2}{3} k\,\delta_{ij} \tag{15}$$

where the effective viscosity $\nu_{eff} = \nu + \nu_t$,ν is the molecular viscosity, ν_t = $C_\mu k^2/\varepsilon$ is the turbulent viscosity, k is the turbulence kinetic energy (TKE), ε is referred to as dissipation rate (DR) of k. We adopt the standard two-equation k - ε model of Launder and Spalding (1974) for turbulence closure:

$$U_j \frac{\partial k}{\partial x_j} = \frac{\partial}{\partial x_i}\left(\frac{\nu_t}{\sigma_k}\frac{\partial k}{\partial x_i}\right) + \tau_{ij}\frac{\partial U_i}{\partial x_j} - \varepsilon \tag{16}$$

$$U_j \frac{\partial \varepsilon}{\partial x_i} = \frac{\partial}{\partial x_i}\left(\frac{\nu_t}{\sigma_\varepsilon}\frac{\partial \varepsilon}{\partial x_i}\right) + C_{1\varepsilon}\frac{\varepsilon}{k}\tau_{ij}\frac{\partial U_i}{\partial x_j} - C_{2\varepsilon}\frac{\varepsilon^2}{k} \tag{17}$$

with the following standard empirical model coefficients: $C\mu$ = 0.09, $C1\varepsilon$ =1.44, $C2\varepsilon$ = 1.92, σk = 1.0, $\sigma\varepsilon$ = 1.3 (Rodi 1980). In the above, the turbulent closure is achieved by computing the turbulence kinetic energy locally; the generation, transport, and decay of turbulence can hence be accounted for. TKE is computed by a transport equation for k, while the equation for ε is by and large empirical. The eddy viscosity $C\mu \bullet k2/\varepsilon$ is then obtained by dimensional analysis.

In addition to the flow equations, the study of puff characteristics necessitates the calculation of a passive scalar field from the tracer mass conservation equation:

$$U_j \frac{\partial C}{\partial x_j} = \frac{\partial}{\partial x_j}\left(\frac{\nu_t}{S_{ct}}\frac{\partial C}{\partial x_j}\right) \tag{18}$$

where Sct stands for the turbulent Schmidt number, taken to be 0.75 as found appropriate for a related problem (Sykes et al. 1986).

4. Numerical Solution

Computational Parameters and Boundary Conditions

In all the results reported herein, the following parameters are adopted: Ua = 0.1887 m/s, Co = 1, θ_0 = 80, $Ujet$ = 1.086 m/s, $Mvo = 3.23 \times 10^{-4}$ m^4/s^2, and $Q = 3.02 \times 10^{-4} m^3/s$. The domain for computation is chosen to be$D \equiv (X1, X2) \times (Y1, Y2) \times (Z1, Z2) = (-0.1, 1.1) \times (-0.4, 0.4) \times (0.0, 0.6)$ with a length of L = 1.2 m, a width of B = 0.8m and a depth of H = 0.6 m. The jet exit $D0$ is given as $x^2 + y^2 \bullet r^2$ on the lower boundary, with a radius $r = 9.48 \times 10^{-3} m$. The computational domain extends for 1.1 m downstream of the jet exit, greater than 55 jet diameters.

All the parameters correspond to or are comparable with those in previous laboratory and numerical experiments of advected line puffs and line puffs. The characteristic length and time scales have values of $Lmv = 9.53 \times 10^{-2} m$, $Lx = 7.06 \times 10^{-3} m$ and $tc = 3.74 \times 10^{-2}$ s, respectively. The downstream outflow boundary corresponds to a dimensionless distance of $x^* > 150$; this is judged to be remote enough for resolving the whole process of spatial evolution of the jet (see later discussion).

By virtue of symmetry, the numerical solution is sought for half of the flow domain, y • 0. On the surface of symmetry at y = 0, V = 0 is imposed, and normal gradient of all other variables are set to zero. On the jet exit at $z = 0$ with x^2+y^2 • r^2, we give $U = Ua$, $V = 0$, $W = Wo$, $C = C0$, $k = k0$, $\varepsilon = \varepsilon_0$. On the inflow boundary at $x = X1$, $U = Ua$, $V = W = C = 0$, $k = ki$, $\varepsilon = \varepsilon_i$ are imposed. On the outflow boundary at $x = X_2$, a zero normal gradient is set for all variables. On the other boundaries, a zero normal gradient is also set for the concentration C and kinetic energy k, with the turbulent dissipation rate ε approximated by conventional procedure based on equilibrium assumption, while stress-free conditions are adopted for velocities as follows: $W = 0$, $\frac{\partial U}{\partial z} = \frac{\partial V}{\partial z} = 0$ for the lower boundary at $z = Z1$ with $x^2 + y^2 > r^2$, and the upper boundary at $Z = Z2$; $V = 0$, $\frac{\partial U}{\partial y} = \frac{\partial W}{\partial y} = 0$ for the lateral boundary at $y = Y2$. The initial vertical velocity and tracer concentration are given the value of $W0 = 1.07$ m/s (according to $Mv0$) and $C0 = 1$ respectively at the jet exit. The values of $k0$ and ε_0 for the jet exit are assumed to be $0.0171 m^2/s^2$ and $0.1298 m^2/s^3$, corresponding to an eddy viscosity estimated by a free shear model $\nu_t = 2\alpha_\nu W_o \gamma$, with $\alpha_\nu \sim 0.01$. For numerical reasons, negligibly small initial values of $(ki, \varepsilon_i) = (5.34 \times 10^{-8}, 2.57 \times 10^{-7})$ are given for the non-turbulent inflow such that $\nu_{ti} < 0.001\nu$. Fluid properties are taken as that of water, i.e., $\rho = 10^3\ kg/m^3$ and $\nu = 10^{-6}\ m^2/s$.

With a moderately high exit to ambient velocity ratio of $rv \equiv W0/Ua = \tan \theta 0 = 5.67$, the lower boundary will not play a notable role so far as the major features and characteristics of the flow and dilution are concerned, according to closely related experimental and numerical studies (Andreopoulos & Rodi 1984; Sykes, Lewellen & Parker 1986). Actually, we do not treat the lower, upper and lateral boundaries as real true walls outside the jet-discharge region; our interest is in the advected line puff remote from the walls, and real walls introduce local boundary layers which need to be resolved by fine numerical meshes. We have no intention to simulate these details. Furthermore, it is found, via tests with other outflow boundary locations of $X2 = 0.8 - 2.0$, that the simplest type of open condition for the outflow boundary is suitable for $X2$ • 1.0.

Computational Procedure

The governing equations are solved numerically using the finite volume method (Patankar 1980) as embodied in the FLUENT code (Fluent Inc. 1995). The equations are discretized on a non-staggered grid on which Cartesian velocities and other variables are defined at the centre of control volumes. The quadratic upwind interpolation (QUICK) approximation of Leonard (1979) is used for spatial discretization. The discretized equations are solved iteratively using the SIMPLEC algorithm for velocity-pressure correction (Van Doormaal &

Raithby 1984). In all calculations, no under-relaxation of pressure is required, while a factor of 0.8 is adopted for velocities and other variables. Convergence is declared when the normalized residuals are less than 5×10^{-4}.

A $92 \times 26 \times 36$ (x-y-z) orthogonal grid is used for the half-geometry of this problem, with the half-jet exit defined over 18 cells. The minimum grid sizes are $\Delta x = 0.0028$, $\Delta y = 0.0028$ and $\Delta z = 0.0037$ respectively; in the region of the growing puff the grid size is no more than 0.006 m. To obtain accurate solutions, 2000 iterations suffice for all cases involved.

Extensive tests have confirmed that the general features and characteristics reported in this paper are unrelated to the numerical procedure adopted. Three simulations with different resolutions on a coarser 72×26×36 grid and two finer 112×26×36 and 92×32×42 grids were made to validate the numerical accuracy. All the contour plots are very similar to the high-resolution results; differences for all the non-dimensional parameters presented are limited to at most six percent. We are hence reasonably assured that the numerical results represent an accurate simulation of the advected line puff. Also, the solution in the asymptotic stage of the puff is rather insensitive to the initial value of k and ε assumed over a reasonable range; this result is similar to the numerical solutions of line puffs (Lee, Rodi & Wong, 1996; Lee & Chen 1998). In any case, the present results have captured the essentially self-similar flow in an interval wherein most of the changes take place.

5. NUMERICAL RESULTS AND DISCUSSION

5.1 Scalar Field

Fig.6 shows contour plots of the computed scalar field in the plane of symmetry at y = 0. The trajectory of the maximum concentration shows a rapid rise to about $z \approx 0.16$ at $x = 0.1$ corresponding to $x^* \approx 14$, whereafter the jet rises very slowly, reaching $z \approx 0.30$ at $x = 1.0$ corresponding to $x^* \approx 142$. The bent-over phase, during which the scalar contours become almost parallel with the mainstream, takes place essentially in the interval of $10 < x^* < 40$.

Fig.7 shows transverse cross sections of the scalar field at different downstream locations. As a salient feature of bent-over jet observed in experimental studies, a double peak concentration distribution can generally be observed in the kidney-shaped sectional puff, which becomes approximately self-similar for $x^* > 20$. This is in sharp contrast to the numerical results for line puffs (Lee, Rodi & Wong 1996; Lee & Chen 1998), where no double peak structure is observed for the asymptotic phase.

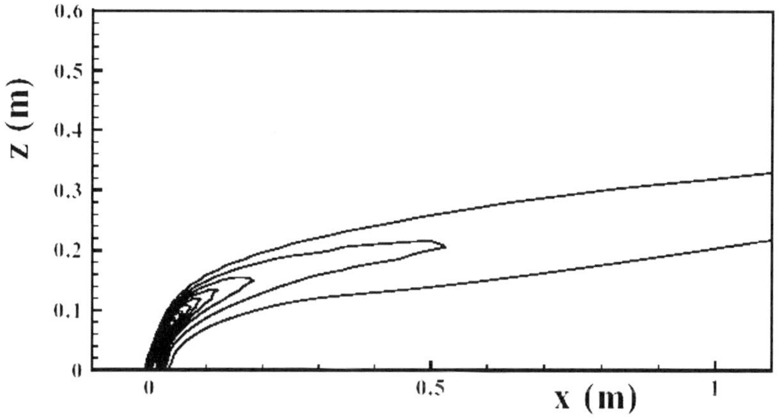

Fig.6 Computed scalar field in the plane of symmetry at plane y=0
(contour interval = 1/30)

i) *Puff aspect ratio*

As a basic parameter to characterize the sectional puff shape, the puff aspect ratio is defined as the ratio of horizontal half-width Rh over vertical half-width Rv of the contour $C = e^{-1}Cm$, with Cm stands for the maximum concentration over the puff section (Fig.5). The results show that the puff aspect ratio is roughly equal to 1.20 for the main stage of $20 < x^* < 70$, but reduces, with a minimum value of 1.05, for the earlier and later stages of $x^* \bullet 10$ and $x^* \bullet 100$. This is consistent with the experimental results of Chu (1996) who found a variation of puff aspect ratio from about 1.05 for the early and later stages to about 1.25 for the bent-over stage, and of Wong (1991) who found a value of 1.23 with probe-based measurements; and comparable to the value of 1.3 suggested by Pratte & Baines (1967) for the developed phase from an analysis of flow visualization. However, no preferred orientation is found in the numerical results for line puffs (Lee, Rodi & Wong 1996; Lee & Chen 1998). Considering the importance of the puff aspect ratio in characterizing the bent-over jet, it is also of interest to note related results on line thermals as follows: analysis of thermal boundaries in the classical study of line thermals (Richards 1965) suggests an averaged aspect ratio of 1.23; analysis of $e^{-1}Cm$ concentration contours measured by Knudsen & Wood (1988) and Gaskin (1995) in the developed phase of advected line thermals with Froude number greater than 1.0 suggests an averaged value of 1.12 and 1.19 respectively; the results of Wong (1991) on advected line thermals suggest a value of 1.0 in the initial phase and a similar value of about 1.2 in the developed phase.

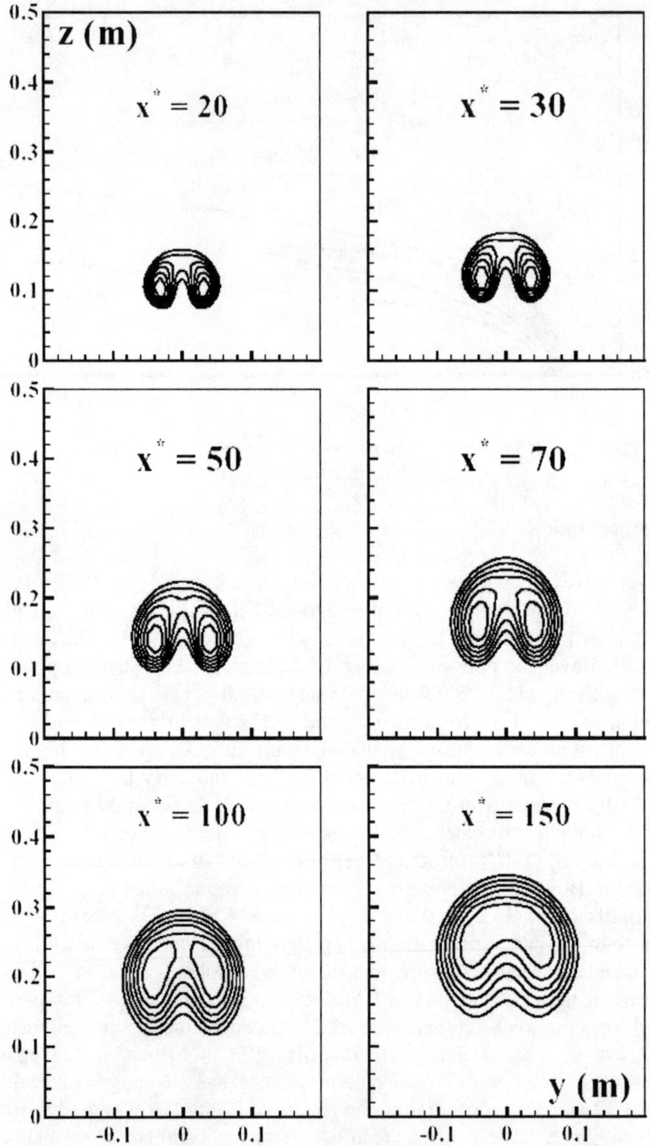

Fig.7 Transverse cross sections of the scalar field at different downstream locations (contour interval = $(1-e^{-1})C_m/6$, outmost concentration contour = $e^{-1}C_m$)

ii) *Maximum concentration ratio*

As observed from the concentration contours, the maximum concentration is not located in the centerline. It is desirable to correlate the centerline maximum concentration *Cc* to the sectional maximum concentration *Cm*, since only centerline concentration was measured in many previous studies. It is found that the ratio of *Cm/Cc* varies, from a smaller value of 1.05 for the initial and later stages of x^* • 10 and x^* • 130, to a higher value of 1.35 for the central phase of 30 • x^* • 120, with an averaged value of 1.2. This is consistent with the range of 1.0 - 1.5 with an average of about 1.2 found by Chu (1996), and compares well with the range of 1.1 to 1.5 found by Wong (1991) in their experiments on advected line puffs. This value is however somewhat lower than a value of about 1.7 given by Fan (1967), based on only two cross-sections of a buoyant jet in a crossflow. In contrast, the ratio is found to be 1.0 for the asymptotic phase in the numerical results for line puffs (Lee, Rodi & Wong 1996; Lee & Chen 1998), as there is no double peak structure and the maximum concentration is located on the centerline.

iii) *Centerline half-width*

Fig.8 shows the centerline half-width *bvc* plotted against centerline trajectory z_c. For z_c/Lmv • 1.2, corresponding to x^* • 10, the results follow the linear similarity law given as Eq.(7) in section 2.3. The corresponding coefficient, referred to as the centerline half-width spreading rate, is found as $C1p = 0.297$. This spreading rate is close to the value of 0.276 experimentally found by Wong (1991), and almost the same as 0.299 by Chu (1996), for advected line puffs; and well within the computed ranges of 0.25 - 0.291 in Lee, Rodi & Wong (1996) and of 0.29 - 0.349 in Lee & Chen (1998), for line puffs. It is also noted that there is some notable effect of virtual origin, due to the special conditions of jet exit.

iv) *Centerline trajectory*

Fig.9 shows the puff trajectory z_c/Lmv plotted against downstream distance *x/Lmv*. The results follow the power similarity law given as Eq.(8). Corresponding coefficient is found as $C2p = 1.56$ and related effect of virtual origin is negligible. This value is very close to the experimental values of 1.56 and 1.63 in Wong (1991) and Chu (1996), respectively, for advected line puffs; and comparable to the value of 1.77 from the buoyant jet experiment of Ayoub (Lee 1989). It also compares well with the computed ranges of 1.51 - 1.59 in Lee & Chen (1998) and 1.66 - 1.70 in Lee, Rodi & Wong (1996)

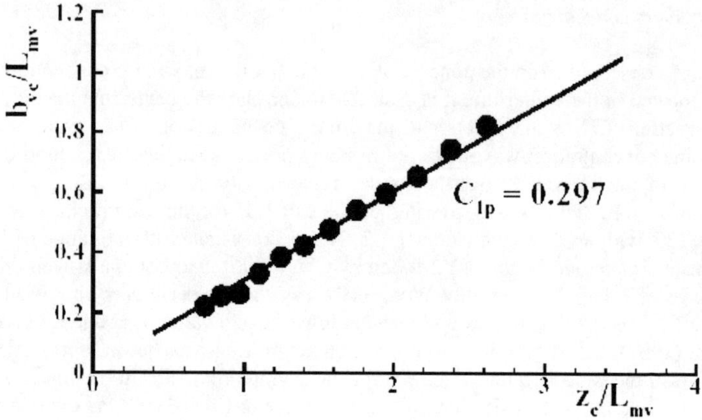

Fig.8 Centerline half-width b_{vc} vs centerline trajectory z_c

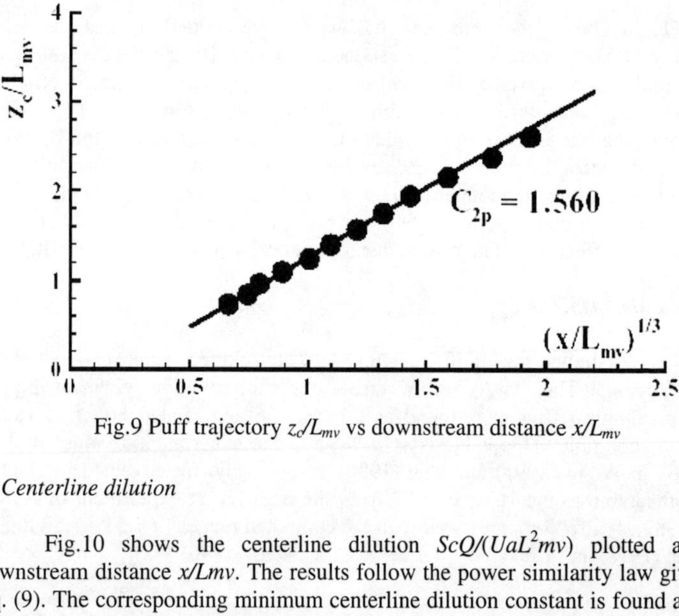

Fig.9 Puff trajectory z_c/L_{mv} vs downstream distance x/L_{mv}

v) *Centerline dilution*

Fig.10 shows the centerline dilution $ScQ/(UaL^2mv)$ plotted against downstream distance x/Lmv. The results follow the power similarity law given as Eq. (9). The corresponding minimum centerline dilution constant is found as $C3p$ = 0.488. No notable effect of virtual origin is observed. This value is comparable to the experimental values of 0.484 in Wong (1991) and 0.409 in Chu (1996), but much higher than the computed range of 0.28 - 0.29 for line puffs in Lee, Rodi &

Wong (1996) and Lee & Chen (1998). The is related to the bifurcation of the scalar field in the cross-section of the bent-over jet in cross-flow.

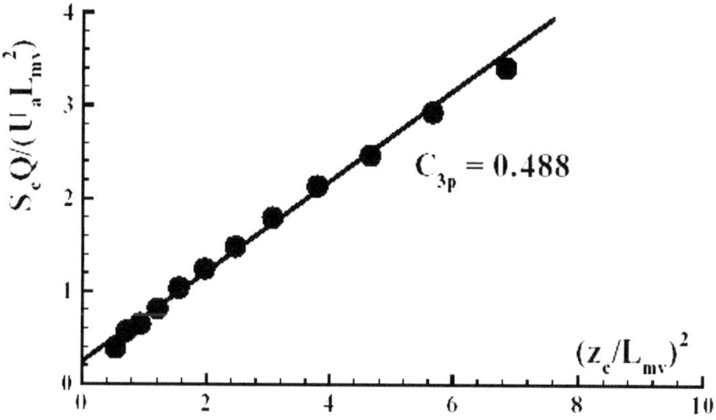

Fig.10 Centerline dilution $S_cQ/(U_aL^2_{mv})$ vs centerline trajectory z_c/L_{mv}

5.2 Velocity Field

Some details of the flow field near the jet exit are illustrated by the velocity-vector plots in the lower horizontal and symmetrical vertical planes (at z=0 and y=0, respectively) shown in Fig.11 and 12. The horizontal section shows a pattern typical of separated flow around a solid cylinder with acceleration around the sides and a reversed flow region extending roughly one diameter downstream. However, the vertical section shows significant upward velocity in the lee of the jet, indicating convergence into the wake; this is the entrainment mechanism which brings environmental fluid into the jet. The resulted secondary flow pattern, which is very similar to that in a line puff (Lee, Rodi & Wang 1996; Lee & Chen 1996), is shown in Fig.13, to illustrate the production of the vortex-pair system in the bent-over phase. The corresponding distributions of vertical velocity and pressure are shown in Fig.14 and 15.

i) *Vertical momentum and added mass effect*

Fig.16 shows the ratio of vertical momentum flux Mv in puff sections over initial vertical momentum flux $Mv0$ plotted against downstream distance x^*. The vertical momentum initially decreases to a value of $0.43Mv0$ at $x.$ = 35, then slightly increases and attains an asymptotic value of $0.46Mv0$ beyond x^* • 70. The reported result is based on the integral of $Mv \equiv \iint_{puff}$ UWdydz taking over the puff defined by the $0.10^{-1}Cm$ contour. Redefining the puff, by the $0.05Cm$ contour produced negligible changes to the asymptotic value, around $0.46Mvo$. The coefficient of 0.46 for x^* > 70 is very close to the experimental value of 0.47 found by Chu (1996) for the developed phase of x^* > 75

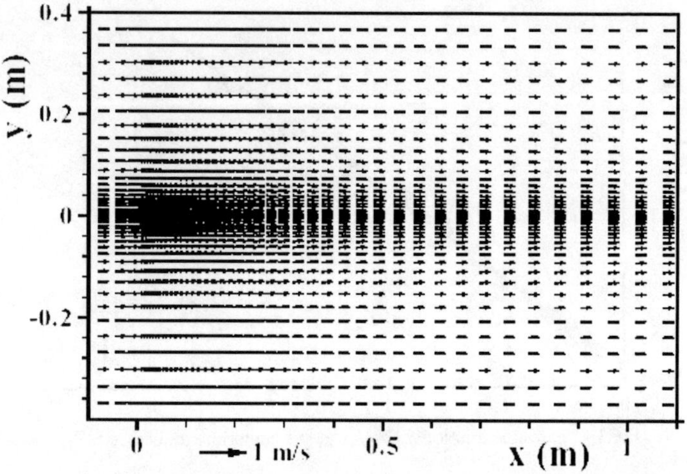

Fig.11 Velocity field in the lower horizontal plane at z = 0

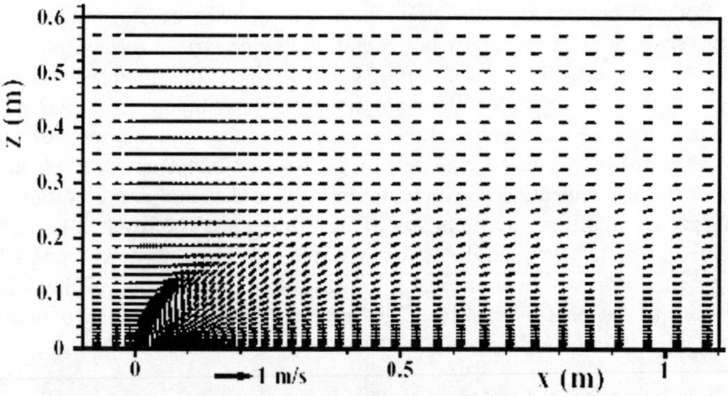

Fig.12 Velocity field in the symmetrical vertical plane at y = 0.

in adevected line puffs, and compares well with computed ranges of 0.483 - 0.513 by Lee, Rodi & Wong and of 0.502 - 0.519 by Lee & Chen (1998) for line puffs. This demonstrates that the vertical momentum of the line puff is conserved once the flow is developed. The advected line puff itself contains only about half of the initial vertical jet momentum, and the another half accounts for the approximately irrotational flow of the surrounding fluid set up by the moving jet. The substantial loss of initial momentum is also supported by a study of vortex rings (Glezer & Coles 1990) in which the remaining dimensionless momentum of the vortex ring was found to be about 0.3. The asymptotic value of 0.5 for the advected puff momentum, a fundamental property, can be interpreted as an added mass of the discharged fluid, with a vertical motion evolving into a transient line puff (Chu 1985; Turner 1986) in a Lagrangian view - implying an added mass coefficient of 1.

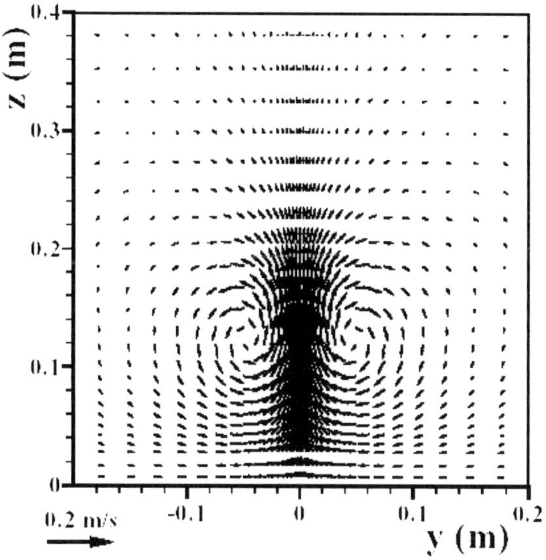

Fig.13 Sectional vortex-pair flow at $x^* = 35$ in the bent-over phase

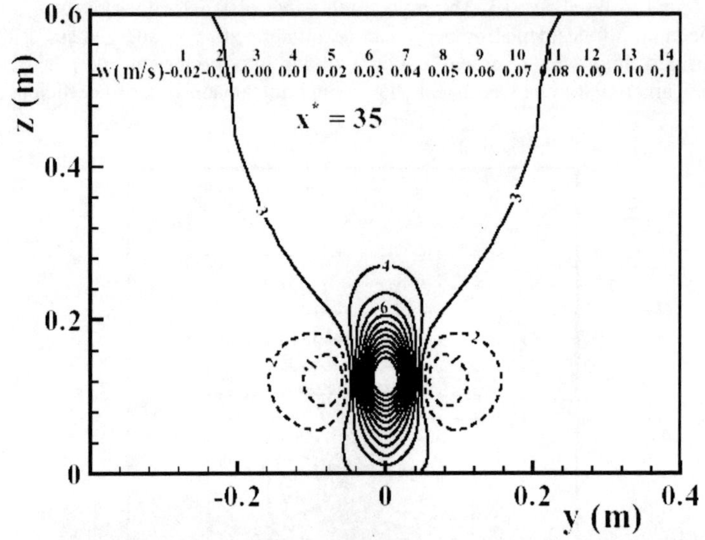

Fig.14 Sectional distribution of the vertical velocity *w* at x^* = 35 in the bent-over phase

ii) *Maximum vertical velocity decay*

Fig.17 shows the result of maximum vertical velocity over the initial vertical velocity of the jet plotted against the downward distance x^*. It is shown that the maximum vertical velocity decay follows the power similarity relation given by Eq.(4). The corresponding coefficient, referred to as the rate of maximum velocity decay, is found to be Cw = 0.75. Effect of virtual origin is negligible. This value of the rate is very close to the experimental result of 0.83 corresponding to Cv = 1.21 found in Chu (1996), but notably greater than the computed values of 0.58 - 0.65 corresponding to Cv = 1.55 - 1.71 (Lee et al 1996) ($Cv = 1/Cw$). This discrepancy means the maximum vertical velocity decays remarkably quicker in the advected line puff than does in the line puff.

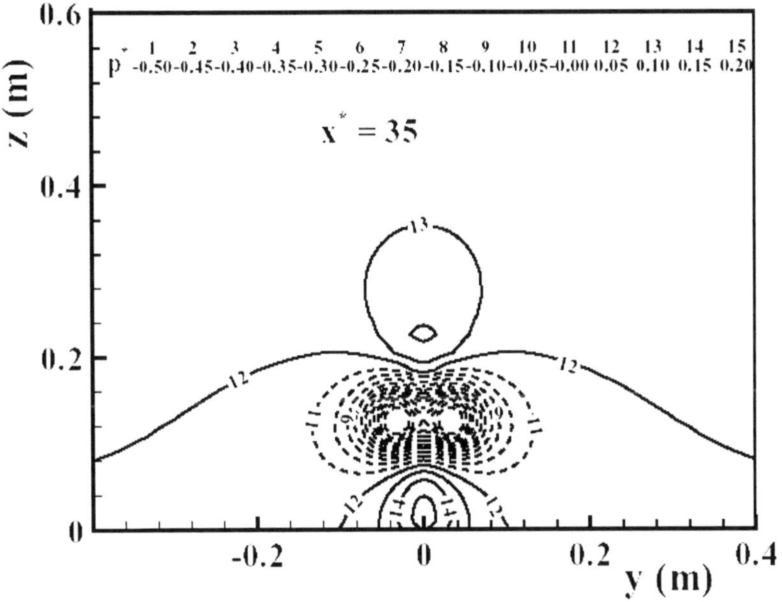

Fig.15 Sectional distribution of dimensionless pressure $p^* = p/(\rho\omega_m^2/2)$ at $x^* = 35$ in the bent-over phase

iii) *Turbulent kinetic energy and dissipation rate*

Similar to the sectional concentration distribution, the turbulent kinetic energy k and dissipation rate ε (hence turbulent viscosity ν_t) both display approximately preserving shapes for t^* • 20. Fig. 18(a) and (b) show the existence of double peaks of high turbulent kinetic energy and its dissipation rate (and hence the turbulent viscosity) located in the middle of both sides of the puff, somewhat similar to that of the scalar concentration field. This is also quite different to the case with the line puff (Lee, Rodi & Wong 1996; Lee & Chen 1998).

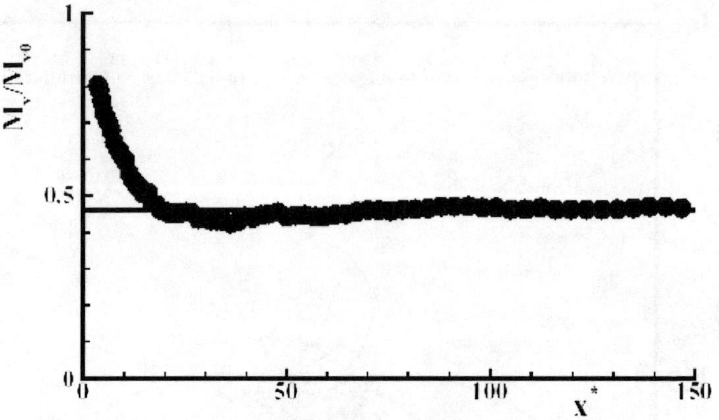

Fig.16 Ratio of vertical momentum flux M_v over initial vertical momentum flux M_{vo} vs downstream distance x^*

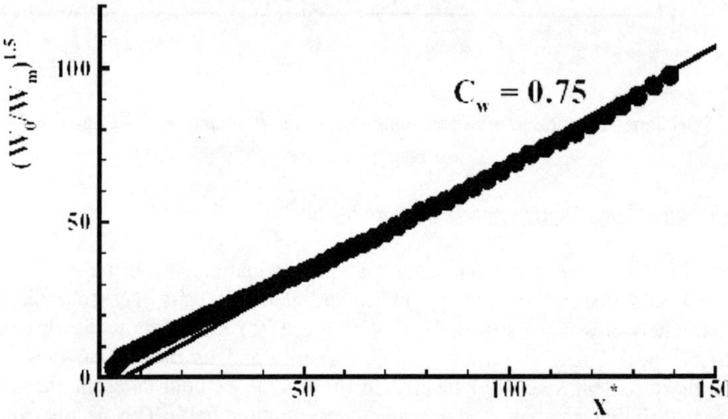

Fig.17 Ratio of maximum vertical velocity to initial vertical jet vs the downstream distance x^*

5.3 Analogy between a line puff and an advected line puff

In Table 1 the numerically derived coefficients in the similarity relations of an advected line puff (3D jet) are summarized and compared with experimental data. Also shown are the corresponding results obtained from turbulence models of a two-dimensional line puff, using either the standard k - ε model (SKE)

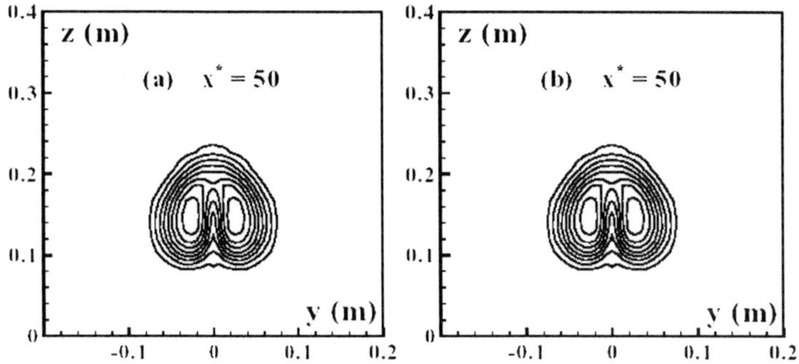

Fig.18 Computed fields of (a) turbulent kinetic energy (k) and (b) its dissipation rate ε at $x^* = 50$ in the bent-over phase (contour interval = one ninth of section maximum value; maximum $k = 3.5 \times 10^{-3}\ m^2/s^2$ and $\varepsilon = 3.6 \times 10^{-3}\ m^2/s$

(Lee, Rodi and Wong 1996) or the Renormalization Group (RNG) model (Lee and Chen 1998). It can be seen that in general almost all the predicted jet properties are in good agreement with experimental data, and certainly within the scatter in the data. It should be pointed out that the experiments by Wong (1991) and Chu (1996) are based on the ALP analog - the data generally exhibit much less scatter than previous investigations; in particular Chu's experiments were performed using non-intrusive LIF as well as LDA techniques.

	Parameter	SKE	RNG	3D JET (ALP)	Experiment
Puff Momentum	M_v / M_{vo}	0.497	0.502	0.46	0.47(Chu 1996) 0.30 (vortex ring)
Max Vertical Velocity	$W_m = C_v M_{lo}^{1/3} t^{-2/3}$	1.61	1.90	1.33	1.21 (Chu 1996)
Trajectory	$z_c = C_{2p} M_{lo}^{1/3} t^{1/3}$	1.70	1.54	1.56	1.44 (Chu 1974) 1.56(Wong 1991) 1.63 (Chu 1996)
Scalar Concentration	$C_o / C_c = C_{3p} z_c^2 / Q_{lo}$	0.28	0.28	0.488	0.484 (Wong 1991) 0.409 (Chu 1996)
	C_m / C_c	1.02	1.02	1.35	1.0-1.56 (Wong 1991) 1.1-1.5 (Chu 1996)
Centerline Half-width	$b_{vc} = C_{1p} z_c$	0.269	0.355	0.297	0.276 (Wong 1991)
Aspect ratio	R_h / R_v	1.03	1.00	1.20	1.23 (Wong 1991) 1.25 (Chu 1996) 1.3 (Pratt& Baines 1967)
Turbulence intensity	u_m / W_m	0.39	0.37	0.38	0.37-0.39* (Chu 1996)

Table 1. Similarity relations derived from numerical solution of Advected Line Puffs (ALP); numbers refer to dimensionless coefficients of equations or ratios. Also shown are experimental

values and numerical results of 2D line puffs using the standard k - ε (SKE) model and the Renormalization Group (RNG) model.

Compared to the present solution, the most significant shortcoming of the 2D line puff models is the inability to predict the bifurcated scalar field in the asymptotic stage ($c_m/c_c \approx 1$, and $R_h/R_v \approx 1$). This immediately leads to an underestimate of the centerline dilution constant. In addition, the velocity decay is significantly under-predicted by the 2D models.

The line puff is an analog to the bent-over jet or ALP. This correspondence in similarity laws is a strict outcome of dimensional analysis, not depending on any assumptions on the details of the problem. But detailed quantitative correspondence requires exact correspondence in governing equations for the two kinds of flows. Actually, by assuming a constant horizontal velocity for the advected line puff, essentially the same 2D equations for the transient line puff would have been obtained from the 3D equations for the 3D advected line puff, if we further assume negligible diffusion in the direction of the crossflow, as in the analytical solution of Yih (1981). The latter assumption enables one to drop all the terms corresponding to diffusion of related quantities (momentum components, scalar concentration, turbulent kinetic energy and dissipation rate), resulting in the so-called "parabolized" Navier-Stokes (PNS) equations (Chen, Gao & Chen 1991). These the two assumptions are further examined for the 3D advected line puff:

i) *Excess horizontal velocity in the advected line puff*

As the exit horizontal velocity of the jet is made equal to the main inflow velocity, the computed averaged horizontal velocity $U_{ave} \equiv \iint_{puff} \mathrm{U dydz} / \iint_{puff} \mathrm{dydz}$ of the puff is expected, due to momentum conservation, to remain equal to Ua, as shown in Fig.19; this is consistent with the LDA measurements of Chu (1996). However, the detailed velocity distribution could be remarkably different from a constant value of Ua, as illustrated by the computed x-velocity distribution in a typical cross-section (Fig.20). Here the relative excess velocity REV is defined as $(U - U_a)/U_a$. In Fig.21, the REV is plotted against the downstream distance x., where $REV_{max} = (U_{max} - U_a)/U_a$ stands for the maximum relative excess velocity; $REV_{min} = (U_{min} - U_a)/U_a$ stands for the minimum relative excess velocity; and REV_t stands for the total relative excess velocity, $(U_{max} - U_{min})/Ua$. It is shown that the sectional velocity difference can be greater than the averaged velocity itself for the initial phase of $x^* \bullet 10$, substantial for the bent-over phase and notable for well-developed phase. A double peak of REV is located at about the same region of scalar concentration maxima, which results in an excess downward advection of the scalar field, along with the averaged advection due to the averaged velocity Ua. This furnishes a major mechanism accounting for the difference between a line puff and an advected line puff.

ii) *Horizontal diffusion in the advected line puff*

For the scalar diffusion, Fig.22 shows a budget between terms of x-diffusion $\frac{\partial}{\partial x}\left(\frac{\nu_t}{S_{ct}}\frac{\partial C}{\partial x}\right)$, y-diffusion $\frac{\partial}{\partial y}\left(\frac{\nu_t}{S_{ct}}\frac{\partial C}{\partial y}\right)$, z-diffusion $\frac{\partial}{\partial z}\left(\frac{\nu_t}{S_{ct}}\frac{\partial C}{\partial z}\right)$ and total diffusion $\frac{\partial}{\partial x}\left(\frac{\nu_t}{S_{ct}}\frac{\partial C}{\partial x}\right)+\frac{\partial}{\partial y}\left(\frac{\nu_t}{S_{ct}}\frac{\partial C}{\partial y}\right)+\frac{\partial}{\partial z}\left(\frac{\nu_t}{S_{ct}}\frac{\partial C}{\partial z}\right)$, in the symmetrical plane of the advected line puff. It is seen that throughout the bent-over phase, the dominant terms are the transverse y-diffusion term and the horizontal x- and vertical z-advection terms (Fig.22 and Fig.23). The assumption of small or negligible streamwise diffusion when compared with advection is confirmed.

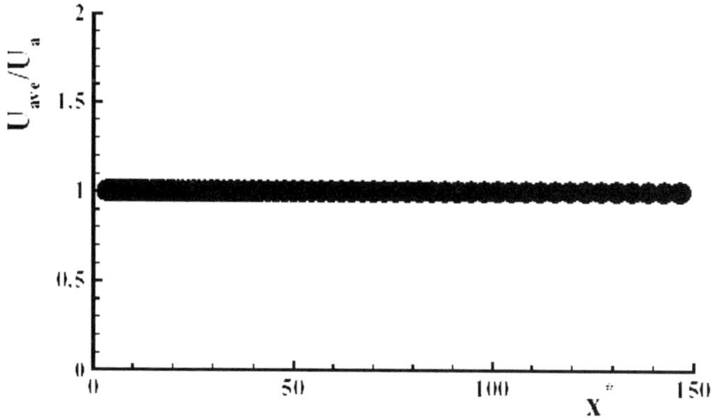

Fig.19 Averaged horizontal velocity vs the downstream distance x^*

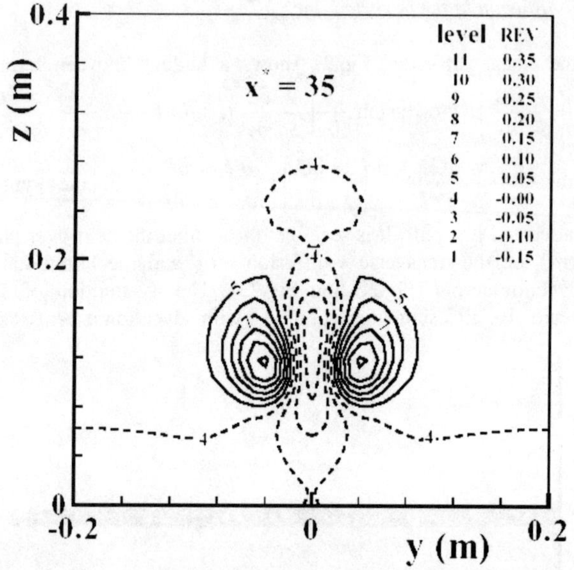

Fig.20 Sectional distribution of relative excess horizontal velocity $(u-u_a)/u_a$ at $x^* = 35$ in the bent-over phase.

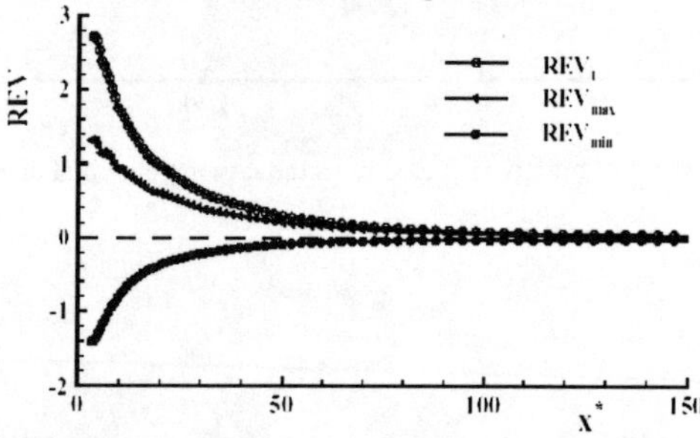

Fig.21 Relative excess horizontal velocity (REV) vs the downstream distance x^*

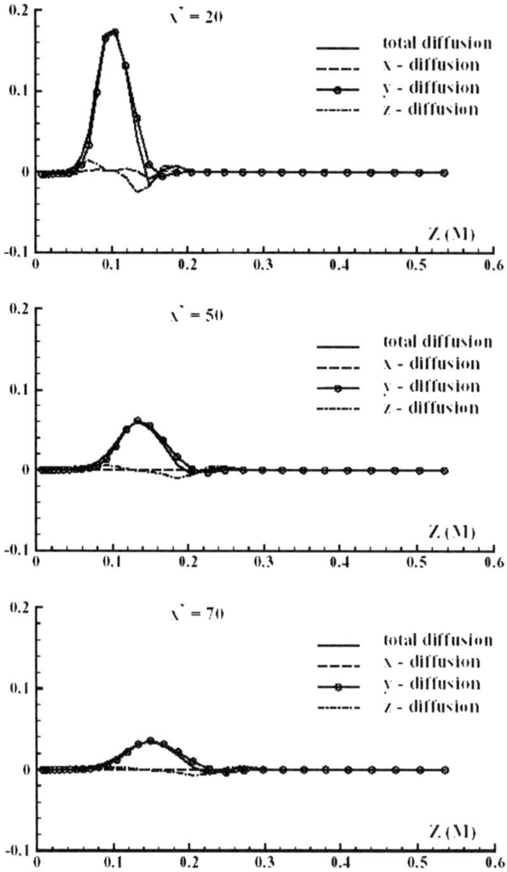

Fig.22 Sectional x-diffusion $\frac{\partial}{\partial x}\left(\frac{vt}{Sct}\frac{\partial C}{\partial x}\right)$, y diffusion $\frac{\partial}{\partial y}\left(\frac{vt}{Sct}\frac{\partial C}{\partial y}\right)$, z-diffusion $\frac{\partial}{\partial z}\left(\frac{vt}{Sct}\frac{\partial C}{\partial z}\right)$ and total diffusion $\frac{\partial}{\partial x}\left(\frac{vt}{Sct}\frac{\partial C}{\partial x}\right)+\frac{\partial}{\partial y}\left(\frac{vt}{Sct}\frac{\partial C}{\partial y}\right)+\frac{\partial}{\partial z}\left(\frac{vt}{Sct}\frac{\partial C}{\partial z}\right)$ in the symmetrical plane at y=0 (S.I. units)

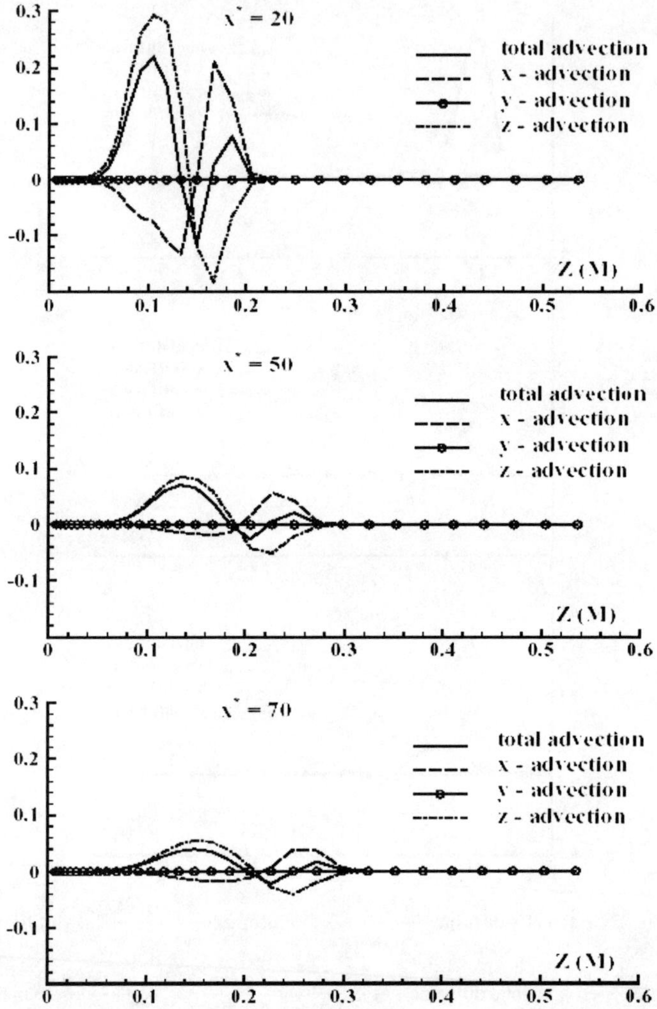

Fig.23 Sectional x-advection $U\dfrac{\partial C}{\partial x}$, y-advection $V\dfrac{\partial C}{\partial y}$, z-advection $W\dfrac{\partial C}{\partial z}$ and total advection $U\dfrac{\partial C}{\partial x}+V\dfrac{\partial C}{\partial y}+W\dfrac{\partial C}{\partial z}$ in the symmetrical plane at y=0 (S.I. units)

6. PRACTICAL APPLICATION

For prediction of jet mixing in a crossflow, the numerical model results can be cast in a more convenient form. For completeness the well-known solution for the MDNF is also included.

Momentum-dominated Near Field (MDNF)

In the absence of body forces, the momentum of a jet in stagnant fluid is conserved. The jet mixes with the surrounding fluid by turbulent entrainment; the streamwise velocity and tracer concentration is self-similar and Gaussian in shape (Fischer *et al.* 1979; Chu and Lee 1996; Lee and Chu 2002). The following relations for the characteristic jet properties can be deduced from the basic equations or by dimensional analysis and verified by experiments.

$$W_m = 7.0 \; M_{vo}^{1/2} \; z^{-1} \tag{19}$$

$$b_c = 0.114 \; z \tag{20}$$

$$S_c = 0.16 \; \frac{M_{vo}^{1/2} \; z}{Q} \tag{21}$$

In the MDNF the jet is only weakly advected; mixing is similar to that of a jet in stagnant fluid. Noting that $\dfrac{dz}{dx} = W/U_a$, and for a Gaussian profile, the ratio of centerline maximum to average velocity is $W_m/\overline{W}$ = 2. The following equations can then be obtained:

z/L_{mv} • 1 (MDNF):

$$\frac{z}{L_{mv}} = 2.65\left(\frac{x}{L_{mv}}\right)^{1/2} \tag{22}$$

$$b = 0.114 \; z \tag{23}$$

$$\frac{S_c Q}{U_a \, L_{mv}^{\;2}} = 0.16\left(\frac{z}{L_{mv}}\right) \quad \text{or} \tag{24}$$

$$S_c = 0.16 \frac{M_{vo}^{1/2} \; z}{Q}$$

where the b is the Gaussian half-width based on the 1/e *Cc* point.

Momentum-dominated Far Field (MDFF)

In the current-dominated regime, based on the ALP numerical model and experimental results, the following equations for the advected line puff are recommended:

<u>*z/Lmv* • 1 (MDFF):</u>

$$\frac{z}{L_{mv}} = 1.56\left(\frac{x}{L_{mv}}\right)^{1/3} \tag{25}$$

$$b = 0.28\,z \tag{26}$$

$$\frac{S_c Q}{U_a\, L_{mv}^{\;2}} = 0.46\left(\frac{z}{L_{mv}}\right)^2 \quad \text{or} \tag{27}$$

$$S_c = 0.46\,\frac{U_a\;z^2}{Q}$$

The jet in crossflow has also been studied using a semi-empirical approach by laboratory and field experiments of a Canadian river (Hodgson and Rajaratnam 1992). It can be shown that the dilution prediction for the asymptotic stage can be expressed as a function of distance downstream (x) from a jet discharge in the following form (Lee 1993):

$$S_c = \frac{C_o}{C_m} = 0.64\left(\frac{W_o\,x}{U_a D}\right)^{2/3} \tag{28}$$

Fig.24 shows the comparison of this equation with the experiments of Hodgson and Rajaratnam and their field dilution data. It is seen the agreement is quite reasonable.

6.1 Application Example

Consider a single sewage jet discharge into a river with Q = 0.0138 *m3/s*. The jet diameter is D = 0.1 m. The river depth is H = 12 *m*, and the design river velocity is Ua = 0.2 *m/s*. The initial density difference (buoyancy) of the effluent is negligible. Estimate the initial dilution, vertical jet velocity, and width at an elevation of z = 12 m. Also estimate the corresponding downstream distance of the jet at this location.

<u>*Solution*</u>: The discharge is modelled as a turbulent jet in crossflow. Compute jet velocity and momentum fluxes and obtain the crossflow momentum length scale *Lmv* as:

$$W_o = \frac{0.0138}{\pi \times 0.1^2/4} = \underline{1.76}\, m/s$$

$$M_{vo} = QW_o = 0.0138 \times 1.76 = \underline{0.0243}\, m^4/s^2$$

$$L_{mv} = \frac{M_{vo}^{1/2}}{U_a} = \frac{\sqrt{0.0243}}{0.2} = \underline{0.78}\, m.$$

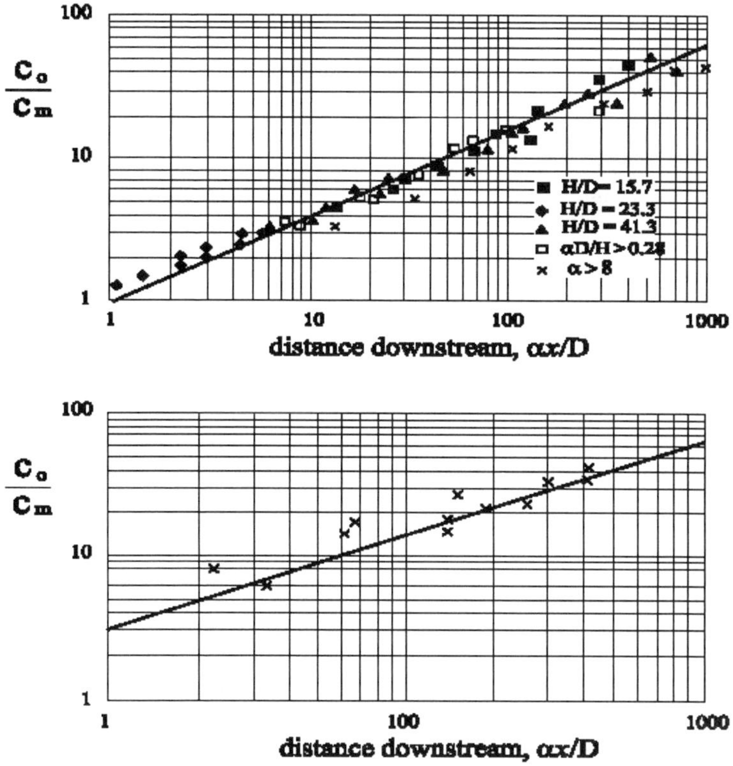

Fig.24 Comparison of Eq.28 with the laboratory and field data (upper/lower plot respectively) of Hodgson and Rajaratnam (1992); $\alpha = W_o/U_a$

At an elevation of z = 12 m, z/Lmv = 12/0.78 = 15.4 >>1, apply the MDFF equations to obtain the dilution, trajectory, and width

$$S_c = 0.46\frac{0.2\times 12^2}{0.0138} = \underline{960}$$

$$2b_{vc} = 2\times 0.28\times 12 = \underline{6.7}\ m$$

This occurs at a distance given by the trajectory equation:

$$x = 0.78 \times (15.4/1.560)^3 = \underline{750}\ m$$

Note the relatively long distance that it takes for the jet to travel to this elevation. Typically a good estimate of the visible width of a jet is approximately $\bullet 2b$; this suggests that the bent-over jet is quite well-mixed over the cross-section at this location. The vertical velocity at this location can also be shown to have decayed to negligible values:

$$Wm = 1.33 \times (0.0243/0.2)^{1/3}/(750/0.2)^{2/3} = \underline{2.73 \times 10^{-3}}\ m/s$$

The efficiency of mixing in a current (moving dilution of $Sc = 960$) can best be appreciated by computing the still water dilution at this elevation (Eq.21):

$$Sc(Ua = 0) = 0.16 \times \frac{0.0243^{1/2} \times 12}{0.0138} = \underline{21.7}!$$

Alternatively, the moving water dilution for this problem can be obtained from the semi-empirical equation based on the field data of Hodgson and Rajaratnam:

$$Sc = 0.64 \times \left(\frac{1.76 \times 750}{0.2 \times 0.1} \right)^{2/3} = \underline{1046}$$

7. CONCLUDING REMARKS

The mixing of a turbulent jet in crossflow is studied using the concept of asymptotic flow regimes. The bent-over phase of a special three-dimensional jet in crossflow, advected line puffs, has been studied using the standard k - ε two equation model. As far as we are aware, this is the first comprehensive study of scalar mixing in the bent-over jet, extending to the asymptotic phase where the analogy of a bent-over jet to an unsteady line puff is valid. The salient features of the predicted flow and scalar mixing rates are well-supported by experiments based on laser-induced fluorescence and laser doppler anemometry. Additional experiments also show the mixing of a vertical jet in crossflow is very similar to that of an advected line puff (in the bent-over phase). Combined with previous numerical studies on line puffs, a better understanding of the bent-over jet flow and mixing characteristics is obtained; simple equations for mixing analysis are developed. The connection between a line puff and an advected line puff is also explored. The main conclusions are as follows:

- The advected line puff is longitudinally characterized by a bent-over phase taking place essentially in the interval of $10 < x^* < 40$, during which the fluid flow and scalar contours become almost parallel with the free stream. The flow is approximately self-similar beyond a dimensionless length of $x^* = 20 - 30$.
- The vertical motion of the advected line puff is characterized by a vortex-pair structure, with an added mass coefficient of approximately 1, very much like that for line puffs.
- The scalar field is sectionally characterized by a kidney-shaped outline containing a double peak of concentration maxima. The puff aspect ratio (defined as horizontal half-width over vertical half-width) and the maximum concentration ratio (defined as sectional maximum concentration over centerline maximum concentration) are all found around 1.2.

- The longitudinal excess velocity of the advected line puff is characterized by a double peak; the effect of transverse diffusion and streamwise/ vertical advection are the significant terms for the bent-over phase.

We have outlined a framework that has proved useful for studying jets in crossflow. The same approach can be used to study plumes in crossflow via the concept of advected line thermals (Wood 1993), for which much more data is available. More details on this can be found in Wood (1993) and in a forthcoming work (Lee and Chu 2002). This Chapter has provided some theoretical basis for an introduction to the structure of jet mixing in a crossflow; the equations provide first order estimates for the jet trajectory and dilution in each of the asymptotic flow regimes. However, for many practical problems we are faced with the prediction of mixing in the transition between regimes. For example, a strongly jet in a weak crossflow (e.g. a 95 percentile value) is often a design condition. For engineering application, a general mathematical model for predicting the characteristic jet properties is often required, especially when buoyancy is included (e.g. Muellenhoff et al. 1985). For an introduction to an educational tool the reader is referred to a recently developed model VISJET based on visualization technology (Lee *et al.* 1999). The model can be downloaded from http://www.aoe-water.hku.hk/visjet; it serves as a starting point for tackling a range of jet and plumes problems.

Acknowledgements:

The project was supported by the Hong Kong Research Grants Council and in part by a grant from the National Science Foundation of China. All computations were performed in the University of Hong Kong. The numerical modeling work on line puffs was originally supported by an Alexander von Humboldt Research Fellowship.

REFERENCES

Ayoub, G.M. (1971). "Dispersion of buoyant jets in a flowing ambient fluid", Ph.D.thesis, Imperial College, University of London.

Andreopoulos,J. & RODI, W. (1984). "Experimental investigation of jets in a cross-flow." *J. Fluid Mech.*, **138**, 93-127.

Chen, G.Q., Gao, Z. & Chen, Y.S. (1991). "Some theoretical aspects of the simplified Navier-Stokes equations." *Science in China*, Ser. A **34**, 943-954.

Chien, C.J. & Schetz, J.A. (1975). "Numerical solution of the three-dimensional Navier-Stokes equations with application to channel flows and a buoyant jet in a cross-flow." *Trans. ASME E: J. Appl. Mech.*, **42**, 575-579.

Chu, P.C.K. (1996). "Mixing of turbulent advected line puffs." Ph.D thesis, University of Hong Kong.

Chu, P.C.K. and Lee, J.H.W. (1996). "Vorticity dynamics of an inverted jet in crossflow." *Proc. Second International Conference on Hydrodynamics, December 1996, Hong Kong*, Vol. 2, 743-748.

Chu, V.H. & Goldberg, M.B. (1974). "Buoyant forced plumes in crossflow." *Proc. ASCE, J. Hydraul. Div.*, **100**, 1203-1214.

Chu, V.H. (1985). "Oblique turbulent jets in a crossflow." *ASCE J. Engr. Mech.*, **111**, 1343-1360.

Chu, V.H. and Lee, J.H.W. (1996). "A general integral formulation of turbulent buoyant jets in crossflow", *Journal of Hydraulic Engineering, ASCE*, **122**, No.1, 27-34.

Demuren, A.O. (1983). "Numerical calculations of steady three-dimensional turbulent jets in cross flow." *Comp. Meth. Appl. Mech. & Engng.*, **37**, 309-328.

Fan, L.N. (1967). "Turbulent buoyant jets into stratified or flowing ambient fluids." *Report No. KH-R-15*, W.M. Keck Lab. of Hydr. and Water Resour., California Institute of Technology, Pasadena, California.

Fischer, H.B. et al. (1979). *Mixing in Inland and Coastal Waters*. Academic Press, New York.

Fric, T.F. and Roshko, A. (1994). "Vortical structure in the wake of a transverse jet", *J. Fluid Mech.*, **279**, 1-47.

Gaskin, S.J. (1995). "Single buoyant jet in a crossflow and the advected line thermal.", Ph.D. thesis, University of Canterbury, Christchurch, New Zealand.

Glezer, A. & Coles, D. (1990). "An experimental study of a turbulent vortex ring." *J. Fluid Mech.*, **211**, 243-283.

FLUENT INC. (1995). "FLUENT (version 4.3) User's Guide." Lebanon, New Hampshire.

Kamotani, Y. & Greber, I. (1972). "Experiments on a turbulent jet in a crossflow." *AIAA Journal*, **10**, 1425-1429.

Keffer, J.F. & Baines, W.D. (1963). "The round turbulent jet in a cross wind." *J. Fluid Mech.*, **15**, 481-496.

Lee, J.H.W. and Neville-Jones, P. (1987) "Initial dilution of horizontal jet in crossflow", *Journal of Hydraulic Engineering, ASCE*, **113**, 615-629.

Lee, J.H.W. (1989). "Note on Ayoub's data of horizontal round buoyant jet in a current", *J. of Hydraulic Engineering, ASCE*, **115**, 969-975.

Lee, J.H.W. (1993). Discussion of "An experimental study of jet dilution in crossflows", *Canadian Journal of Civil Engineering*, **20**, 1073-1076.

Lee, J.H.W. Rodi, W. & Wong, C.F. (1996). "Turbulent momentum line puffs." *ASCE J. Engr. Mech.*, **122**, 19-29.

Lee, J.H.W. & Chen, G.Q. (1998). "A numerical study of turbulent line puffs via the renormalization group (RNG) k._model.", *Int. J. Numer. Meth. Fluids*, **26**, 217-234.

Lee, J.H.W., Li, L., and Cheung, V. (1999). "A semi-analytical self-similar solution of a bent-over jet in crossflow", *Journal of Engineering Mechanics, ASCE*, **125**, 733-746.

Lee, J.H.W., Cheung, V., Wang, W.P., and Cheung, S.K.B. (1999). "Lagrangian modeling and visualization of rosette outfall plumes", *Proc. Hydroinformatics2000, University of Iowa, July 23-27, 2000*, (CDROM)

Lee, J.H.W. and Chu, V.H., *Turbulent jets and plumes - a Lagrangian approach*, Kluwer Academic Publishers (to appear in 2002).

Leonard, B.P. (1979). "A stable and accurate convective modelling procedure based on quadratic upstream interpolation.", *Comp. Meth. Appl. Mech. Engr.*, **19**, 59-98.

Moussa, Z.M., Trischka, J.W. & Eskinazi, S. (1977). "The near field in the mixing of a round jet with a cross-stream.", *J. Fluid Mech.*, **80**, 49-80.

Patankar, S.V., Basu, D.K. & Alpay, S.A. (1977). "Prediction of the three dimensional velocity field of a deflected turbulent jet.", *Trans. ASME I: J. Fluids Engng.*, **99**, 758-762.

Patankar, S.V. (1980). *Numerical Heat Transfer and Fluid Flow*, Hemisphere Publishing Co., New York.

Pratte, B.D. & Baines, W.D. (1967). "Profiles of the round turbulent jet in a cross flow.", *J. Hydr. Div., ASCE*, **93**, No. HY6, 53-64.

Rajaratnam, N. (1974), *Turbulent jets.*, Elsevier, New York.

Richards, J.M. (1965). "Puff motions in unstratified surroundings." *J. Fluid Mech.*, **21**, 97 106.

Richards, J.M. (1963). "Experiments on the motions of isolated cylindrical thermals through unstratified surroundings.", *Int. J. Air. Water Poll.*, **7**, 17-34.

Rodi, W. (1980). *Turbulence Models and Their Application in Hydraulics*. IAHR, Delft, Netherlands.

Scorer, R.S. (1978). *Environmental Aerodynamics*, Ellis Horwood, Chichester.

Sykes, R.I., Lewellen, W.S. & Parker, S.F. (1986). "On the vorticity dynamics of a turbulent jet in a crossflow." *J. Fluid Mech.*, **168**, 393-413.

Tsang, G. (1971). "Laboratory study of line thermals." *Atmos. Envir.*, **5**, 445-471.

Turner, J.S. (1986). "Turbulent entrainment: the development of the entrainment assumption, and its application to geophysical flows." *J. Fluid Mech.*, **173**, 431- 471.

Van Doormaal, J.P. et al. (1984). "Enhancement of the SIMPLE method for predicting incompressible fluid flows." *Num. Heat Transfer*, **7**, 147-163.

Wong, C.F. (1991). "Advected line thermals and puffs.", M.Phil. thesis, University of Hong Kong.

Wood, I.R. (1993). "Asymptotic solutions and behavior of outfall plumes," *J. of Hydraulic Engr., ASCE,* **119**, 555-580.

Wright, S.J. (1977) "Effects of ambient crossflow and density stratification on the characteristic behavior of round turbulent buoyant jets," *Report No. KH-R-36,* W.M. Keck Lab. of Hydr. and Water Resour., California Inst. of Tech., Pasadena, Calif.

Wright, S.J. (1977). "Mean behavior of buoyant jet in a crossflow." *Proc. ASCE, J. Hydraul. Div.*, **103**, 499-513.

Yih, C.S. (1981). "Similarity solutions for turbulent jets and plumes." *ASCE J. Engr. Mech.*, **107**, 455-478.

Chapter 4

MULTI-PHASE PLUMES IN UNIFORM, STRATIFIED, AND FLOWING ENVIRONMENTS

Scott A. Socolofsky[1], Brian C. Crounse[2], and E. Eric Adams[3]

ABSTRACT

Multi- phase plumes in uniform, stratified, and flowing environments have been studied through both laboratory and limited field experiments and through numerical modeling. Of particular interest to the authors is the behavior of multi-phase plumes in the deep ocean, with applications ranging from carbon sequestration to the fate of oil released from an oil-well blowout. Here, we review the pertinent literature and present some initial experimental and analytical results of our own. We start with a description of a multi-phase plume in a uniform ambient, then build in complications of stratification and crossflow. Laboratory experiments are presented for a 2.4 m deep stagnant, stratified ambient and a 0.7 m deep uniform crossflow. The experiments supplement observations of plume type from the literature and quantify entrained fluid volume fluxes in stratification to help validate numerical models. Our theoretical analysis includes a re-examination of the governing dimensionless parameters, with specific focus on deep water applications, and presentation of a double plume model, which incorporates the effects of plume peeling and bubble dissolution.

1. INTRODUCTION

Multi-phase plumes are buoyancy driven flows where the buoyancy is provided by a continuous release of an immiscible dispersed phase, such as gas bubbles, liquid droplets or solid particles. Some environmental applications include air bubble plumes used for reservoir destratification (Schladow 1993, Lemckert & Imberger 1993), reaeration (Wüest, et al. 1992), ice prevention in harbors (Baddour 1994, McDougall 1978), and contaminant containment (Milgram 1983); continuous particle clouds resulting from the release of dredged

[1] Res. Assoc., Inst. for Hydromech., Univ. of Karlsruhe, 76128 Karlsruhe, Germany, socolofs@alum.mit.edu
[2] Engineer, Carlisle & Company, 30 Monument Sq., Concord MA 01742, USA, brian.crounse.94@alum.dartmouth.org
[3] Sr. Res. Engr. and Lecturer, Dept. of Civ. & Envirn. Engrg., Mass. Inst. of Tech., Rm. 48-325, Cambridge, MA 02139, USA, eeadams@mit.edu

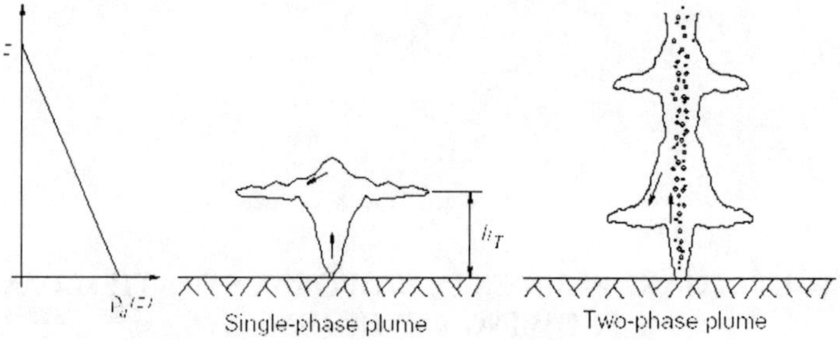

Fig. 1 Schematic of single- and two phase plumes in stratification.

sediments (Koh & Chang 1973, HAVIS Environmental 1994); liquid CO_2 plumes for deep-ocean carbon sequestration (Liro et al. 1992, Adams et al. 1997, Alendal & Drange 2000); and deep sea blowouts of oil and gas (Yapa & Zheng 1997*a*, 1997*b*, Johansen 1999). Our recent research has focused on the latter two applications.

While these plumes result from a range of dispersed phases, we will adopt the general term bubble plume and focus mainly on bubble plumes in water. Physically, a bubble plume is described by the release conditions of the dispersed phase as well as the ambient environmental conditions. Important parameters describing the release include the composition and physical characteristics of the dispersed phase (e.g. density ρ_b, viscosity μ_b, and surface tension σ_b), its flow rate, and the geometry of the release. Pertinent environmental conditions include ambient density stratification and currents.

Bubble plumes are similar in many ways to single-phase plumes, such as sewage plumes in sea water or heated water plumes resulting from the use of once-through cooling in electric power production, but they also have some important differences. The main difference between single- and multi-phase plumes results from the discrete nature of the buoyant dispersed phase. In the case of a single-phase plume, the buoyancy is well mixed with the bulk fluid--the advection of buoyancy is controlled by the motion of the fluid. In contrast, the bubbles themselves comprise the buoyancy in a bubble plume; they exhibit a slip velocity, u_s, relative to the bulk plume fluid, and their distribution over the plume cross section is controlled both by bubble dynamics as well as motion of the bulk plume fluid. Sec. 2 discusses the complications of a dispersed phase in detail for a uniform ambient.

In the presence of ambient stratification, the dynamics of single- and multi-phase plumes are significantly altered. Fig. 1 depicts classic single- and two-phase plumes in stratification. As the single-phase plume rises, it loses buoyancy relative to the environment. Because of its excess momentum, however, it initially overshoots a point of neutral buoyancy, becomes negatively buoyant, and eventually falls back, or traps, to form an intrusion. From dimensional analysis a prediction for the plume trap height, h_T, should have the form

$$h_T = 2.8\left(\frac{B}{N^3}\right)^{1/4} \tag{1}$$

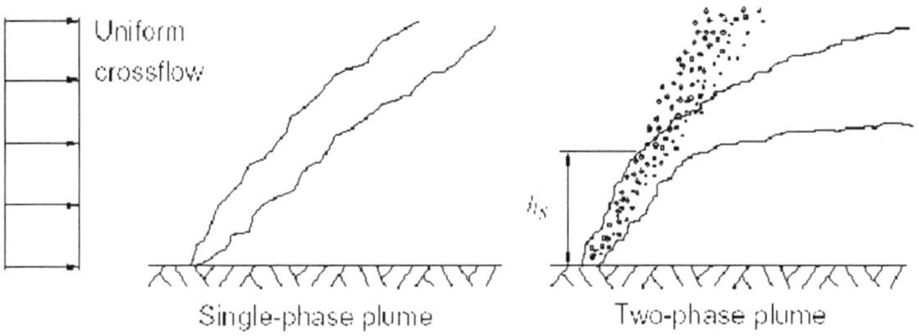

Fig. 2. Schematic of single- and two-phase plumes in a uniform crossflow.

where $B = gQ_b(\rho - \rho_b)/\rho$ is the total kinematic buoyancy flux, Q_b is the volume flux of effluent at the source, ρ is the ambient density, $N = [-(g/\rho)(\partial\rho/\partial z)]^{1/2}$ is the Brunt-Vaisälä buoyancy frequency, and 2.8 is an empirical constant (Fischer et al. 1979, Turner 1986, Crawford & Leonard 1962). This relationship has been verified from laboratory scales to the scales of forest fires and volcanic eruptions (Turner 1986).

Contrast this with a bubble plume. As fluid is entrained and lifted by the bubbles, it becomes negatively buoyant but will continue to rise as long as the bubble drag force exceeds the negative buoyancy of the fluid itself (McDougall 1978). Eventually, the plume will reach a point where the bubbles can no longer support the negative buoyancy of the water. At this height, some portion of the plume water will separate, or peel, from the plume and form a negatively buoyant outer plume (Asaeda & Imberger 1993). If there is no ambient current, this downward plume will surround the upward plume and interact with it. Eventually, the peeled fluid will fall to a point where it is neutrally buoyant, significantly below the height at which it left the upward plume, and intrude as a horizontal density current (Asaeda & Imberger 1993, Lemckert & Imberger 1993, McDougall 1978). Meanwhile, unless their slip velocity is very small, the buoyant bubbles will continue to rise past the peeling location, carrying some fraction of the plume water with them. These peeling events will continue at higher elevations, until either the surface is reached, the dispersed phase reaches neutral buoyancy, or the dispersed phase dissolves (Wüest et al. 1992, Socolofsky & Adams 2000a, Crounse 2000). Sec. 3 discusses stratification in detail and presents both laboratory experiments and numerical results demonstrating these effects.

In a crossflow, the velocity difference between the slipping bubbles and the rising entrained fluid can lead to another type of separation. Fig. 2 shows single- and two-phase plumes in a uniform crossflow. If the crossflow is strong enough to push fluid out of the plume and advect it downstream, the bubbles will separate from the entrained fluid, at some height, h_S, forming a bubble column that rises as a result of the bubble slip velocity alone. Above h_S, fluid advected into the front of the bubble column by the current is lifted a short distance as it interacts with the buoyant bubbles before it is released in the lee of the plume (Socolofsky & Adams 2000*c*, Hugi 1993). Sec. 4 discusses the effects of crossflow in more detail and presents the results of some laboratory experiments and dimensional analysis.

2. SIMPLE BUBBLE PLUMES

This section describes the behavior of axisymmetric bubble plumes in a stagnant, unstratified environment, which we will call simple bubble plumes. By definition, the initial momentum for a bubble plume is negligible. We limit our discussion further to plumes where the volume fraction of the dispersed phase is low, or dilute, so that the plume behavior is controlled by fluid forces, rather than particle collisions (Crowe et al. 1998) and to fully turbulent, steady, unbounded flows. We will describe bubble plume behavior in terms of a one-dimensional integral model. Although such models are approximate, they are useful both heuristically and as a quantitative tool.

2.1 Theory

2.1.1 Similarity

Looking first to simple single-phase plumes, one important characteristic is that they are self-similar. Physically, self-similar behavior indicates that the flow has established a sort of moving equilibrium, in which the evolution of the flow is self-governing (Townsend 1976, Turner 1986, Fischer et al. 1979). Quantitatively, this means that time-average cross-sectional profiles of plume quantities (taken normal to the mean flow) maintain a fixed, near-Gaussian shape and a constant spreading rate (Fischer et al. 1979). Thus, plume properties are fully described by a characteristic centerline value and a characteristic radius, both functions of height. The self-similar region of a simple plume is known as the zone of established flow (Morton et al. 1956, Fischer et al. 1979, Turner 1986).

The advantage of assuming self-similarity is that it reduces the complexity of the mathematical model. Since characteristic plume variables change only with height, a plume may be modeled by a one-dimensional, i.e. ordinary, set of differential equations. Assumption of similarity also allows us to ignore some details of the turbulent flow.

In contrast to single-phase plumes, bubble plumes are not strictly self-similar because the ratio of the decreasing continuous phase fluid velocity to the constant bubble slip velocity is variable. Nonetheless, invoking the self-similarity assumption for the analysis of simple bubble plumes yields useful results. The implications of this assumption will be revisited in Sec. 2.2.

2.1.2 Flux Model

The descriptive variables of a simple bubble plume are its width, $b(z)$, the time-averaged velocity profile of the entrained fluid, $u(z,r)$, and the fraction of the plume cross-sectional area occupied by bubbles, or void fraction, $C(z,r)$. Fig. 3 shows typical profiles of u and C at two heights in a simple bubble plume. Other variables which are pertinent include the density of the bulk fluid, ρ_w, the density of the bubbles, $\rho_b(z)$, the characteristic bubble radius, $r_b(z)$, and the slip velocity of the bubbles, $u_s(z)$. We define z as the vertical distance above the plume origin.

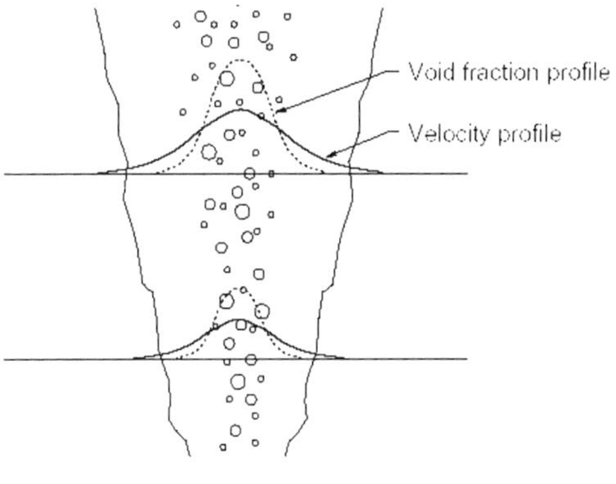

Fig. 3. Schematic of a simple bubble plume with profiles of velocity and void fraction.

These variables may be integrated through a plane normal to the mean flow to give fluxes of volume, mass, momentum and buoyancy. By employing the laws of mass and momentum conservation, along with a closure scheme, we arrive at a complete set of ordinary differential equations which describe the evolution of plume characteristics with height.

Two volume fluxes are defined: the flux of plume water, Q_p, and the flux of the dispersed phase, Q_b. The plume water flux is

$$Q_p(z) = \int_0^\infty 2\pi r(1 - C(z,r))u(z,r)dr. \tag{2}$$

The definition of Q_b is complicated by the bubble slip velocity. Typically, the transport velocity for the bubbles is assumed to be the sum of the plume fluid rise velocity and an additional bubble slip velocity, such that $u_b = u + u_s$ (Kobus 1968, McDougall 1978, Milgram 1983, Asaeda & Imberger 1993). Making this assumption,

$$Q_b(z) = \int_0^\infty 2\pi r C(z,r)(u(z,r) + u_s(z,r))dr. \quad (3)$$

The mass fluxes of plume water and the dispersed phase are $\rho_w Q_p$ and $\rho_b Q_b$, respectively.

The total kinematic momentum flux, M, is defined by

$$M(z) = \gamma \int_0^\infty 2\pi r \left[u^2(z,r)(1 - C(z,r)) + \frac{\rho_b(z)}{\rho_w}(u(z,r) + u_s(z,r))^2 C(z,r) \right] dr \quad (4)$$

where γ, an amplification term defined in Milgram (1983), accounts for the fact that use of the mean velocity, u, in Eq. (4) implicitly ignores turbulent momentum transport. Momentum amplification is discussed in Sec. 2.2.

As the driving force for the plume, the density difference between the dispersed phase and the ambient is $\Delta\rho_b = \rho_w - \rho_b$, which is positive for rising bubbles. As the void fraction is small, the Boussinesq approximation is applicable, so that the reduced gravity is defined as $g' = g\Delta\rho_b/\rho_w$ (Liro et al. 1994). In an unstratified environment, ρ_w is constant.

With these definitions, the total kinematic buoyancy flux, B, of a simple bubble plume is defined as,

$$B(z) = \int_0^\infty 2\pi r C(z,r) g \frac{\Delta\rho_b(z)}{\rho_w}[u(z,r) + u_s(z,r)]dr = g' Q_b(z) \quad (5)$$

Another useful quantity involving buoyancy is the integrated buoyant force, $\hat{B}_b$, acting on a unit height of the plume,

$$\hat{B}_b(z) = \int_0^\infty 2\pi r C(z,r) g \Delta\rho_b(z) dr \quad (6)$$

Note that this is a force per unit height, unlike the preceding flux quantities.

Evaluation of these integrals requires selection of a shape for the profiles depicted in Fig. 3. As noted, simple single-phase plumes exhibit near-Gaussian property profiles. Here, we assume that simple bubble plumes do as well. This assumption is supported by experimental results and is discussed further is Sec. 2.2. Thus,

$$u(z,r) = U_m(z) e^{-r^2/b^2} \quad (7)$$

$$C(z,r) = C_m(z)e^{-r^2/(\lambda_1 b)^2} \tag{8}$$

where U_m and C_m are centerline values. Because the bubble column does not occupy the full plume width, λ_1 is introduced, defined as the spreading ratio of the bubbly region relative to the plume width ($0 < \lambda_1 \leq 1$). $1/\lambda_1^2$ is a turbulent bubble Schmidt number that would be constant for a strictly similar flow (Ditmars & Cederwall 1974).

With the profiles described by Eqs.(7) and (8), the volume flux of the plume water becomes

$$Q_p = \pi U_m b^2 \left(1 - C_m \frac{\lambda_1^2}{\lambda_1^2 + 1}\right)$$

Because the bubbles are dilute, such that $C_m << 1$, the volume flux reduces to

$$Q_p(z) \approx \pi U_m b^2 \tag{9}$$

To evaluate Q_b, we make the assumption that u_s is independent of the radial location. If the distribution of u_s for a given application were available, its variation across the cross-section could be included. For now, the volume flux of the dispersed phase becomes

$$Q_b(z) = \frac{\pi \lambda_1^2 b^2 C_m}{\lambda_1^2 + 1}(U_m + U_b) \tag{10}$$

where we define $U_b = (1 + \lambda_1^2)u_s$ for convenience (McDougall 1978).

The momentum flux equation is simplified by considering the magnitude of each term in Eq. (4). The first term describes the momentum flux associated with the mean bulk fluid flow, while the second term describes that of the mean bubble flow. Taking the ratio of these two terms indicates that the momentum flux of the fluid is $\rho_w(1 - C)/\rho_b C$ times the momentum flux of the bubbles. Limiting to dilute plumes where $C << 1$, we may neglect the bubble momentum term. For air bubbles in water, ρ_w/ρ_b is also large, further supporting this approximation. Thus, dropping the bubble momentum flux term, and letting $1 - C(z, r) \approx 1$, the momentum flux equation becomes

$$M(z) \approx \frac{\gamma}{2}\pi b^2 U_m^2 \tag{11}$$

The buoyancy flux can be written immediately as $g'Q_b(z)$; thus, the remaining integral is the buoyant force, $\hat{B}_b$, which becomes

$$\hat{B}_b(z) = \pi g \lambda_1^2 b^2 C_m g \Delta\rho_b = \frac{1+\lambda_1^2}{U_m + U_b} Q_b g \Delta\rho_b \tag{12}$$

We see here that the slip velocity, which appears in the denominator, acts to reduce the buoyant force at a given elevation.

2.1.3 Governing Differential Equations

The evolution of the plume in space is controlled by two physical processes: the turbulent entrainment of ambient water and the buoyant forcing of the bubbles. In this section the governing dynamic equations are derived by invoking volume and momentum conservation. Because bubble evolution can be de-coupled from the dynamic equations, the bubbles themselves are treated in a separate bubble sub-model.

Volume conservation is governed by turbulent entrainment: the buoyant bubbles induce a turbulent flow, which in turn causes eddies from the plume to engulf ambient liquid and mix it into the plume (Turner 1986). The most successful method to date for quantifying the rate of entrainment is the entrainment hypothesis, formally introduced by Morton et al. (1956). This hypothesis states that the mean entrainment velocity across a shear flow boundary, perpendicular to the direction of flow, is proportional to a characteristic velocity of the flow (Turner 1986). Thus, the entrainment volume flux is the entrainment velocity multiplied by the surface area of the flow. Adoption of this hypothesis leads to

$$\frac{dQ_p}{dz} = 2\pi b \alpha u \tag{13}$$

where α is a turbulent entrainment coefficient. It is constant for single-phase jets and plumes, but varies with local plume conditions for simple bubble plumes (see Sec. 2.2.4).

Momentum conservation represents a force balance: buoyant forces cause the bubbles to rise relative to the bulk fluid; the subsequent drag on the bubbles effectively transfers the buoyant forcing to the bulk fluid. Thus, momentum conservation can be expressed using the integrated buoyant force:

$$\frac{dM}{dz} = \frac{\hat{B}_b(z)}{\rho_w} = \pi \lambda_1^2 b^2 g' C_m \tag{14}$$

Together, Eqs. (13) and (14) describe the dynamics of a simple bubble plume. Due to dissolution and changes in hydrostatic pressure, the bubbles undergo additional transformation.

Bubble plume applications in which bubble dissolution is significant include lake aeration (Wüest et al. 1992) and deep-ocean CO_2 sequestration (Liro et al. 1992). Bubble dissolution is dependent on factors such as bubble size. If the flux of the number of bubbles, N_b, is assumed to be constant in the zone of established flow, then the average bubble mass may be defined as $m_b = \rho_b Q_b(z)/N_b$; a characteristic bubble radius is then $r_b = [3m_b/(4\pi\rho_b)]^{1/3}$. The time rate of change of the mass of a single bubble is commonly modeled as

$$\frac{dm_b}{dt} = -4\pi r_b K \rho_b (C_s - C_\infty) \tag{15}$$

where C_s is the solubility and, thus, surface concentration of the dispersed phase, C_∞, for the substance in question, and K is a mass transfer coefficient. K is a function of bubble characteristics; its determination is examined in detail in Clift et al. (1978).

As the effective velocity of the dispersed phase is $(U_m + U_b)/(1 + \lambda_1^2)$, Eq. (15) can be written with respect to distance by invoking the chain rule:

$$\frac{d(\rho_b Q_b)}{dz} = -4\pi N_b K r_b \rho_b \frac{1+\lambda_1^2}{U_m + U_b}(C_s - C_\infty) \tag{16}$$

Often, the flux of the dissolved phase of the solute is also of interest and must be tracked separately (see Sec. 3).

While dissolution decreases the bubble volume flux with height, bubble expansion due to decreasing pressure acts to increase it. In common reservoir applications the ideal gas law together with adiabatic expansion can be used to obtain

$$Q_b(z) = \frac{H_A}{H_T - z} \rho_{bA} \tag{17}$$

where dissolution is neglected. H_A is the atmospheric pressure head and ρ_{bA} is the gas density at the reservoir surface. $H_T = H_A + H$ is the total pressure head, where H is the depth of the release. In the deep ocean H is of the order of 1000 m and compressibility effects can generally be ignored.

If dissolution is negligible, Eq. (17) represents the typical bubble evolution. A more flexible approach, which allows for dissolution, is to track the mass flux of the bubbles via Eq. (16), and then to calculate their volume flux after determining their density via an appropriate equation of state. Hence, Eqs. (13) and (14) together with either Eqs. (16) or (17) represent the governing equations. They may be expressed in terms either of the integral variables Q_p, Q_b, and M, or the local variables U_m, b, and C_m, by use of the following relationships:

$$U_m = \frac{2}{\gamma}\frac{M}{Q_p} \tag{18}$$

$$b = \left(\frac{\gamma Q_p^{\ 2}}{2\pi M} \right)^{1/2} \tag{19}$$

$$C_m = \frac{Q_b \left(1 + \lambda_1^{\ 2}\right)}{\pi \lambda_1^{\ 2} b^2 \left(U_m + U_b\right)} \tag{20}$$

2.2 Observations

The previous section presented a theoretical framework for the analysis of simple bubble plumes. This section presents experimental data and discusses several of the assumptions and coefficients underlying this framework.

2.2.1 Profiles and Similarity

The assumption of Gaussian profiles for both the continuous and dispersed phases of dilute bubble plumes is supported by experimental evidence over a wide range of experimental conditions and fluids (Kobus 1968, Milgram 1983, Tacke et al. 1985).

Strict similarity for bubble plumes requires that λ_1 and α be constant. Experiments show, however, that both these parameters vary with height, indicating that simple bubble plumes are not strictly self-similar (Wilkinson 1979, Milgram 1983). Fortunately, model results are not sensitive to reasonable variation in λ_1 (Liro et al. 1991). They are more sensitive to α, but reasonable results are obtained using representative average or height-dependent values of α (Milgram 1983, Turner 1986).

2.2.2 Release Conditions

The governing equations, as stated, do not apply near the point of gas release (zone of flow establishment or ZFE), where the gas fraction is high and the flow changes rapidly with height. One may avoid analyzing this zone by assigning a virtual release point, such that the virtual plume is identical to the observed plume in the zone of established flow. The height of the virtual release depends on the geometry of the release, and generally lies below the actual release point for a bubble plume (Kobus 1968).

Liro et al. (1991) describes a fairly simple technique for assigning initial conditions above the ZFE: the height of the ZFE is assumed to be approximately five times the diameter of the release port, and the location of the virtual release is assumed to lie an equal distance below the actual port. Applying analytical solutions for a single-phase plume between the virtual origin and the top of the ZFE yields

$$b = \frac{6}{5}\alpha z_0$$

$$U_m = \left(\frac{25 g Q_b \left(1 + \lambda_1^{\ 2}\right)}{24\alpha^2 \pi z_0} \right)^{1/3}$$

where z_0 is the distance from the virtual release point to the top of the ZFE. While this method is approximate, it provides reasonable estimates of initial conditions when buoyancy, rather than momentum, dominates the flow conditions. Because buoyant plumes are mainly sensitive to the initial buoyancy flux, efforts to model the ZFE more accurately do not yield significant dividends. For plumes where momentum dominates the ZFE, as may be the case in an oil well blowout, an alternate approach may be required (see e.g. Yapa & Zheng 1997*a*).

2.2.3 Bubble Characteristics

Significant physical parameters describing bubbles include the slip velocity, u_s, and the spreading rate of the bubble column, λ_1.

In a stagnant environment u_s depends on factors such as bubble size and shape, buoyancy, viscosity, and surfactant concentration (Clift et al. 1978). Bubble shape changes with increasing diameter, progressing from spherical through elliptical to spherical cap. Other effects, such as temperature, also affect rising bubbles (Leifer et al. 2000).

The stable bubble size for a given release is affected by the dispersed phase flow rate and the salinity of the receiving water. For air bubbles in fresh water, the bubble size increases with airflow rate, and is fairly independent of the diffuser geometry (Kobus 1968, Milgram 1983, Tacke et al. 1985). In sea water, however, small bubbles are more stable, and diffuser geometry can play a role, even at low exit velocity (Beyersdorf 1997). In our experiments a diffuser that produced 2 mm bubbles in fresh water produced 0.5 mm bubbles in sea water for the same gas flow rate (Socolofsky 2000). This change in size reduced the slip velocity from approximately 20 cm/s to 7 cm/s.

Given certain bubble properties, other environmental factors can affect the slip velocity. In a bubble plume the flow field surrounding a bubble is complicated by wakes from neighboring bubbles and by the turbulence generated by the plume's shear flow. Many investigators, such as Ditmars & Cederwall (1974), Milgram (1983), Tacke et al. (1985) and Brevik & Killie (1996) have assumed that these turbulent conditions would alter the quiescent water terminal velocity; hence, they chose to treat the slip velocity as a free parameter. Others, such as Liro et al. (1991) and Wüest et al. (1992), kept the quiescent terminal velocity for u_s, omitting it as a calibration parameter. The limited experimental data of Chesters et al. (1980), Roig et al. (1998) and Leitch & Baines (1989) show that the observed slip velocities are indeed comparable to the stagnant water terminal velocity, implying that turbulence in dilute plumes does not significantly alter the bubble terminal velocity.

The bubble core spreading rate is not well understood. Experimental observations indicate that λ_1 is approximately constant with height, but varies widely by interpretation. Milgram (1983) claimed that the small values (in the range of 0.3) of λ_1 reported by Ditmars & Cederwall (1974) were artifacts of plume wandering. Plume wandering occurs when a bubble plume meanders about its centerline because of recirculation currents set up in confined basins. Ignoring this experimental artifact leads to an overestimate of the plume width and a corresponding underestimate of λ_1. When plume wandering was accounted for, Milgram reported that λ_1 should be in the range of 0.7-0.8. However, Chesters et al. (1980) report a value of λ_1 = 1. Socolofsky & Adams (2000*a*), accounting for bubble wander, showed that λ_1 can range from 0.5 to 1.0 depending on B and u_s. Fortunately, plume predictions are only weakly dependent on λ_1 (Liro et al. 1992).

2.2.4 Turbulent Entrainment

The turbulent entrainment of an integral model cannot be determined from first principles. Rather, a closure scheme must be assumed. As discussed previously, most investigators have adopted the entrainment assumption, which states that the effective inward (entrainment) velocity over a defined flow interface is proportional to the mean characteristic velocity of that flow (Turner 1986).

Turner (1986) and Fischer et al. (1979) discuss turbulent entrainment in the context of single phase jets and plumes in detail. The most notable result for simple single phase jets and plumes is that the entrainment coefficient, α, is constant. For round jets (no buoyant effects), $\alpha_j = 0.054$, while for round plumes $\alpha_p = 0.083$. The difference between α_j and α_p must be due to differences in the character of turbulence between the two flows, implying that buoyancy affects entrainment rate. The constant values of α are evidence of the flows' self-similarity.

In contrast, the entrainment rate for simple bubble plumes is not universally constant. This fact was first reported in Ditmars & Cederwall(1974), who found that plume average values of α ranged from 0.04 to 0.08 in Kobus's(1968) experiments, increasing with higher gas flow rates. In other experiments, Tacke et al.(1985) reported entrainment coefficients in the range of 0.075 to 0.13.

The most detailed analysis of α for simple bubble plumes is reported by Milgram (1983), who found that local values of α ranged from 0.037 to 0.165. He suggested an empirical formula for α:

$$\alpha(F_B) = K\frac{F_B}{A + F_B} \tag{21}$$

where $K = 0.165$ and $A = 7.598$. F_B is a bubble Froude number, defined as

$$F_B = C_m^{2/5} \frac{L_M}{L_D} \tag{22}$$

where $L_M = \left(Q_b^2 / gC_m^2\right)^{1/5}$ and $L_D = \left(\sigma_b / g(\rho_w - \rho_b)\right)^{1/2} / C_m^{1/3}$. Milgram suggested that L_M is a characteristic bubble mixing length, and L_D is a characteristic distance separating bubbles. He argued that an increase in the bubble Froude number enhances turbulent entrainment by increasing turbulence near the entrainment interface. Fig. 4 shows all available data for entrainment coefficient as a function of bubble Froude number (local values of α were not reported by Tacke et al. (1985)). The solid line in Fig. 4 plots Eq. (21).

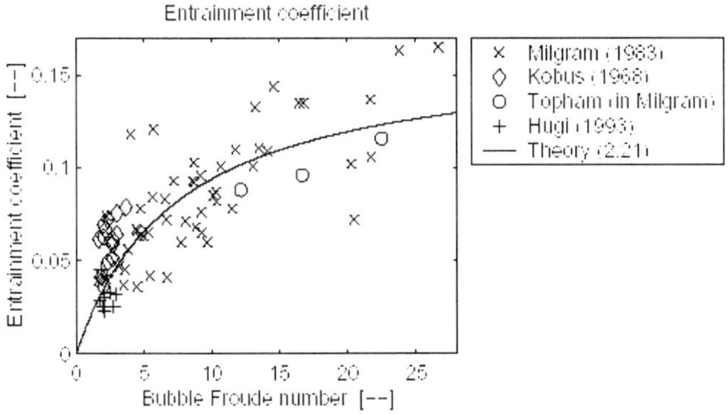

Fig. 4. Correlation of entrainment coefficient with bubble Froude number.

Other closure schemes have been investigated. For instance, Brevik & Killie (1996) and Brevik & Kluge (1999) proposed that the dominant Reynolds stress could be assumed to be self-similar, so that the spreading rate could be defined by an integral constant. Unfortunately, the integral constant, which must be determined experimentally, varies with physical parameters, such as gas flow rate, similarly to α. Hence, alternative schemes do not offer significant advantages over the entrainment assumption.

2.2.5 Momentum Amplification

Introduced in Eq. (4), γ accounts for the fact that the plume is turbulent. The instantaneous vertical plume velocity may be decomposed into mean and turbulent quantities: $u = \overline{u} + u'$. Values of $\gamma > 1$ account for the momentum flux associated with $\overline{u'u'}$.

Milgram (1983) presents an analysis of the momentum amplification effect. There, Milgram correlated γ with a phase distribution number, N_P, that describes the coherency of the bubble column, such that

$$\gamma(N_P) = 1.07 + \frac{D_1}{N_P^{D_2}} \tag{23}$$

where $N_P = L_V / L_D$ and $L_V = U_m^2 / gC_m$. Correlation with the available data gave $D_1 = 977$ and $D_2 = 1.5$.

A slow stream of isolated bubbles has a low N_P, which gives a high value for γ. Leitch & Baines (1989) also report that turbulent momentum flux increases with decreasing airflow rate, though they found discrepancies between their data and Eq. (23). In the opposite limit, as $N_P \to \infty$, $\gamma \to 1.07$, the value for a single-phase plume.

3. BUBBLE PLUMES IN STRATIFICATION

This section incorporates the complications of stratification into our description of axisymmetric bubble plumes. We assume a linear stratification and a stagnant ambient for our analytical results; however, the numerical model described at the end of this section can incorporate arbitrary stratification and any density feedback due to dissolution of the dispersed phase. The final section addresses crossflows.

3.1 Theory

3.1.1 Dimensional Analysis

There are two independent techniques to derive a set of governing non-dimensional numbers. First, if the important physical parameters can be fully listed, the Buckingham Π theorem can be used to find a set of non-dimensional numbers (Fischer et al. 1979). Second, if the governing differential equations can be written, the non-dimensional numbers can be derived by normalizing the governing equations. Here, we start with the first method in order to introduce the physics of stratification independent of an analytical model. Later, we will use the non-dimensional groups obtained here to help normalize the governing equations.

We begin with a single-phase plume in stratification where the important independent parameters are the buoyancy flux, B, and the stratification frequency, N, and the desired dependent variable is the trap height, h_T (refer to Fig. 1). Normalizing h_T by B and N, we define the first non-dimensional group, π_1, as

$$\pi_1 = \frac{h_T}{(B/N^3)^{1/4}} \tag{24}$$

which gives the ratio of the trap height to the characteristic length scale of a stratified single-phase plume, $l_C = (B/N^3)^{1/4}$. From Eq. (1) we have $\pi_1 = 2.8$.

Turning to multi-phase flow, we consider a sediment plume where expansion and dissolution are negligible and where particle size and slip velocity can be accurately controlled. This introduces two-phase plume physics without the complications of bubble expansion. Several sediment characteristics are important, including size, density, shape, and possibly cohesion. Since u_s is itself a function of these parameters, we assume that the slip velocity incorporates the important characteristics for describing the simple two-phase plume. Thus, a second, non-dimensional group, π_2, may be written as

$$\pi_2 = \frac{u_s}{(BN)^{1/4}} \tag{25}$$

where π_2 is the ratio of the slip velocity to a characteristic plume fluid velocity, $u_C = (BN)^{1/4}$.

The Buckingham Π theorem states that π_1 should depend on π_2. Reingold (1994) first attempted to relate h_T to u_s; however, she did not make use of π_2. Additional data, together with the data from Reingold (1994), are presented in Fig. 5. The relationship plotted in the figure is give by

$$\pi_1 = 2.8 - 0.27\pi_2 \tag{26}$$

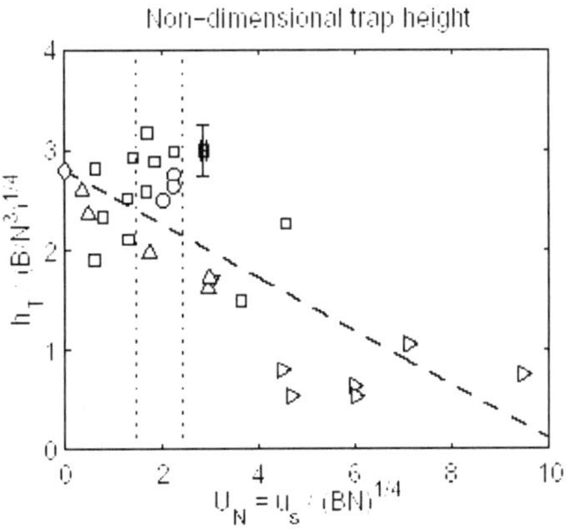

Fig. 5. Dependence of trap height on slip velocity. Sediment and bubble experiments by Reingold (1994) are Δ and ∇, respectively. ▷ are both field and laboratory data from Lemckert & Imberger (1993). O are laboratory experiments by Asaeda & Imberger (1993). □ are our laboratory experiments, and ◊ is the single-phase value from Eq. (1).

The trend of reduced h_T for increasing π_2 yields physical insight. Because water that peels from a two-phase plume loses its buoyancy when the bubbles continue to rise, it will trap lower than for a single phase plume, where the fluid and buoyancy are advected together into the intrusion layer. Completing the dimensional analysis, we consider the pressure effects resulting in bubble expansion. For ideal gas behavior, the important parameter would be the total pressure head, H_T. The only other independent non-dimensional group, π_3, is

$$\pi_3 = \frac{H_T}{\left(B / N^3\right)^{1/4}} \tag{27}$$

which is the ratio of H_T to the natural length scale of the plume, l_C.

There are thus three non-dimensional groups describing multi-phase plumes in stratification. The group π_1 comes from single-phase plumes. Introducing the dispersed phase gives π_2, and we have $\pi_1 = f(\pi_2)$ when compressibility effects are negligible.

Finally,when compressibility is important, we have all three dimensionless groups, and expect $\pi_1 = f(\pi_2, \pi_3)$. Since π_2 is the dominant parameter representing the non-dimensional slip velocity, we give it the name $U_N \equiv \pi_2$.

3.1.2 Similarity

Bubble plumes have already been shown to lack strict similarity due to variations in α and λ_1 (see Sec. 2). Stratification adds other effects that degrade the self-similarity of both single-phase and bubble plumes.

Stratification affects self-similarity in two ways: it affects entrainment, and it causes peeling and trapping. For a single-phase plume, the entrainment coefficient may be correlated with the plume Richardson number, R_P, given by

$$R_P = \frac{QB^{1/2}}{M^{5/4}} \tag{28}$$

Alternatively, we could use the plume Froude number, $F_P \sim 1/R_P$. For a single-phase plume in an unstratified environment, the plume Richardson number is a constant, $R_P = 0.557$ (Fischer et al. 1979). In stratification, R_P is not constant, but changes due to entrainment of ambient fluid. The entrainment for a bubble plume in a stratified environment should be similarly affected. This further complicates the correlations for α for bubble plumes stated in Sec.2.2.4. Hussain & Narang (1984) proposed a complex expression for entrainment into a bubble plume in a stratified environment, but did not offer experimental verification. Currently, there is no experimental correlation for entrainment of a bubble plume in stratification.

The velocity profile of a self-similar flow has a consistent shape at all heights. This condition is violated in stratification because of peeling events. The velocity profiles at a peeling event, and at the depth where an outer plume intrudes, are fundamentally different from those found elsewhere in the plume. Thus, the plume as a whole is clearly not self-similar.

Nonetheless, the tools granted by the similarity assumption, namely the integral model and the entrainment assumption, may still be used to provide insight into the dynamics of stratified bubble plumes. This is possible because plume properties are weakly dependent on variations in α due to stratification (Turner 1986) and because the plume can be decomposed into rising and descending flows (Asaeda & Imberger 1993). However, the equations from Sec. 2 must be expanded to account for the density effects of stratification.

3.1.3 Flux Model

The ambient fluid density, presumed constant in Sec. 2, is described in stratification by the variable $\rho_a(z)$. As the plume entrains water from this stratification profile, the density of the plume water, ρ_w, varies with height. In order to properly account for the buoyancy of the plume water, we define the density defect $\Delta\rho_w = \rho_w - \rho_a$. The variation of $\Delta\rho_w$ requires that new buoyancy flux and force terms be added to those in Sec. 2.1.2.

In general, determination of the density of the plume fluid in a stratified environment requires knowledge of the concentrations, and thus fluxes, of the stratifying agents, e.g. temperature and/or dissolved solutes (salinity):

$$J(z) = \int_0^\infty 2\pi r(1 - C(z,r))\rho_w c_p(z) T(z,r) u(z,r) dr \tag{29}$$

$$S(z) = \int_0^\infty 2\pi r(1 - C(z,r)) s(z,r) u(z,r) dr \tag{30}$$

J is the heat energy flux, S is the salt flux, c_p, approximated as a constant in most cases, is the specific heat of the plume fluid, T is the temperature of the plume fluid, and s is the salinity of the plume water.

If the fluxes in Eqs. (29) and (30) are tracked, water density can be calculated from an equation of state (Gill 1982). However, if the difference between the minimum and maximum ambient densities is small, water density may be approximated as a linear function of salinity and temperature. In this case, it is adequate to track the conservation of density defect, $\rho_r - \rho_w$, where ρ_r is the reference density. The buoyancy flux of the plume water is then defined as

$$B_w(z) = \int_0^\infty 2\pi r(1 - C(z,r)) g \frac{\rho_r - \rho_w(z,r)}{\rho_r} u(z,r) dr \tag{31}$$

To evaluate this integral, we introduce a new profile (assumed Gaussian) which gives the difference in density between the ambient, ρ_a, and the plume fluid:

$$\Delta\rho_w(z,r) = \Delta\rho_{w,m}(z)e^{-r^2/(\lambda_2 b)^2}$$

where $\Delta\rho_{w,m}$ is the centerline value of $\Delta\rho_w$ and λ_2 is the spreading ratio of density defect to velocity (Fischer et al. 1979). Evaluating Eq. (31) in terms of $\Delta\rho_w$ gives the new plume buoyancy flux

$$B_w(z) = \frac{gU_m b^2}{\rho_r}\left[(\rho_r - \rho_a) - \frac{\lambda_2{}^2\Delta\rho_{w,m}}{1+\lambda_2{}^2}\right] \quad (32)$$

which will be used to form a buoyancy conservation equation.

For the plume force balance, the buoyant force of the bubbles, $\hat{B}_b$, from Sec. 2 is now offset by a negative buoyant force of the entrained water, $\hat{B}_w$, written per unit height as

$$\hat{B}_w = \int_0^\infty 2\pi rg(1 - C(z,r))\rho_w(r,z)dr \approx \pi\lambda_2{}^2 b^2 g\Delta\rho_{w,m} \quad (33)$$

3.1.4 Governing Differential Equations

With the additions of the previous section, the governing equations can be derived for the stratified case. The volume (or mass) conservation equation, Eq. (13), is unchanged from Sec. 2.1.3. The governing equation for momentum gains a term from Eq. (33), due to the forcing of the negatively buoyant plume fluid. Thus, Eq. (14) becomes

$$\frac{dM}{dz} = \frac{\hat{B}_b - \hat{B}_w}{\rho_w} = \frac{\pi g b^2}{\rho_w}\left(\lambda_1{}^2 C_m \Delta\rho_b - \lambda_2{}^2\Delta\rho_{w,m}\right) \quad (34)$$

A new equation represents the conservation of buoyancy flux, given by

$$\frac{dB_w}{dz} = 2\pi bU_m \alpha g\frac{\rho_r - \rho_a(z)}{\rho_r}. \quad (35)$$

Equation (35) indicates that the buoyancy flux of the plume water changes due to entrainment of ambient water that itself has a density defect relative to the reference density. Together with Eq. (13) and a bubble dynamics sub-model, Eqs. (34) and (35) govern a bubble plume in stratification.

More generally, the buoyancy conservation equation can be replaced by the conservation of heat and salinity flux, which follow directly from Eq. (13):

$$\frac{dJ}{dz} = 2\pi b\alpha U_m \rho_w c_p T_a(z) + \frac{dW_b}{dz}\Delta H_{sol} \quad (36)$$

$$\frac{dS}{dz} = 2\pi b\alpha U_m s_a(z) \tag{37}$$

where T_a and s_a are the temperature and salinity of the ambient fluid at a given height and H_{sol} is the heat of solution for the dissolving dispersed phase. Tracking the fluxes of both heat and salinity (plus any other solutes of interest) is sensible when there are multiple stratifying agents or when dissolution of the dispersed phase affects density.

To non-dimensionalize the governing equations, it is useful to recast them as functions of the governing variables U_m, b, C_m, $\Delta\rho_{w,m}$, and $\Delta\rho_b$. The set of dimensionless variables are formed from l_C and u_C. For deep water plumes, where $H >> H_A$, the length scale for calculating derivatives is chosen based on the trap height, so the combination $(B/N^3)^{1/4}$ becomes the dominant length scale. This contrasts with McDougall (1978), who used H as his normalizing length scale. The other normalization variables in McDougall (1978) will be retained, however, yielding

$$z = \left(\frac{B}{N^3}\right)^{1/4} Z; \qquad b = 2\alpha\left(\frac{B}{N^3}\right)^{1/4} B_N$$

$$U_m = U_b V; \qquad C_m = X$$

$$\frac{\Delta\rho_{w,m}\, g}{\rho_r} = \frac{U_b^{\,2}}{\lambda_2^{\,2}}\left(\frac{N^3}{B}\right)^{1/4} G_w; \qquad \frac{\Delta\rho_b\, g}{\rho_r} = \frac{U_b^{\,2}}{\lambda_1^{\,2}}\left(\frac{N^3}{B}\right)^{1/4} G_b$$

where Z, B_N, V, X, G_w, and G_b are the non-dimensional variables for elevation, width, plume fluid velocity, bubble concentration, plume fluid buoyancy, and bubble buoyancy, respectively.

Inserting the non-dimensional variables, Eqs. (13) and (34) become

$$\frac{d}{dZ}\left[VB_N^{\,2}\right] = VB_N \tag{38}$$

$$\frac{d}{dZ}\left[VB_N\right] = \frac{B_N}{\gamma V}\left(XG_g - G_w\right) \tag{39}$$

respectively. If the buoyancy flux Eq. (32) is substituted into Eq. (35) and expanded, the latter equation becomes

$$\frac{d}{dZ}\left[\frac{VB_N^{\,2}G_w\left(1+\lambda_1^{\,2}\right)^2}{\left(1+\lambda_2^{\,2}\right)}\right] = \frac{-VB_N^{\,2}}{U_N^{\,2}} \tag{40}$$

which gives the conservation of the plume fluid buoyancy, independent of the bubbles. The final equation represents buoyancy changes due to bubble expansion, neglecting dissolution. The gas volume flux can be expressed as a bubble buoyancy flux term by multiplying Q_b by $g(\rho_r - \rho_b)/\rho_r$. Applying conservation of buoyancy to the bubble volume flux and assuming the ideal gas law yields

$$\frac{d}{dZ}\left[(V+1)B_N{}^2 XG_b\left(1+\lambda_1{}^2\right)^2\right]=\frac{M_{HT}}{\left(1+Z/(P_{HT})^{1/4}\right)^2} \tag{41}$$

where M_{HT} and P_{HT} were defined in Asaeda & Imberger (1993). P_{HT} is just $\pi_3{}^4$. Asaeda & Imberger (1993) primarily use a different form of P_{HT}, called P_N, in their analyses:

$$P_N = \frac{N^3H^4}{B} \tag{42}$$

In deep water $H_A << H$ and $P_N \approx P_{HT}$. M_{HT} combines the bubble slip velocity and expansion effects and can be written as a function of U_N and P_{HT}, namely

$$M_{HT} = \frac{P_{HT}{}^{-1/4}}{4\pi\alpha^2 U_N{}^3} \tag{43}$$

Asaeda & Imberger (1993) also primarily use a different form of M_{HT}, called M_H, in their analyses:

$$M_H = \frac{B}{4\pi\alpha^2 H u_s{}^3} \tag{44}$$

In the literature, variations of P_N and M_H have been used extensively (e.g. Asaeda & Imberger 1993, McDougall 1978, Schladow 1993, Lemckert & Imberger 1993); whereas, U_N is a new parameter. This came about because researchers have been primarily interested in reservoir destratification, where bubble expansion is a significant physical process.

The non-dimensional governing equations Eqs. (38)-(41) together with the dimensional analysis help identify the proper use of these three parameters. U_N enters the conservation of fluid buoyancy equation and the dimensional analysis when the dispersed phase is introduced. It is the fundamental parameter describing the effects of a dispersed phase on plume dynamics. Both P_{HT} and M_{HT} enter with bubble expansion since they are both dependent on the water depth. Because $M_{HT} = f(P_{HT}, U_N)$ we propose that P_{HT} and U_N should be used as the two governing non-dimensional variables for two-phase plumes.

3.2 Observations

3.2.1 Plume Classification

Asaeda & Imberger (1993) and Lemckert & Imberger (1993) present a wide range of laboratory and field experimental observations. They were primarily interested in the efficiency of reservoir destratification and showed that plumes having one intermediate peel and one surface peel were the most efficient at reservoir mixing. Based on their experimental observations, Asaeda & Imberger (1993) defined three distinct modes of two-phase plume behavior in linear stratification. Shown schematically in Fig. 6 together with Type 1* to be described shortly, these modes were called Types 1, 2, and 3. The Type 1 plumes have no intermediate intrusions, detraining all of the entrained fluid in the surface radial jet. The Type 2 plumes have one or more intermediate intrusions, but each intrusion is a distinct layer. The Type 3 plumes appear to lose entrained fluid randomly, forming a continuous structure of sub-surface intrusions.

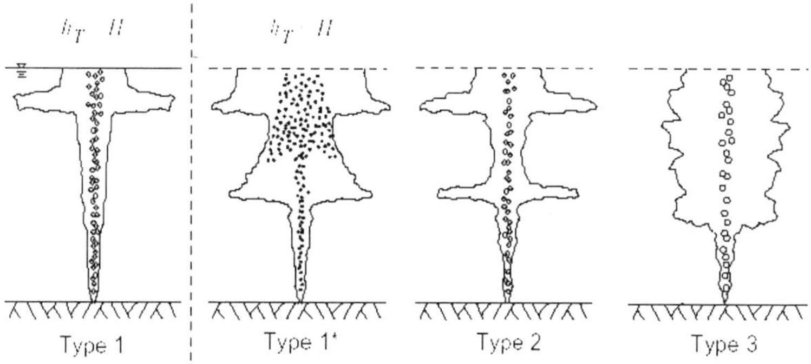

Fig. 6. Schematic of the plume type classification

The Type 1* plumes occur when the dispersed phase has a low enough slip velocity compared to the turbulence at the plume peel that the bubbles peel with the fluid. In most of the experiments reported in the literature, the bubble slip velocity is about 20 cm/s, too high to be affected by the peeling fluid. However, our experiments with oil, fine air bubbles and sediment (u_s = 3 - 8 cm/s) show that Type 1* plumes exist and differ from Type 2 plumes: the bubble core is more spread out above the first peel, and fluid is re-entrained out of the intrusion and carried upward by the peeled bubbles as they reform the secondary plumes.

The Type 3 plumes actually result from inefficient peeling events, where only a portion of the entrained fluid is detrained. In this case, detraining fluid may have been lifted a significant elevation, and may be denser than intrusions below it. This instability allows Type 3 peels to overlap, yielding the random, continuous-peeling nature of the Type 3 plume. In contrast, Type 1* and 2 plumes detrain most of their water at each peel. As a result, the peels do not overlap but remain localized at separate intrusion levels. Because the plume source originates as if it were above a peel with complete detrainment, the first peel in any plume must be Type 1* or 2.

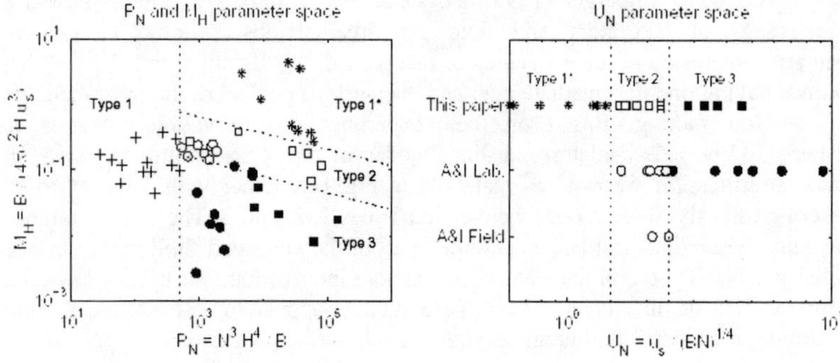

Fig. 7. Correlation of plume type with the governing non-dimensional parameters. Circles and pluses are from Asaeda & Imberger (1993); squares and stars are the current authors. Pluses are Type1 plumes, stars are Type 1* plumes, open symbols are Type 2 plumes and filled symbols are Type 3 plumes. The Asaeda & Imberger (1993) data include field and laboratory experiments.

Using P_N and M_H, Asaeda & Imberger (1993) were able to predict plume Types 1, 2, and 3. The plot to the left in Fig. 7 shows the plume type parameter space defined by Asaeda & Imberger (1993). Fig. 7 contains all of the data in Asaeda & Imberger's (1993) paper, along with data from our own experiments.

As discussed in the previous section, we prefer to use P_N and U_N and to avoid M_H. If we neglect bubble expansion, then we can limit ourselves further to U_N alone. The plot to the right in Fig. 7 shows an alternate plume-type prediction scheme using U_N. Type 1* plumes are generated for U_N from zero up to about 1.5. Type 2 plumes exist for U_N between about 1.5 and 2.4. For U_N greater than about 2.4 plumes are Type 3. The Type 1 plumes defined by Asaeda & Imberger (1993) do not plot in this parameter space since the reservoir depth is neglected. Also, in the case of air bubbles where the reservoir depth is of the order of H_A, bubble expansion would be significant, and P_N should be added to the analysis. As illustrated in the figure, however, the existing data are well represented using U_N alone.

3.2.2 Experiments

Our experiments with two-phase plumes in stratification have addressed two issues. First, plumes were observed over a wide range of the non-dimensional parameter space in order to broaden our understanding of plume typology; these data were presented in the previous section. Second, as discussed in this section, measurements were made to quantify the induced flow field to provide validation data for numerical models.

These experiments were conducted in the Parsons Laboratory at MIT using a 1.2 m square by 2.4 m deep tank. The tank is stratified using the two-tank method (Asaeda & Imberger 1993), creating a linear profile with a stratification

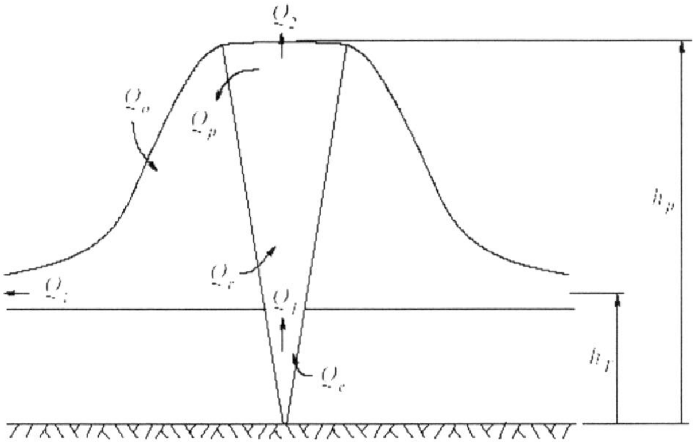

Fig. 8. Schematic model of a plume peel showing definitions of the plume volume fluxes and characteristic heights.

frequency of N, typically 0.3 s^{-1}. Multiple dispersed phases were investigated, including air, oil and sediment (which forms an inverted plume). The flow field was visualized by injecting Rhodamine 6G dye at the diffuser source and illuminating a slice of the plume through the centerline with a LASER light sheet (called the LASER induced fluorescence (LIF) method). Quantitative measurements included plume length scales from the visualization and pre- and post-experiment profiles of salinity and dye tracer. Additional details of the experiments are presented in Socolofsky (2000) and Socolofsky & Adams (2000*a*, 2000*b*).

To interpret the raw density and dye concentration profiles, a conceptual model of a plume peel was created (see Fig. 8). Two characteristic heights and seven volume fluxes are defined. The two heights are measured from the dye concentration profiles (h_T is the center of mass of the intrusion, and h_P is the point of the minimum dye concentration between peels). The power of this conceptual model is in dealing with the exchange of fluid between the downdraught and upward moving plumes. In this model, this exchange is simplified into the recirculating flow, Q_r, which is calculated from volume conservation as described in detail below.

Because velocity measurements in stratified ambients are difficult to obtain, we instead determine plume volume fluxes. Following Leitch & Baines

(1989) and Baines & Leitch (1992), the net volume flux integrated across the plume is

$$Q(z) = A\frac{\partial \rho_a / \partial t}{\partial \rho_a / \partial z} \tag{45}$$

where A is the cross-sectional area of the tank and t and z are the time and spatial coordinates. Using this equation, the flux at the bottom of the peel, Q_1, and the flux at the top of the peel, Q_2, are directly computed. One more flux, $Q_1 - Q_i$, is obtained by taking the maximum negative (downward) flux in the peeling region. Three more equations follow from conservation of volume; from Fig. 8

$$Q_e = Q_1 \tag{46}$$

$$Q_1 + Q_r = Q_2 + Q_P \tag{47}$$

$$Q_p + Q_o = Q_r + Q_i \tag{48}$$

A final equation uses the dye concentration profile, taken at the end of the experiment. By integrating the dye profile, the mass of dye at each level can be calculated. The fraction of dye injected that remains trapped in the intrusion, f^* (measured from the dye profiles), is assumed to equal the fraction of plume fluid that peels, f; from Fig. 8

$$f = \frac{Q_p}{Q_p + Q_2} \approx f^* \tag{49}$$

which gives us a closed set of equations.

This simple model was applied to the first peel of a sample Type 1* plume with an airflow rate of 3 mL/s and bubble slip velocity of 7 cm/s. The entrainment below the shrouded region was 300 ± 30 mL/s, and the net upward flux through the peel was 100 ± 30 mL/s. 93% of the plume fluid peeled, which resulted in 1000 ± 300 mL/s peeling, 870 ± 300 mL/s recirculating, and 580 ± 80 mL/s intruding. Socolofsky & Adams (2000*b*) presents a more general approach where the volume fraction of fluid peeled is not assumed equal to f^*, but rather an optimization scheme is used to calculate the seven flow rates and their associated tracer and salinity fluxes. For this example with a large f^*, the results are largely unchanged and the errors are reduced. Sec. 3.3 presents a comparison of these fluxes with results from an integral model.

3.3 Models

The governing equations presented in the previous sections are useful for analytical purposes. To form a complete model formulation, one must add a mechanism for determining behavior at a peeling event. Though peeling can be added, models of this type, called single plume models, predict behavior of the rising plume while neglecting the dynamics of the intrusion flow. An alternative type of formulation, called a double-plume model, explicitly models the dynamics of the intrusion flow as well as the rising inner plume. This section reviews the relative merits of different models.

3.3.1 Model Formulations

Single-plume models are attractive because they represent initial value problems. All that is required in addition to the flux definitions and governing equations of Sec. 2.1 and 3.1 is a scheme for describing peeling events. Without such a scheme, numerical integration of the governing equations cannot continue past the first peeling event. The actual dynamics of peeling events, however, are not at present well understood, so the approaches used to date are approximate.

In the models described by Liro et al. (1992) and Thorkildsen et al. (1994), peeling occurs at the height where the net buoyancy of the plume, i.e. the right-hand side of Eq. (34), becomes negative. At this point, a certain fraction of volume and momentum flux is removed from the plume, so that the plume again becomes positively buoyant. Integration then proceeds to the next peeling event. The fate of the water lost from a given peeling event is not explicitly modeled.

Schladow (1993), in a study of the evolution of stratification of a reservoir mixed by an air bubbler, described a model in which the governing equations are integrated until the momentum flux approaches zero. Schladow assumed that all of the fluid leaves the plume at this point and ultimately intrudes into the environment at its depth of neutral buoyancy. His combined plume-reservoir model successfully predicted the time for reservoir turnover due to bubbling.

Double-plume models have been formulated by McDougall (1978), Asaeda & Imberger (1993), and Crounse (2000). To account for the dynamics forced by stratification, each of these models decomposes the bubble plume into inner and outer plumes. The inner and outer plumes are described by separate volume (Q_i and Q_o, not related to the fluxes in the previous section), momentum (M_i and M_o), and buoyancy (B_i and B_o) fluxes. Because the property profiles are no longer expected to be Gaussian, these models cast these fluxes in terms of simpler top-hat profiles. For example, the inner plume profiles are

$$u_i(z,r) = u_i(z), \quad r \le b_i$$

$$C(z,r) = C(z), \quad r \le \lambda_1 b_i$$

$$\rho_i(z,r) = \rho_i(z), \quad r \le \lambda_2 b_i$$

where b_i is the top-hat inner plume width. Defining $\Delta\rho_i = \rho_i - \rho_a$, the fluxes for the inner plume become

$$Q_i = \pi u_i b_i^2 \tag{50}$$

$$M_i = \gamma \pi u_i^2 b_i^2 \tag{51}$$

$$B_i = \pi u_i b_i^2 g \frac{\Delta \rho_i}{\rho_r} \tag{52}$$

For simplicity, this definition of B_i expresses buoyancy flux in reference to ρ_a, unlike Eq. (32). Top-hat and Gaussian flux quantities are interchangeable using

$$b_i = \sqrt{2}b \quad u_i = \frac{U_m}{2}, \quad \alpha_i = \sqrt{2}\alpha$$

The flux expressions for the outer, intruding plume are identical to the inner plume, except that the outer velocity is u_o (positive for upward-flowing and negative for downward-flowing outer plumes), the density defect is $\Delta\rho_o$, and the area of the outer plume is $\pi(b_o - b_i)^2$, rather than πb_i^2.

McDougall (1978) proposed a model which splits the plume into two coflowing annular plumes. The inner plume consists of a rising flow of bubbles and water, while the outer plume consists only of water. These plumes interact through mixing across the plume boundaries. McDougall reasoned that the mixing between the plumes, and between the outer plume and the ambient, could be parameterized with the entrainment assumption, so that the entrainment fluxes are defined as

$$E_i = 2\pi b_i \alpha_i (u_i - u_o) \tag{53}$$

$$E_o = 2\pi b_i \alpha_o u_o \tag{54}$$

$$E_a = 2\pi b_o \alpha_a u_o \tag{55}$$

where E_i is the entrainment flux from the outer plume to the inner plume, E_o is the entrainment flux from the inner to the outer plume, and E_a is the entrainment flux from the ambient environment to the outer plume, all per unit height. α_i, α_o and α_a are entrainment coefficients governing the three entrainment fluxes. Just as entrainment transported volume and associated buoyancy from the ambient to the plume flow in single-plume models, these entrainment fluxes transport volume and associated buoyancy and momentum between the inner plume, the outer plume, and the ambient environment. The mixing assumptions are based in part on Morton (1962), who applied the entrainment assumption to coaxial, coflowing single-phase jets. This particular application of the entrainment hypothesis has not been clearly verified by experiment. Although this model can produce peeling events, which occur when the outer plume momentum approaches zero, it does not model the intrusion flow which originates at a peeling event.

Whereas McDougall's (1978) plumes were coflowing, Asaeda & Imberger (1993) formulated a double-plume model which incorporates counterflowing plumes. In this model, the inner plume encompasses all of the upward-moving fluid and bubbles (i.e. both McDougall's inner and outer upward-flowing plumes are treated together), while the outer plume represents the descending intrusion flow. Mixing between the inner and outer plumes and the ambient environment are modeled with relationships identical to those of McDougall (1978), except that in this case $u_o < 0$, so the entrainment fluxes Eqs. (53)-(55) become

$$E_i = 2\pi b_i \alpha_i (u_i - u_o) \quad (56)$$

$$E_o = 2\pi b_i \alpha_o u_o \quad (57)$$

$$E_a = 2\pi b_o \alpha_a u_o \quad (58)$$

Asaeda & Imberger (1993) found that $2\alpha_i \approx \alpha_o \approx \alpha_a$, which also agrees with McDougall (1978), produced the best fit of the model to their experimental data for trap height and number of sub-surface intrusions.

To solve the model, the governing equations of the inner plume are integrated from the virtual origin to the point where the momentum flux of the inner plume approaches zero. At this point, Asaeda & Imberger approximate the peeling process by assuming that 100 percent of the inner plume fluid exits the plume and begins to descend. The governing equations of the outer plume are integrated downward from this point until the plume reaches neutral buoyancy, at which point the outer plume is assumed to intrude into the ambient environment. As the properties of the inner plume had been initially calculated without the presence of the outer plume, the process of integration of the inner plume, and then the outer plume, must be repeated until the predicted flows converge. Once this is achieved, a new inner plume is initialized above the previous peel location, and the process is repeated until the surface is reached.

The Asaeda & Imberger model formulation accounts for the fact that the location of the intrusion is often significantly below the depth of the peeling event. The assumption that all of the plume fluid detrains at a peeling event is reasonable, as the actual percentage for a typical Type 2 plume has been observed to be approximately 90 percent (Socolofsky & Adams 2000*b*). However, 100 percent peeling dictates that the outer plume intrudes at some depth between peeling events, and cannot overlap a lower outer plume segment. Hence, this approach does not apply to Type 3 plumes.

A more general method for modeling the peeling process is to treat detrainment as a process analogous to turbulent entrainment, so that the flux of fluid out of the inner plume is expressed in terms of local plume conditions. Crounse (2000) proposed the detrainment equation

$$E_p = \varepsilon \left(\frac{u_b}{u_i} \right)^2 \left(\frac{B_i}{u_i^{\ 2}} \right) \quad (59)$$

where ε is an empirical parameter. This exact relationship has not been experimentally confirmed, but it reflects the observed physical process. The percentage of volume flux which is lost over a Type 2 peeling event can be matched by varying ε. Furthermore, use of this equation allows simulation of Type 3-like behavior.

Taking into account all of the discussed interactions between the inner and outer plumes, the governing equations for a double-plume model can be derived from the equations presented in Sec. 2.1.2 and 3.1.3. For example, the volume flux equations become

$$\frac{dQ_i}{dz} = E_i + E_o + E_p \tag{60}$$

$$\frac{dQ_o}{dz} = -E_i - E_o - E_a - E_p \tag{61}$$

for the inner and outer plumes, respectively.

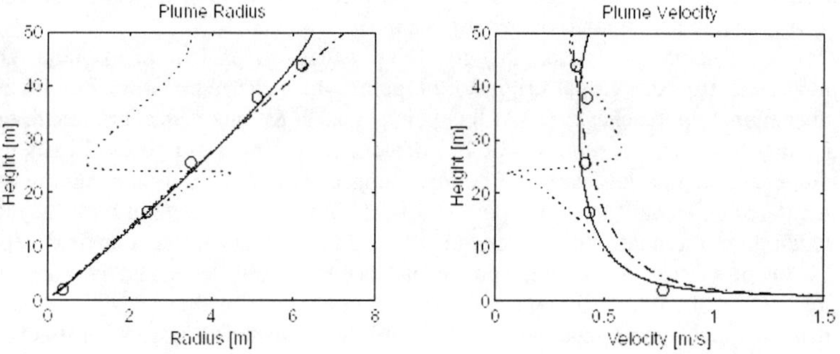

Fig. 9. Radius and velocity profiles for three different modeled plumes. The single-phase plume is represented by a dash-dot line, the simple bubble plume by a solid line, and the stratified bubble plume by a dotted line. Data for a simple bubble plume from Milgram (1983) are indicated by O. All cases have a source buoyancy flux of B =0.19m^4/s^3.

Treatment of peeling as a continuous process requires a different iteration scheme than that of Asaeda & Imberger (1993). Given initial conditions, the inner plume is integrated upward until either the bubbles have fully dissolved, or a surface is reached. Then, starting from the top of the plume, outer plume segments are initiated wherever the peeling flux E_p is significant. The outer plume is then integrated downward until the plume intrudes. Then, the next lower outer plume is initiated, and so on down the plume. As a consequence, overlapping outer plumes completely mix. This process is then iterated until the fluxes between the inner and outer plumes are consistent. Convergence is readily achieved for Type 2 plumes with a limited number of peels, but becomes more difficult as the plume approaches Type 3 behavior.

3.3.2 Model Results

Fig. 9 illustrates model results for three different plumes having identical initial buoyancy fluxes. The three cases modeled are a single-phase plume, a bubble plume in a homogeneous environment, and a bubble plume in a linearly stratified environment ($N = 0.04\ s^{-1}$, $U_N = 1.2$). Experimental data from Milgram (1983) for a simple bubble plume are plotted for comparison.

The modeled single-phase and two-phase simple plumes are rather similar. The plume radii are both nearly linear with height. The velocity of the simple bubble plume is slightly lower than that of the single phase plume at lower elevations because the effective buoyant force per unit height is smaller in the bubble plume than in the simple plume. As the bubbles expand with height, the buoyancy flux increases and the fluid accelerates. The simple two-phase plume model results match the data from Milgram.

The only difference between the two modeled bubble plumes is the presence of stratification. The properties of the stratified plume are initially similar to the other plumes, but diverge as the plume fluid is arrested by buoyant forces and subsequently detrains. Above the peel the plume reforms, but is significantly smaller than the simple bubble plume at all points above the plume.

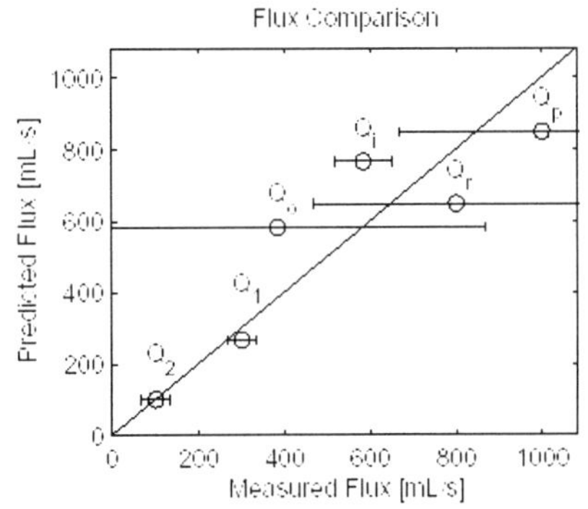

Fig. 10. Comparison of model results with experimental fluxes.

Fig. 10 compares model results with experimental flux calculations as described in Sec. 3.2.2. All of the predicted fluxes lie within the range of experimental error except for Q_i, the flux of fluid intruding into the ambient. Because the experiment was conducted in a finite tank, it is possible that the experimental value of Q_i is suppressed by blockage due to the boundary.

Fig. 11 shows model results for continuous releases of liquid CO_2 in a stagnant, stratified environment, such as might occur in an oceanic CO_2

sequestration scheme. The only difference among the three cases presented is the dissolution rate of the CO_2 droplets, which profoundly affects the plume structure. The dissolution can be strongly affected by the formation of a hydrate film on the surface of the CO_2 droplets, a process that is not completely understood (Wong & Hirai 1997). Not only does the dissolution lead to reduced positive buoyancy with height, but the dissolved CO_2 also increases the density of the plume water. Thus, the effects of increased dissolution are amplified.

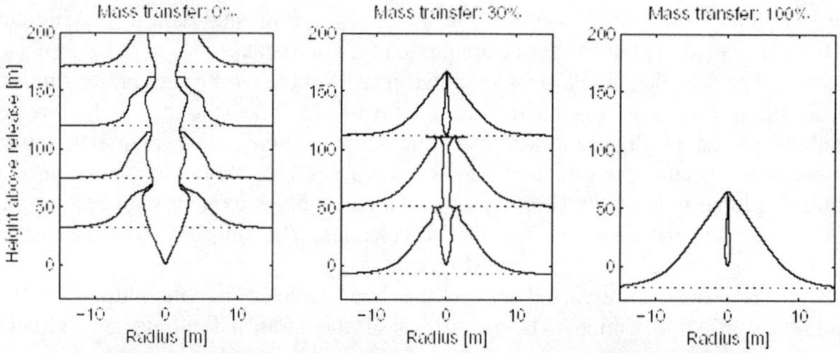

Fig. 11. Profiles of CO_2 droplet plumes. Shown are the inner and outer plume radii of three plumes resulting from the release of buoyant CO_2 droplets ($B = 10^{-3}$ m^4/s^3; initial diameter = 0.4 cm) at 800 meters depth with a typical ocean stratification ($N = 0.003$ s^{-1}). The mass transfer rate is varied between zero and 100 percent of that predicted by Eq. (15). The case without mass transfer, which continues to the water surface, has been truncated for clarity.

3.3.3 Alternative Models

The advantages of integral models are that the governing equations allow insight into the dynamics of the flow, they are computationally efficient, and they produce reasonably accurate results in many cases. However, it is clear that integral models gradually lose their validity as the plume structure becomes less self-similar, due to factors such as stratification.

Integral models are not the only tool for modeling multi-phase flows: more rigorous computational fluid dynamic codes, which solve the 3-D Navier-Stokes equations, may also be used. They generally follow one of two approaches. In the Lagrangian/Eulerian approach, individual particles of the dispersed phase are tracked as they advect through a fluid modeled by the Navier-Stokes equations. In the Eulerian/Eulerian approach, both phases are described by the Navier-Stokes equations (Jakobsen et al. 1997). These types of models have been mostly applied to confined multi-phase flows such as bubble columns (e.g. Lapin & Lübbert 1994, Jakobsen et al. 1997), although some investigators have examined bubble plumes as well (Alendal & Drange 2000, Chen et al. 2000, Sato & Hama 2000).

4. BUBBLE PLUMES IN CROSSFLOW

Returning to bubble plumes in unstratified environments, this section addresses the complications of a uniform, flowing ambient, or crossflow. We use dimensional analysis to classify the possible range of behavior of bubble plumes in a crossflow and present the results of some of our experiments.

4.1 Theory

Crossflows affect a wide range of plume properties, changing the basic plume dynamics. Even in the case of single-phase plumes, crossflows enhance entrainment, deflect the plume centerline, can deform the plume shape into a vortex pair, and can cause fluid to leak in the downstream wake of the plume (Fischer et al. 1979, Davidson & Pun 1999, Yappa & Zheng 1997*a*). Bubble plumes are affected similarly, and have additional complications due to the slip velocity of the dispersed phase.

4.1.1 Analysis of Single-Phase Jets and Plumes

As a basis for understanding bubble plumes in a crossflow, we first look at single-phase jets and plumes. Davidson & Pun (1999) studied the effects of crossflow on a single-phase momentum jet. Because the mean jet, entrainment, and turbulent velocities all decrease with height above the source, the effect of a uniform crossflow on the jet increases with height. Following Davidson & Pun (1999), three modes of behavior can be anticipated based on the height, z, relative to a characteristic jet length scale in a crossflow,

$$l_{jc} = \frac{M_0^{1/2}}{u_\infty} \tag{62}$$

where M_0 is the jet momentum flux and u_∞ is the crossflow velocity. For $z/l_{jc} < 0.14$ the jet is Gaussian, adequately described by the stagnant water equations, but deflected by the vector sum of the crossflow and the mean vertical jet velocity. At greater elevations, the jet begins to leak fluid in the downstream wake, but concentration and velocity profiles remain reasonably Gaussian. However, for $z/l_{jc} > 1$, the Gaussian model begins to fail, and the jet becomes strongly affected by the current.

The length scale, l_{jc}, is proportional to the height where the jet entrainment velocity, u_e, balances with the crossflow velocity. Fischer et al. (1979) report that u_e varies with height as

$$u_e = 7\alpha_j \frac{M_0^{1/2}}{z} \tag{63}$$

Combining Eqs. (62) and (63), we find that the entrainment velocity equals the crossflow velocity at $z/l_{jc} = 7\alpha_j = 0.4$, a value between the point where leakage begins and where the Gaussian structure breaks down (Davidson & Pun 1999).

Pun & Davidson (1999) investigated buoyant plumes and showed that they behave similarly. A crossflow is considered weak when the Gaussian model still applies. When the Gaussian model is no longer adequate, the crossflow is considered strong, and Hugi (1993) reported that a plume in a strong crossflow resembles a line thermal. Applying similar scaling principles to buoyant single-phase plumes, Pun & Davidson (1999) find a characteristic length scale for plumes in crossflow,

$$l_{pc} = \frac{B}{u_\infty^{\ 3}} \qquad (64)$$

Using the results of Davidson & Pun (1999), $z/l_{pc} = 0.003$ gives the point where fluid begins to leak, $z/l_{pc} = 0.06$ gives the point where the entrainment velocity equals the crossflow velocity, and Pun & Davidson (1999) report $z/l_{pc} = 0.5$ is the transition height at which the plume resembles a line thermal.

4.1.2 Complications Due to Slip Velocity

Because the fluid and the bubbles move at different velocities, the crossflow can create a separation between the rising plume fluid and the bubbles, as depicted in Fig. 2. In the simplest model, the bubble trajectory is given by the vector sum of the crossflow velocity and the bubble rise velocity. This introduces two important effects. First, this leads to a fractionation of the bubbles: fast rising bubbles stay in front, and slow rising bubbles move to the back of the bubble plume (Hugi 1993). Second, since the vertical velocity of the bubbles is the plume fluid velocity plus the bubble slip velocity, the trajectory of the bubble column is steeper than the trajectory of the entrained fluid. Unless the bubble core turbulence is sufficient to keep the entrained plume fluid with the bubbles, the fluid and the bubbles will separate, in which case the traditional view of an integral model will no longer hold.

Hugi (1993) conducted a series of experiments in a 3 m deep tank with air flow rates of 1 to 9 mL/s, slip velocities of 17 to 30 cm/s, and crossflow velocities of 1.2 to 7.9 cm/s. The crossflow was simulated by towing the bubble source. From visual observation, bubble fractionation was verified. Lagrangian integration of LDV measurements taken at a point as the plumes passed by indicted that fluid entrained in the front of the plume was indeed ejected in the lee of the plume after being elevated a short distance, indicating separation. Dye injected with the bubble plume also confirmed separation. Hugi (1993) concluded that coherent, self-similar plumes would not form in a crossflow because of the separation between the rising bubbles and the entrained fluid.

We have conducted similar experiments in a 0.8 m square cross-section flume. Our experiments confirmed Hugi's (1993) observation that the fluid and bubbles can separate, but suggest that there is a critical height, h_S, associated with the onset of separation. To analyze our experiments, we define two non-dimensional velocities: $u_s/(B/z)^{1/3}$ and $u_\infty/(B/z)^{1/3}$, which are the slip and crossflow velocities normalized by the characteristic plume fluid rise velocity. Substituting h_S for z and invoking the Buckingham Π theorem, the separation height should be derived from

$$\frac{u_\infty}{(B/h_S)^{1/3}} = f\left(\frac{u_s}{(B/h_S)^{1/3}}\right) \tag{65}$$

This relationship is calibrated to our experiments in Sec. 4.2.

4.1.3 Complications due to Stratification

In the presence of both a crossflow and stratification, there are competing forces tending to break down plume similarity. The dimensional analysis presented previously can help identify the dominant processes. If the height h_T for plume trapping due to stratification is significantly below the height h_S where currents cause separation, then stratification is dominant, and the plume can be modeled as in Sec. 3. In the case where the heights are reversed, the crossflow would dominate, and stratification can be ignored.

4.1.4 Models

Despite the potential loss of similarity, integral models have been applied to study multi-phase plumes in crossflows. Yapa & Zheng (1997*a*) presents a model of an ocean blowout and Yapa & Zheng (1997*b*) validates the model to some shallow field data in the North Sea. Their model implicitly assumes that bubbles or droplets do not separate from the rising plume fluid prior to trapping. Socolofsky & Adams (2000*c*) presents a

Experiment I.D.	Flow Rate			u_∞	h_S
	Air [mL/s]	Oil [mL/s]	Alcohol [mL/s]	[cm/s]	[cm]
Exp-B8	33	0	0	2	-
Exp-C8	17	17	0	5	-
Exp-C19	0	0	2.5	5	-
Exp-C14	0	10	0	5	-
Exp-B3	3	0	0	5	16
Exp-B10	3	0	0	10	7
Exp-C16	10	10	0	10	28
Exp-C15	4	0	2.5	5	15

Table 1. Experimental parameters for selected crossflow experiments.

modeling algorithm using h_S. The multi-phase plume is integrated as a mixture up to the separation height, taken as the lower of h_S or h_T. Above this transition, the entrained fluid is modeled as a buoyant, single-phase jet in the lee of the bubble column, where the initial momentum is the momentum of the mixed model at the separation height. The bubbles are modeled above the separation height as the vector sum of the bubble slip velocity and the crossflow. An alternate model, presented by Johansen (1999), accounts for the separation directly: when the slip velocity of the bubbles is great enough that they would be lost on the leading edge of the plume, the droplets are ejected from the integral model and tracked separately.

4.2 Observations

The experiments confirmed the presence of fractionation and separation. For plumes with high buoyancy and low crossflow velocity, separation did not occur by the time the bubbles reached the water surface, and the situation is classified as a weak crossflow. In other experiments, separation did occur and the conditions are classified as strong crossflow. Table 1 presents the parameters of the experiments shown in the following figures.

4.2.1 Weak Crossflows

In weak crossflows, some entrained fluid stays with the bubble plume from the injection point to the flume surface. Fig. 12 shows four representative experiments in weak crossflows.

While major separation between the lightest dispersed phase and the other components of the plume does not occur before the plume reaches the surface, two forms of detachment, or leakage, are observed. First, as reported by Davidson & Pun (1999) for single-phase jets, some entrained fluid leaks into the downstream wake. Comparing frames (a.) and (b.) in Fig. 12 to frames (c.) and (d.), the detachment is much greater for air bubble plumes than for the oil or alcohol plumes, even though the crossflow velocity was greater for the oil and alcohol plumes. This is explained by the fact that bubbles with higher slip velocities advect much faster than their accompanying entrained fluid (Leitch & Baines 1989). The second form of leakage is seen by the fractionation of the bubbles and droplets in the crossflow, leaking smaller bubbles into the downstream wake. Frame (b.) in Fig. 12 is the most striking example of fractionation, where the air bubbles lead in the front and the oil bubbles fall to the back of the plume, but dye marking the entrained fluid is present throughout the plume.

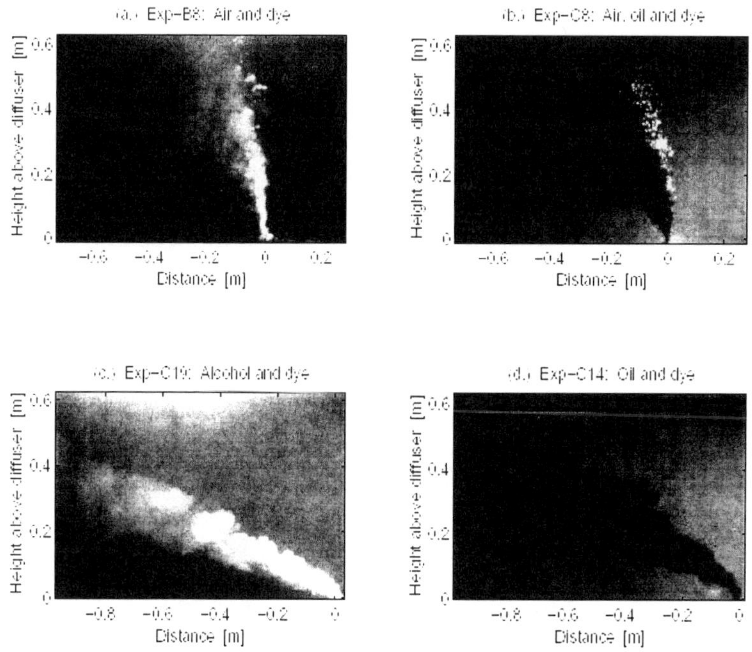

Fig. 12. Experiments showing multi-phase plumes in weak crossflows. Experimental conditions are summarized in Tab. 1.

4.2.2 Strong Crossflows

In strong crossflows there is significant separation between the dominant dispersed phase and the entrained fluid and the separated fluid rises independently in the far field. Fig. 13 shows four representative experiments in strong crossflows. Frames (a.) and (b.) in Fig. 13 are for two-phase air-bubble plumes and frames (c.) and (d.) are multi-phase alcohol and air and oil and air plumes, respectively.

For the air-bubble plumes, complete separation occurs between the entrained fluid and the rising bubble column. Dye injected near the release point separates from the bubble column, but continues to rise in the far field even though the dye and entrained fluid are neutrally buoyant due to acceleration within the bubble column. This indicates that, beyond the point of separation, the injected dye tracer behaves like a momentum jet.

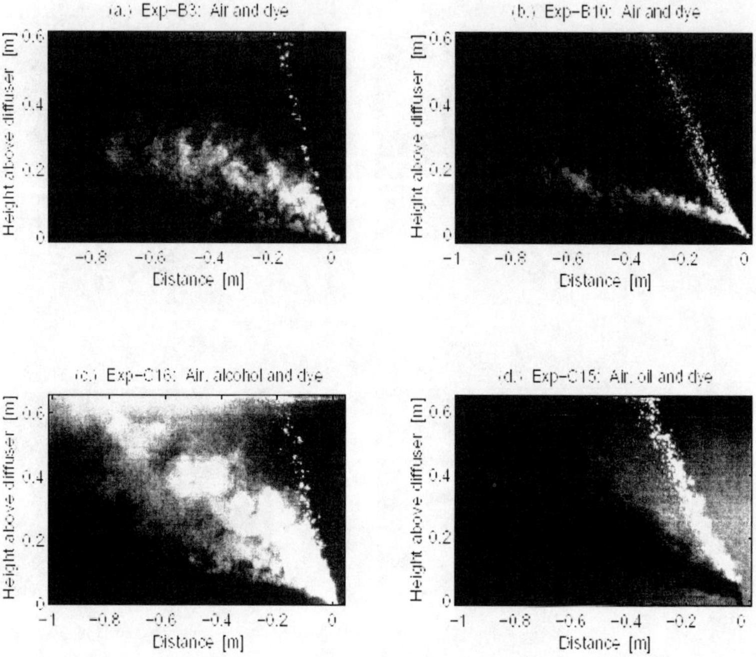

Fig. 13. Experiments showing multi-phase plumes in strong crossflows. Experimental conditions are summarized in Tab. 1.

Detachment is also observed throughout the mixed and separated plume regions. Above the separation height, the trajectory of the bubble column appears linear, represented by the vector sum of the group rise velocity of the bubbles and the crossflow velocity. This further indicates that the bubble column in a strong crossflow is not plume-like above the separation height since the downstream coordinate of a pure plume should vary as the $^{4}/_{3}$- power of height above the diffuser (Fischer et al. 1979).

For the multi-phase plumes in frames (c.) and (d.), complete separation occurs between the air bubbles and the other dispersed phase, but further separation is not observed between the separated oil and alcohol plumes and their entrained fluid. Following the description above, the separated oil and alcohol plumes are accelerated in the plume region before they separate; thus, they should be represented as buoyant momentum jets in the far field. Fractionation and leakage remain as characteristic features of these plumes.

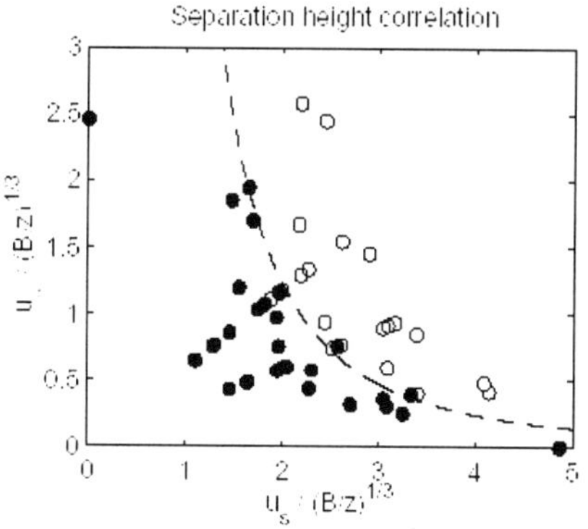

Fig. 14. Transition height correlation for multi-phase plumes in a crossflow. Filled circles indicate heights below which the phases remain mixed; open circles indicate heights above which one or more phases have separated. The dashed line plots the relationship in Eq. (66).

4.2.3 Separation Height

From a suite of 25 experiments (which includes those presented in this section) Socolofsky & Adams (2000*c*) calibrated Eq. (65) as shown in Fig. 14, yielding,

$$\frac{u_\infty}{(B/h_S)^{1/3}} = 6.32\left(\frac{u_s}{(B/h_S)^{1/3}}\right)^{-2.36} \tag{66}$$

5. SUMMARY

The defining character of a multi-phase plume is the relative independence of the dispersed phase from the surrounding fluid. In this chapter we have summarized this characteristic in the parameter u_s, the slip velocity. The main effects of u_s are to erode the self-similarity and to reduce the effective buoyant force of the multi-phase plume.

In homogeneous environments, the bubble motion leads to a variable entrainment coefficient, α, and variable spreading ratio for the bubbly core, λ_1. Although non-constant α and λ_1 imply the loss of strict similarity, an integral model, such as described in Sec. 2, continues to perform well due to relative insensitivity of the model to variation in these parameters over their expected range of environmental values (Turner 1986, Liro et al. 1992).

Stratification breaks down similarity in two different ways. First, the variation of α is enhanced by stratification. Second, the stratification causes a separation between the bubbles and plume fluid at a peel, completely eliminating self-similarity.

In addition to its effects on self-similarity, u_s in stratification affects other plume properties. First, because the peeled water loses its buoyancy when the bubbles continue to rise, the intrusion layers form much lower than predicted by trap height equations for single-phase plumes. Second, the magnitude of u_s for a given release changes the efficiency of plume peeling. Higher u_s causes greater leakage and lower efficiency. As the efficiency decreases or when density changes caused by dissolution increase the tendency for intrusions to overlap, plumes tend toward Type 3 behavior. The effects of u_s in stratification can be analyzed using U_N and P_N. Together these non-dimensional numbers correlate with the variable trap height of multi-phase plume intrusions and with the overall plume efficiency or typology.

Two types of integral models were discussed to deal with the effects of stratification. Single-plume models were applied as didactic tools to illustrate the interplay of forces in the bubble plume. For numerical modeling, double-plume models capture more of the flow characteristics. These models address the break down in similarity at a peel directly by separating the plume peeling and intrusion flows from the upward-moving, nearly self-similar plume core.

In a crossflow, similarity is degraded due to advection of the entrained plume fluid away from the bubbles by the crossflow itself. In weak crossflows, the bubbles fractionate, leaving small bubbles in the wake of the plume. Under these conditions, integral models incorporating the crossflow were shown to apply. As the crossflow increases, however, the bubbles separate completely from the entrained fluid, and self-similarity is lost. When separation occurs, the bubbles should be modeled individually, and the ejected fluid can be treated as a buoyant momentum jet.

ACKNOWLEDGMENTS

This study was supported by the MIT Sea Grant College Program, the National Energy Technology Laboratory of the U.S. Department of Energy, and the Deep Spills Task Force, comprised of the Minerals Management Service of the U.S. Department of Interior and a consortium of 12 member oil companies of the Offshore Operator's Committee.

REFERENCES

Adams, E. E., Caulfield, J. A., Herzog, H. J. and Auerbach, D. I. (1997). Impacts of reduced pH from ocean CO_2 disposal: Sensitivity of zooplankton mortality to model parameters. *Waste Management* **17**:375-380.

Alendal, G. and Drange, H. (2000). Two-phase, near-field modeling of purposefully released CO_2 in the ocean. *J. Geophy. Res.: Oceans (submitted).*

Asaeda, T. and Imberger, J. (1993). Structure of bubble plumes in linearly stratified environments. *J. Fluid Mech.* **249**:35-57.

Baddour, R. E. (1994). Thermal-saline bubble plumes. *Recent Research Advances in the Fluid Mechanics of Turbulent Jets and Plumes.* Editors: P. A. Davies and M. J. V. Neves, Kluwer Academic Publishers, The Netherlands:117-129.

Baines, W. D. and Leitch, A. M. (1992). Destruction of stratification by bubble plumes. *J. Hydr. Engrg.* **118**(4):559-577.

Beyersdorf, J. (1997). Verhalten von Luftblasen und Sedimenten in Blasensäulen in abhängigkeit vom Saltzgehalt im Wasser, Mitteilungen, Heft 79, Franzius-Institut f. Wasserbau u. Küsteningenieurwesen, Universität Hannover.

Brevik, I. and Killie, R. (1996). Phenomenological description of the axisymmetric air-bubble plume. *Int. J. Multiphase Flow* **22**(3):535-549.

Brevik, I. and Kluge, R. (1999). On the role of turbulence in the phenomenological theory of plane and axisymmetric air-bubble plume. *Int. J. Multiphase Flow* **25**:87-108.

Chen, B., Masuda, S., Nishio, M., Someya, S. and Akai, M. (2000). A numerical prediction of plume structure of CO_2 in the ocean-a near field model. In *5th Int. Conf. Greenhouse Gas Control Tech., Cairns, Australia*, IEA Greenhouse Gas R&D Program.

Chesters, A., van Doorn, M. and Goossens, L. H. J. (1980). A general model for unconfined bubble plumes from extended sources. *Int. J. Multiphase Flow* **6**:499-521.

Clift, R., Grace, J. R. and Weber, M. E. (1978). *Bubbles, Drops, and Particles*, Academic Press, New York, NY.

Crawford, T. V. and Leonard, A. S. (1962). Observations of buoyant plumes in calm stably stratified air. *J. App. Met.* p. 254.

Crounse, B. C. (2000). Modeling buoyant droplet plumes in a stratified environment, MS thesis, Dept. of Civ. Env. Engrg., MIT, Cambridge, MA.

Crowe, C., Sommerfield, M. and Tsuji, Y. (1998). *Multiphase flows with droplets and particles*, CRC Press.

Davidson, M. J. and Pun, K. L. (1999). Weakly advected jets in cross-flow. *J. Hydr. Engrg.* **125**(1):47-58.

Ditmars, J. D. and Cederwall, K. (1974). Analysis of air-bubble plumes. In *Proc. of 14th Int. Conf. Coastal Engrg., Copenhagen*, ASCE:2209-2226.

Fischer, H. B., List, E. G., Koh, R. C. Y., Imberger, J. and Brooks, N. H. (1979). *Mixing in Inland and Coastal Waters*, Academic Press, New York, NY.

Gill, A. (1982). *Ocean-Atmosphere Dynamics*, Academic Press, New York, NY.

HAVIS Environmental (1994). Mixing zone simulation model for dredge overflow and discharge into inland and coastal waters, Prepared for U.S. Army COE Waterways Experiment Station, Technical Report DACW39-93-C-0109, HAVIS Environmental, Fort Collins, CO.

Hugi, C. (1993). Modelluntersuchungen von Blasenstrahlen für die Seebelüftung, Ph.D. Thesis, Inst. f. Hydromechanik u. Wasserwirtschaft, ETH, Zürich.

Hussain, N. A. and Narang, B. S. (1984). Simplified analysis of air-bubble plumes in moderately stratified environments. *J. Heat Trans.* **106**:543-551.

Jakobsen, H. A., Sannaes, B. H., Grevskott, S. and Svendsen, H. F. (1997). Modeling of vertical bubble-driven flows. *Ind. Eng. Chem. Res.* **36**:4052-4074.

Johansen, Ø. (1999). DeepBlow - a Lagrangian plume model for deep water blowouts. In *Proc. 3rd Int. Marine Environ. Modelling Seminar*, Lillehammer.

Kobus, H. E. (1968). Analysis of the flow induced by air-bubble systems. In *Proc. 11th Int. Conf. Coastal Engrg.*, London, ASCE:1016-1031.

Koh, R. C. Y. and Chang, Y. C. (1973). Mathematical model for barged ocean disposal of waste. *Technical Report 660/2-73-029*, U.S. EPA, Washington, D.C.

Lapin, A. and Lübbert, A. (1994). Numerical simulation of the dynamics of two-phase gas-liquid flows in bubble columns. *Chemical Engineering Science* **49**(21):3661-3674.

Leifer, I., Patro, R. and Bowyer, P. (2000). A study on the temperature variation of rise velocity for large clean bubbles. *J. Atm. and Ocean Tech., (accepted)* .

Leitch, A. M. and Baines, W. D. (1989). Liquid volume flux in a weak bubble plume. *J. Fluid Mech.* **205**:77-98.

Lemckert, C. J. and Imberger, J. (1993). Energetic bubble plumes in arbitrary stratification. *J. Hydr. Engrg.* **119**(6):680-703.

Liro, C. R., Adams, E. E. and Herzog, H. J. (1991). Modeling the release of CO_2 in the deep ocean. *Technical Report MIT-EL 91-002*, Energy Laboratory, MIT, Cambridge, MA.

Liro, C. R., Adams, E. E. and Herzog, H. J. (1992). Modeling the release of CO_2 in the deep ocean. *Energy Conserv. Mgmt.* **33**(5-8):667-674.

McDougall, T. J. (1978). Bubble plumes in stratified environments. *J. Fluid Mech.* **85**(4):655-672.

Milgram, J. H. (1983). Mean flow in round bubble plumes. *J. Fluid Mech.* **133**:345-376.

Morton, B. R. (1962). Coaxial turbulent jets. *J. Heat and Mass Transfer* **5**:955-965.

Morton, B. R., Taylor, S. G. I. and Turner, J. S. (1956). Turbulent gravitational convection from maintained and instantaneous sources. *Proc. of the Royal Soc.* **A234**:1-23.

Pun, K. L. and Davidson, M. J. (1999). On the behavior of advected plumes and thermals. *J. Hydr. Res.* **37**(4):519-540.

Reingold, L. S. (1994). An experimental comparison of bubble and sediment plumes in stratified environments. MS thesis, Dept. of Civ. Envir. Engrg., MIT, Cambridge, MA.

Roig, V., Suzanne, C. and Masernat, L. (1998). Experimental investigation of a turbulent bubbly mixing layer. *Int. J. Multiphase Flow* **24**(1):35-54.

Sato, T. and Hama, T. (2000). Numerical simulation of dilution process in CO_2 ocean sequestration. In *5th Int. Conf. Greenhouse Gas Control Tech.*, Cairns, Australia, IEA Greenhouse Gas R&D Program.

Schladow, S. G. (1993). Lake destratification by bubble-plume systems: Design methodology. *J. Hydr. Engrg.* **119**(3):350-368.

Socolofsky, S. A. (2000). Laboratory Experiments of Multi-phase Plumes in Stratification and Crossflow, Ph.D. Thesis, Dept. of Civ. Env. Engrg., MIT, Cambridge, MA (in prep.).

Socolofsky, S. A. and Adams, E. E. (2000*a*). Multi-phase plumes in stratification: Dimensional analysis. *J. of Fluid Mechanics* (submitted).

Socolofsky, S. A. and Adams, E. E. (2000*b*). Multi-phase plumes in stratification: Liquid volume fluxes. *J. of Fluid Mechanics* (submitted).

Socolofsky, S. A. and Adams, E. E. (2000*c*). Multi-phase plumes in uniform and stratified crossflow. *J. of Hydraulic Research* (submitted).

Tacke, K. H., Schubert, H. G., Weber, D. J. and Schwerdtfeger, K. (1985). Characteristics of round vertical gas bubble jets. *Metal. Trans. B* **16B**(2):263-275.

Thorkildsen, F., Alendal, G. and Haugan, P. M. (1994). Modelling of CO_2 droplet plumes, Technical Report 96, Nansen Environmental and Remote Sensing Center, Adv. Griegsvei 3A, N - 5037 Solheimsviken, Norway.

Townsend, A. A. (1976). The structure of turbulent shear flow. Cambridge University Press.

Turner, J. S. (1986). Turbulent entrainment: the development of the entrainment assumption, and its application to geophysical flows. *J. Fluid Mech.* **173**:431-471.

Wilkinson, D. L. (1979). Two-dimensional bubble plumes. *J. Hydr. Div.* **105**(HY2):139-154.

Wong, C. S. and Hirai, S. (1997). Ocean storage of carbon dioxide--a review of ocean carbonate and CO_2 hydrate chemistry. Technical report, IEA Greenhouse Gas R&D Prog., Cheltenham, UK.

Wüest, A., Brooks, N. H. and Imboden, D. M. (1992). Bubble plume modeling for lake restoration. *Water Resour. Res.* **28**(12):3235-3250.

Yapa, P. D. and Zheng, L. (1997*a*). Simulation of oil spills from underwater accidents I: Model development. *J. Hydr. Res.* **35**(5):673-687.

Yapa, P. D. and Zheng, L. (1997*b*). Simulation of oil spills from underwater accidents II: Model verification. *J. Hydr. Res.* **35**(5):688-697.

Chapter 5

TURBULENT TRANSPORT PROCESSES ACROSS NATURAL STREAMS

Vincent H. Chu[1]

ABSTRACT

The transport of pollutant, particulates, heat, salinity and/or moisture across natural streams is dependent on vertical and horizontal turbulence of very different length scales. The vertical turbulence is not directly significant in the transport process because its length scale, limited by the depth of the stream, is small compared with the width of the stream. The horizontal turbulence, with a length scale comparable to the width of the flow, is effective in the transport across the streams. However, whether or not the horizontal turbulence can be maintained, by the energy supply from the mean flow, is determined by the frictional energy dissipation and the related process of the momentum exchange by the vertical turbulence. In this chapter, the respective roles of the horizontal and vertical turbulence on the transport process are considered. LES and RANS simulation models of the horizontal turbulence are developed based on a sub-depth-scale model of the vertical turbulence. The results of a series of numerical simulations conducted using these models are verified with laboratory and field data.

1. INTRODUCTION

Natural streams in the atmosphere, oceans, lakes and rivers are wide compared with the depth of the streams. Heat, mass and momentum are transported across the width of these streams by horizontal turbulence that can be hundreds of times greater in width than the depth of the streams. In the Lower St. Lawrence River Estuary, horizontal turbulence as large as 30 km has been observed in an area of the estuary where the depth is 300 m (see Figure 1). The Gulf Stream in the Atlantic has a width of 60 km and depth of about 600 m. The Jet Stream in the atmosphere has a vertical extent of about 10 km and a width exceeding 1000 km. In all these examples, the horizontal length scale of the

[1]Professor, Department of Civil Engineering and Applied Mechanics, McGill University, Montreal, Canada H3A 2K6, vincent.chu@mcgill.ca

turbulent flows is two orders of magnitude greater than the depth of the flows. These flows are *shallow* in the sense that the depth is small compared with the width of the flow. Large-scale and small-scale turbulence can coexist in these *shallow quasi-two-dimensional* (SQ2D) turbulent flows. The large-scale turbulent motion is two-dimensional because the motion is confined in a predominantly horizontal direction. However, the small-scale turbulent motion is not confined and is free in all directions.

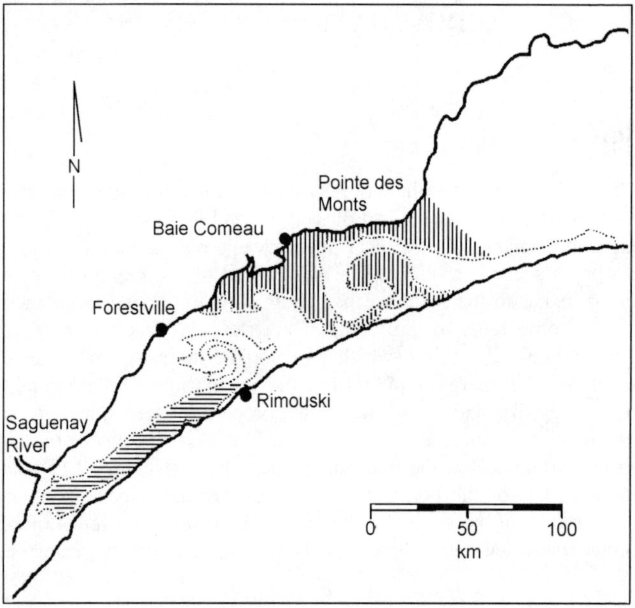

Figure 1: Eddies produced by instability of the Gaspe Current along the south shore of the Lower St. Lawrence River Estuary. The temperature ranges from 5 °C in the horizontally shaded region to 15 °C in the vertically shaded region. The circulation in these eddies is quite weak but the exchanges of salinity and pollutants over tens of kilometres of distance between the south shore and the north shore are affected by the circulation. Reproduced from Mertz and Gratton (1990).

Pollutant, particulates, heat, salinity and/or moisture are transported across these natural streams by turbulent motion of many different length scales. The greater the length scale of the turbulence the greater the contribution of the turbulence to transport processes since the turbulent diffusion coefficient is proportional to the length and velocity of the turbulent motion. The wake of the grounded tanker Argo Merchant on Nantucket shoals as shown in Figure 2 is an example of the SQ2D turbulent flows. The large-scale turbulence eddies produced by the momentum defect in the wake has a horizontal length scale ℓ_h of 700 m (2000 ft). The small-scale turbulence produced by vertical shear has a length scale ℓ_v equal to the water depth, which in this area of the Nantucket shoals is only 18 m (50 ft). The horizontal to vertical length scale ratio, ℓ_h / ℓ_v, of the turbulent motion in this example is 40 to 1.

The turbulent diffusivity, D_T, of the SQ2D turbulent flow is composed of two parts: $D_{Th} \sim u_h \ell_h$ and $D_{Tv} \sim u_v \ell_v$, associated with horizontal and vertical turbulence, respectively. If the velocity stays constant, i.e., if $u_h \sim u_v$, the turbulent diffusivity coefficient would be linearly dependent on the length scale of the turbulent motion: $D_{Th} / D_{Tv} \sim \ell_h / \ell_v$. If the energy cascade rate is constant, i.e., if $u_h^3 / \ell_h = u_v^3 / \ell_v$, the turbulent diffusivity would follow the Richardson's four-third-power-law: $D_{Th} / D_{Tv} \sim \ell_h^{4/3} / \ell_v^{4/3}$. Measurements in the open waters have suggested a relation somewhere between the linear and the four-third-power law (Okubo, 1968; Okubo and Ozimidov, 1970).

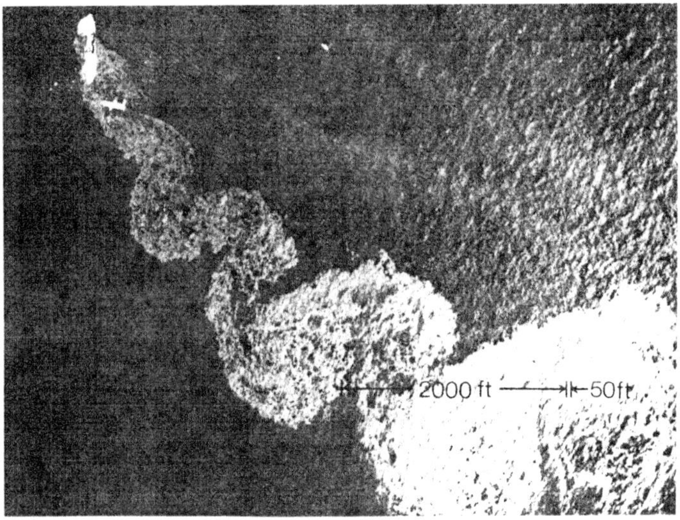

Figure 2: The wake of the grounded tanker Argo Merchant on Nantucket shoals. Large-scale and small-scale turbulence eddies in the wake are made visible by the crude oil on the sea surface. Water depth in this area of the Nantucket shoals is about 18 m (50 ft) as marked. Reproduced from *Album of Fluid Motion* (1982).

The diffusivity, D_{Tv}, associated with the vertical turbulence is proportional to the local shear velocity, u_*, and the water depth, h (that is $u_v = u_*$; $\ell_v = h$). In the absence of the horizontal turbulence in wide open-channel flow of uniform depth, the dimensionless coefficient, $D_{Tv} / u_* h$, was determined from measurements to be a constant ranging from 0.08 to 0.13 (Fisher et al., 1979; Nokes and Wood, 1988). The contribution to diffusion by the horizontal turbulence is not as well understood (Rastogi and Rodi, 1978; Chu and Babarutsi, 1988; Holley, 1996). The part of the diffusivity due to horizontal turbulence, D_{Th}, can be significantly greater than the part due to the vertical turbulence, D_{Tv}, if the horizontal length scale, ℓ_h, is large compared with the vertical length scale, ℓ_v. Sediments brought into suspension in the main channel of a river could be deposited along the shore if the horizontal turbulence is there to transport the materials across the stream (Schmidt et al., 1999; Wiele et al., 1999). Without the horizontal turbulence, pollutants, ice and debris entering the shore from overland would be trapped along the coasts; wastewater discharged into the river would stay relatively unmixed as it flows down the rivers (Chu, Gehr and Leduc, 1991).

EXAMPLE 1.1: *Find the lateral spreading of the oil in the wake of the grounded tanker Argo Merchant on Nantucket shoals assuming that the transport of the oil across the wake is due to the vertical turbulence only.*

The turbulent diffusion coefficient associated with the vertical turbulence is $D_{Tv} = 0.13 u_* h$ where $u_* = \sqrt{c_f / 2}\, U$ is the shear velocity, c_f the bottom-friction coefficient, U the velocity and h the depth of the flow. Given this diffusion coefficient D_{Tv}, the concentration profile of the contaminant from a point source is Gaussian:

$$C = C_m \exp\left[-\frac{y^2 U}{4 D_{Tv}\, x} \right]$$

where y is the lateral coordinate in the direction across the stream. For a water depth of $h = 18$ m and an estimated friction coefficient $c_f = 0.006$, the width of this Gaussian profile at a longitudinal distance $x = 1500$ m from the source is

$$\sigma = \sqrt{2 D_{Tv} \frac{x}{U}} = \sqrt{0.26 \sqrt{\frac{c_f}{2}}\, U h \frac{x}{U}} = \sqrt{0.26 \sqrt{\frac{0.006}{2}} \times 18 \times 1500} = 19.3 \text{ m}$$

This estimate of the width is an order of magnitude smaller than the observation. As shown in Figure 2, the wake width should be about 700 m at a distance of 1500 m downstream from the source. The width is under estimated in the above calculations because the assumption made in the calculations is incorrect. The horizontal turbulence not only is not negligible but is in this case the dominant contribution to the transport of the contaminant across the wake.

The length scale of the horizontal turbulence may be large. Its energy however is limited due to the existence of an energy dissipation mechanism by friction (Chu, Wu and Khayat, 1991; Babarutsi and Chu, 1998; Altai, Zhang and Chu, 1999). Therefore, to reproduce the correct transport processes across the natural streams, the horizontal turbulence must be simulated in a separated model from the vertical turbulence. We will show in this chapter how separate models of the horizontal and vertical turbulence are developed and how these models are applied to calculate the transports across the streams.

The organization of this chapter is as follows. Section 2 describes how a sub-depth-scale average is made so that the horizontal and vertical turbulence can be modeled separately. Section 3 provides a sub-depth-scale model and Section 4 a sub-grid-scale model. Large Eddy Simulations (LES) were carried out using these models. Model verifications were conducted using the experimental results obtained in the turbulent flow of a shallow open-channel expansion. The overall effects of the friction on the turbulent energy budget are considered in Section 5. The experimental data are analyzed to show how the length scale of the horizontal turbulence is determined by the relative intensity of the vertical turbulence. Section 6 will introduce a two-length-scale RANS (Reynolds Averaged Navier Stokes) turbulence model. Calculations are conducted using the model to show that the conventional single-length-scale turbulence models are biased toward the vertical turbulence and therefore not suitable for general simulation of the transport processes across the natural streams. The chapter ends in Sections 7 and 8 with two modeling examples encountered in practice.

2. SUB-DEPTH-SCALE AVERAGING

An important step in the development of a model for the horizontal turbulence is the sub-depth-scale average, which is a filtering operation to remove the vertical turbulence. The visual perception of the horizontal turbulence in the wake of the tanker Argo Merchant as shown in the aerial photograph in Figure 2 is obtained from such a filtering operation. Turbulence in the sub-depth-scale is visually removed from the aerial photo as the 18 m sub-depth-scale eddies are barely recognizable in the photograph. The enormous difference in the time and the length scales between the horizontal and vertical turbulence in the natural streams has made it necessary to model the two separately.

The formulation is carried out here for incompressible turbulent flow in a shallow open channel, in which the water depth, h, is small compared with the horizontal length scale of the flow and is gradually varying in the horizontal direction. The horizontal co-ordinates are x_i $(i = 1, 2)$ and the vertical co-ordinate is z. Average elevation of the free surface is located at $z = 0$ and the channel bed at $z = -h(x_i)$ (see Figure 3).

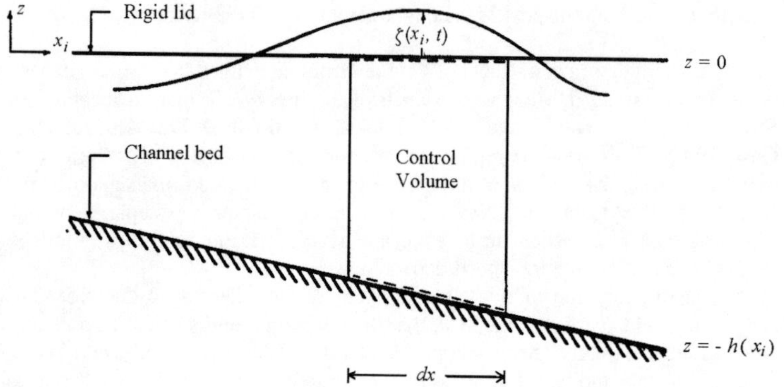

Figure 3: The control volume bounded by the free water surface and the channel bed in an open channel flow of variable depth.

The horizontal velocity components, u_i, are separated into two parts, namely, the depth-independent part, $\tilde{u}_i(x_i,t)$, and the sub-depth-scale fluctuation, $u_i''(x_i,z,t)$, as follows (see Figure 4): $u_i(x_i,z,t)=\tilde{u}_i(x_i,t)+u_i''(x_i,z,t);\ i=1,2$. The depth-independent part, $\tilde{u}_i\ (x_i,t)$, associated with the horizontal turbulence, is obtained by the sub-depth-scale average operation

$$\tilde{u}_i = \langle\langle u_i \rangle\rangle = \frac{1}{h}\int_{-h}^{\zeta} \langle u_i \rangle dz\,;\ \ \mathrm{i}=1,2 \tag{1}$$

where ζ is the water surface elevation. The sub-depth-scale averages, $\tilde{u}_i = \langle\langle u_i \rangle\rangle$, denoted by the tilde symbol or the double bracket operator, have two operations. First, the velocity u_i is averaged over time to give $\langle u_i \rangle$. The time average, $\langle u_i \rangle$, is subsequently averaged over the depth to give $\tilde{u}_i$ or $\langle\langle u_i \rangle\rangle$. The angle bracket, $\langle\ \rangle$, denotes a short-time average over a period of time that is long compared with the time scale of the vertical turbulence but short compared with the time scale of the horizontal turbulence. The double bracket, $\langle\langle\ \rangle\rangle$, denotes the sub-depth-scale averaging operation as an alternate to the operation associated with the tilde symbol.

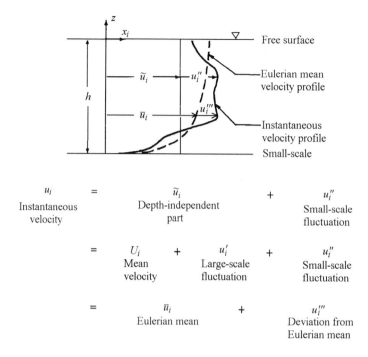

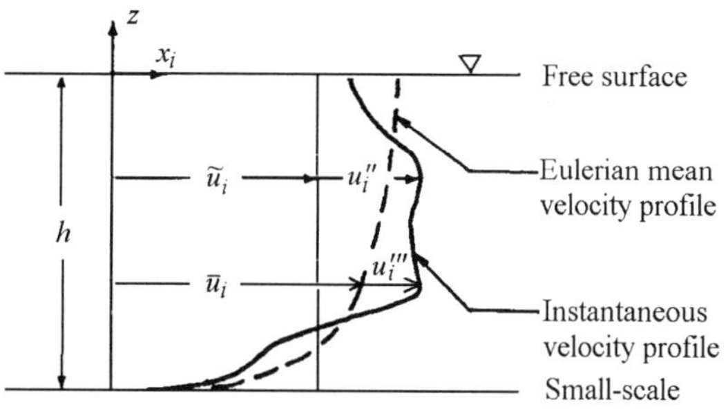

u_i Instantaneous velocity	=		$\tilde{u}_i$ Depth-independent part		+	u_i'' Small-scale fluctuation
	=	U_i Mean velocity	+	u_i' Large-scale fluctuation	+	u_i'' Small-scale fluctuation
	=		$\bar{u}_i$ Eulerian mean	+		u_i''' Deviation from Eulerian mean

Figure 4: The velocity, u_i, is separated into a depth-independent part, $\tilde{u}_i$, and a small-scale fluctuation, u_i''. This division of the turbulent motions by its scales is to be distinguished from the conventional approach of partitioning the instantaneous component, u_i, into the Eulerian mean, $\bar{u}_i$, and the deviation, u_i''', from the mean value.

The vertical velocity component is also separated into two parts, but in a slightly different manner, as follows: $w(x_i, z, t) = \langle w\,(x_i, z, t)\rangle + w''(x_i, z, t)$. The time average, $\langle w\rangle$, is the part associated with the horizontal turbulence while the fluctuating w'' with vertical turbulence. Since the depth and the roughness vary gradually across the open channel flow, $\langle w\rangle$ is negligible, but w'' is not. The vertical momentum equation, after neglecting the terms associated with $\langle w\rangle$, is

$$\frac{\partial\,\langle w''^2\,\rangle}{\partial z} \approx -\frac{1}{\rho}\frac{\partial\langle p\rangle}{\partial z} - g, \tag{2}$$

in which p = pressure, ρ = fluid density, and g = gravity constant. Integration of this equation with respect to the vertical coordinate, z, gives the nearly hydrostatic pressure variation with the depth as follows: $\langle p \rangle \approx -\rho g(z-\zeta) - \rho\langle w''^2 \rangle$. The continuity and horizontal momentum equations are obtained by balancing the fluxes in and out of the control volume as shown in Figure 3:

$$\frac{\partial \zeta}{\partial t} + \frac{\partial}{\partial x_j}\int_{-h}^{\zeta} u_j dz = 0 \tag{3}$$

$$\rho\frac{\partial}{\partial t}\int_{-h}^{\zeta} u_i dz + \rho\frac{\partial}{\partial x_j}\int_{-h}^{\zeta} u_i u_j dz = -\frac{\partial}{\partial x_i}\int_{-h}^{\zeta} p dz + p_b\frac{\partial h}{\partial x_i} - \tau_{bi} \tag{4}$$

where p_b is the pressure and τ_{bi} the friction stresses exerted on the channel bed. The short-time averaging operation, $\langle\ \rangle$, is performed on every term of these equations. The result of the averaging operation is the following set of equations:

$$\frac{\partial \zeta}{\partial t} + \frac{\partial h\tilde{u}_j}{\partial x_j} = 0 \tag{5}$$

$$\frac{\partial \tilde{u}_i}{\partial t} + \tilde{u}_j\frac{\partial \tilde{u}_i}{\partial x_j} = -g\frac{\partial \zeta}{\partial x_i} - \frac{\langle \tau_{bi}\rangle}{\rho h} + \frac{1}{\rho h}\frac{\partial h\tilde{\tau}_{ij}}{\partial x_j} \tag{6}$$

where $\tilde{\tau}_{ij} = \rho(\langle\langle w''^2 \rangle\rangle\delta_{ij} - \langle\langle u_i''u_j''\rangle\rangle)$ is the Reynolds stresses associated with the sub-depth-scale velocity fluctuations and the double brackets denote the sub-depth-scale operation. In deriving these equations, a rigid-lid approximation was made to evaluate the high-order nonlinear terms.

3. SUB-DEPTH-SCALE MODEL

The Reynolds stresses $\tilde{\tau}_{ij}$ and the bed-friction stresses $\langle \tau_{bi} \rangle$ are the basic elements that determine the depth-averaged flow and the structures of the horizontal turbulence. In the present sub-depth-scale model, a quadratic relation with the local velocity, $\tilde{u}_i$, is assumed for the friction stresses, $\langle \tau_{bi} \rangle$, as follows:

$$\frac{\langle \tau_{bi}\rangle}{\rho} = \frac{c_f}{2}\tilde{u}_i\sqrt{\tilde{u}_l\tilde{u}_l} \tag{7}$$

The proportionality constant c_f is the bed-friction coefficient. The Reynolds stress, $\tilde{\tau}_{ij} = \rho(\langle\langle w''^2 \rangle\rangle\delta_{ij} - \langle\langle u_i''u_j''\rangle\rangle)$, is assumed to be linearly related to the mean velocity gradient through a Boussinesq type of hypothesis as follows:

$$\tilde{\tau}_{ij} = \rho\nu_{SD}\left(\frac{\partial \tilde{u}_i}{\partial x_j} + \frac{\partial \tilde{u}_j}{\partial x_i}\right) - \rho\left[\nu_{SD}\frac{\partial \tilde{u}_k}{\partial x_k} + \frac{1}{2}\langle\langle u_k''''u_k'' - w''\rangle\rangle\right]\delta_{ij} \tag{8}$$

In this equation, the turbulent viscosity ν_{SD} is introduced to represent the energy dissipation effect by turbulence of the sub-depth scales. The sub-depth-scale viscosity is proportional to the shear velocity $\tilde{u}_*$ and the water depth h:

$$\nu_{SD} = c_\nu u_* h \tag{9}$$

where $u_* = \sqrt{\tau_b / \rho}$ is the friction velocity derived from the friction stress $\tau_b = \sqrt{\langle \tau_{bk} \rangle \langle \tau_{bk} \rangle}$ that acts on the channel bottom. The value of the coefficient is $c_\nu \cong 0.13$ for straight and wide open-channel flow (Nokes and Wood, 1988). The turbulent pressure term is $\rho(\frac{1}{2}\langle\langle u_k" u_k" \rangle\rangle - \langle\langle w_k"^2 \rangle\rangle) \approx 4.0 \rho u_*^2$ for logarithmic velocity profile. Sub-depth-scale viscosity, ν_{SD}, is a parameter in the model characterizing the horizontal exchange of momentum by the sub-depth-scale turbulence.

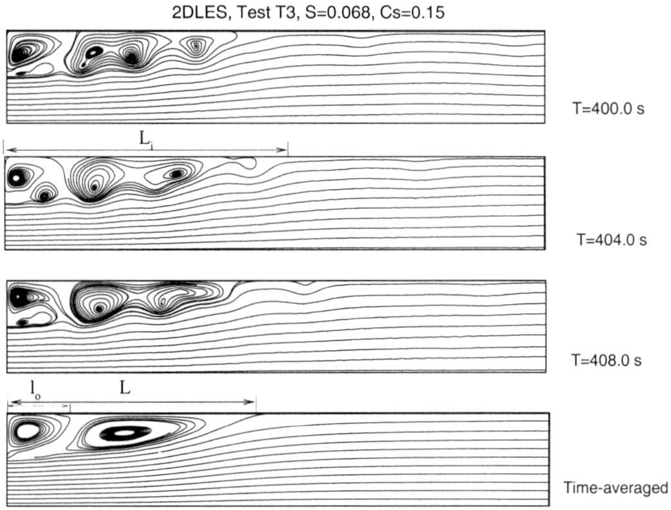

Figure 5: Instantaneous streamlines at time t = 400 s, 404 s and 408 s, and the time-averaged streamlines, of the flow in the channel expansion; Test T3, water depth h = 3.50 cm.

4. LARGE EDDY SIMULATION

Computations are conducted to simulate the horizontal turbulence using the model equations, Eqs. (5) and (6), derived from sub-depth-scale averaging. The re-circulation in the sudden widening of an open-channel flow is simulated using the conditions of the laboratory experiments conducted by Barbarutsi and Chu (1991). The width of the flows changes from 30.5 cm to 61.0 cm. Figure 5 shows the streamline patterns in the open channel of water depth h = 3.50 cm. The first three flow patterns shown in the figure are the instantaneous streamlines obtained at time $t = 400$ s, 404 s and 408 s. The bottom of the figure is the streamlines obtained by averaging over a period of time. The time-averaged patterns show the turbulent flows in the channel expansion to be consisting of a main re-circulating zone and a secondary eddy. The time-average flow patterns obtained for four tests are given in Figure 6. The depth of the open-channel flow in these four tests, T1, T2, T3 and T4, are h =1.45 cm, 2.50 cm, 3.50 cm and 6.90 cm, respectively.

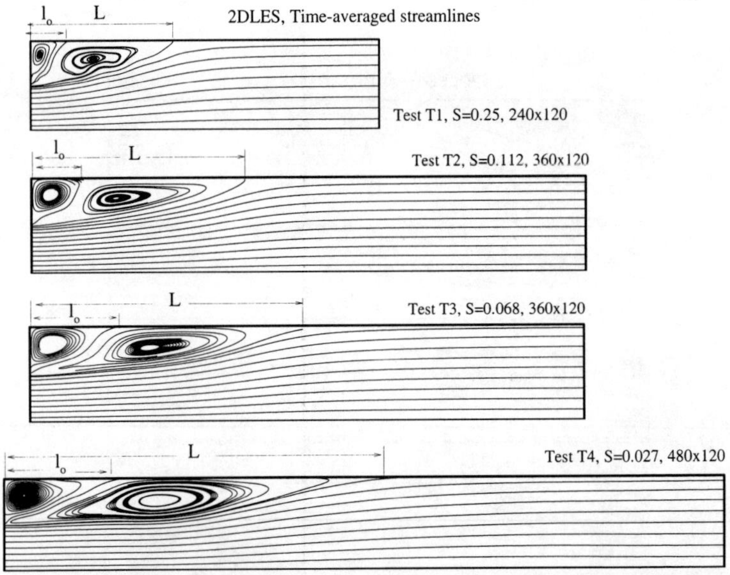

Figure 6: The streamlines obtained from long-time average of the instantaneous profiles; (a) Test T1, $\tilde{S} = 0.253$, $\ell / L = 0.22$, $L/d = 2.78$, (b) Test T2, $\tilde{S} = 0.112$, $\ell / L = 0.23$, $L/d = 4.42$, (c) Test T3, $\tilde{S} = 0.0685$, $\ell / L = 0.25$, L/d = 5.43, (d) Test T4, $\tilde{S} =$ 0.0267, $\ell / L = 0.26$, $L/d = 7.51$.

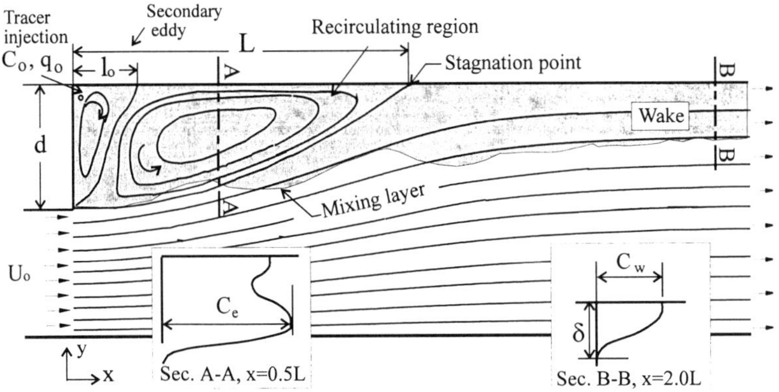

Figure 7: Plan-view of the re-circulating flow in an open channel due to a sudden widening of the flow in the transverse direction.

The length of the main re-circulating eddy, L, the length of the counter rotating secondary eddy, ℓ, the visual wake width, δ_v, for each test are obtained from the numerical simulations for comparison with the laboratory observations. During the laboratory experiments, dye was injected from a point at the corner of the channel expansion. The entrainment into the mixing layer and the re-attached of the layer onto the sidewall set up the re-circulating flow. Part of the dye moved upstream from the point of re-attachment. The rest moved downstream to form a wake. Figure 7 shows the flow and the dye concentration distribution in the re-circulating flow. The inserts to the figure show the dye concentration profiles in the center of the re-circulating zone and in the wake. The conditions of the numerical simulations were the same as the laboratory experiments conducted by Barbarutsi and Chu (1991). The numerical simulations were carried out using a two-dimensional grid on the horizontal plane. In addition to the sub-depth-scale viscosity ν_{SD}, a sub-grid-scale eddy viscosity ν_{SG} is needed to account for the energy dissipation at the sub-grid-scales. The best-known model for the sub-grid-scale viscosity is the Smagorinsky model.

$$\nu_{SG} = C_S^2 \Delta^2 \sqrt{S_{ij} S_{ij}} \; ; \quad i, j = 1, 2, 3 \tag{10}$$

The two-dimensional version of this model is

$$\nu_{2DSG} = C_S^2 \Delta^2 \left[\left(\frac{\partial \tilde{u}}{\partial x} \right)^2 + \left(\frac{\partial \tilde{v}}{\partial y} \right)^2 + \frac{1}{2} \left(\frac{\partial \tilde{v}}{\partial x} + \frac{\partial \tilde{u}}{\partial y} \right)^2 \right]^{\frac{1}{2}} \tag{11}$$

in which C_S = Smagorinsky parameter and $\Delta = \sqrt{\Delta x^2 + \Delta y^2}$; $\tilde{u}$ and $\tilde{v}$, are components of the velocity on the *x* and *y* plane. If the grid is large compared with the water depth (i.e., if $\Delta x, \Delta y >> h$), the total sub-depth and sub-grid viscosity is the sum of the sub-depth-scale component given by Eq. (9) and the sub-grid component by Eq. (11) as follows: $\nu_T = \nu_{SD} + \nu_{2DSG}$. However, if the grid were significantly smaller than the depth of the water (i.e., if $\Delta x, \Delta y << h$), the three-dimensional model would be more correct:

$$\nu_{3DSG} = C_S^2 \Delta_3^2 \sqrt{S_{ij} S_{ij}} \tag{12}$$

In this three-dimensional model, $\Delta_3 = \sqrt{\Delta x^2 + \Delta y^2 + \Delta z^2}$ and

$$S_{ij} = \frac{1}{2}\left(\frac{\partial u_i}{\partial x_j} + \frac{\partial u_j}{\partial x_i}\right); \quad \text{i, j} = 1, 2, 3 \tag{13}$$

The three-dimensional strain rate tensor has nine components:

$$S_{ij} = \begin{vmatrix} S_{11} & S_{12} & S_{13} \\ S_{21} & S_{22} & S_{23} \\ S_{31} & S_{32} & S_{33} \end{vmatrix} \tag{14}$$

Not all the components are determined by a two-dimensional simulation but the values of some of the components can be estimated from the two-dimensional calculation of $\tilde{u}$ and $\tilde{v}$ using the following procedure. The two-dimensional strain rate tensor has only four components: S_{11}, S_{12}, S_{21} and S_{22} . Based on isotropic assumption, the remaining components of the three-dimensional strain tensor ($S_{13}, S_{23}, S_{31}, S_{32}$ and S_{33}) are related to the two-dimensional components as follows:

$$S_{13} = S_{31} = S_{23} = S_{32} = \frac{1}{2}(S_{12} + S_{21}), \quad S_{33} = \frac{1}{2}(S_{11} + S_{22})$$

With these relations, the three-dimensional sub-grid-scale viscosity is given by the approximate formula

$$\nu_{3DSG} \approx C_S^2 \frac{3\Delta^2}{2} \sqrt{\frac{3}{2}\left[\left(\frac{\partial \tilde{u}}{\partial x}\right)^2 + \left(\frac{\partial \tilde{v}}{\partial y}\right)^2 + \left(\frac{\partial \tilde{v}}{\partial x} + \frac{\partial \tilde{u}}{\partial y}\right)^2\right]} \tag{15}$$

Hence, if the size of the grid is small compared with the depth of the flow, the sub-grid-scale viscosity would be given by Eq. (15). There would be no sub-depth-scale viscosity and the total for the sub-grid model is $\nu_T = \nu_{3DSG}$.

Nassiri (1999) used the following interpolation formula for the sub-depth and sub-grid scale viscosity:

$$\nu_T = \nu_{2DSG} + \frac{\Delta^{\frac{4}{3}}}{\Delta^{\frac{4}{3}} + h^{\frac{4}{3}}}\nu_{SD} + \frac{h^{\frac{4}{3}}}{\Delta^{\frac{4}{3}} + h^{\frac{4}{3}}}\left(\nu_{3DSG} - \nu_{2DSG}\right) \qquad (16)$$

According to this formula, depending on the size of the grid, the sum of the sub-depth and sub-grid viscosity would be $\nu_T = \nu_{SD} + \nu_{2DSG}$ if $\Delta >> h$, or $\nu_T = \nu_{3DSG}$ if $\Delta << h$. The results of Nassiri's LES are presented in Figures 5 and 6. The instantaneous velocity of the turbulent flow, $\tilde{u}_i$, was obtained by numerical solution of Eqs. (5) and (6) using the viscosity given by the interpolation formula Eq. (16).

The significant features of the mean flow in the channel expansion are the re-circulating eddy, the secondary eddy and the wake as delineated in Figure 7. These are correctly simulated by the LES if the optimal value of the Smagorinsky parameter C_S is selected for the simulation. However, the optimal value is not a constant but varies in the range from $C_S = 0.125$ to $C_S = 0.200$ depending on the size of the grid and also the depth of the water. These dependencies of the parameters on the sub-depth and sub-grid scales are partly affected by the numerical methods, which could have a significant impact on the results of the simulation. Advection is generally dominant over the diffusion in the simulation of turbulent flow using LES method. Numerical scheme must be correctly selected to minimize the effects of the numerical oscillations and diffusion (see Chu and Altai, 1998, 2000, 2001; and Altai and Chu, 2001).

5. EFFECT OF FRICTION AND VERTICAL TURBULENCE

It is clear from the LES results obtained of the shallow re-circulating turbulent flow in the channel expansion that friction plays a significant role in the development of the horizontal turbulent motion. In this section, this friction effect is further examined based on the experimental observations obtained for several shallow turbulent flows including the re-circulating flows in the channel expansion.

5.1 Re-circulating Flow in the Channel Expansion

The laboratory experiment by Babarutsi, Ganoulis and Chu (1989) and Babarutsi and Chu (1991) was conducted in an open-channel expansion. Figure 8 shows how the dye concentration distributions in the flow are dependent on the friction effects.

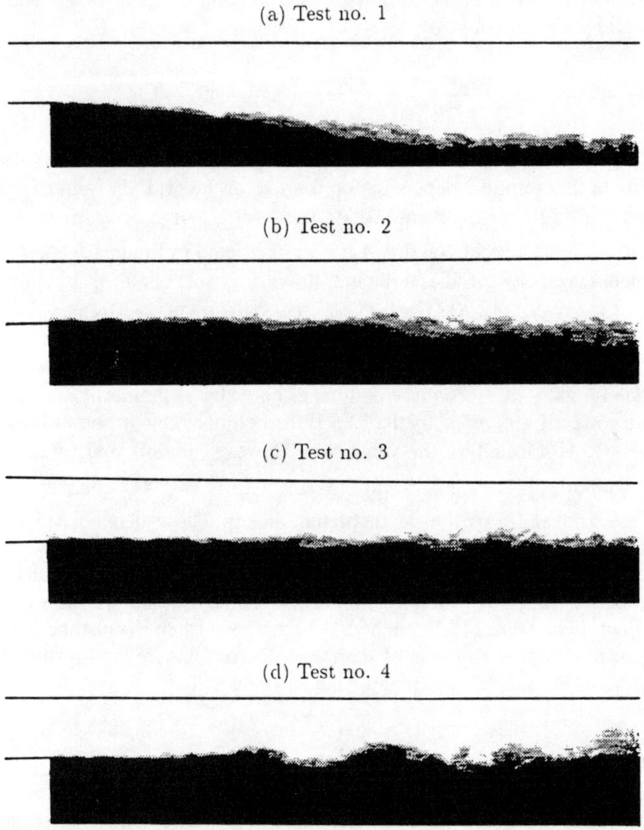

Figure 8: Images of the dye in the re-circulating open-channel flow obtained using an overhead camera above the channel by Babarutsi, Ganoulis and Chu (1989). (a) Test T1, $\tilde{S} = 0.253$, $L = 0.82$ m, $\ell = 0.175$ m, $\delta_v = 0.15$ m; (b) Test T2, $\tilde{S} = 0.112$, $L = 1.42$ m, $\ell = 0.365$ m, $\delta_v = 0.21$ m; (c) Test T3, $\tilde{S} = 0.0685$, $L = 1.64$ m, $\ell = 0.435$ m, $\delta_v = 0.27$ m; (d) Test T4, $\tilde{S} = 0.0267$, $L = 2.14$ m, $\ell = 0.502$ m, $\delta_v = 0.40$ m. The dye was injected into the re-circulating zone at the right-hand corner the sudden expansion.

Babarutsi, Ganoulis and Chu (1989) correlated the friction effect on the flow with an overall bed-friction number

$$\tilde{S} = \frac{c_f d}{2h} \tag{17}$$

where c_f is the bed-friction coefficient, d the width of the channel expansion and h the water depth. For the four tests, T1, T2, T3 and T4, The bed-friction numbers are $\tilde{S}$ = 0.253, 0.112, 0.0685, and 0.0267 with the water depths of the flows h = 1.45 cm, 2.50 cm, 3.50 cm and 6.90 cm, respectively. The friction effect, inversely proportional to the depth of the water, is most significant in test T1 and least significant in test T4. As the depth of the flow increases from h = 1.45 cm in test T1 to a depth of h = 6.90 cm in test T4, while the bed-friction number decreases from $\tilde{S}$ = 0.253 to 0.0267, the friction effect becomes progressively less significant compared with the lateral exchange of momentum by the horizontal turbulence.

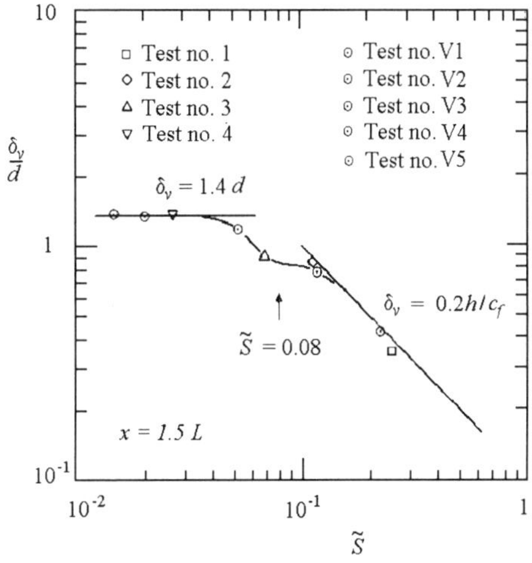

Figure 9: Variation of the visual wake width δ_v / d with the overall bed-friction number $\tilde{S} = c_f\, d/(2h)$. The width δ_v is obtained at a distance $x = 1.5L$ from the channel expansion and L is the length of the main re-circulating eddy.

The visual wake width δ_v was determined from the concentration profiles shown in Figure 8, as a measure of the length scale of the horizontal turbulence. The relative width of the wake δ_v/d, which decreases from 1.3 to 0.3 as the bed-friction number increases in value from $\tilde{S} \approx 0.027$ to $\tilde{S} \approx 0.025$ (Figure 9). The maximum dye concentration in the wake, c_m, increase with the bed-friction number (Figure 10). The concentration is inversely related to the width of the wake because the product of velocity, concentration and width gives approximately the mass flux. High mixing rate would lead to a large width and consequently low concentration in the wake. According to these observations, horizontal turbulence would be the dominant effect on mixing when friction effect is negligible. The results of the laboratory experiments as shown in Figures 8, 9 and 10 are consistent with the LES of the horizontal turbulence using the sub-depth-scale model (see Figures 6 and 7).

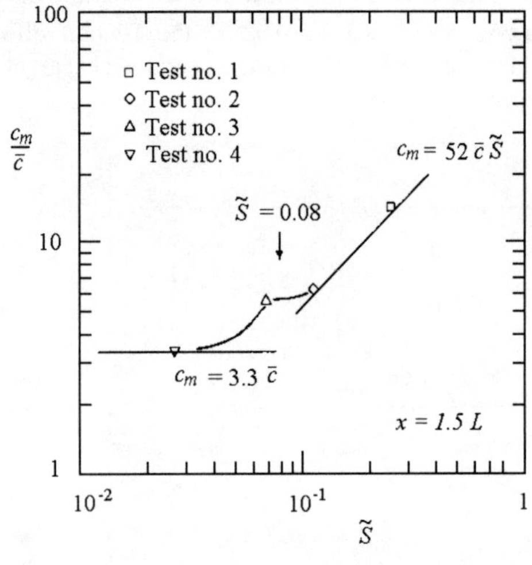

Figure 10: Variation of the dye concentration in the wake $c_m/\bar{c}$ with the overall bed-friction number $\tilde{S} = c_f\, d/(2h)$; c_m is the maximum concentration in the profile and $\bar{c} = c_o q_o/(U_o dh)$ is the average value of the profile across the channel.

5.2 Island Wakes

The re-circulating flow in the channel expansion (Figure 8) is similar in its mixing characteristics to the re-circulating flow in the wake bubbles of islands (Figure 9). Figure 9 shows the aerial photograph of the wakes down streams of two islands in Rupert Bay, Quebec (Ingram and Chu, 1987). The turbidity provides the evidence of mixing in the wakes. The small island has a turbid wake, while the large island has a clear water wake characterized by low turbulence and low turbidity. Despite the very different island diameters (D = 280 m and 560 m), the lengths of the two wake bubbles are approximately the same [L = 290 m and 270 m as marked in Figures 9(a) and 9(b), respectively]. In these wakes of the island, friction is the dominant effect as the diameter of the island is huge compared with the water depth. The friction length scale, $2h/c_f$, is the dominant length scale and the lengths of the wake bubbles are directly proportional to this friction length scale.

The re-circulation flow in the wake bubbles of the islands is not exactly the same as the re-circulating flow in the laboratory channel expansion but the data of both flows are similarly dependent on the friction effect and follow the same trend as shown in Figure 11. Due to interaction of the mixing layers developed along the two sides of the islands, the length of the island wake bubbles is slightly (about 60%) shorter than the length of re-circulation eddy in the channel expansion. For comparison in the figure, the data of the island wakes are normalized by $D/2$ while the laboratory data of the channel expansion are normalized by the width of the expansion, d.

EXAMPLE 5.1: *Estimate the length of the re-circulation region in the wake of an island assuming that the diameter of the island is* D = *560 m, the water depth around the island* $h \approx 2$ *m and the friction coefficient* c_f = *0.0062.*

The estimate will be made in this example based on the island-wake data shown in Figures 11 and 12. If friction is the dominant effect, the length of the re-circulating zone, L, would be directly proportional to the depth of water, h, inversely proportional to the friction coefficient, c_f, and would not be dependent of the size of the island. On the other hand, if the friction effect were negligible, the length of the re-circulating region would be dependent only on the diameter of island. For the island wake in this example,

$$\tilde{S} = \frac{c_f D}{4h} \approx \frac{0.0062 \times 560}{4 \times 2} = 0.43 . \tag{18}$$

Friction in this case is the dominant effect. The best fit of the data in the friction dominant regime of Figure 12 (for $\tilde{S} \geq 0.2$) gives $2L/D = 0.4/\tilde{S}$; that is

$$L = 0.8 \frac{h}{c_f} \tag{19}$$

It follows from this equation that $L \approx 0.8 \times 2/0.0062 = 260$ m. This estimate of the length of the re-circulation region in the island wake is consistent with the aerial photograph of the flow around the island shown in Figure 11.

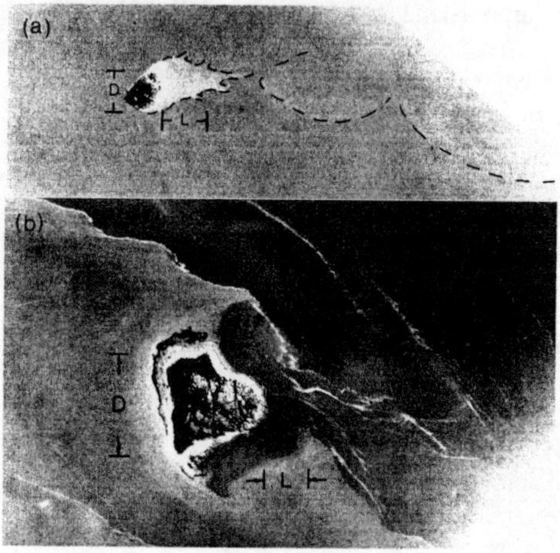

Figure 11: Aerial photographs of flows around islands in Rupert Bay (Ingram and Chu, 1987). (a) D = 280 cm, h = 3 m, $c_f = 0.0056$, L = 290 m, $\tilde{S} = c_f\, D/(4h) = 0.13$; (b) D = 560 m, h = 2 m, $c_f = 0.0062$, L = 270 m, $\tilde{S} = c_f\, D/(4h) = 0.43$.

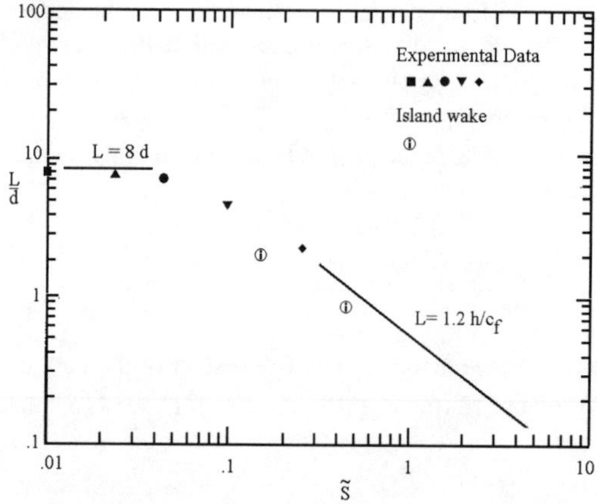

Figure 12: Data of L/d versus $\tilde{S} = c_f\, d/(2h)$ for the re-circulation flow in the channel expansion. The island wake data are included. The wake bubble length to island diameter ratios, $2L/D$, are plotted with the island bed-friction number, $\tilde{S} = c_f\, d/(2h)$, for comparison.

EXAMPLE 5.2: *Untreated wastewater is discharged into the wake of an island through a 0.3 m diameter pipe at a rate of* $q_0 = 0.01\ m^3/s$. *The water depth around the island is h = 2 m. The coliform counts of the effluent at the outfall is* c_0 *= 400000 counts per liter. The effluent re-circulates in the back of the island and then forms a plume in the wake of the island. Assume complete mixing of the effluent in the wake. Estimate the width of the plume and the coliform counts in the wake. The velocity of the current around the island is* $U_0 \approx 1\ m/s$.

There is no data available for the wake width of island. However, the results obtained from the channel-expansion experiment are used to make an estimate of the width. The best fit of the data shown in Figures 9 and 10 gives the following asymptotic formulae for the width and the maximum concentration in the friction dominant regime of the wake:

$$\delta_v = 0.2\frac{h}{c_f}, \qquad c_m = 52\,\bar{c}\,\tilde{S} = 26\frac{q_0 c_f}{U_0 h^2} c_0 \tag{20}$$

(See Babarutsi and Chu (1991) for details of these formulae.) Since h = 2 m and c_f = 0.0062, the half width of the wake is

$$\delta_v = 0.2 \times \frac{2}{0.0062} = 64.5 \text{ m.}$$

The coliform counts in the wake is

$$c_m = 26 \times \frac{0.01 \times 0.0062}{1 \times 2^2} \times 400000 \times 10^3 = 161200 \text{ counts/s.}$$

The minimum dilution ratio is c_0/c_m = 400000/161200 = 2.48. These estimates are based on the assumptions that the effluent is mixed thoroughly in the re-circulation region and that the effluent plume has the same width as the wake of the island. The coliform concentration distribution in the wake is Gaussian with a maximum concentration c_m and a half-width δ_v defined at a location where the concentration $c = \frac{1}{2}c_m$.

5.3 Energy Consideration

A more general assessment of the friction effect is obtained based on energy consideration. Production of turbulence energy by the horizontal turbulence must be balanced by the dissipation. Chu, Wu and Khayat (1991) estimated the frictional dissipation of the horizontal turbulence using a linear theory in a parallel flow and found the rate to be given by

$$F' = \frac{c_f U}{2h}(2\overline{u'^2} + \overline{v'^2}). \tag{21}$$

The production rate by transverse shear is

$$P' = -\overline{u'v'}U_y \tag{22}$$

where u' and v' are the velocity fluctuation components of the horizontal turbulence, U the mean velocity and $U_y = dU/dy$ the velocity gradient in the transverse direction of the mean flow. These expressions for the dissipation and production terms, F' and P' are obtained for nearly parallel flows $U(y)$. The ratio of the dissipation term F' to the production term P' is the flux bed-friction number

$$S_f = \frac{F'}{P'} = \frac{2\overline{u'^2} + 2\overline{v'^2}}{-\overline{u'v'}}S \tag{23}$$

where S is the gradient bed-friction number

$$S = \frac{c_f}{2h}\frac{U}{U_y}. \tag{24}$$

If $\overline{v'^2} \approx 0.44\overline{u'^2}$ and $-\overline{u'v'} \approx 0.3\overline{u'^2}$ (that is assuming the same relations as in a free jet), the term $[-\overline{u'v'}] / [2\overline{u'^2} + \overline{v'^2}]$ would have an estimate value of about 0.12. Therefore, the gradient bed-friction number and the flux bed-friction number are related through the relation $S = 0.12\ S_f$. Beside the frictional dissipation, the nonlinear cascade process also dissipates the energy of the horizontal turbulence by transferring its energy to turbulence of smaller length scales at a rate equal to ε'. For maintenance of the horizontal turbulence, the energy production, P', must be greater than the combined dissipation, F' and ε'; i.e.,

$$F' + \varepsilon' < P'. \tag{25}$$

Dividing this inequality by P', gives

$$S_f = \frac{F'}{P'} < 1 - \frac{\varepsilon'}{P'}. \tag{26}$$

This then lead to the necessary condition

$$S < 0.12(1 - \frac{\varepsilon'}{P'}). \tag{27}$$

for maintenance of the horizontal turbulence. According to this condition, the critical value of the gradient bed-friction number should be less than $S_c = 0.12$ if the dissipation due to nonlinear cascade ε' were ignored. This results obtained from the energy consideration is consistent with the hydrodynamic stability analysis by Chu, Wu and Khayat (CWK, 1991) and Alavian and Chu (AC, 1985). CWK analyzed the stability of 'inviscid' and parallel flows with hyperbolic-tangent and hyperbolic-secant transverse velocity profiles, and found the critical values to be $S_c = 0.120$ and 0.145, respectively. AC included the turbulent viscosity in their analysis and found the critical value to be $S_c = 0.06$.

The term U/U_y in Eq. (24) is a length scale associated with the width of the transverse velocity profile. In correlating the experimental data of the open-channel flow expansion, this length scale is taken to be the width of the channel expansion, d. The overall bed-friction number $\tilde{S} = c_f d/(2h)$ (Eq. (17)) is the gradient bed-friction number S (Eq. (24)) obtained by setting the length scale $U/U_y = d$. In correlating data for the island wake, this length scale is the diameter of the island, D.

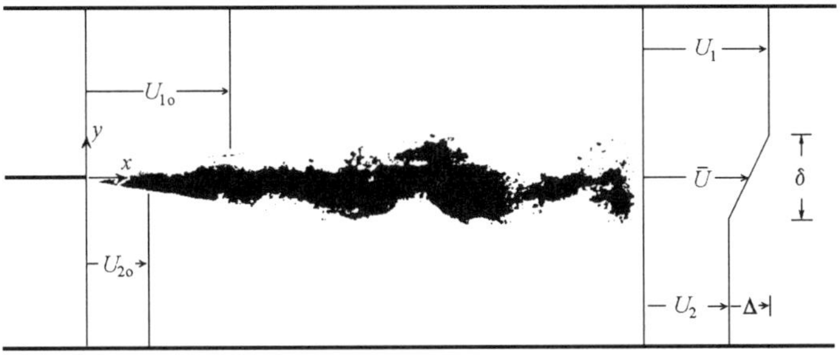

Figure 13: Transverse mixing layer in shallow open-channel flow. This is a top view of the flow obtained by an overhead camera. Dye is introduced along the sides of the splitter plate. The water depth of the open-channel flow for this test is $h = 2.64$ cm; $h/c_f = 400$ cm; $x = 0$ to 150 cm.

5.4 Mixing-layer Experiments

The first series of experiments to study the friction effects on horizontal turbulence was the mixing-layer experiments conducted by Chu and Babarutsi (1988). Figure 13 shows the set up of the experiments from a top view of the open-channel flow. The open-channel flow is divided into two streams of different velocities U_1 and U_2 by a splitter plate. Horizontal turbulence is produced by the transverse shear downstream of the splitter plate. To study the friction effect, velocity and turbulent intensity were measured across the mixing layer. The width of the mixing layer, δ, were determined from the measurements using the definition

$$\delta = \frac{U_1 - U_2}{(dU/dy)_{\max}} = \frac{\Delta}{(dU/dy)_{\max}} \tag{28}$$

where $U(y)$ was the velocity and $(dU/dy)_{max}$ its transverse gradient at the inflection of the profile. Figures 14 and 15 show the data obtained from the measurements. The dimensions of the horizontal turbulence is proportional to the width, δ, but the friction effect imposes a limit on the length scale of the turbulent eddies. Two velocity scales are associated with the mixing layer: they are the average velocity of the dominant eddies in the mixing layer $\overline{U} = (U_1 + U_2)/2$ and the velocity difference across the mixing layer $\Delta = U_1 - U_2$. The width data are normalized in the figure by the friction length scale, $2h/c_f$, the average velocity $\overline{U}$ and the velocity difference across the mixing layer Δ. Figure 14 shows the normalized width of the mixing layer. Initially the width increases at a rate

$$\frac{\overline{U}}{\Delta}\frac{d\delta}{dx} = 0.12 \tag{29}$$

In the far field, the width approaches the asymptote

$$\frac{c_f \delta_{\max} \overline{U}_0}{h\Delta_0} = 0.065 \tag{30}$$

as friction becomes the dominant effect in the region. The subscript 'o', in $\overline{U}_0$ and Δ_0, denote the conditions at the splitter plate. The maximum width of the mixing layer, δ_{max}, is inversely related to and determined by the value of friction coefficient, c_f.

Figure 15 shows the intensity of the horizontal turbulence, $\sqrt{u'^2}_{\max}$. The relative intensity would be a constant, that is $\sqrt{u'^2}_{\max}/\Delta \approx 0.24$, if the friction effect were negligible. In shallow waters, this relative intensity, $\sqrt{u'^2}_{\max}/\Delta$, decrease with the distance, x, from the splitter plate. The intensity of the horizontal turbulence is reduced directly as a result of the friction effect.

EXAMPLE 5.3: *The velocity changes from U_1 = 0.9 m/s to U_2 = 0.1 m/s across the transverse mixing layer at the confluence of two rivers where the water depth h = 1.3 m. Estimate the width of the mixing layer assuming that the value of the friction coefficient is (a) c_f = 0.006, (b) c_f = 0.0006.*

For this example, $\Delta_0 = U_1 - U_2 = 0.8$ m/s, $\overline{U}_0 = (U_1 + U_2)/2 = 0.5$ m/s. The maximum width of the mixing layer is inversely proportional to the value of the friction coefficient according to Eq. (30). If the value of the friction coefficient were $c_f = 0.006$,

$$\delta_{\max} = \frac{0.065 \times 1.3 \times 0.8}{0.006 \times 0.5} = 22.5 \text{ m}.$$

On the other hand, if the value were ten times smaller, i.e., if $c_f = 0.0006$,

$$\delta_{\max} = \frac{0.065 \times 1.3 \times 0.8}{0.0006 \times 0.5} = 225 \text{ m}.$$

Friction therefore is the deciding effect that determines the length scale of the horizontal turbulence. Without friction, the length scale of the horizontal turbulence would continue to increase with distance from the confluence, bounded only by the width of the river. What should be the value of friction coefficient? The answer to this question will be attempted in a later section.

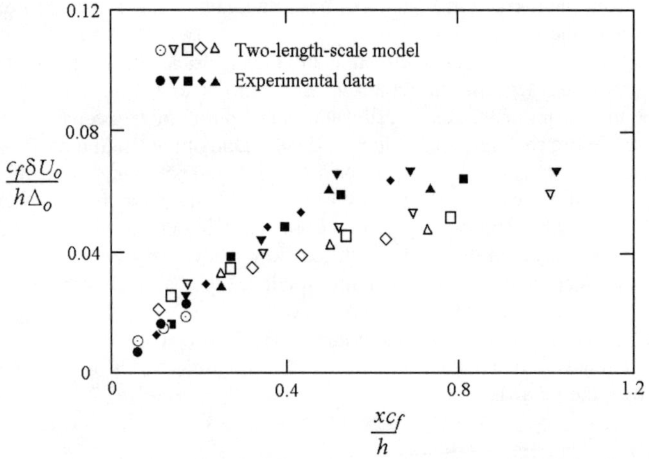

Figure 14: Width of the mixing layer, δ, plotted with distance from the splitter plate, *x*. The experimental data obtained by Chu and Babarutsi (1988) are denoted by the solid symbols. The simulation data obtained using the two-length-scale turbulence model are denoted by the open symbols.

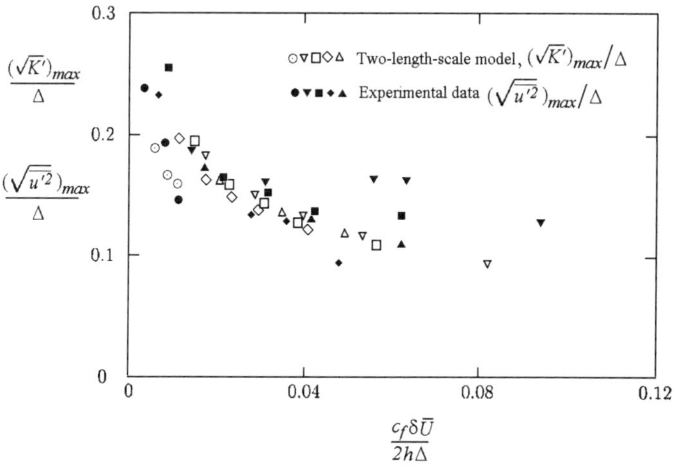

Figure 15: RMS (root-mean-square) velocity fluctuations in the mixing layer, $\sqrt{\overline{u'^2}}_{max}/\Delta$, plotted against the gradient bed-friction number.

The aerial photograph in Figure 16 shows the development of a transverse mixing layer in the shallow water at the confluence of two rivers. The mixing layer spreads out in the horizontal direction and then approaches a width of about 140 m as the water depth in the region varies from 0.8 to 1.7 m. The friction coefficient at the confluence can be estimated to be approximately equal to 0.001 using the procedure shown in Example 5.1. Such value of the friction coefficient is small for open-channel flow of uniform density. The river confluence is located in a shallow estuary (Rupert Bay, Quebec) where the flow is stably stratified due to the intrusion of salinity from the sea into the estuary. The transverse mixing layer in the stably stratified flow is affected by the *interfacial* friction coefficient, which could be quite small as the vertical turbulence is suppressed by the stable stratification.

Recently, the mixing-layer experiments are repeated in a large channel facility by Uijttewaal and Tukker (1998). The asymptotic width of the mixing layer based on their data is

$$\frac{c_f \delta_{max} \overline{U}_0}{h\Delta_0} = 0.05, \tag{31}$$

which is slightly smaller than the width given by Eq. (30). The important conclusion obtained by Chu and Babarutsi (1988), that the length scale of the horizontal turbulence is inversely proportional to the friction coefficient, nevertheless is confirmed.

Figure 16: Aerial photograph of transverse shear flow at confluence of Nottaway River and Broadback River in Rupert Bay, Quebec. Large-scale turbulence eddies are visible between fast-flow and turbid water of Nottaway (shown in upper part of photograph) and slow and clear water of Broadback; flow directions are from left to right in photograph.

5.5 Friction Coefficients and Vertical Turbulence

The friction coefficient is merely a parameter introduced in the depth-averaged formulation to characterize the vertical exchange of momentum by the vertical turbulence. We found from the mixing-layer experiment that the length scale of the horizontal turbulence δ_{max} is inversely proportional to the value of the friction coefficient c_f (Eq. (30)). A strong exchange of momentum by vertical turbulence would lead to a large friction coefficient and consequently relatively insignificant transport across the stream by the horizontal turbulence. Conversely, if the friction coefficient was small, for example, due to the suppression of the vertical turbulence in a stably stratified flow, the horizontal turbulence would play a more significant role in the transport process. The concept that the existence of the horizontal turbulence is inversely related to the vertical turbulence is fundamental to understanding the transport processes across the natural streams.

In open-channel flow of uniform density, the value of the friction coefficient is a function of the roughness of the channel bed. For fully rough bed,

$$c_f = 0.028(\frac{k_s}{h})^{\frac{1}{3}} \tag{32}$$

where k_s is the sand-grain roughness and h is the depth of the open channel flow (see e.g. Henderson 1966). If the channel bed were not a fully rough surface, the friction coefficient would be dependent on also the Reynolds number. The A.S.C.E. Task Force on Friction Factor in Open Channel (1963) recommends the following formula:

$$\frac{1}{\sqrt{c_f}} = -4.0 \log\left(\frac{k_s}{12h} + \frac{1.25}{\mathrm{Re}\sqrt{c_f}}\right) \tag{33}$$

where k_s = equivalent sand-grain roughness, h = water depth and $\mathrm{Re} = 4Uh/\nu$ = Reynolds number.

5.6 Density Stratification

In natural streams, the vertical turbulence is often affected by the presence of density stratification. The intensity of the vertical turbulence is enhanced in unstable stratification and reduced in stable density stratification. The relation between the interfacial friction coefficient and vertical turbulence in stratified flow can be determined by correlating the Reynolds stress with the intensity of the turbulence as follows. The friction stress $\langle \tau_b \rangle$ in Eq. (6), representing the vertical exchange of momentum near the bed, is equal to the Reynolds stress $\rho\, \overline{u_b'''w_b'''}$ in the fully developed turbulent flow near the bed. The friction coefficient by definition is

$$c_f = \frac{2\langle \tau_b \rangle}{\rho \tilde{u}^2} = \frac{2\overline{u_b'''w_b'''}}{\tilde{u}^2} \tag{34}$$

where u_b''' and w_b''' are the velocity fluctuations of the turbulent flow near the bed. The triple prime symbol denotes the velocity fluctuations associated with the vertical turbulence (see also Figure 4). The numerical relation between c_f and $\rho\, \overline{u_b'''w_b'''}$ depends on density stratification. The following example shows this relation in an open-channel flow of uniform density on a smooth surface.

EXAMPLE 5.4: *Find the friction coefficient of an open-channel flow of uniform density on a smooth surface. Use the following values for the Reynolds stresses obtained from laboratory measurements near the channel bed:* $\overline{u_b'''w_b'''}/(\sqrt{\overline{u_b'''^2}}\sqrt{\overline{w_b'''^2}}) \approx 0.44$, $\sqrt{\overline{u_b'''^2}}/\tilde{u} \approx 0.15$, *and* $\sqrt{\overline{w_b'''^2}} \approx 0.05$ *where* $\tilde{u}$ *is depth-average velocity of the open-channel flow.*

Use Eq. (34) for the relation between the friction coefficient and the intensity of the vertical turbulence:

$$c_f = \frac{2\overline{u_b'''w_b'''}}{\sqrt{\overline{u_b'''^2}}\sqrt{\overline{w_b'''^2}}}\frac{\sqrt{\overline{u_b'''^2}}}{\tilde{u}}\frac{\sqrt{\overline{w_b'''^2}}}{\tilde{u}} \tag{35}$$

Substituting for the values given,

$$c_f = 2\times 0.44\times 0.15\times 0.05 = 0.0066 \tag{36}$$

This value of the friction coefficient is typical of the open-channel flow of uniform density over a smooth bed at moderate Reynolds number. In the presence of a stable density stratification, the value of the friction coefficient could significantly be reduced due to suppression of the Reynolds stresses, $\rho\, \overline{u_b'''w_b'''}$, $\rho\, \overline{u_b'''^2}$ and $\rho\, \overline{w_b'''^2}$ by the buoyancy forces. The mixing layer at the confluence of rivers in Figure 16 is an example. The water depth in the area of the confluence varying from 0.8 to 1.7 m is quite small compared to the 100 m size of the eddies across the width of the mixing layer. Knowing the asymptotic width of the mixing layer $\delta_{max} \approx 100$ m, and assuming that the depth, the velocity difference and the average velocity across the mixing layer are h = 1.3 m, $\Delta_0 = 0.8$ m/s and $\overline{U}_0 = 0.5$ m/s, the formula in Eq. (30) is used to estimate the friction coefficient

$$c_f = 0.065\frac{h}{\delta_{max}}\frac{\Delta_0}{\overline{U}_0} = \frac{0.065\times 1.3\times 0.8}{100\times 0.5} = 0.0014$$

This value of the friction coefficient is quite small compared with the value for open-channel flow of uniform density. The presence of salinity and stable density stratification in this area of the Rupert Bay probably has suppressed the intensity of the vertical turbulence. Layers form in a stably density stratified flow and, under this condition, the interfacial friction coefficient would be the more appropriate parameter to characterize the vertical exchange of momentum.

The inter-connecting relations between the density stratification, interfacial friction, interfacial waves and the vertical turbulence have been examined by many previous investigations (see, e.g., Turner, 1973, Gargett et al., 1984, Chu and Baddour, 1984, and Rodi, 1987). Many important transport processes in the natural streams, including those of the jet stream in the atmosphere and the gulf stream of the Atlantic ocean, are affected by density stratification (Garvine, R. W., 1974; Pringree, 1978, 1979; Simpson, 1981). The length scale and intensity of the horizontal turbulence in these stratified flows are closely related to the exchange of momentum by the vertical turbulence. We know how the horizontal turbulence is dependent on interfacial friction coefficient. However, the exact connection between the horizontal and the vertical exchanges is not entirely understood.

6. TWO-LENGTH-SCALE RANS MODEL

In a one-layer system, the horizontal turbulence can be computed by the LES method as described in section 4 or computed using a RANS (Reynolds Average Navier Stokes) model to be explained in this section. Using the LES method, the instantaneous velocity components $\tilde{u}_i$ are computed first. The mean flow components U_i are subsequently determined by time averaging the instantaneous components. In the RANS model, the mean flow is directly simulated using a closure model. Chu and Babarutsi (1989) introduced a two-length-scale RANS model for the horizontal turbulence. In this model, the mean velocity components are obtained from the numerical solution of the continuity and momentum equations:

$$\frac{\partial hU_k}{\partial x_k} = 0 \tag{37}$$

$$\frac{\partial U_i}{\partial t} + U_j \frac{\partial U_i}{\partial x_j} = -\frac{\partial P}{\partial x_i} + \frac{1}{h}\frac{\partial}{\partial x_j}[\nu_T h(\frac{\partial U_i}{\partial x_j} + \frac{\partial U_j}{\partial x_i})] - \frac{1}{h}\frac{\partial}{\partial x_i}[\nu_T h(\frac{\partial U_\ell}{\partial x_\ell})] - \frac{c_f U_i}{2h}\sqrt{U_\ell U_\ell} \tag{38}$$

where U_i is mean velocity and ν_T = eddy viscosity associated with the turbulent motion. The turbulent viscosity ν_T has two parts. The part due to vertical turbulence is $c_v U_* h$. The part due to horizontal turbulence is $c_\mu K'^2/\varepsilon'$. The sum of the two is

$$\nu_T = c_v U h + c_\mu \frac{K'^2}{\varepsilon'} \tag{39}$$

in which $U_* = \sqrt{\tau/\rho} = \sqrt{c_f U_\ell U_\ell / 2}$ is the shear velocity. The turbulent kinetic energy, $K' = \frac{1}{2}\overline{u'_k u'_k}$, and the energy cascade rate, ε', are the part associated with the horizontal turbulence. The K'-equation for the horizontal turbulence,

$$\frac{\partial K'}{\partial t} + U_\ell \frac{\partial K'}{\partial x_\ell} = \frac{\partial}{\partial x_\ell}(\frac{\nu_T}{\sigma_K}\frac{\partial K'}{\partial x_\ell}) + P' + F' - \varepsilon' \tag{40}$$

is energy equation obtained from the Reynolds averaging. The ε'-equation,

$$\frac{\partial \varepsilon'}{\partial t} + U_\ell \frac{\partial \varepsilon'}{\partial x_\ell} = \frac{\partial}{\partial x_\ell}(\frac{\nu_T}{\sigma_\varepsilon}\frac{\partial \varepsilon'}{\partial x_\ell}) + [c_{1\varepsilon}(P' + c_{3\varepsilon}F') - c_{2\varepsilon}\varepsilon']\frac{\varepsilon'}{K'} \tag{41}$$

is a model equation based on an assumed anology between the effects of friction and buoyancy. The production and friction terms, P' and F', in the energy and dissipation equations, Eqs. (40) and (41), are

$$P' = \tau_{ij}\frac{\partial U_i}{\partial x_j} = \nu_T\left(\frac{\partial U_i}{\partial x_j}+\frac{\partial U_j}{\partial x_i}\right)\frac{\partial U_i}{\partial x_j} - \nu_T\left(\frac{\partial U_\ell}{\partial x_\ell}\right)^2 - K'\left(\frac{\partial U_\ell}{\partial x_\ell}\right) \quad (42)$$

$$F' = \overline{f'_\ell u'_\ell} = -\frac{c_f}{2h}\left[\overline{u'_\ell u'_\ell}\sqrt{\overline{U_m U_m}} + \overline{u'_\ell u'_n}\frac{\overline{U_\ell U_n}}{\sqrt{\overline{U_m U_m}}}\right] \quad (43)$$

where U_i = mean velocity, $u'_i = \tilde{u} - U_i$ = velocity fluctuations associated with the large-scale turbulence, and f'_i = friction-force fluctuations. The derivations for the expressions of P' and F' are given in Babarutsi (1991). The expressions for nearly parallel flow have been given previously in Eqs. (21) and (22). Beside the energy dissipation by friction, the energy of horizontal turbulence is passed on to the smaller scales by nonlinear process at a rate equal to ε'. The nonlinear cascade process is quite complicated and is determined by the model equation, Eq. (41). The values of the coefficients selected for the model are

$$c_v = 0.08,\ c_\mu = 0.09,\ c_{1\varepsilon} = 1.44,\ c_{2\varepsilon} = 1.92,\ \sigma_k = 1.0,\ \sigma_\varepsilon = 1.3,\ c_{3\varepsilon} = 0.8 \quad (44)$$

These are the coefficients used in standard $k-\varepsilon$ model except the coefficient $c_{3\varepsilon}$ which is introduced here for the body-force effect.

The above model equation for ε' is identical in form to the model equation used by Hossian (1980) and Rodi (1985) for buoyancy effect in gravity stratified shear flow. In modeling of the buoyancy effect, the term F' is the work done by the buoyancy force. The model coefficient of $c_{3\varepsilon} = 0.8$ was used by Hossian and Rodi in their simulation of the buoyancy effect in gravity stratified shear flow. The rational for selecting the model coefficients for the shallow shear flow the same as the gravity stratified flow is based on the assumed analogy between buoyancy and friction.

Computations using the RANS model have been conducted for a number of SQ2D turbulent flows. Chu and Babarutsi (1989) and Babarutsi and Chu (1998) used the model to calculate the transverse development of the mixing layers. Figures 14 shows the results obtained for the width of the mixing layers δ and Figure 15 the results for the intensity of the horizontal turbulence K'. The model produced a some what smaller growth rate in the initial development close to the splitter plate and also slightly smaller asymptotic width. The model prediction for the intensity of the horizontal turbulence follows closely the trend of the experimental data.

Simulations using the two-length-scale RANS model also have been conducted to calculate re-circulating flows in the open-channel flow expansion by Babarutsi, Nassiri and Chu (1996) and starting jets of small depth by Altai, Zhang and Chu (1999). The results are generally consistent with the experimental observations. The readers are referred to the original publications for details of these simulations. We conclude this chapter by two examples encountered in practice. The first example is concerned with mathematical model of the wastewater effluent in the St. Lawrence River while the second example is concerned with a physical model of cooling water discharge from a power plant into a coastal water body.

7. WASTE WATER EFFLUENT IN THE ST. LAWRENCE RIVER

The MUC (Montreal Urban Community) sewage treatment plant discharges its effluent into the St. Lawrence River at an outfall located at the northeast end of the Vaches Island. The effluent enters the river horizontally through a 6.1 m diameter circular pipe in a direction nearly perpendicular to, and at a location approximately 170 m off the centerline of the seaway's shipping lane (see the map in Figure 17 and the river cross-section in Figure 18). The main channel of the river is along the shipping lane of the St. Lawrence Seaway. The depth of the seaway, maintained through regular dredging, is about 11 to 15 m and the width 244 m. Nearly 80% of the river flows along the main shipping channel and 20% along the shore. Strong lateral exchanges of mass and momentum take place between the main channel and the shallow shore. Turbulence is generated on the one hand through bed-friction stresses and on the other hand through lateral exchange of momentum between the fast flow in the main channel and the slower flow near the shore; both play a significant role on the dilution of the effluent.

The dilution characteristics of the effluent plume in the river was determined mathematically, using a version of the two-length-scale model described in the previous section, over a distance of about 2 km downstream of the outfall. The model was then verified by a Rhodamine dye test carried out on May 17, 1989. Dye concentration profiles were obtained from river sampling at two river cross sections: section B and section C at 1.07 km and 2.14 km downstream of the outfall respectively. Figures 19 and 20 shows the profiles at the two river cross sections and the comparison with the predictions by the mathematical model.

In the mathematical model, the dye concentration distribution was obtained by solving numerically the diffusion equation

$$\frac{\partial c}{\partial x} = \frac{\partial}{\partial \psi}[m_x h^2 U D_T \frac{\partial c}{\partial \psi}] \tag{45}$$

in which c = dye concentration, ψ = stream function co-ordinate, x = longitudinal co-ordinate along the stream tube, U = velocity component in the longitudinal direction, m_x = the metric coefficient related to curvature of the stream tubes, and D_T = turbulent diffusion coefficient (or the transverse mixing coefficient). This is the diffusion equation derived by Yotsukura and Sayre (1976), using the stream function as the transverse co-ordinate, and it has been used quite successfully by several numerical simulations of transverse mixing in natural rivers (see Harden and Shen, 1979, and Lau and Krishnappan, 1981).

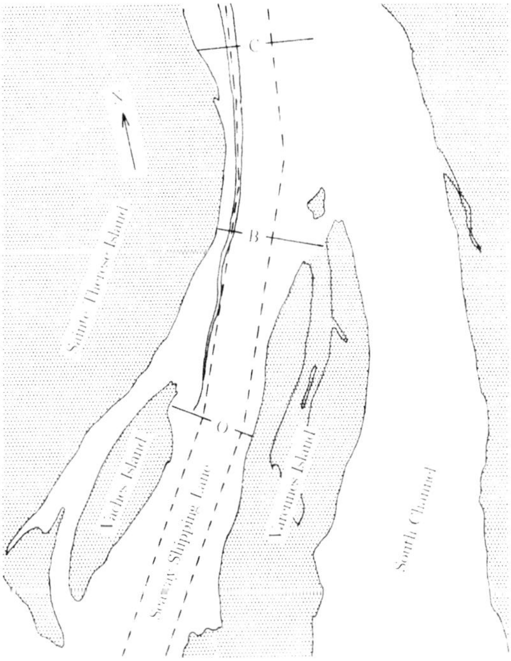

Figure 17: Topographic map showing locations of the outfall (section O), section B, section C and the effluent plume predicted by the mathematical model.

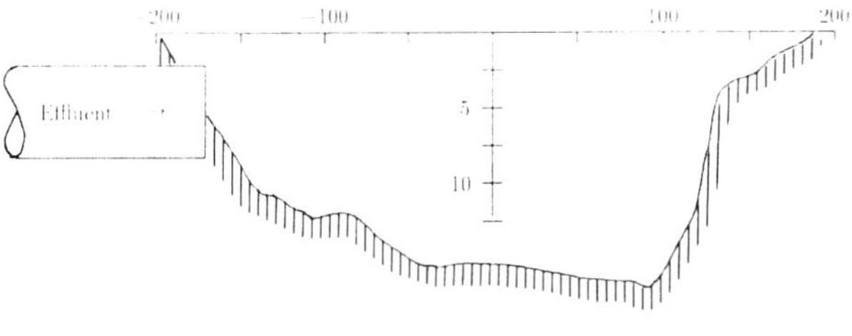

Figure 18: River cross section O at the outfall. The dimensions are in meters.

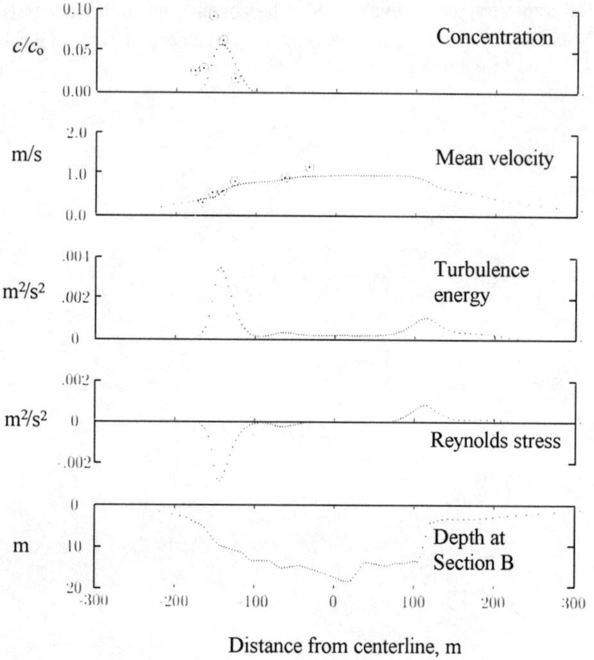

Figure 19: Concentration, mean velocity, turbulent intensity, Reynolds stress and topographic profiles at section B, 1070 m downstream of the outfall.

In most of the previous simulations of mixing processes in rivers, it was necessary to correlate the turbulent diffusion coefficient empirically with the flow and cross-sectional geometry of the river. It is difficult to do the same in the present situation because the effluent plume is located in a region of rapidly changing water depth. In the present simulation, the turbulent diffusion coefficient was obtained using a two-length-scale Reynolds-stress turbulence model by Babarutsi and Chu (1993). In this model, the diffusion coefficient is given by

$$D_T = \frac{1}{\sigma_T}(c_v u_* h + c_\varepsilon \frac{K'}{\varepsilon'}\overline{v'v'}) \tag{46}$$

where $u_* = \sqrt{\tau_0/\rho}$ = local bed-friction velocity, h = water depth, $K' = \frac{1}{2}(\overline{u'u'} + \overline{v'v'})$ turbulence energy, ε' = energy dissipation rate, and $\overline{v'v'}$ = one component of the Reynolds-stress tensor; the coefficients $c_v = 0.08$ and $c_\varepsilon = 0.15$, and the turbulent Schmidt number σ_T is assumed to be equal to 0.7.

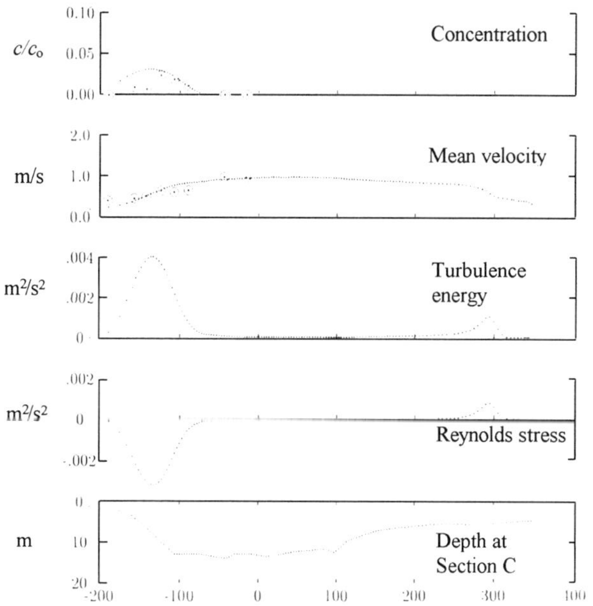

Figure 20: Concentration, mean velocity, turbulent intensity, Reynolds stress and topographic profiles at section C, 2140 m downstream of the outfall.

The turbulence diffusion coefficient as given by Eq. (46) has two parts, the turbulence diffusion due to vertical turbulence and the turbulence diffusion due to horizontal turbulence. The Reynolds-stress turbulence model calculates the part due to horizontal turbulence. The governing equations for the components of the Reynolds-stress tensor, $\overline{u'u'}$, $\overline{u'v'}$ and $\overline{v'v'}$, are:

$$U\frac{\partial \overline{u'u'}}{\partial x} + V\frac{\partial \overline{u'u'}}{\partial y} = \frac{1}{h}\frac{\partial}{\partial y}[h(c_s \frac{K'}{\varepsilon'}\overline{v'v'} + v_s)\frac{\partial \overline{u'u'}}{\partial y}] + P'_{11} + \Pi_{11} - F_{11} - \varepsilon' \tag{47}$$

$$U\frac{\partial \overline{u'v'}}{\partial x} + V\frac{\partial \overline{u'v'}}{\partial y} = \frac{1}{h}\frac{\partial}{\partial y}[h(c_s \frac{K'}{\varepsilon'}\overline{v'v'} + v_s)\frac{\partial \overline{u'v'}}{\partial y}] + P'_{12} + \Pi_{12} - F_{12} \tag{48}$$

$$U\frac{\partial \overline{v'v'}}{\partial x} + V\frac{\partial \overline{v'v'}}{\partial y} = \frac{1}{h}\frac{\partial}{\partial y}[h(c_s \frac{K'}{\varepsilon'}\overline{v'v'} + v_s)\frac{\partial \overline{v'v'}}{\partial y}] + P'_{22} + \Pi_{22} - F_{22} - \varepsilon' \tag{49}$$

in which (U,V) and (u',v') are the mean and fluctuating velocity components in the longitudinal, x, and transverse, y, directions, respectively; $\nu_s = c_v u_* h$ is the eddy viscosity due to the vertical turbulence. The production terms, the pressure-strain term, and the bed-friction terms are as follows:

$$P'_{11} = -2\overline{u'v'}\frac{\partial U}{\partial y},\ P'_{12} = -\overline{v'v'}\frac{\partial U}{\partial y},\ P'_{22} \simeq 0,\ P' = \frac{P'_{11} + P'_{22}}{2} \tag{50}$$

$$\Pi_{11} = -c_1\frac{\varepsilon}{K'}(\overline{u'u'} - K') - \gamma(P'_{11} - P') + c_3(F_{11} - F) \tag{51}$$

$$\Pi_{12} = -c_1\frac{\varepsilon}{K'}(\overline{u'v'}) - \gamma(P'_{12}) + c_3(F_{12}) \tag{52}$$

$$\Pi_{22} = -c_1\frac{\varepsilon}{K'}(\overline{v'v'} - k') - \gamma(P'_{22} - P') + c_3(F_{22} - F) \tag{53}$$

$$F_{11} = \frac{4c_f U}{2h}\overline{u'u'},\ F_{12} = \frac{3c_f U}{2h}\overline{u'v'},\ F_{22} = \frac{2c_f U}{2h}\overline{v'v'},\ F = \frac{F_{11} + F_{22}}{2} \tag{54}$$

The dissipation rate, which is reduced due to the confinement effect, is given by $\varepsilon' = c_{cp}\varepsilon$, where c_{cp} is a model coefficient and ε is determined by the standard model equation

$$U\frac{\partial \varepsilon}{\partial x} + V\frac{\partial \varepsilon}{\partial y} = \frac{1}{h}\frac{\partial}{\partial y}[h(c_s\frac{K'}{\varepsilon}\overline{v'v'} + \nu_s)\frac{\partial \varepsilon}{\partial y}] + c_{1\varepsilon}\frac{\varepsilon}{K'}P' - c_{2\varepsilon}\frac{\varepsilon^2}{K'} \tag{55}$$

The model coefficients of the Reynolds-stress model, as given in Table 1, were obtained from the calibration of the model against the data obtained from laboratory investigations.

Table 1: Coefficients of the Reynolds Stress Model

$c_{1\varepsilon}$	$c_{2\varepsilon}$	c_1	c_3	γ	c_s	c_ε	c_{cp}	c_v
1.44	1.92	1.5	-2.0	0.72	0.25	0.15	0.88	0.08

The mean velocity profile in the river was calculated using the procedure described in Yotsokura and Sayre (1976). In this procedure, the pressure is assumed constant across the river and the local velocity depends on the local water depth through the Manning equation. The Manning coefficient $n = 0.025$ was used in the present calculation.

The Reynolds stresses and the dye concentration were calculated using a forward-marching finite difference scheme of Patankar and Spalding (1970). The concentration equation and the Reynolds-stress were expressed in term of the stream-function co-ordinates as Eq. (45). The total volume of the river flow, estimated to be 3483 m^3/s on May 17, is divided into 500 stream-tubes; i.e., $\Delta\psi =$ 6.97 m^3/s. For stability of the numerical computation, $\Delta x = 3.05$ m. The results of the numerical computation are presented in Figures 19 and 20. The dye concentrations of the river samples are higher than predicted by the model at section B, but lower than predicted at section C. This difference is, of course, a reflection of the uncertainty of the field measurements and the imperfection of the mathematical model. It is also a reflection of the variability of the effluent plume, which is not entirely steady but meandering along the edge of the shipping channel. The location and width of the plume obtained from the numerical simulation is superimposed on the topographic map in Figure 17. The width of the plume is defined as the width where the concentration, c, is greater then one-half of the maximum, c_{max}, along the centerline. Despite the uncertainty associated with the field observation and the imperfection of mathematical model, a general agreement is obtained between the simulation and the field observation. In this reach of St. Lawrence River, the turbulent diffusion by the vertical turbulence is comparable in order of magnitude to the diffusion due to horizontal turbulence. According to the numerical calculations, the maximum diffusion coefficient associated with the horizontal turbulence (that is the second term of Eq. (46)) is three times greater than the part associated with the vertical turbulence (the first term of Eq. (46)).

It is interesting to observe that the MUC effluent plume is confined in a narrow corridor of about 100 m in the middle of the St. Lawrence River. The concern expressed by the residences for possible contamination in the regions along the south shore is unwarranted.

8. PHYSICAL MODEL OF COOLING WATER EFFLUENT

A 1:75 undistorted scale model of the Diablo Canyon Power Plant was constructed in 1974 to study the cooling water discharge into a shallow rocky cove located on bluff overlooking the Pacific Ocean west of San Luis Obispo, California (Ryan et al., 1987). The cooling water system withdraws 57 m^3/s through a shoreline intake and discharges into the Diablo Cove as shown in Figure 21. The temperature rise at the outfall is 11.1 °C. The discharge forms a turbulent jet, which was observed to meander and have a tendency to attach onto the side of the cove. This had led to a pool of water with a higher temperature on the right hand side of the cove as shown in Figure 22. The width of the warm water pool is about 700 ft (prototype) while the water depth in this area of the cove is less than 20 ft. The turbulent jet is in shallow water as the depth is very small compared with the width of the flow. The mixing process in this case is primarily dependent on the horizontal turbulence, the existence of which is linked closely to the exchange of momentum by the vertical turbulence.

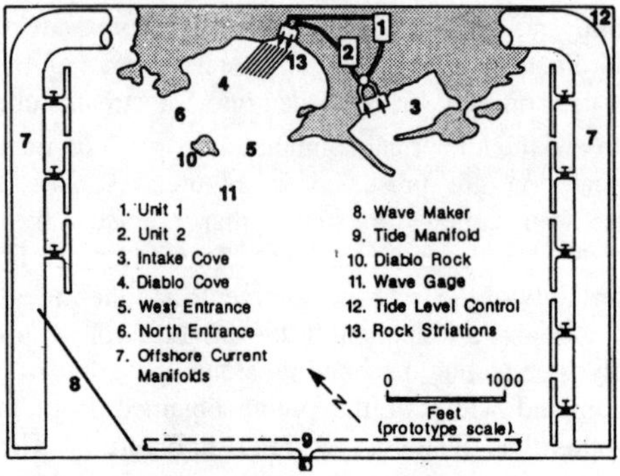

Figure 21: Layout of the physical model of the Diablo Canyon Power Plant, San Luis Obispo, California. Reproduced from Ryan, Tu, Ismail and Wiegal (1987).

Field tests were performed from late 1984 to 1986 in an attempt to verify the prediction by the physical model. The temperature in the field was found to be significantly higher than the prediction. Large meander of the jet discharge was observed in the model but was not in the field. The physical model had incorrectly predicted the temperature because the transport by the horizontal turbulence was much too high in the model.

The original model, constructed in 1974, was based on the available 50 ft grid data, and had a relatively smooth bottom. In contrast, the actual cove bottom was very rugged, with rock striations near the discharge structure with a height of several feet. With refinement of the bathymetry and the increase of the bottom roughness, Ryan et al. (1987) was able to reproduce the correct temperature distribution in the *revised physical model* in agreement with the observations in the field tests (see Figure 22). However, the main reason for the agreement between the revised model results and the field observations according to Ryan et al. (1987) was the increase of the bottom roughness in the revised model. Even after the best available bathymetry was included in the revised model, large meander of the jet discharge was observed leading to excess entrainment of cold ambient water into the plume. To increase the bottom roughness, 5-mm size gravel was scattered randomly on the bottom of the revised model. The roughness eliminates the meander and consequently produces the correct temperature distribution in the model. In the 1:75 model, the 5 mm gravel is equivalent to a 75×5 mm = 0.375 m rough elements in the prototype. This size of the rough element in the revised model is consistent with the 0.3 m to 0.6-m size rocks being observed in the rugged bottom of the Diablo Cove. The model law was based on the Froude similarity. To correctly reproduce the transverse mixing process, similitude in bottom roughness between the model and the prototype also must be maintained.

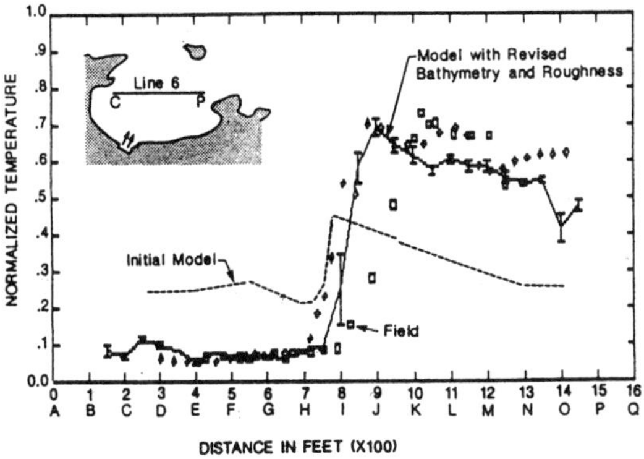

Figure 22: Temperature profiles obtained in the initial model and in the revised model compared with the field data.

The model tests of the Diablo Canyon Power Plant have demonstrated the significance of the vertical turbulence in the overall transport process across the natural streams. The greater bottom roughness in the revised physical model has led to greater intensity of the vertical turbulence and an increase in bed friction. With the increase in bed friction, the transveıse shear flow becomes more stable. The horizontal turbulence becomes less intense. The net result of the greater intensity of the vertical turbulence, and less intensity of the horizontal turbulence, is overall reduction of the mixing of the effluent plume with its surrounding waters and hence higher temperature in the plume as observed in the revised model in agreement with the field data.

The roughness may have produced greater intensity of the vertical turbulence but have eliminated the meandering of the effluent. So mixing is reduced with the increase in the roughness on the bed of the revised model. This observation is contrary to common sense, which may lead us to the wrong conclusion that increase in roughness should produce more intense mixing and greater dilution of the effluent with its surrounding waters.

9. CONCLUSION

We have shown in this chapter how, and why it is necessary, to partition the turbulence in natural streams into a horizontal component and a vertical component so that the components may be simulated by separate models. Although the direct contribution to transport across the stream is small, the vertical turbulence is significant for its effect on the horizontal turbulence. LES and RANS models were developed to calculate the transports of mass and momentum across the width of the streams by horizontal turbulence. Simulations were successfully conducted using these models for processes across shallow turbulent flows such as jets, wakes, mixing layers and re-circulating flows.

An inverse relation was observed between the horizontal and vertical turbulence. The length scale of the horizontal turbulence is inversely proportional to the friction coefficients - a parameter characterizing the momentum exchange by the vertical turbulence. In the model tests conducted for the Diablo Canyon Power Plant, the increase in the bottom roughness was observed to have a negative effect on mixing. The intensification of the vertical turbulence due to the roughness had produced an opposite effect on the horizontal turbulence. The observation is contrary to common sense but is fundamental in the concept of mixing across the natural streams.

In many previous studies of the mixing processes in rivers, estuaries and oceans, longitudinal and transverse diffusion coefficients were introduced and semi-empirically determined by correlating the mixing coefficients with the current structures and the geometry of the streams (see Harleman, 1966, Okubo and Ozimidov, 1970, and Fisher 1979). The correlations were often conducted for the purpose of one-dimensional calculations as described in textbooks such as Thomann and Mueller (1987) and Chapra (1997). But natural streams are wide compared with its depth. Important transport processes in the streams are not generally describable by one-dimensional models. (See the examples shown in Figures 1, 2, 11, 17 and 21.) With the advance of computation methods, direct simulations of turbulence in the turbulent flow can be conducted for the transports of matters such as heat, salinity, ices, pollutants and sediments across the natural streams. The LES and RANS models of horizontal turbulence and vertical turbulence are developed to provide more realistic description and better understanding of these processes.

REFERENCES

Alavian, V. (1984) "The stability of turbulent exchange flows in shallow waters," M. Eng. Thesis, Department of Civil Engineering and Applied Mechanics, McGill University, Montreal, Canada.

Alavian, V. and Chu, V. H. (1985) "Turbulent exchange flow in a shallow compound channel," *Proc. 21th Congress of Int'l Assoc. Hydraulic Res.*, Vol.3, pp. 446-451.

Altai, W. and Chu, V. H. (2000) "Large eddy simulation by Lagrangian block method," *Computational Fluid Dynamics 2000*, N. Satofuka Ed., Springer-Verlag, New York, pp. 467-472.

Altai, W., Zhang, J.-B. and Chu, V. H. (1999) "Shallow turbulent flow simulation using two-length-scale model," *J. Engineering Mechanics,* Vol. 125, No. 7, pp. 780-788.

Babarutsi, S. (1991) *Modelling Quasi-two-dimensional Turbulent Shear Flow,* Ph. D. thesis, Department of Civil Engineering and Applied Mechanics, McGill University, Montreal, Canada.

Babarutsi, S. and Chu, V. H. (1991) "Dye-Concentration Distribution in Shallow Re-circulating Flows," *J. Hydr. Div.*, ASCE, Vol. 117, No. 5, pp. 643-659.

Babarutsi, S. and Chu, V. H. (1993) "Reynolds-stress model for quasi-two-dimensional turbulent shear flow," in *Engineering Turbulence Modelling and Experiments 2*, Rodi, W. and Martelli, F. Editors, Elsevier Science Publishers, pp.93-102.

Babarutsi, S., Ganoulis, J. and Chu, V. H. (1989) "Experimental Investigation of Shallow Recirculating Flows," *J. Hydr. Div.*, ASCE, Vol. 115, pp. 906-924.

Babarutsi, S. and Chu, V. H. (1998) "Modeling Transverse mixing layer in shallow open-channel flows," *J. Hydraulic Engineering*, ASCE, Vol. 124, No.7, pp. 718-727.

Chapra, S. C. (1997) *Surface Water-Quality Modeling*, McGraw Hill, New York, 844pp.

Chu, V. H. and Altai, W. (1998) "Generalized second moment method for advection and diffusion processes," *Proc. 2nd Int. Symp. Envir. Hydraulics*, Balkema, The Nethelands, pp. 357-362.

Chu, V. H. and Altai, W. (2000) "Lagrangian block method," *Computational Fluid Dynamics 2000*, N. Satofuka Ed., Springer-Verlag, New York, pp. 771-771.

Chu, V. H. and Altai, W. (2001) "Simulation of Shallow Transverse Shear Flow by Generalized Second Moment Method," *J. Hydr. Res.* , Vol. 39, No. 6, pp. 575-582.

Chu, V. H. and Babarutsi, S. (1989) "Modelling of the turbulent mixing layers in shallow open channel flows," *Proc. 23rd Congress of Int'l Assoc. Hydr. Res.*, Vol. A, pp. 191-198.

Chu, V.H. and Babarutsi, S. (1988) "Confinement and bed-friction effects in shallow turbulent mixing layers," *J. Hydraulic Engineering*, Vol. 114, pp. 1257-1274.

Chu, V. H. and Baddour, R. E. (1984) "Turbulent gravity-stratified shear flow," *J. Fluid Mech.*, Vol. 138, pp. 353-378.

Chu, V. H., Wu, J.-H. and Khayat, R. E. (1991) "Stability of transverse shear flows in shallow open channels," *J. Hydraulic Engineering*, Vol. 117, No.10, pp. 1370-1388.

Chu, V. H., Gehr, R. and Leduc, R. (1991) "Dilution of MUC wastewater treatment plant effluent in the St. Lawrence River," *Proc. of the 1991 CSCE Annual Conference*, Vancouver, British Columbia, Vol. 4, pp. 239-248.

Fischer, et al. (1979) *Mixing in Inland and Coastal Waters*, chapter 5, Academic Press, pp. 104-147.

Garvine, R. W. (1974) "Frontal structure of a river plume," *J. Geophysical Res.*, Vol. 79, pp. 2251-2259.

Harden, T. O. and Shen, H. T. (1979) "Numerical simulation of mixing in rivers," *J. of Hydraulics Div.*, Vol. 104, No. HY4, pp. 393-408.

Harleman, D. R. F. (1966) "Pollution in Estuaries," chapter 14, in *Estuary and Coastal Hydrodynamics*, A. T. Ippen edited, McGraw Hill, New York, pp. 630-647.

Holly, E. R. (1996) *Diffusion and Dispersion*, in chapter 4, *Environmental Hydraulics*, V. P. Singh and W. H. Hager editors, pp. 111-151.

Hossain, W. S. (1980) "Mathematische Modellierung von turbulenten Auftriebsströmungen," Ph.D. Thesis, University of Karlsruhe, West Germany.

Lau, Y. L. and Krishnappan, B. G. (1981) "Modelling transverse mixing in natural streams," *J. of Hydraulics Div.*, ASCE, Vol. 107, No. HY2, pp. 209-226.

Ingram, R. G. and Chu, V. H. (1987) "Flow around islands in Rupert Bay: An investigation of the bed-friction effect," *J. Geophy. Res. Ocean*, Vol. 92, No. C13, pp. 14521-14534.

Kraichnan, R. H. (1967) "Inertial range in two-dimensional turbulence," *Phys. of Fluids*, Vol. 10, No. 7, pp. 1417-1423.

Mertz, G. and Gratton, Y. (1990) "Topographic waves and topographically induced motions in the St. Lawrence Estuary," *Coastal and Estuarine Studies*, Vol. 39 M. I. El-Sabh and N. Silverberg Eds., Oceanography of a Large-Scale Estuarine System - The St. Lawrence, Chapter 5, pp. 94-108, Springer-Verlag, New York.

Mertz, G., Gratton, Y. and Gagne, J. A. (1990) "Properties of unstable waves in the Lower St. Lawrence Estuary," *Atmosphere-Ocean*, Vol. 28, pp. 230-240.

Nassiri, M. (1999) *Two-dimensional Simulation Models of Shallow Re-circulating Flows*, Ph. D. thesis, Department of Civil Engineering and Applied Mechanics, McGill University, Montreal, Canada H3A 2K6.

Noke, R. I. and Wood, I. R. (1988) "Vertical and lateral dispersion: some experimental results," *J. Fluid Mech.*, Vol. 187, pp. 373-394.

Patankar, S. V. and Spalding, D. B. (1970) *Heat and Mass Transfer in Boundary Layers*, Intertext Publishers, London, England.

Okubo, A. (1968) "Some remarks on the importance of the shear effect on horizontal diffusion," *J. Ocean Soc. Japan*, Vol. 24, pp. 60-69.

Okubo, A. and Ozimidov, R. V. (1970) "Empirical dependence of the horizontal diffusion coefficient in the ocean on the scale of phenomenon question," *Izv. Atmos. Oceanic Phys.*, Vol. 6, pp. 308-309.

Pingree, R. D. (1978) "Cyclonic eddies and cross frontal mixing," *J. Mar. Biol. Assoc. U. K.*, Vol. 58, pp. 955-963.

Pingree, R. D. (1979) "Baroclinic eddies bordering the Celtic Sea in late summer," *J. Mar. Biol. Assoc. U. K.*, Vol. 59, pp. 689-698.

Rodi, W. (1979) *Turbulent models and their application in hydraulics - a state of the art review*, IAHR publications, 104 pp.

Rodi, W. (1985) "Calculation of stably stratified shear-layer flows with a buoyancy-extended κ-ε turbulence model," in *Turbulence and Diffusion in Stable Environments*, Edited by J. C. R. Hunt, Clarendon Press, Oxford.

Rodi, W. (1987) "Examples of calculation methods for flow and mixing in stratified fluids," *J. Geophy. Res. Oceans*, Vol. 92, no. C5, pp. 5305-5328.

Rastogi, A. K. and Rodi, W. (1978) "Predictions of Heat and Mass Transfer in Open Channels," *J. Hydraulic Engineering*, Vol. 104, pp.397-420.

Ryan, P. J., Tu, S. W., Ismail, N. M. and Wiegel, R. L. (1987) "Physical model verification of a coastal discharge," *Proc. of the 1987 ASCE National Conference on Hydraulic Engineering*, Williamsburg, pp. 1106-1111.

Schmidt, J. C., Gram, P.E., and Leschin, M. F. (1999) "Variation in the magnitude and style of deposition and erosion in three long (8-12 km) reaches as determined by photographic analysis," in *The Controlled Flood in Grand Canyon*, Geophysical Monographs 110, American Geophysical Union, Washington D. C.

Simpson, J. H. (1981) "The shelf sea fronts: Implications of their existence and behaviour," *Philos. Trans. R. Soc. London*, Ser. A, Vol. 302, pp. 531-546.

Thomann, R. V. and Mueller, J. A. (1987) *Principles of Surface Water Quality Modeling and Control*, Harper & Row, Cambridge, 644 pp.

Turner, J. S. (1973) *Buoyancy Effects in Fluids*, Cambridge University Press.

Uijttewaal, W.S.J. and Tukker, J. (1998) "Development of quasi two-dimensional structures in a shallow free-surface mixing layer," *Exp. Fluids*, Vol. 24, pp. 192-200.

Wiele, S. M., Andrews, E.D., and Griffin, E. R. (1999) "The effect of sand concentration on deposition rate, magnitude, and location in the Colorado River below the Little Colorado River," in *The Controlled Flood in Grand Canyon*, Geophysical Monographs 110, American Geophysical Union, Washington D. C.

Yotsukura, N., and Sayre, W. W. (1976) "Transverse mixing in natural stream," *Water Resource Res.*, Vol. 12, pp. 695-704.

Zhang, J.-B. (1996) *Turbulence Measurements in Shallow Shear Flow using Video Imaging Method*, Ph.D. thesis, Department of Civil Engineering and Applied Mechanics, Montreal, Canada,183 pp.

Chapter 6

THREE-DIMENSIONAL HYDRODYNAMIC AND SALINITY TRANSPORT MODELING IN ESTUARIES

Keh-Han Wang[1]

ABSTRACT

Hydrodynamic and salinity transport modeling is important to the study of an estuary's dynamic environment. This chapter presents a comprehensive overview of the development of a three-dimensional hydrodynamic and salinity transport model. The fundamental equations that govern fluid flows, movement of free surface, and salinity/temperature transport are formulated. In addition, coordinate transformation techniques and applications of the boundary conditions for estuary modeling are described in detail. Model equations are solved by using an implicit, finite-difference scheme. A widely used mode-splitting technique is introduced to simplify the numerical procedure of computing three-dimensional velocity components. Description of a second-order turbulence closure model for the determination of the vertical eddy and mixing coefficients is provided. Model application to the hydrodynamic studies of selected estuaries is presented. The effects of wind stress, fresh water inflow and tidal oscillation on the flow circulation and salinity change are analyzed and discussed.

1. INTRODUCTION

The coastal zone and estuaries are regions receiving considerable attention as a result of increasing utilization of their resources. In addition to the effects of a number of physical mechanisms, social, ecological, and economic issues make the study of an estuarine system even more complicated. Estuaries can be a nursery and adult habitat for commercial fishery species, provide flood control for coastal communities, support pollution control for the surrounding coastal environments, serve as a transportation route, and be a recreation area. Occasionally, oil spills, thermal discharges, and transport of polluted material from point/nonpoint sources, coastal dredging and offshore drilling may significantly affect the natural system. A better understanding of the basic flow circulation and associated transport mechanism in estuaries becomes essential to assist in maintaining a rich

[1] Associate Professor, Department of Civil and Environmental Engineering, University of Houston, Houston, TX 77204-4791, e-mail:khwang@uh.edu.

and healthy ecosystem and preventing the further deterioration of the environment. With knowledge of fluid movement and mixing capability, predictions of the transport of pollutants and salinity can be carried out and the water quality issue can be addressed. The environmental impact resulting from any physical changes of the system can also be studied. Rational planning and management decisions can then be established after the evaluation of alternative methods for salinity and quality control.

The physical behavior of estuaries is a complex and dynamic process. It involves factors including astronomical tides, wind, freshwater inflows, return flows, waste loadings of all types, bathymetry, evaporation, rainfall, and runoff from adjacent watershed areas. With such a large number of factors and the complicated dynamic interactions between factors, the need for comprehensive analyses of the system is evident. Hydrodynamic, salinity, and pollutant transport modeling can provide solutions for the evaluation of estuarine and coastal systems. The models would define the salinity distribution and provide an economical way to explore relationships between circulation, water quality, and ecology using physical circulation patterns as an input.

Transport phenomenon in estuaries is a complicated process, involving various constituents. In estuaries that are not excessively polluted, salinity is a major component affecting the ecosystem and a major concern for preserving such estuaries. In this chapter, transports of salinity and temperature, which affect the density of the water, are addressed. For a large estuarine system, salt water and fresh water mix, creating a brackish habitat for organisms like oysters, shrimp, and coastal fish. Changes in salinity level and possible salt water intrusion may have substantial impact on the life cycle of fisheries and may directly affect water resources management. For example, it is widely known that oyster productivity is highly influenced by estuarine salinity (Butler 1954; Gunter and Menzel 1957; Menzel et al. 1966; Chatry et al. 1983). Oyster larvae prefer a salinity of less than 15 ppt (parts per thousand) while mature oysters can tolerate salinity between 15 to 25 ppt. Other common species (brown shrimp, white shrimp, bay anchovy, sand seatrout, redfish, etc.) also require low salinity for larval stages to survive.

Productivity is also associated with the exclusion of marine predators that are intolerant of low and fluctuating salinity. Predator encroachment into estuaries during dry periods when the salinity rises has been noticed in many areas adjacent to the Gulf of Mexico and in Chesapeake Bay. Studies have suggested that marine predators (e.g. oyster drill, crown conch, stone crab, etc.) cannot penetrate estuaries where the salinity is less than 15 ppt (St. Amant 1957, Manzi 1970). Therefore, the naturally balanced brackish system provides healthy habitats for fishery species. Freshwater diversion for inland use or widening and deepening of a ship channel may influence the physical characteristics of the estuary and may change the salinity distributions and the circulatory patterns. These changes may reduce the ability of an estuary for maintaining its healthy and natural functions.

Hydrodynamic and salinity transport modeling is important to the studies of an estuary's dynamic system. Numerous models have been developed over the past three decades, in which various hydrodynamic models as solutions of practical problems have been reported. Simplified two-dimensional models using alternating direction implicit (ADI) method (Shankar and Masch 1970, Reid and Whitaker 1981) have been employed at the early stage of model development to provide numerical solutions for studies of oceans and estuaries. More robust two-dimensional flow simulators, which are economically competitive with ADI methods, were later developed and applied. These methods include semi-implicit (Backhaus 1983, Casulli 1990) as well as fully implicit (Stelling et al. 1986, Wilders et al. 1988, Evans 1989) splitting methods. With the limitation of spatial dimensions, a two-dimensional model can produce only the free surface and the depth-averaged velocity components. For many applications, this information is insufficient to describe the actual variation of an estuary system; computation of velocity components as a function of vertical direction is required. The flow velocities as well as the transport and mixing phenomena are indeed three dimensional in most estuaries.

The development of a comprehensive three-dimensional hydrodynamic and salinity/temperature transport model was reported as early as 1973 by Leendertse et al. and later extended by Leendertse and Liu (1977). Since then, various three-dimensional hydrodynamic and transport models have been developed to fulfill required applications to water bodies of interest. Each model solves essentially the same flow equations and transport equations but uses different numerical methods and computational structures in either Cartesian or curvilinear coordinate systems. In terms of computational structures, equations may be solved either directly or by mode-splitting techniques. When mode splitting is used, care must be exercised to ensure the consistency of the physical quantities derived from the external (two-dimensional vertically integrated equations) and internal (three-dimensional equations) modes. Both finite-difference and finite-element models have been developed and applied to practical applications. Other critical components affecting the performance of a three-dimensional model include the implementation of a numerical scheme (either semi-implicit or fully implicit) and the development of a coupled turbulence closure model for the determination of vertical diffusion and mixing coefficients.

For finite-difference codes, selected three-dimensional models and literature, which describe model development and applications, are listed here.

Coordinate system	Models
Cartesian	Rand Model (Leendertse and Liu 1977) EHSM3D (Sheng 1983) TRIM_3D (Casulli and Cheng 1992, Casulli et al. 1993)
Orthogonal curvilinear and vertically stretched	ECOM and POM Models (Blumberg and Mellor 1987) ECOMsi (Blumberg et al. 1996) EFDC (Hamrick, J. 1993 and Hamrick and Wu 1997)
Non-orthogonal curvilinear and vertically stretched	CH3D (Sheng 1987, Johnson et al. 1993, Wang 1992) Huang and Spaulding's model (1994)

At the early stage of three-dimensional model development, Cheng and Smith (1989) also provided a detailed survey and comparison of various three-dimensional numerical estuarine models. As the application of three-dimensional hydrodynamic and transport models for estuarine and coastal studies becomes increasingly important, a series of national and international biennial conferences on estuarine modeling have been held during the past 10 years (Proceedings of Estuarine and Coastal Modeling 1989–1999). The development, application, calibration, validation, and visualization of predictions from estuarine and coastal models have been addressed and discussed in those conferences.

For the United States, a selected list of critical estuaries along the East, West and Gulf of Mexico coastlines have received extensive modeling and monitoring efforts since the establishment of National Estuary Program in 1987 (EPA 1992). Many scientific articles have emphasized the application of hydrodynamic and transport models in assisting the establishment of a better and comprehensive management plan for each estuary. Examples showing estuarine modeling studies include Chesapeake Bay (Kim et al. 1989, Wang 1992, Johnson et al. 1993, Hamrick 1993), San Francisco Bay (Smith and Cheng 1989, Smith 1994, Casulli and Cheng 1992), Galveston Bay (Wang 1994, Schmalz 1996), Tampa Bay (Hess 1994, Vincent et al. 1995), Delaware Bay (Kim et al. 1994), and San Diego Bay (Wang et al. 1998).

A hydrodynamic and transport model is a simulation tool that uses a numerical scheme to solve the basic flow and transport equations to predict fluid flow, salinity/temperature distribution, and concentration level in the study area. The models require appropriate initial and boundary conditions as described by the hydrodynamic quantities of tide, wind, and river inflows, as well as the transport quantities of salinity, temperature and concentration. This chapter provides a detailed description of a finite-difference numerical circulation and salinity/temperature transport model. Since the coordinate system selected for the model's development is the generalized curvilinear coordinate system, the model is a curvilinear three-dimensional multi-layered estuarine and coastal model. A turbulence closure model is also introduced and incorporated with the flow and transport modules to provide a realistic parameterization of the vertical mixing processes. Model equations are the usual Reynolds-averaged Navier-Stokes (RANS) equations, and salinity and temperature transport equations. The flow equations include the nonlinear, diffusion, baroclinic and coriolis terms, where the baroclinic effect is calculated by integrating horizontal density gradients from the surface to the bottom that provides a more rigorous definition of the density throughout the system. The boundary conditions due to the external forces of wind, tides, and freshwater inflows are formulated and discussed. The variables solved include three-dimensional flow velocities, free-surface elevation, salinity, and temperature. Vertical eddy and mixing coefficients are calculated based on a simplified second-order turbulence closure model. A typical mode-splitting technique is introduced in this chapter to allow the free-surface elevation (including the vertically averaged velocity components) to be calculated separately from the three-dimensional velocity field.

This model's performance has been tested in two major estuaries: Chesapeake Bay and Galveston Bay. Results showing the curvilinear coordinates, the time variation of the velocity field, free-surface elevation, and the salinity distribution are presented and discussed.

2. BASIC EQUATIONS

The basic equations that govern the three-dimensional flow field are the Reynolds-averaged Navier-Stokes equations. Employing the eddy-viscosity concept and including the Coriolis effect, the continuity equation and momentum equations are written in the Cartesian coordinates as

$$\frac{\partial u}{\partial x}+\frac{\partial v}{\partial y}+\frac{\partial w}{\partial z}=0, \tag{1}$$

$$\frac{\partial u}{\partial t}+\frac{\partial uu}{\partial x}+\frac{\partial uv}{\partial y}+\frac{\partial uw}{\partial z}=fv-\frac{1}{\rho}\frac{\partial p}{\partial x}+\frac{\partial}{\partial x}(\upsilon_H\frac{\partial u}{\partial x})+\frac{\partial}{\partial y}(\upsilon_H\frac{\partial u}{\partial y})+\frac{\partial}{\partial z}(\upsilon_v\frac{\partial u}{\partial z}), \tag{2}$$

$$\frac{\partial v}{\partial t}+\frac{\partial uv}{\partial x}+\frac{\partial vv}{\partial y}+\frac{\partial vw}{\partial z}=-fu-\frac{1}{\rho}\frac{\partial p}{\partial y}+\frac{\partial}{\partial x}(\upsilon_H\frac{\partial v}{\partial x})+\frac{\partial}{\partial y}(\upsilon_H\frac{\partial v}{\partial y})+\frac{\partial}{\partial z}(\upsilon_v\frac{\partial v}{\partial z}), \tag{3}$$

$$\frac{\partial w}{\partial t}+\frac{\partial uw}{\partial x}+\frac{\partial vw}{\partial y}+\frac{\partial ww}{\partial z}=-\frac{1}{\rho}\frac{\partial p}{\partial z}-g+\frac{\partial}{\partial x}(\upsilon_H\frac{\partial w}{\partial x})+\frac{\partial}{\partial y}(\upsilon_H\frac{\partial w}{\partial y})+\frac{\partial}{\partial z}(\upsilon_v\frac{\partial w}{\partial z}). \tag{4}$$

For large-scale flow simulation, it is reasonable to assume that vertical acceleration is negligible compared to the vertical pressure gradient (Boussinesq assumption). Therefore, the vertical momentum equation (Eq. 4) after simplification becomes

$$-\frac{1}{\rho}\frac{\partial p}{\partial z}=g. \tag{5}$$

In the equations described above, x and y are the horizontal coordinates and z points vertically upward. The notations u, v and w are the velocity components along the x, y and z directions, respectively (See Fig. 1); ρ = fluid density; and g = gravitational acceleration. The Coriolis terms (fv in Eq. (2) and $-fu$ in Eq. (3)) are included in the basic equations to reflect the effect of Earth's rotation on fluid motion. For a large-scale simulation in a domain with long wave motion, the Coriolis acceleration becomes either significant or equivalently important, as the wave frequency is of the same order as f. Typically, the Coriolis acceleration can produce significant effects in tidal waves. The Coriolis parameter, f, is defined as $2\ \Omega \sin \phi$, where ϕ is the Earth's latitude measured positive and negative in the northern and southern hemispheres, respectively, and Ω is the rotational speed of Earth, $\Omega = 7.292 \times 10^{-5}\ s^{-1}$. The vertical and horizontal turbulent eddy coefficients are denoted by υ_V and υ_H, which may vary in time and space. It is also assumed that the only effect of the density variation is on the gravitational body force term. The pressure at depth z can be obtained by integrating equation (5) from z to free-surface ζ, and is given as

$$p = \rho_0 g\zeta + g\int_z^0 \rho(x,y,z',t)dz' , \tag{6}$$

where ρ_0 is a reference fluid density.

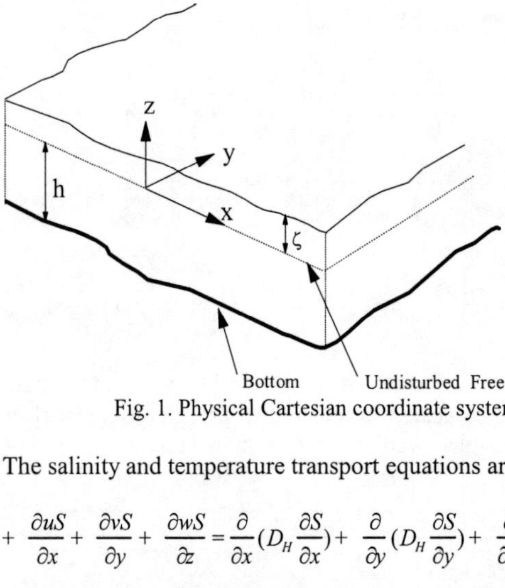

Fig. 1. Physical Cartesian coordinate system.

The salinity and temperature transport equations are expressed as

$$\frac{\partial S}{\partial t} + \frac{\partial uS}{\partial x} + \frac{\partial vS}{\partial y} + \frac{\partial wS}{\partial z} = \frac{\partial}{\partial x}(D_H\frac{\partial S}{\partial x}) + \frac{\partial}{\partial y}(D_H\frac{\partial S}{\partial y}) + \frac{\partial}{\partial z}(D_v\frac{\partial S}{\partial z})], \tag{7}$$

$$\frac{\partial T}{\partial t} + \frac{\partial uT}{\partial x} + \frac{\partial vT}{\partial y} + \frac{\partial wT}{\partial z} = \frac{\partial}{\partial x}(D_H\frac{\partial T}{\partial x}) + \frac{\partial}{\partial y}(D_H\frac{\partial T}{\partial y}) + \frac{\partial}{\partial z}(D_v\frac{\partial T}{\partial z})], \tag{8}$$

where S = salinity and T = temperature. D_H is the horizontal diffusion coefficient and D_v represents the vertical turbulent mixing coefficient, which can be evaluated based on a turbulence closure model. A similar transport equation can be formulated for modeling transport of pollutants. Since density is a function of salinity and temperature, various forms of equations of state can be used in the model. For example, Eckart (1958) described the following to relate density with the salinity and temperature

$$\rho = \frac{5890 + 38T - 0.375T^2 + 3S}{5890.72 + 37.774T - 0.33625T^2 - 1.706S - 0.1TS} \quad . \tag{9}$$

3. COORDINATE TRANSFORMATION AND TRANSFORMED EQUATIONS

In actuality, most flow domains are irregular and water depths are non-uniform. The irregularity of the flow regions along the horizontal and vertical directions makes the numerical approach a challenging task. For the horizontal domain, the irregular boundary complicates the application of the boundary condition. The numbers of grid points also need to increase considerably to be able to model the actual physical boundary, which will substantially increase the computational time. In the vertical direction, flow equations can be formulated along the standard z plane. Then, the unequal vertical grid cells in deep and shallow waters require extensive interpolations to exchange quantities of physical variables around grid points throughout the computation. All the difficulties resulting from the irregularity of the study domain may be resolved by using coordinate transformations.

3.1 Vertical Coordinate Transformation

A vertically stretched coordinate or so-called σ-grid (Phillips 1957, Sheng 1983), namely

$$\sigma = \frac{z - \zeta}{H}, \quad H = h + \zeta, \tag{10}$$

may be used to resolve bathymetry gradually with an equal number of vertical grid points in deep and shallow waters (Fig. 2). H represents the total water depth, h is the undisturbed water depth. The transformed coordinate σ ranges from $\sigma = 0$ at $z = \zeta(x,y,t)$ to $\sigma = -1$ at $z = -h(x,y)$. The relationships linking derivatives in the old coordinate system (x,y,z,t) to those in the new system (x', y', σ, t') with $x' = x$, $y' = y$, and $t' = t$ can be obtained by applying the chain rules of partial derivative. After omitting the prime notation used in the new coordinate system, we have

$$\begin{aligned} \frac{\partial G}{\partial x} &= \frac{\partial G}{\partial x} - \frac{(1+\sigma)}{H}\frac{\partial \zeta}{\partial x} - \frac{\sigma}{H}\frac{\partial h}{\partial x} \\ \frac{\partial G}{\partial y} &= \frac{\partial G}{\partial y} - \frac{(1+\sigma)}{H}\frac{\partial \zeta}{\partial y} - \frac{\sigma}{H}\frac{\partial h}{\partial y} \\ \frac{\partial G}{\partial t} &= \frac{\partial G}{\partial t} - \frac{(1+\sigma)}{H}\frac{\partial \zeta}{\partial t} \end{aligned} \tag{11}$$

$$\frac{\partial G}{\partial z} = \frac{1}{H}\frac{\partial G}{\partial \sigma}.$$

Here, G is an arbitrary field variable. For the stretched vertical coordinate, a new vertical velocity ω can be defined as

$$\omega = \frac{D\sigma}{dt} = \frac{1}{H}[w - (1+\sigma)\frac{D\zeta}{dt} - \sigma\,(u\frac{\partial h}{\partial x} + v\frac{\partial h}{\partial y})], \tag{12}$$

where $D(\ )/dt$ represents a total derivative. It should be noted that with the application of the kinematic free-surface and bottom boundary conditions,

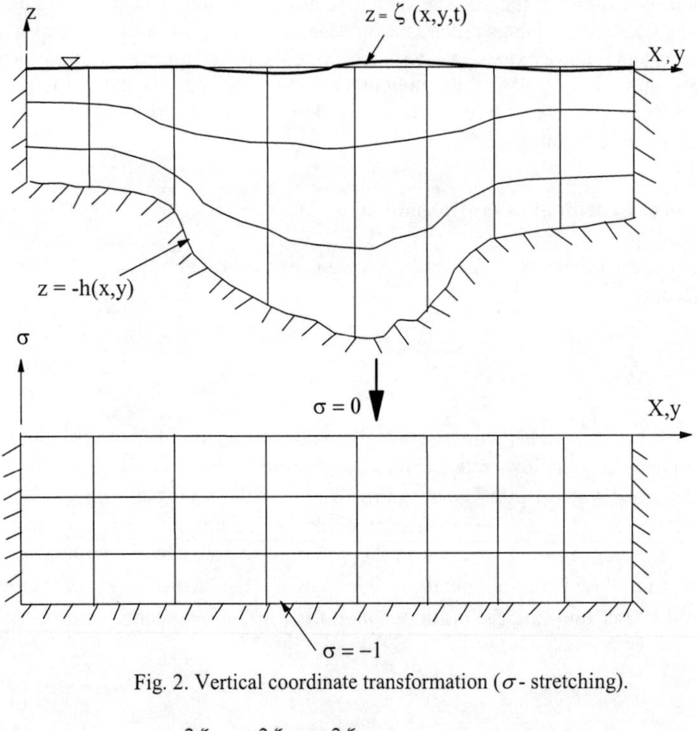

Fig. 2. Vertical coordinate transformation (σ - stretching).

$$\frac{\partial \zeta}{\partial t} + u\,\frac{\partial \zeta}{\partial x} + v\,\frac{\partial \zeta}{\partial y} = w \quad \text{at } z = \zeta, \tag{13}$$

$$u\frac{\partial h}{\partial x} + v\frac{\partial h}{\partial y} = -\,w \qquad at\ z = -h, \tag{14}$$

the vertical velocity ω satisfies

$$\omega\ (x,y,0,t) = 0, \tag{15a}$$

$$\omega\ (x,y,-1,t) = 0. \tag{15b}$$

Using Eqs. (11) and (12), the basic equations (1)-(3) can be reduced to

$$\frac{\partial \zeta}{\partial t} + \frac{\partial Hu}{\partial x} + \frac{\partial Hv}{\partial y} + \frac{\partial H\omega}{\partial \sigma} = 0\ , \tag{16}$$

$$\begin{aligned}\frac{\partial Hu}{\partial t} + \frac{\partial Huu}{\partial x} + \frac{\partial Huv}{\partial y} + \frac{\partial Hu\omega}{\partial \sigma} &= H\ f\ v - H\frac{1}{\rho_0}\frac{\partial p}{\partial x} \\ &+ H\,[\frac{\partial}{\partial x}(\upsilon_H \frac{\partial u}{\partial x}) + \frac{\partial}{\partial y}(\upsilon_H \frac{\partial u}{\partial y}) + \frac{\partial}{\partial z}(\upsilon_v \frac{\partial u}{\partial z})]\end{aligned} \tag{17}$$

$$\begin{aligned}\frac{\partial Hv}{\partial t} + \frac{\partial Huv}{\partial x} + \frac{\partial Hvv}{\partial y} + \frac{\partial Hv\omega}{\partial \sigma} &= -H\ f\ u - H\frac{1}{\rho_0}\frac{\partial p}{\partial y} \\ &+ H\,[\frac{\partial}{\partial x}(\upsilon_H \frac{\partial v}{\partial x}) + \frac{\partial}{\partial y}(\upsilon_H \frac{\partial v}{\partial y}) + \frac{\partial}{\partial z}(\upsilon_v \frac{\partial v}{\partial z})]\end{aligned} \tag{18}$$

Following Eq. (6), the pressure gradient along the *x* and *y* directions can be expressed in terms of the gradient of free-surface elevation and the effect of density gradients as

$$\frac{\partial p}{\partial x} = \rho_0 g \zeta_x + gH\int_\sigma^0 \rho_x d\sigma + gH_x[\int_\sigma^0 \rho d\sigma + \rho\sigma], \tag{19}$$

$$\frac{\partial p}{\partial y} = \rho_0 g \zeta_y + gH\int_\sigma^0 \rho_y d\sigma + gH_y[\int_\sigma^0 \rho d\sigma + \rho\sigma]. \tag{20}$$

The first term of the right-hand sides of Eqs. (19) and (20) represents the barotropic effect, whereas the second and the third terms reflect the baroclinic effect due to density gradients.

3.2 Curvilinear Coordinate Transformation and Contravariant Velocities

For the horizontal domain, a curvilinear coordinate transformation technique (Thompson and Johnson 1985) can be used to generate curvilinear grids to best represent the geometrically complex water bodies. One advantage of adopting the curvilinear coordinate transformation is that the number of curvilinear grid points is generally much less than that of the Cartesian grid system having the same level of representation of the fluid domain. This coordinate transformation is performed by solving a set of Poisson's equations which relate the curvilinear coordinates in the physical plane, x and y, with the uniformly spaced coordinates in a transformed plane ξ and η (Fig. 3). To perform the computation in the transformed plane, the curvilinear-coordinate based hydrodynamic and transport equations need to be formulated.

Along irregular fluid-solid interfaces, the application of the boundary conditions is complicated and potential errors may be introduced. To improve the numerical procedure of modeling the boundary conditions at the lateral fluid-solid interfaces or open boundaries, contravariant velocities may be used. The physical velocity components u and v along the x and y directions, respectively, can be transformed into contravariant velocities q_1 and q_2 using

$$q_1 = \xi_x u + \xi_y v = \frac{y_\eta}{J} u - \frac{x_\eta}{J} v, \tag{21}$$

and

$$q_2 = \eta_x u + \eta_y v = \frac{-y_\xi}{J} u + \frac{x_\xi}{J} v. \tag{22}$$

where q_1 and q_2 are perpendicular to the η and ξ lines, respectively, as shown in Fig. 3. The subscripts of ξ_x, ξ_y, η_x, η_y, x_ξ, and y_ξ denote partial differentiation.

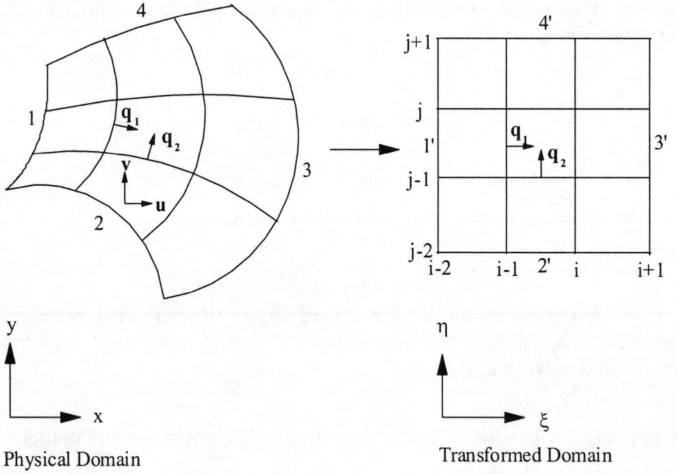

Fig. 3. Curvilinear coordinate system and contravariant velocity transformation.

3.3 Transformed Equations

Applying the above described coordinate and velocity transformations, the three-dimensional equations of motion written in terms of the horizontal contravariant velocities (q_1, q_2) and vertical velocity ω in the transformed coordinates (ξ, η, σ) are formulated from Eqs.(16) -(18) as

$$\frac{\partial \zeta}{\partial t} + \frac{1}{J}\left[\frac{\partial}{\partial \xi}(JHq_1) + \frac{\partial}{\partial \eta}(JHq_2)\right] + H\frac{\partial \omega}{\partial \sigma} = 0, \tag{23}$$

$$\begin{aligned}
\frac{1}{H}\frac{\partial Hq_1}{\partial t} = &- g\left(g^{11}\frac{\partial \zeta}{\partial \xi} + g^{12}\frac{\partial \zeta}{\partial \eta}\right) + \frac{g_{12}}{J} f q_1 + \frac{g_{22}}{J} f q_2 \\
&- \frac{1}{HJ^2}\left\{ y_\eta \left[\frac{\partial}{\partial \xi}\left(x_\xi HJq_1^2 + x_\eta HJq_1q_2\right) + \frac{\partial}{\partial \eta}\left(x_\xi HJq_1q_2 + x_\eta HJq_2^2\right)\right]\right. \\
&\left. - x_\eta \left[\frac{\partial}{\partial \xi}\left(y_\xi HJq_1^2 + y_\eta HJq_1q_2\right) + \frac{\partial}{\partial \eta}\left(y_\xi HJq_1q_2 + y_\eta HJq_2^2\right)\right] + J^2\frac{\partial Hq_1\omega}{\partial \sigma}\right\} \\
&- \frac{g}{\rho_0}\left[H\int_\sigma^0\left(g^{11}\frac{\partial \rho}{\partial \xi} + g^{12}\frac{\partial \rho}{\partial \eta}\right)d\sigma + \left(g^{11}\frac{\partial H}{\partial \xi} + g^{12}\frac{\partial H}{\partial \eta}\right)\left(\int_\sigma^0 \rho d\sigma + \sigma\rho\right)\right] \\
&+ \frac{1}{H^2}\frac{\partial}{\partial \sigma}\left(\upsilon_v \frac{\partial q_1}{\partial \sigma}\right) + \upsilon_H \text{ (Diffx)},
\end{aligned} \tag{24}$$

$$\begin{aligned}
\frac{1}{H}\frac{\partial Hq_2}{\partial t} = &- g\left(g^{21}\frac{\partial \zeta}{\partial \xi} + g^{22}\frac{\partial \zeta}{\partial \eta}\right) - \frac{g_{11}}{J} f q_1 - \frac{g_{21}}{J} f q_2 \\
&- \frac{1}{HJ^2}\left\{ - y_\xi \left[\frac{\partial}{\partial \xi}\left(x_\xi HJq_1^2 + x_\eta HJq_1q_2\right) + \frac{\partial}{\partial \eta}\left(x_\xi HJq_1q_2 + x_\eta HJq_2^2\right)\right]\right. \\
&\left. + x_\xi \left[\frac{\partial}{\partial \xi}\left(y_\xi HJq_1^2 + y_\eta HJq_1q_2\right) + \frac{\partial}{\partial \eta}\left(y_\xi HJq_1q_2 + y_\eta HJq_2^2\right)\right] + J^2\frac{\partial Hq_2\omega}{\partial \sigma}\right\} \\
&- \frac{g}{\rho_0}\left[H\int_\sigma^0\left(g^{21}\frac{\partial \rho}{\partial \xi} + g^{22}\frac{\partial \rho}{\partial \eta}\right)d\sigma + \left(g^{21}\frac{\partial H}{\partial \xi} + g^{22}\frac{\partial H}{\partial \eta}\right)\left(\int_\sigma^0 \rho d\sigma + \sigma\rho\right)\right] \\
&+ \frac{1}{H^2}\frac{\partial}{\partial \sigma}\left(\upsilon_v \frac{\partial q_2}{\partial \sigma}\right) + \upsilon_H \text{ (Diffy)},
\end{aligned} \tag{25}$$

Here, $J = x_\xi y_\eta - x_\eta y_\xi$ denotes the Jacobian of transformation, g_{ij} is the metric tensors, and g^{ij} denotes the conjugate metric tensors. The expressions g_{ij} and g^{ij} are summarized in the Appendix. The complete horizontal diffusion terms, Diffx and Diffy, in Eq. (24) and Eq. (25), respectively, are also shown in the

Appendix. The contravariant velocities can be calculated from the above equations and then transformed back to the physical velocity components for analyses.

The computation of the unknown ζ may be decoupled from the three-dimensional equations by a mode splitting technique. This technique permits the external free-surface elevation to be calculated from the vertically integrated equations. Once the free-surface elevation ζ is determined, the horizontal velocity field can be evaluated from the horizontal momentum equations (24) and (25). The vertical velocity w along the z direction can then be computed from

$$w = H\omega + (1+\sigma)\frac{D\zeta}{Dt} + \sigma\left(q_1\frac{\partial h}{\partial \xi} + q_2\frac{\partial h}{\partial \eta}\right) \quad , \tag{26}$$

where ω is determined from Eq. (23) as

$$\omega = -\frac{1+\sigma}{H}\frac{\partial \zeta}{\partial t} - \frac{1}{H}\int_{-1}^{\sigma}\frac{1}{J}\left[\frac{\partial}{\partial \xi}(JHq_1) + \frac{\partial}{\partial \eta}(JHq_2)\right]d\sigma . \tag{27}$$

The vertically-integrated equations are obtained by integrating the three-dimensional continuity and momentum equations (Eqs. (23)-(25)) from bottom to the free surface. Those equations are given as

$$\frac{\partial \zeta}{\partial t} + \frac{1}{J}\left[\frac{\partial}{\partial \xi}(JU) + \frac{\partial}{\partial \eta}(JV)\right] = 0, \tag{28}$$

$$\begin{aligned}
\frac{\partial U}{\partial t} = & -Hg\left(g^{11}\frac{\partial \zeta}{\partial \xi} + g^{12}\frac{\partial \zeta}{\partial \eta}\right) + \frac{g_{12}}{J}fU + \frac{g_{22}}{J}fV + \tau_{\xi s} - \tau_{\xi b} \\
& - \int_{-1}^{0}\frac{1}{HJ^2}\left\{y_\eta\left[\frac{\partial}{\partial \xi}\left(x_\xi HJq_1^2 + x_\eta HJq_1q_2\right) + \frac{\partial}{\partial \eta}\left(x_\xi HJq_1q_2 + x_\eta HJq_2^2\right)\right]\right. \\
& \left. - x_\eta\left[\frac{\partial}{\partial \xi}\left(y_\xi HJq_1^2 + y_\eta HJq_1q_2\right) + \frac{\partial}{\partial \eta}\left(y_\xi HJq_1q_2 + y_\eta HJq_2^2\right)\right] + J^2\frac{\partial Hq_1\omega}{\partial \sigma}\right\}d\sigma \\
& - \frac{gH}{\rho_0}\int_{-1}^{0}\left[H\int_{\sigma}^{0}\left(g^{11}\frac{\partial \rho}{\partial \xi} + g^{12}\frac{\partial \rho}{\partial \eta}\right)d\sigma + \left(g^{11}\frac{\partial H}{\partial \xi} + g^{12}\frac{\partial H}{\partial \eta}\right)\left(\int_{\sigma}^{0}\rho d\sigma + \sigma\rho\right)\right]d\sigma \\
& + \int_{-1}^{0} H\upsilon_H\,(\text{Diffx})\,d\sigma,
\end{aligned} \tag{29}$$

$$\frac{\partial V}{\partial t} = -Hg\left(g^{21}\frac{\partial \zeta}{\partial \xi} + g^{22}\frac{\partial \zeta}{\partial \eta}\right) - \frac{g_{11}}{J}fU - \frac{g_{21}}{J}fV + \tau_{\eta s} - \tau_{\eta b}$$

$$-\int_{-1}^{0} \frac{1}{HJ^2}\left\{ -y_\xi \left[\frac{\partial}{\partial \xi}\left(x_\xi HJq_1^2 + x_\eta HJq_1q_2\right) + \frac{\partial}{\partial \eta}\left(x_\xi HJq_1q_2 + x_\eta HJq_2^2\right)\right] \right.$$

$$\left. + x_\xi \left[\frac{\partial}{\partial \xi}\left(y_\xi HJq_1^2 + y_\eta HJq_1q_2\right) + \frac{\partial}{\partial \eta}\left(y_\xi HJq_1q_2 + y_\eta HJq_2^2\right)\right] + J^2 \frac{\partial Hq_2\omega}{\partial \sigma} \right\} d\sigma$$

$$-\frac{gH}{\rho_0}\int_{-1}^{0}\left[H\int_{\sigma}^{0}(g^{21}\frac{\partial \rho}{\partial \xi} + g^{22}\frac{\partial \rho}{\partial \eta})d\sigma + (g^{21}\frac{\partial H}{\partial \xi} + g^{22}\frac{\partial H}{\partial \eta})(\int_{\sigma}^{0}\rho d\sigma + \sigma\rho)\right] d\sigma$$

$$+ \int_{-1}^{0} H\upsilon_H \,(\text{Diffy})\, d\sigma, \qquad (30)$$

where $\tau_{\xi s}$ and $\tau_{\eta s}$ are the shear stresses acting on the free surface along the ξ and η directions, respectively. $\tau_{\xi b}$ and $\tau_{\eta b}$ represent the bottom stresses. The vertically-integrated velocities U and V are expressed as

$$(U,V) = H\int_{-1}^{0}(q_1, q_2)d\sigma \qquad (31)$$

The free-surface elevation and vertically integrated velocities are determined from Eqs. (28)-(30).

The salinity and temperature transport equations in the transformed coordinate system can be expressed as

$$\frac{1}{H}\frac{\partial HS}{\partial t} = \frac{1}{H^2}\frac{\partial}{\partial \sigma}\left(D_v \frac{\partial S}{\partial \sigma}\right) - \frac{1}{HJ}\left[\frac{\partial}{\partial \xi}\left(JHq_1 S\right) + \frac{\partial}{\partial \eta}\left(JHq_2 S\right)\right]$$

$$-\frac{1}{H}\frac{\partial H\omega S}{\partial \sigma} + D_H\left[g^{11}\frac{\partial^2 S}{\partial \xi^2} + 2g^{12}\frac{\partial^2 S}{\partial \xi \partial \eta} + g^{22}\frac{\partial^2 S}{\partial \eta^2}\right], \qquad (32)$$

$$\frac{1}{H}\frac{\partial HT}{\partial t} = \frac{1}{H^2}\frac{\partial}{\partial \sigma}\left(D_v \frac{\partial T}{\partial \sigma}\right) - \frac{1}{HJ}\left[\frac{\partial}{\partial \xi}(JHq_1 T) + \frac{\partial}{\partial \eta}(JHq_2 T)\right]$$

$$-\frac{1}{H}\frac{\partial H\omega T}{\partial \sigma} + D_H\left[g^{11}\frac{\partial^2 T}{\partial \xi^2} + 2g^{12}\frac{\partial^2 T}{\partial \xi \partial \eta} + g^{22}\frac{\partial^2 T}{\partial \eta^2}\right]. \qquad (33)$$

The temporal and spatial variations of salinity and temperature can be simulated by solving the above two equations. The density can then be updated using the equation of state (9). This approach allows the effects of salinity and temperature to be applied to the computation of three-dimensional fluid velocities, as the baroclinic terms are recalculated basing on the newly updated density distribution. The transport equations are coupled with the three-dimensional flow equations.

4. BOUNDARY CONDITIONS

4.1 Boundary Conditions at the Free Surface

For flow field computation, appropriate free-surface boundary conditions need to be specified to model forces applied on the surface. Considering wind effect, the boundary conditions at the free surface are wind stresses calculated by means of the following quadratic stress law

$$\rho_0 \upsilon_v \left(\frac{\partial u}{\partial z}, \frac{\partial v}{\partial z}\right) = (\tau_{sx}, \tau_{sy}) = \rho_a C_{da} (u_w^2 + v_w^2)^{1/2} (u_w, v_w) , \quad (34)$$

where τ_{sx} and τ_{sy} are the wind stresses along the x and y directions respectively, ρ_a is the air density, C_{da} is the drag coefficient, u_w and v_w are wind velocity components at a certain height (e.g. 10 m) above the water surface. Using Garratt's formula (Garratt, 1977), C_{da} can be estimated as

$$. C_{da} = 0.001\left[0.75 + 0.067 ({u_w}^2 + {v_w}^2)^{1/2}\right] \quad (35)$$

The units used for wind speed are meters per second. To use Eq. (34) in a curvilinear coordinate system, u_w and v_w need to be transformed into wind velocities along the ξ and η directions, respectively.

4.2 Boundary Conditions at the Bottom

As the bottom shear stress is proportional to u_*^2 , the boundary conditions for the fluid-solid interfaces at the bottom require that the shear stresses satisfy a quadratic stress law. Here, u_* is the friction (or shear) velocity. The conditions are given as

$$\rho\ \upsilon_v \left(\frac{\partial u}{\partial z}, \frac{\partial v}{\partial z}\right) = (\tau_{bx}, \tau_{by}) = \rho C_d\ (u_1^2 + v_1^2)^{1/2} (u_1, v_1) \quad , \quad (36a)$$

where τ_{bx} and τ_{by} are bottom shear stresses along the x and y directions, respectively, u_1 and v_1 are horizontal velocities at the first layer above the bottom, and C_d is the drag coefficient. The drag coefficient can be evaluated based on the logarithmic velocity profile near the bottom as shown below

$$C_d = \left[\frac{\kappa}{\ln\left(\frac{\Delta z_1}{2z_0}\right)} \right]^2 , \tag{36b}$$

where κ is the von Karman constant, Δz_1 is the thickness of the bottom layer, and z_0 represents roughness height. In the transformed coordinates, Eq. (36a) can be rewritten as

$$\rho \upsilon_v (\frac{\partial u}{\partial \sigma}, \frac{\partial v}{\partial \sigma}) = (\tau_{b\xi}, \tau_{b\eta}) = \rho C_d (g_{11} q_{11}^2 + 2g_{12} q_{11} q_{21} + g_{22} q_{21}^2)^{1/2} (q_{11}, q_{21}), \tag{37}$$

where q_{11} and q_{21} are the contravariant velocity components at the first grid point above the bottom.

4.3 Boundary Conditions along the Lateral Boundaries

Along the lateral open boundaries, the normal velocity component either is prescribed (river inflows) or is assumed to be of zero slope depending on the given physical forcings, while the surface elevation is specified at those boundaries. For the coastal open boundaries, the input of the surface elevation can be either the measured tidal elevation or the tide described by the summation of the tidal constituents,

$$\varsigma = \sum_{n=1}^{N} H_n \cos(\frac{2\pi t}{T_n} + \beta_n) , \tag{38}$$

where H_n, T_n, and α_n are the mean amplitude, period and phase angle of tidal constituent *n*, respectively. Harmonic analysis can provide such tidal constituent data. Along the lateral solid boundaries, velocities are set to be equal to zero.

4.4 Boundary Conditions for the Computation of Salinity and Temperature

Boundary conditions for the salinity and temperature transport equations are that, at the bottom and lateral solid boundaries, the normal derivatives of temperature and salinity are zero. At the free surface, the salinity flux is zero and the temperature gradient is balanced by the heat flux in and out of the surface. Along the lateral open boundaries, a one-dimensional convective open boundary condition can be applied to control the salinity and temperature change according to inflow and outflow.

5. NUMERICAL METHODS

Different numerical methods can be applied to solve the hydrodynamic and transport equations. Described in this section is an implicit, finite-difference scheme used to solve the three-dimensional hydrodynamic and transport equations. For numerical formulation, a staggered grid system is used in both the horizontal and vertical directions of the computational domain (Fig. 4). The horizontal velocity components are set at the boundaries of each grid cell whereas the salinity, temperature, and free-surface elevation are set at the center of the grid cell. The mode-splitting technique decouples the computation of the external free-surface elevation from the full three-dimensional equations.

5.1 External Mode

The external mode of the flow, which is described by the free-surface elevation and the vertically integrated velocities, can be calculated from the vertically integrated equations (Eqs. (28)-(30)) by using the ADI (Alternating Direction Implicit) method. The

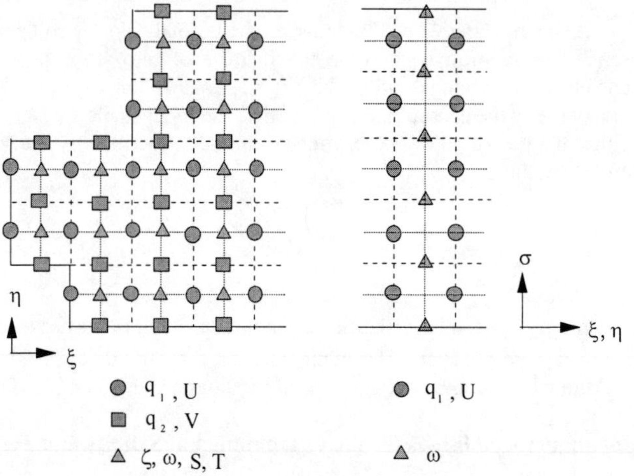

Fig. 4. Staggered grid system in horizontal and vertical directions.

finite-difference forms for solving the external mode (vertically integrated equations) are written as

x-sweep :

$$\zeta^{\overline{n+1}} + \alpha_1 \delta_\xi (JU)^{\overline{n+1}} = \zeta^n - \alpha_1 \delta_\eta (JV)^n \quad , \tag{39}$$

$$\alpha_2 g^{11} \delta_\xi \zeta^{\overline{n+1}} + (1+\alpha_3)(JU)^{\overline{n+1}} = (JU)^n - \alpha_2 g^{12} \delta_\eta \varsigma^n + JF^n \Delta t \quad , \tag{40}$$

y-sweep :

$$\zeta^{n+1} + \alpha_1 \delta_\eta (JV)^{n+1} = \zeta^{\overline{n+1}} + \alpha_1 \delta_\eta (JV)^n \quad , \tag{41}$$

$$\alpha_2 g^{22} \delta_\eta \zeta^{n+1} + (1+\alpha_4)(JV)^{n+1} = (JV)^n - \alpha_2 g^{21} \delta_\xi \varsigma^n + JG^n \Delta t \tag{42}$$

The vertically- integrated velocity U at $n+1$ time step can be updated from the corrector

$$U^{n+1} = \frac{1}{J}[(JU)^{\overline{n+1}} + \frac{\alpha_2}{1+\alpha_3} g^{11}(-\delta_\xi \varsigma^{n+1} + \delta_\xi \varsigma^{\overline{n+1}})] \tag{43}$$

Superscript n represents time step and $\overline{n+1}$ denotes the intermediate time step between time step n and time step $n+1$. The following symbols are used in the above finite difference equations:

$$\alpha_1 = \frac{\Delta t}{J} \quad , \tag{44a}$$

$$\alpha_2 = HgJ\Delta t \quad , \tag{44b}$$

$$\alpha_3 = (-\frac{g_{12}}{J} f + C_{\tau\xi})\Delta t \quad , \tag{44c}$$

$$\alpha_4 = (\frac{g_{21}}{J} f + C_{\tau\eta})\Delta t \quad , \tag{44d}$$

$$C_{\tau\xi} = \rho C_{d\xi}(g_{11}U^2 + 2g_{12}UV + g_{22}V^2)^{1/2} \quad , \tag{44e}$$

$$C_{\tau\eta} = \rho C_{d\eta}(g_{11}U^2 + 2g_{12}UV + g_{22}V^2)^{1/2} \quad , \tag{44f}$$

$$\delta_\xi () = [()_{i+1,j} - ()_{i,j}] / \Delta\xi \quad , \tag{44g}$$

and

$$\delta_\eta () = [()_{i,j+1} - ()_{i,j}] / \Delta\eta \quad , \tag{44h}$$

where $\Delta\xi = \Delta\eta = 1$. Here F^n represents the combined $g_{22} fV / J$, $\tau_{\xi s}$, and three integration terms of convection, baroclinic and diffusion representations in Eq. (29) at $n\Delta t$ time step, while G^n denotes equivalent terms in Eq. (30). By solving Eqs. (39)-(43) (tri-diagonal matrix equations) with the above-described boundary conditions, the free-surface elevation and vertically integrated velocities U and V can be determined.

5.2 Internal Mode

After the determination of free-surface elevation from the procedure of external mode described above, three-dimensional velocities can then be obtained by solving three-dimensional equations (23)-(25). This numerical procedure is referred as internal mode. Simplifying the expressions of Eqs. (24) and (25), we have

$$\frac{1}{H}\frac{\partial Hq_1}{\partial t} = B_x + \frac{1}{H^2}\frac{\partial}{\partial \sigma}[\upsilon_v \frac{\partial q_1}{\partial \sigma}] \quad , \tag{45}$$

$$\frac{1}{H}\frac{\partial Hq_2}{\partial t} = B_y + \frac{1}{H^2}\frac{\partial}{\partial \sigma}[\upsilon_v \frac{\partial q_2}{\partial \sigma}] \quad , \tag{46}$$

where B_x and B_y represent all the terms (except vertical diffusion terms) on the right hand side of equations (24) and (25).

The finite difference equations are:

$$(Hq_1)^{n+1} - (\frac{\Delta t}{H})^n \delta_\sigma(\upsilon_v \delta_\sigma q_1{}^{n+1}) = (Hq_1)^n + \Delta t(HB_x)^n \tag{47}$$

$$(Hq_2)^{n+1} - (\frac{\Delta t}{H})^n \delta_\sigma(\upsilon_v \delta_\sigma q_2{}^{n+1}) = (Hq_2)^n + \Delta t(HB_y)^n \tag{48}$$

where

$$\delta_\sigma() = [()_{i,j,k+1} - ()_{i,j,k}]/\Delta\sigma \, . \tag{49}$$

The subscripts i, j represent a fixed horizontal position and k is a typical vertical layer. The vertical diffusion terms in the above equations are treated implicitly to ensure numerical stability. It is also important to ensure the velocity closure condition is satisfied, i.e.

$$H[\int_{-1}^{0} q_1 d\sigma] - U = 0 \quad , \tag{50}$$

$$H[\int_{-1}^{0} q_2 d\sigma] - V = 0 \quad . \tag{51}$$

Equation (47) forms a system of algebraic equations for q_1 at every vertical layer. The solutions can be obtained by using a standard matrix technique. Similarly, q_2 can be determined by solving Eq. (48). Once the horizontal velocity components

q_1 and q_2 are calculated, the vertical velocity w can then be evaluated from Eq. (26).

Following a similar approach, the finite-difference forms of salinity and temperature transport equations ((32) and (33)) are

$$(HS)^{n+1} - (\frac{\Delta t}{H})^n \delta_\sigma (D_v \delta_\sigma q S^{n+1}) = (HS)^n + \Delta t (HS_r)^n \quad , \qquad (52)$$

$$(HT)^{n+1} - (\frac{\Delta t}{H})^n \delta_\sigma (D_v \delta_\sigma q T^{n+1}) = (HT)^n + \Delta t (HT_r)^n \quad , \qquad (53)$$

where S_r and T_r represent the convection and horizontal diffusion terms in Eqs. (32) and (33), respectively. It should be noted that the finite-difference schemes for modeling convection terms are crucial for the computation of salinity and temperature values. The higher-order scheme (e.g. QUICKEST, Leonard 1979) may substantially improve the accuracy of the numerical computation.

The use of a σ-stretched coordinate system, in general, can overcome the difficulty encountered in the z-plane models, such as three-dimensional numerical formulations under nonuniform vertical layers and simulation with input of large-amplitude tidal oscillation. However, occasional problems may arise from the use of σ-grid model. One problem is associated with potential errors occurring in the computation of pressure gradients (the baroclinic terms appeared in Eqs. (24) and (25)) in regions of large bottom slope. Haney (1991) suggested that using nonuniform sigma intervals or reducing cell spacing in regions with steep topography might alleviate the problem. Sheng, Lee, and Wang (1990) also reported the problem, but recommended an interpolation procedure to calculate the pressure gradients equivalent to Eqs. (19) and (20) along grid points at the same vertical level. Mellor and Blumberg (1985) proposed a set of new diffusion terms in a sigma-coordinate system so that the bottom boundary layers can be realistically modeled when bottom topographical slopes are large. For salinity transport modeling, the exchange through advection of lighter, higher layers in shallow water with adjacent denser, lower layers in deep water may over transport the salinity from deep water to shallow water (Sheng, Lee, and Wang 1990). Finite difference expressions for convection terms need to be carefully selected and tested to alleviate the problem.

6. SIMPLIFIED SECOND-ORDER CLOSURE MODEL

The vertical turbulent eddy coefficient υ_v and vertical mixing coefficient D_v can be determined by either using the empirical formulas or by solving the equations from a turbulence closure model. For a stratified flow, the Munk-Anderson type empirical formula for υ_v and D_v (Sheng 1983) are

$$\upsilon_v = \upsilon_{v0} \phi_1 (R_i) \quad , \qquad (54)$$

$$D_v = D_{v0}\phi_2(R_i) \quad , \tag{55}$$

where R_i is the Richardson Number

$$R_i = \frac{-\dfrac{gH}{\rho_0}\dfrac{\partial\rho}{\partial\sigma}}{\left(\dfrac{\partial u}{\partial\sigma}\right)^2 + \left(\dfrac{\partial v}{\partial\sigma}\right)^2} \quad . \tag{56}$$

Here, υ_{v0} and D_{v0} are eddy and mixing coefficients in the absence of any density stratification and ϕ_1 and ϕ_2 are stability functions. Munk and Anderson (1948) provided the following formula

$$\phi_1 = (1 + 10R_i)^{-1/2} \quad , \tag{57}$$

$$\phi_2 = (1 + 3.33R_i)^{-3/2} \quad . \tag{58}$$

However, the stability functions may vary with different study areas.

A better approach for obtaining the eddy coefficients is to solve the correlation equations of turbulent variables. For a stratified fluid system, a second-order turbulence closure model described in terms of Reynolds stress correlation, density fluctuation correlation and correlation of velocity and density fluctuations is given as (Donaldson, 1973)

$$\begin{aligned}\frac{D\overline{u'_i u'_k}}{Dt} &= -\overline{u'_j u'_k}\frac{\partial u_i}{\partial x_j} - \overline{u'_j u'_k}\frac{\partial u_k}{\partial x_j} - \frac{1}{\rho_0}(g_i\overline{u'_k\rho'} + g_k\overline{u'_i\rho'}) \\ &+ \frac{\partial}{\partial x_j}\left[\Lambda q\left(\frac{\partial\overline{u'_i u'_j}}{\partial x_k} + \frac{\partial\overline{u'_j u'_k}}{\partial x_i} + \frac{\partial\overline{u'_k u'_i}}{\partial x_j}\right)\right] + \frac{\partial}{\partial x_i}\left(\Lambda q\frac{\partial\overline{u'_j u'_k}}{\partial x_j}\right) + \frac{\partial}{\partial x_k}\left(\Lambda q\frac{\partial\overline{u'_j u'_i}}{\partial x_j}\right) \\ &- \frac{q}{\Lambda}\left(\overline{u'_i u'_k} - \frac{\delta_{ik}}{3}q^2\right) + \upsilon_0\frac{\partial^2\overline{u'_i u'_k}}{\partial x_j^2} - 2\upsilon_0\frac{\overline{u'_i u'_k}}{\lambda^2} \quad ,\end{aligned} \tag{59}$$

$$\frac{D\overline{\rho'^2}}{Dt} = -2\overline{u'_j\rho'}\frac{\partial\rho}{\partial x_j} + \frac{\partial}{\partial x_j}\left(\Lambda q\frac{\partial\overline{\rho'^2}}{\partial x_j}\right) + \upsilon_0\frac{\partial^2\overline{\rho'^2}}{\partial x_j^2} - 2\upsilon_0\frac{\overline{\rho'^2}}{\lambda^2} \quad , \tag{60}$$

$$\frac{D\overline{u'_k\rho'}}{Dt} = -\overline{u'_j u'_k}\frac{\partial \rho}{\partial x_j} - \overline{u'_j \rho'}\frac{\partial u_k}{\partial x_j} - \frac{1}{\rho_0} g_k \overline{\rho'^2}$$
$$+\frac{\partial}{\partial x_j}\left[\Lambda q\left(\frac{\partial \overline{u'_j\rho'}}{\partial x_k} + \frac{\partial \overline{u'_k\rho'}}{\partial x_j}\right)\right] + \frac{\partial}{\partial x_k}\left(\Lambda q \frac{\partial \overline{u'_j\rho'}}{\partial x_j}\right)$$
$$-\frac{q}{\Lambda}\overline{u'_k\rho'} + \upsilon_0 \frac{\partial^2 \overline{u'_k\rho'}}{\partial x_j^2} - 2\upsilon_0 \frac{\overline{u'_k\rho'}}{\lambda^2} \quad , \tag{61}$$

where a bar over a quantity denotes a time average and a prime denotes the fluctuating part of a physical quantity, υ_0 is the kinematic viscosity, $q = \left(\overline{u'_m u'_m}\right)^{1/2}$, and $q^2/2$ represents the turbulent kinetic energy. Λ and λ are length scales. A simplified second order closure model can be constructed by assuming super-equilibrium shear flow condition: (1) $D(\ )/Dt \approx 0$ as the correlation of turbulent variables is in local equilibrium, (2) the effect of diffusive terms is neglected, (3) $\lambda^2 \approx \upsilon_0 \Lambda / bq$, and (4) $\partial(\ )/\partial z >> \partial(\ )/\partial x$ or $\partial(\ )/\partial y$. The constant b can be set at 0.125. Equations (59)-(61) are reduced to a set of algebraic equations, which are

$$(2b+1)\frac{q}{\Lambda}\overline{u'u'} + 2\overline{u'w'}\frac{\partial u}{\partial z} - \frac{q^3}{3\Lambda} = 0, \tag{62}$$

$$(2b+1)\frac{q}{\Lambda}\overline{v'v'} + 2\overline{v'w'}\frac{\partial v}{\partial z} - \frac{q^3}{3\Lambda} = 0, \tag{63}$$

$$(2b+1)\frac{q}{\Lambda}\overline{w'w'} + \frac{2g}{\rho_0}\overline{w'\rho'} - \frac{q^3}{3\Lambda} = 0, \tag{64}$$

$$(2b+1)\frac{q}{\Lambda}\overline{u'v'} + \overline{v'w'}\frac{\partial u}{\partial z} + \overline{u'w'}\frac{\partial v}{\partial z} = 0, \tag{65}$$

$$(2b+1)\frac{q}{\Lambda}\overline{u'w'} + \overline{w'w'}\frac{\partial u}{\partial z} + \frac{g}{\rho_0}\overline{u'\rho'} = 0, \tag{66}$$

$$(2b+1)\frac{q}{\Lambda}\overline{v'w'} + \overline{w'w'}\frac{\partial v}{\partial z} + \frac{g}{\rho_0}\overline{v'\rho'} = 0, \tag{67}$$

$$\overline{w'\rho'}\frac{\partial \rho}{\partial z} + b\frac{q}{\Lambda}\overline{\rho'^2} = 0, \tag{68}$$

$$\overline{u'w'}\frac{\partial \rho}{\partial z} + \overline{w'\rho'}\frac{\partial u}{\partial z} + A\frac{q}{\Lambda}\overline{u'\rho'} = 0 \quad , \tag{69}$$

$$\overline{v'w'}\frac{\partial \rho}{\partial z} + \overline{w'\rho'}\frac{\partial v}{\partial z} + A\frac{q}{\Lambda}\overline{v'\rho'} = 0 \quad , \tag{70}$$

and

$$\overline{w'w'}\frac{\partial \rho}{\partial z} + \frac{g}{\rho_0}\overline{\rho'^2}\frac{\partial u}{\partial z} + A\frac{q}{\Lambda}\overline{w'\rho'} = 0 \ . \tag{71}$$

Here, $A = 1+2b$. The correlation of $\overline{w'\rho'}$, $\overline{u'w'}$, and $\overline{v'w'}$ can be determined by solving the above model equations. The turbulence eddy coefficient, υ_v, and vertical mixing coefficient, D_v, can then be evaluated by the following formulas

$$\upsilon_v = -\frac{\overline{u'w'}}{\partial u / \partial z} = \frac{(A+\Psi)\Lambda}{A(A-\psi)q}\overline{w'w'} \quad , \tag{72}$$

$$D_v = -\frac{\overline{w'\rho'}}{\partial \rho / \partial z} = \frac{b\Lambda}{A(b-\psi)q}\overline{w'w'} \quad , \tag{73}$$

where

$$\overline{w'w'} = \frac{q^2}{3(A-2\Psi)} \quad , \tag{74a}$$

$$\Psi = \frac{\psi}{1-\psi / b} \quad , \tag{74b}$$

$$\psi = \frac{g\Lambda^2}{\rho_0 A q^2}\frac{\partial \rho}{\partial z} \quad , \tag{74c}$$

$$q = \sqrt{\left(\frac{\partial u}{\partial z}\right)^2 + \left(\frac{\partial v}{\partial z}\right)^2}\,\Lambda Q \ , \tag{74d}$$

$$Q = \sqrt{\frac{-\Phi A^2 + \sqrt{\Phi^2 A^4 - 12A^4 b^2[R_i^{\,2}(3bA + 6b^2 + b) + R_i(b-A)]}}{6A^4 b^2}} \ , \tag{74e}$$

and

$$\Phi = 3AbR_i + 9b^2 R_i - b + bR_i \quad . \tag{74f}$$

For modeling in a vertically stretched coordinate system, the above equations need to be transformed into a σ plane for computation. The determination of D_v and υ_v depends heavily on the velocity and density gradients along the vertical direction and the turbulent length scale. The input parameters for turbulence closure model need to be calibrated using field data. The above described second-order turbulence closure model can be extended and improved by solving additional turbulent kinetic energy and turbulent macroscale equations (Schmalz, Jr. 1996, Hamrick and Wu 1997).

7. CASE STUDIES

Any hydrodynamic and transport model must be calibrated and validated before its use in practical application. Field data collected from the study area can be used to examine the accuracy of model predictions and fine tune the model parameters. For the above-described three-dimensional hydrodynamic and transport model, model simulations have been carried out at water bodies of interest to test the performance of the model. Selected model results for Chesapeake Bay and Galveston Bay are presented in the following.

7.1 Chesapeake Bay

Chesapeake Bay (Fig. 5) is the largest and most highly productive estuary in the United States. The bay lies along the coastlines of Maryland and Virginia. Ten major rivers discharge freshwater into the estuary. Model simulations have been performed to calculate free-surface elevation, velocity field and salinity distribution using 1983 (September) data. The data include river inflows, wind stresses and tidal elevations measured at the bay entrance. The model's performance can be evaluated by comparing model predictions to field measurements. The physical curvilinear grid system of the Chesapeake Bay is shown in Fig. 6 to illustrate the fitness of grid lines with the domain boundaries. To fit the physical coordinates to the boundaries, skewed grid cells may be generated. Having grid lines nearly orthogonal can alleviate potential numerical errors. Detailed input data and model results for the Chesapeake Bay studies can be found in Wang (1992). In this chapter, example plots showing the predicted free-surface elevation, velocity filed and salinity variation in the bay are presented.

General circulation patterns (27-day averaged velocity field) near the surface and bottom of the entire Chesapeake Bay are shown in Fig. 7. Fig. 7 shows that the bottom fluid flows in the opposite direction of the surface fluid near the central part of the bay. Three-dimensional velocity structures can be clearly depicted in the velocity vector plots. To demonstrate the model's performance, the computed fluid velocities and free-surface elevations are compared with the field measurements. Fig. 8 shows the comparison between the simulated surface currents (u and v) and measured data at Bay Bridge and comparisons of bottom velocities are plotted in Fig. 9. Model predictions show good agreement with the field data. Although a slight phase shift exists between the model results and the measured data, the magnitude of the flow velocities is well predicted.

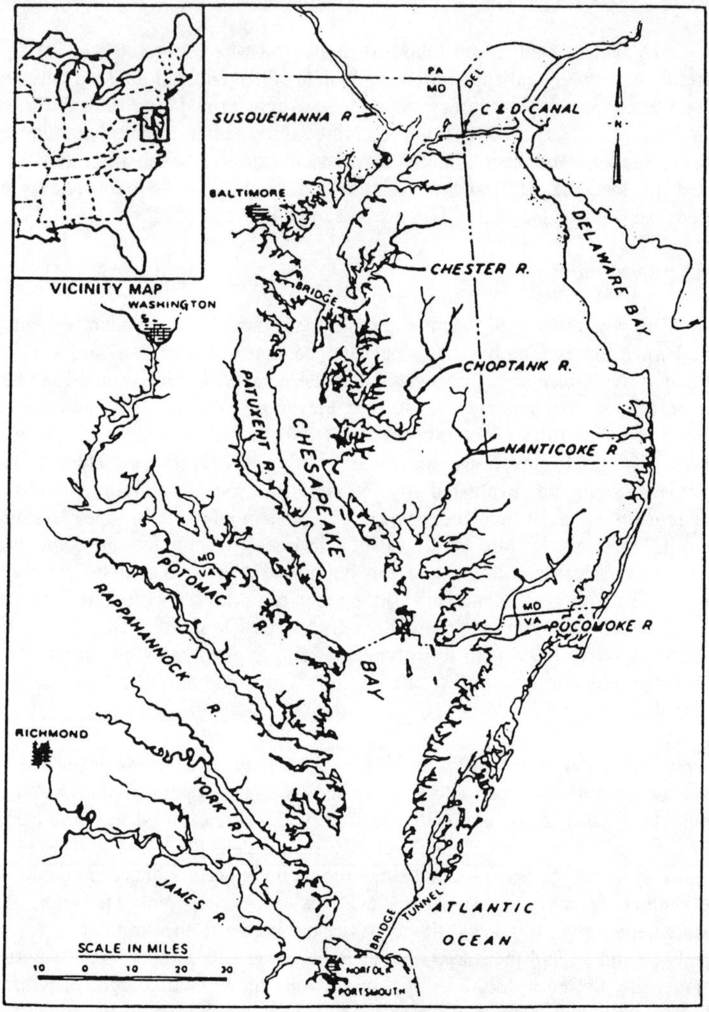

Fig. 5. Chesapeake Bay.

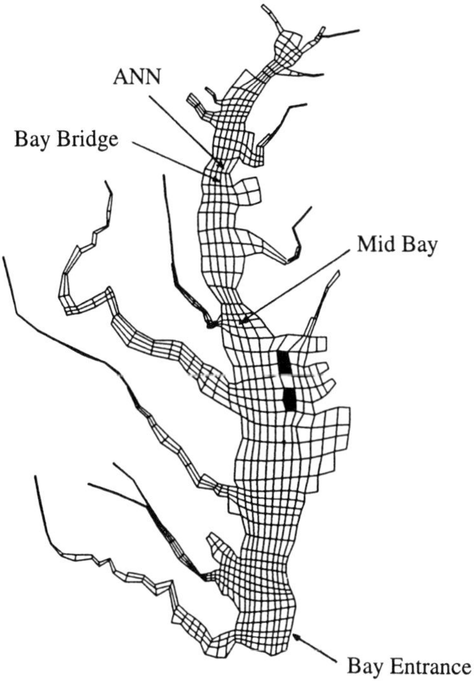

Fig. 6. Curvilinear grid system of Chesapeake Bay.

Figure 10 shows the computed and measured free-surface elevations at station ANN. The agreement again shows that the model is capable of reproducing the free-surface elevations and fluid velocities under the prescribed external driving mechanisms.

The time variation of the surface and bottom salinity at Mid Bay is plotted in Fig. 11 to show the phenomenon of wind-induced destratification resulting from strong winds acting on the water column after day 21. During the strong wind period, the field measurements show a large oscillation of the salinity of the bottom layer. Mid Bay station is located in the middle of the Bay along the deep ship channel. There is a large bottom slope along the transverse direction. The timing, duration and direction (southeastward wind) of the strong wind events cause interaction of fluid flows within the geometry of the bay and control the mixing processes between the deeper and shallower waters. The wind-generated internal velocity shear produces a strong vertical instability to enhance the processes of stratification and destratification of the water column as reflected by the oscillation of the bottom salinity. Without large oscillation, the model predicts wind-induced destratification as strong winds blow across the water column.

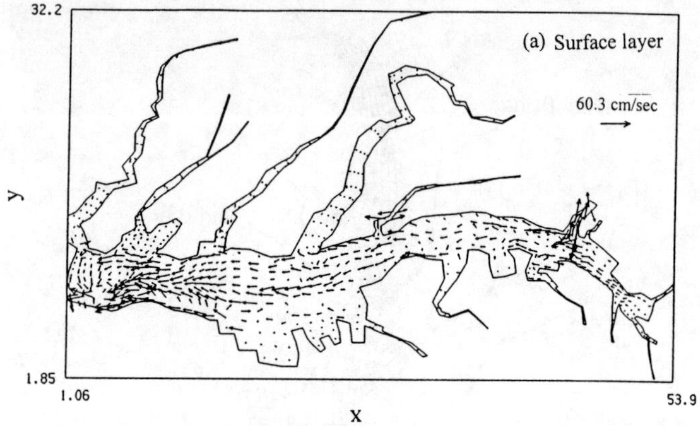

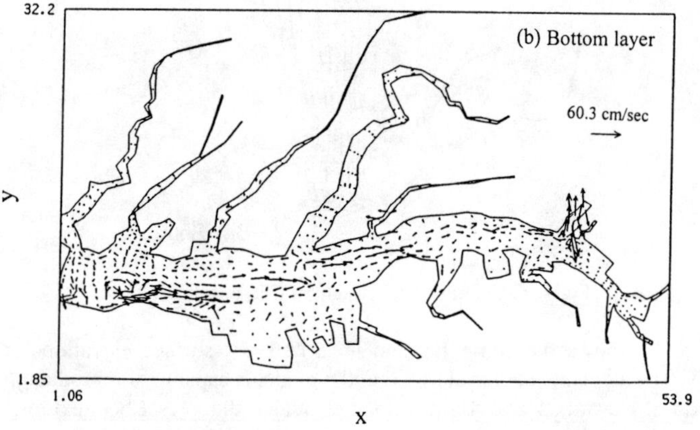

Fig. 7. Predicted general circulation of Chesapeake Bay (a) surface layer (b) bottom layer.

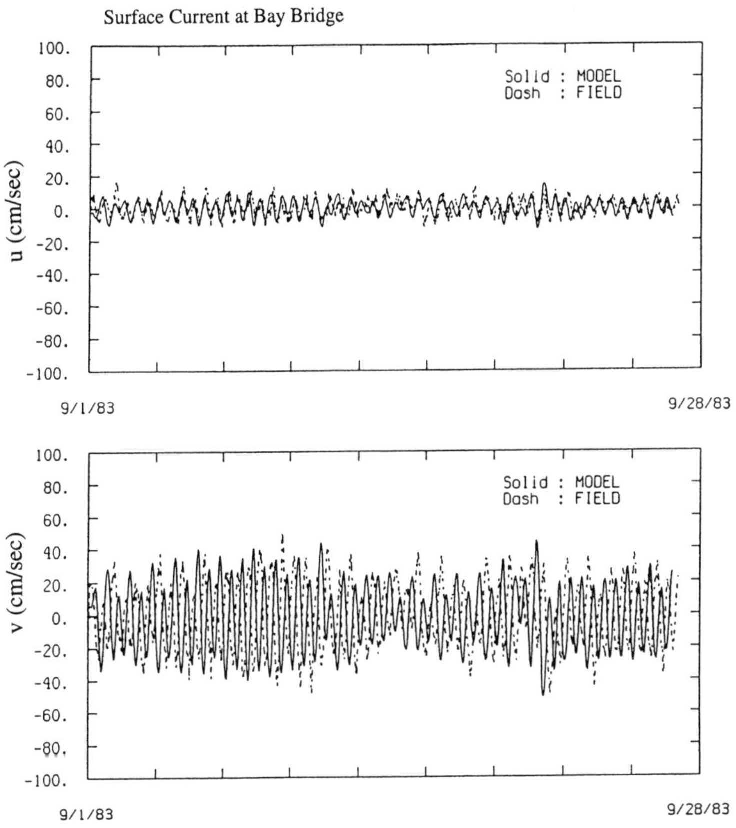

Fig. 8. Comparisons of simulated surface current and measured data at Bay Bridge in Chesapeake Bay.

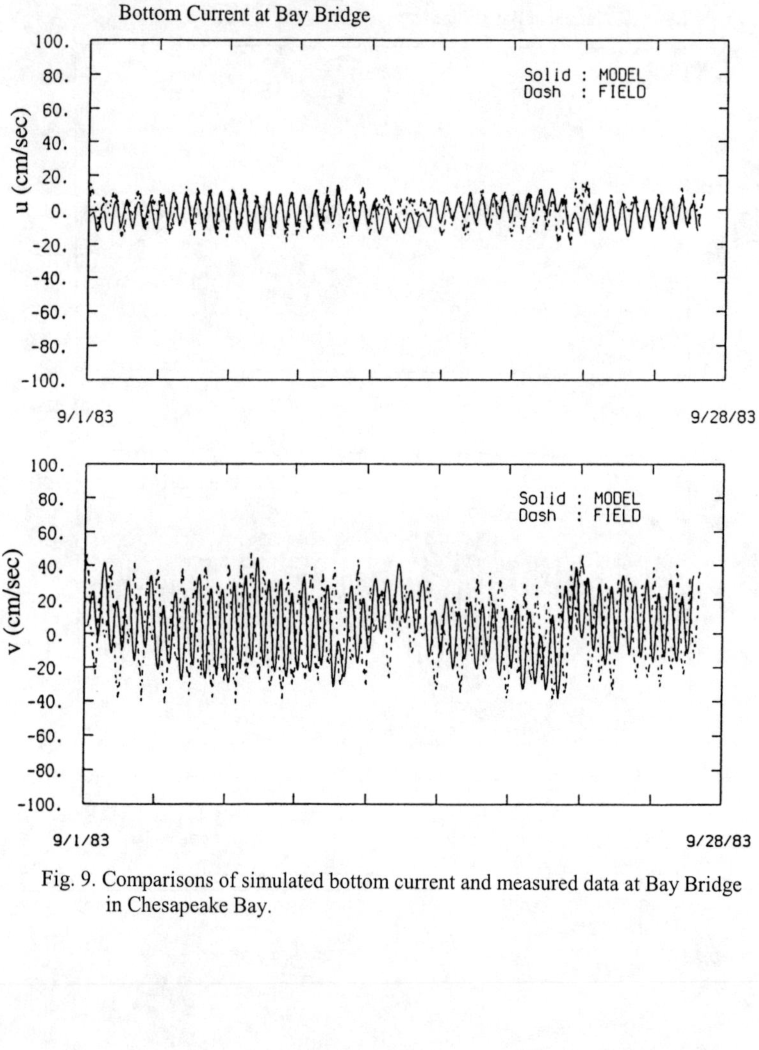

Fig. 9. Comparisons of simulated bottom current and measured data at Bay Bridge in Chesapeake Bay.

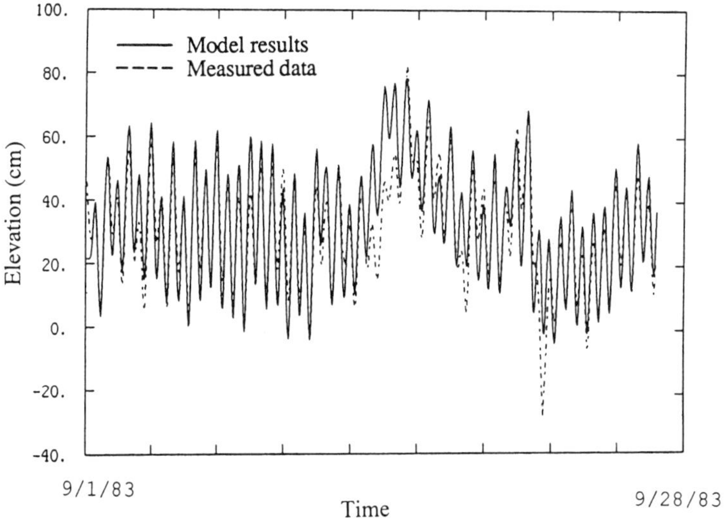

Fig. 10. Comparisons of simulated tidal elevations and measured data at station ANN.

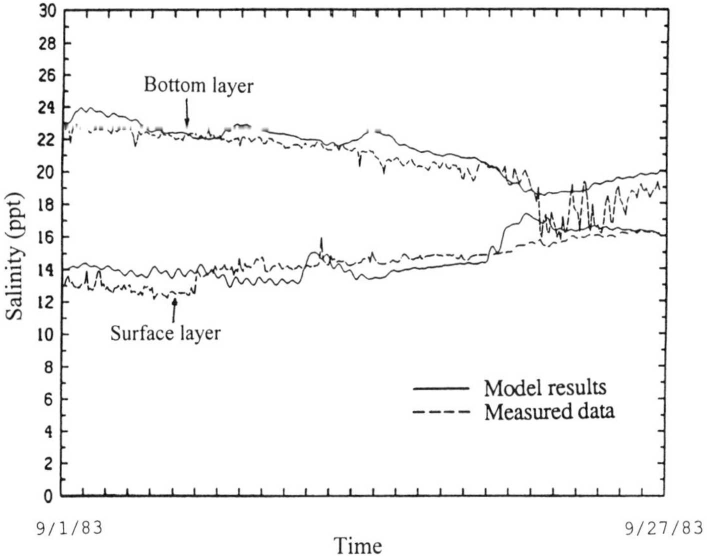

Fig. 11. Time variation of simulated and measured salinity at Mid Bay in Chesapeake Bay.

7.2 Galveston Bay

Galveston Bay (Fig. 12) is the seventh-largest estuary in the Unites States. The bay, which lies southeast of the Houston metropolitan area in the northwest Gulf of Mexico, is also one of the most highly productive estuaries. Numerical investigation of Galveston Bay has been conducted to define the flow circulation and examine the effect of freshwater inflows on the salinity variation. A typical curvilinear grid system of Galveston Bay is plotted in Fig. 13.

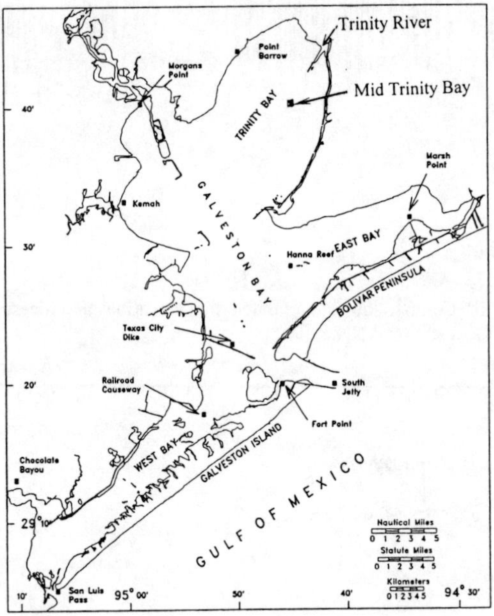

Fig. 12. Galveston Bay.

The results presented here were obtained from monthly simulations using inputs of tides and river inflows measured at open boundaries from Dec. 9,1970 to January 7, 1971. Detailed input data and model results for the Galveston Bay studies can be found in Wang (1994). Example plots showing a snapshot of surface and bottom flow circulation at time = 10 days are presented in Fig. 14. It can be noticed that, at several locations inside the bay, the flow directions between the surface and bottom fluids are different. Recirculation zones can be identified in the bay area. Three-dimensional velocity structures are again revealed in the Galveston Bay system. Comparisons between simulated free-surface elevations and field measurements at Hanna Reef are provided in Fig. 15. The results indicate good model performance in predicting free-surface elevation.

With continuous freshwater inflows, the salinity can be maintained at a low level. Fig. 16 shows the freshwater induced destratification (or mixing) over the entire water depth at the location of Mid Trinity Bay, which has a large influx of freshwater from the Trinity River. Freshwater inflow plays an important role in the salinity distribution in the bay.

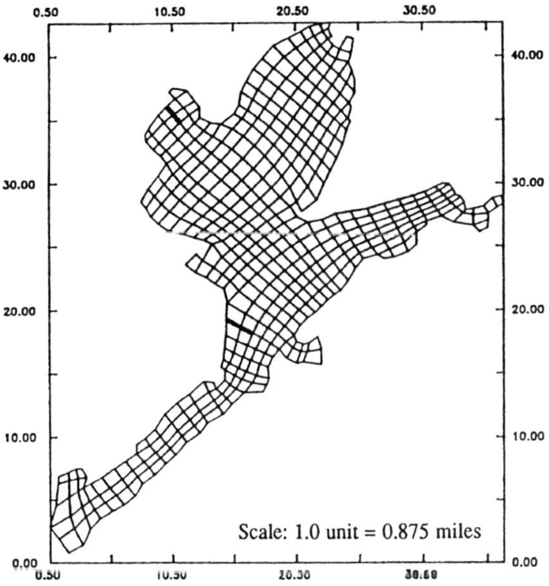

Fig. 13. Galveston Bay curvilinear grids.

8. SUMMARY

Hydrodynamic and transport modeling is important to a variety of studies of estuaries and coastal waters. These include studying circulation patterns, salinity levels, flushing capabilities, data synthesis, physical mechanisms, impact analyses, and resource management. Different types of estuarine models may be developed to provide solution tools for a particular problem of interest. For a comprehensive investigation of an estuarine system, in general, the model features must include: fluid velocity, salinity, and temperature that are three-dimensional and time dependent; free-surface elevations; nonlinear horizontal and vertical advection; vertical diffusion and mixing; horizontal density gradient; and variable grid spacing to resolve the irregular domain boundaries and bottom topography. The model equations and numerical features of time-differencing scheme, spatial-differencing scheme, horizontal and vertical grid structures, and equation solvers need to be constructed for the establishment of a model. It is clear that no two models are identical in all numerical features. Moreover, there exists no universal model that can be applied for all estuarine and coastal waters without modification or customization.

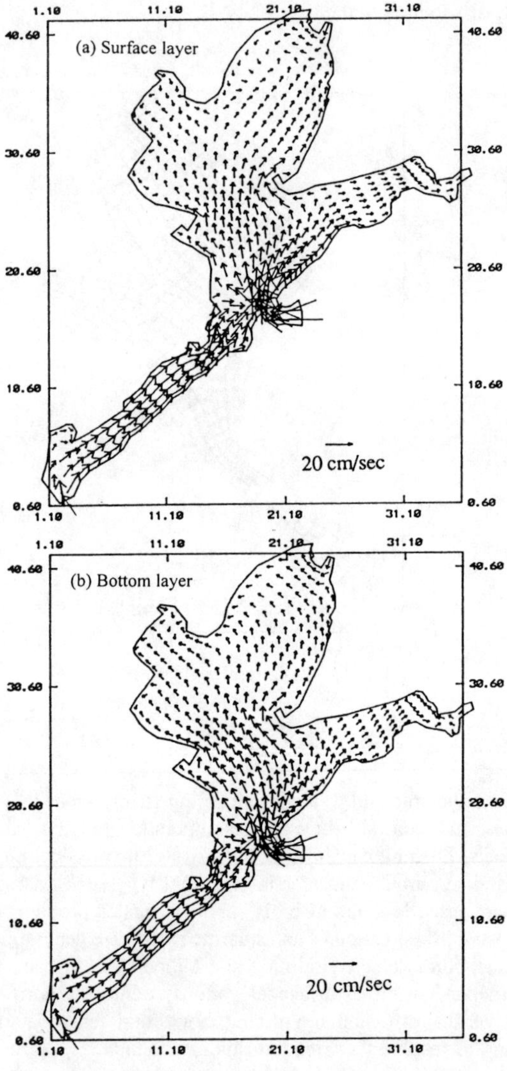

Fig. 14. Predicted circulation patterns of Galveston Bay at time= 10 days. (a) surface layer (b) bottom layer.

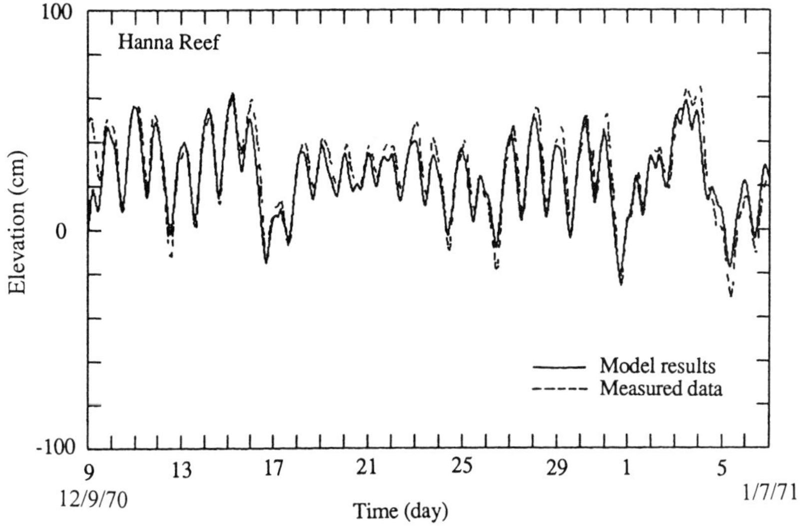

Fig. 15. Comparisons of simulated tidal elevations and measured data at Hanna Reef in Galveston Bay.

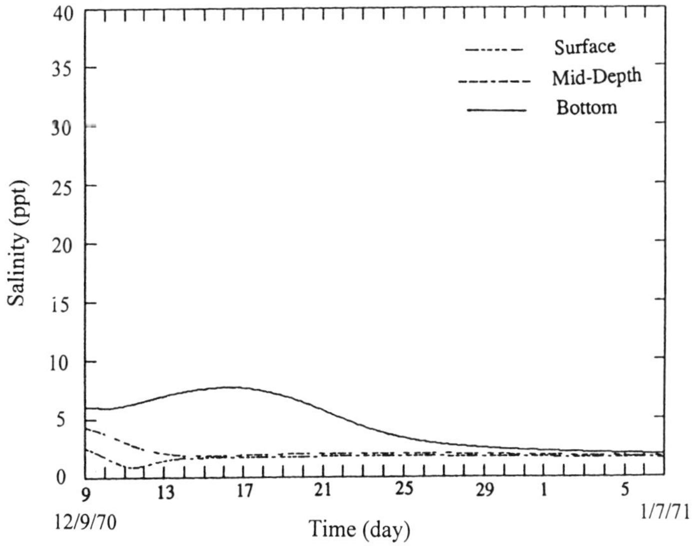

Fig. 16. Time variation of simulated salinity at Mid Trinity Bay in Galveston Bay.

This chapter provides a systematic overview of the development of a three-dimensional hydrodynamic and salinity/temperature transport model. The fundamentals of model equations, coordinate transformations, and boundary conditions are presented in detail. The numerical technique introduced in this chapter is an implicit finite difference scheme. Higher-order differencing schemes may be considered to improve the performance of the model. Another important and challenging feature related to estuarine modeling is the determination of the vertical eddy and mixing coefficients. The second-order correlation equations of turbulent variables and concepts of turbulence closure modeling is introduced in this chapter. A simplified second-order closure model can be constructed by assuming super-equilibrium shear flow condition. The procedure for deriving formulations of vertical eddy and mixing coefficients is reported. Continuing research efforts should be carried out to improve currently existing turbulence closure models. Finally, modeling studies of two major estuaries, Chesapeake Bay and Galveston Bay, are briefly summarized in the chapter. The results showing the effects of wind stresses and freshwater inflows on the flow circulation and salinity distribution are presented and discussed.

9. REFERENCES

Backhaus, J.O. (1983). A semi-implicit scheme for the shallow water equations for application to shelf sea modeling. Continental Shelf Res., **2**, pp. 243-254.

Blumberg, A.F. and Mellor, G.L. (1987). A description of a three-dimensional coastal ocean circulation model. *Three-Dimensional Coastal Ocean Models*, Editor: Norman S. Heaps, American Geophysical Union, pp.1-16.

Blumberg, A.F., Ji, Z.G., and Ziegler, C.K. (1996). Modeling outfall plume behavior using far field circulation model. *J. of Hydraulic Engrg*., ASCE, **122(11)**, pp. 610-616.

Butler, P.A. (1954). Summary of our knowledge of the oyster in the Gulf of Mexico. U.S. Fish and Wildlf. Serv. Fish. Bull., **55**, pp. 479-489.

Casulli, V. (1990). Semi-implicit finite difference methods for the two-dimensional shallow water equations. *J. Comput. Phys*., **86**, pp. 56-74.

Casulli, V. and Cheng, R.T. (1992). Semi-implicit difference methods for three-dimensional shallow water flow. *Int. J. for Numer. Methods in fluids,* **15**, pp. 629-648.

Casulli, V. , Bertolazzi, E., and Cheng, R.T. (1993). TRIM_3D- A three-dimensional model for accurate simulation of shallow water flow. Proceedings of 1993 National Conference on Hydraulic Engineering, ASCE, San Franciso, CA, pp. 1988-1993.

Chatry, M., Dugas, R.J., and Easley, K.A. (1983). Optimum salinity regime for oyster production on Louisiana's state seed grounds. *Contr. to Mar. Sci*., **26**, pp. 81-94.

Cheng, R.T. and Smith, P.E. (1989). A survey of three-dimensional numerical estuarine models. Proceeding of the 1st Estuarine and Coastal Modeling, ASCE, pp. 1-15.

Donaldson, C. duP. 1973. Construction of a dynamic model of the production of atmospheric turbulence and the dispersal of atmospheric pollutants. AMS Workshop on Micrometeorology, pp. 313-392.

Eckart, C. (1958). Properties of water. Part II – the equation of state of water and seawater at low temperature and pressure. *Am. J. Sci.*, **256**, pp. 225-240.

Environmental Protection Agency (EPA) (1992). The National Estuary Program after four years – a report to Congress. EPA report 503/9-92/007.

Evans, G.P., Mollowney, B.M., and Spoel N.C. (1989). Two-dimensional modeling of the Bristol Channel, UK. Proceedings of the 1st Estuarine and Coastal Modeling, ASCE, pp. 331-340.

Garratt, J.R., (1977). Review of drag coefficients over oceans and continents. *Monthly Weather Review*, **105**, pp.915-929.

Gunter, G. and Menzel, R.W. (1957). The crown conch, Melongena Corona, as a predator upon the Virginia oyster. Nautilis, **70**, pp. 84-87.

Hamrick, J.H. (1993). Three-dimensional variable resolution hydrodynamic and transport modeling of the Chesapeake Bay system. Proceedings of National Conference on Hydraulic Engineering, pp. 2110-2115.

Hamrick, J.H. and Wu, T.S. (1997). Computational design and optimization of the EFDC/HEM3D surface water hydrodynamic and eutrophication models. *Next Generation Environmental Models and Computational Methods*, G. Delic and M.F. Wheeler, eds., Society for Industrial and Applied Mathematics (SIAM), Philadelphia.

Haney, R.L. (1991). On the pressure gradient force over steep topography in sigma coordinate ocean models, J. of Physical Oceanography, **21**, pp 610-619.

Hess, K.W. (1994). Tampa Bay Oceanography Project: Development and application of the numerical circulation model. NOAA, National Ocean Service, Office of Ocean and Earth Sciences, NOAA Technical Report NOS OES 005.

Huang, W. and Spaulding, M. (1994). Three-dimensional modeling of circulation and salinity in Mt. Hope Bay and the lower Taunton River. Proceedings of the 3rd International Conference on Estuarine and Coastal Modeling, pp. 500-508.

Johnson, B. H., Kim, K.W., Heath, R.E., Bernard, B.B., and Butler, H.L. (1993). Validation of three-dimensional hydrodynamic model of Chesapeake bay. *J. of Hydraulic Engineering*, ASCE, **119**(1), pp. 2-20.

Kim, K.W., Johnson, B.H., and Sheng, Y.P. (1989). Modeling of wind-mixing and fall turnover event on Chesapeake Bay. Proceedings of the 1st Estuarine and Coastal Modeling Conference, pp. 172-181.

Kim, K.W., Johnson, B.H., and Gebert, J. (1994). Verification of a three-dimensional model of Delaware Bay. Proceedings of National Conference on Hydraulic Engineering, ASCE, pp. 574-578.

Leendertse, J.J., Alexander, R.C., and Liu, S-K (1973). A three-dimensional model for estuaries and coastal seas: Volume I, Principles of computation: R-1417-OWRR, RAND Corp, Santa Monica, CA.

Leendertse, J.J. and Liu, S-K (1977). A three-dimensional model for estuaries and coastal seas: Volume IV, Turbulent energy computation: R-2187-OWRT, RAND Corp, Santa Monica, CA.

Leonard, B.P. (1979). A stable and accurate convective modeling procedure based on quadratic upstream interpolation. *J. of Computer Methods in Applied Mechanics and Engineering*, **19**, pp. 59-98.

Manzel, R.W., Hulings, N.C., and Hathatway, R.R. (1966). Oyster abundance in Apalachicola Bay, Florida in relation to biotic associations influenced by salinity and other factors. Gulf Resources Reports, **2**, pp. 73-96.

Manzi, J.J. (1970). Combined effects of salinity and temperature on the feeding, reproductive, and survival rates of Eupleura caudata and Urosalpinx cinera. Biological Bulletin, **138**, pp. 35-46.

Mellor, G.L. and Blumberg, A.F. (1985). Modeling vertical and horizontal diffusivities with the sigma coordinate system. Monthly Weather Review, **113**, pp. 1379-1383.

Munk, W.H. and Anderson, E.P. (1948). Notes on the theory of the thermocline. *J. of marine Research*. **1**, pp. 276-295.

Phillips, N.A. (1957). A coordinate system having some special advantages for numerical forecasting. *J. of Meteor.*, **14**, pp. 184-185.

Reid, R.O. and Whitaker, R.E. (1981). Numerical model for astronomical tides in the Gulf of Mexico, Vol. I: theory and application, Texas A&M University, College of Geosciences, Colloge Station, TX.

Schmalz, R.A., Jr. (1996). National ocean service partnership: DGPS – supported hydrosurvey, water level measurement, and modeling of Galveston Bay. NOAA Technical Report NOS OES 012.

Shankar N.J. and Masch, F.D. (1970). Influence of tidal inlets on salinity and related phenomena in estuaries. Tech. Rep. HYD 16-7001, CRWR 49, The University of Texas at Austin.

Sheng, Y.P., (1983). Mathematical modeling of three-dimensional coastal currents and sediment dispersion: Model Development and Application, Technical Report CERC-83-2, U.S. Army Engineer Waterways Experi. Station, Vicksburg, MS.

Sheng, Y.P., (1987). On modeling three-dimensional estuarine and marine hydrodynamics. *Three-Dimensional Models of Marine and Estuarine Dynamics*, J.C.J. Nihoul and B.M. Jamart (editors), Elsevier Oceanography Series, Amsterdam, Holland.

Sheng, Y.P., Lee, H.K., and Wang, K.H. (1989). On numerical strategies of estuarine and coastal modeling. Proceedings of the 1st Estuarine and Coastal Modeling Conference, pp. 291-301.

Smith, P.E. and Cheng, R.T. (1989). Recent progress on hydrodynamic modeling of SanFranciso Bay, California. Proceedings of the 1st Estuarine and Coastal Modeling Conference, pp. 502-510.

Smith, P.E. (1994). San Francisco Bay test case for 3-D model verification. Proceedings of National Conference on Hydraulic Engineering, ASCE, pp. 885-889.

St. Amant, L.S. 1957. The southern oyster drill. La. Wildlf. Fish. Comm. 7th biennial report, pp. 81-85.

Stelling, G.S., Wiersma, A.K., and Willemse, J.B.T. (1986). Practical aspects of accurate tidal computations. *J. of hydraulic Div.*, ASCE, **112**, pp. 802-817.

Thompson J.F. and Johnson, B.H. (1985). Development of an adaptive boundary-fitted coordinate code for use in coastal and estuarine areas", Technical Report HL-85-5, U. S. Army Engineer Waterways Experiment Station, Vicksburg, MS.

Vincent, M., Galperin, B., and Luther, M. (1995). Development and application of a real-time three-dimensional hydrodynamic model of Tampa Bay, Florida. Proceedings of the Symposium on Coastal and Ocean Management, pp. 145-146.

Wang, K.H. (1992). Three-dimensional circulation modeling of the coastal and ocean environments", ASCE Civil Engineering in the Oceans V Conference, College Station, Texas, pp. 637-651.

Wang, K.H. (1994). Characterization of circulation and salinity change in Galveston Bay. *J. of Engineering Mechanics*, ASCE, **120**(3), pp. 557-579.

Wang, P.F., Cheng, R.T., Richter, K., Gross, E.S., Sutton, D., and Gartner, J.W. (1998). Modeling tidal hydrodynamics of San Diego Bay, California. J. of American Water Resources Association, **34**, pp. 1123-1140.

Wilders, P., van Stijn, Th. L., Stelling, G.S., and Fokkema, G.A. (1988). A fully implicit splitting method for accurate tidal computations. *Int. J. of Numer. Methods in Engrg.*, **26**, pp. 2707-2721.

10. APPENDIX

Expressions of the metric tensors:

$$g_{22} = x_\eta^{\,2} + y_\eta^{\,2} \qquad g_{11} = x_\xi^{\,2} + y_\xi^{\,2} \qquad (75a,b)$$

$$g_{12} = g_{21} = x_\xi x_\eta + y_\xi y_\eta \qquad g^{22} = g_{11}/J^2 \qquad (75c,d)$$

$$g^{11} = g_{22}/J^2 \qquad g^{12} = g^{21} = -g_{12}/J^2 \qquad (75e,f)$$

The horizontal diffusion term in Equation (24):

$$\text{Diffx}=\frac{1}{J}\Big\{\frac{1}{J^2}[g_{22}(y_\eta u_{\xi\xi}-x_\eta v_{\xi\xi})-2g_{12}(y_\eta u_{\xi\eta}-x_\eta v_{\xi\eta})+g_{11}(y_\eta u_{\eta\eta}-x_\eta v_{\eta\eta})]$$
$$+\frac{1}{J^3}\{\delta[y_\xi(y_\eta u_\eta-x_\eta v_\eta)-y_\eta(y_\eta u_\xi-x_\eta v_\xi)] \tag{76}$$
$$+\gamma[x_\eta(y_\eta u_\xi-x_\eta v_\xi)-x_\xi(y_\eta u_\eta-x_\eta v_\eta)]\}\Big\}$$

The horizontal diffusion term in Equation (25):

$$\text{Diffy}=\frac{1}{J}\Big\{\frac{1}{J^2}[g_{22}(-y_\xi u_{\xi\xi}+x_\xi v_{\xi\xi})-2g_{12}(-y_\xi u_{\xi\eta}+x_\xi v_{\xi\eta})$$
$$+g_{11}(-y_\xi u_{\eta\eta}+x_\xi v_{\eta\eta})]+\frac{1}{J^3}\{\delta[y_\xi(-y_\xi u_\eta+x_\xi v_\eta)-y_\eta(-y_\xi u_\xi+x_\xi v_\xi)] \tag{77}$$
$$+\gamma[x_\eta(-y_\xi u_\xi+x_\xi v_\xi)-x_\xi(-y_\xi u_\eta+x_\xi v_\eta)]\}\Big\}$$

where

$$u = x_\xi q_1 + x_\eta q_2 \tag{78a}$$
$$v = y_\xi q_1 + y_\eta q_2 \tag{78b}$$
$$\delta = g_{22}x_{\xi\xi} - 2g_{12}x_{\xi\eta} + g_{11}x_{\eta\eta} \tag{78c}$$
$$\gamma = g_{22}y_{\xi\xi} - 2g_{12}y_{\xi\eta} + g_{11}y_{\eta\eta} \tag{78d}$$

Chapter 7

FLUID FLOWS AND REACTIVE CHEMICAL TRANSPORT IN VARIABLY SATURATED SUBSURFACE MEDIA

Gour-Tsyh Yeh[1], Ming-Hsi Li[2], Malcolm D. Siegel[3]

ABSTRACT

The couplings among fluid flow, advective and dispersive/diffusive transport, and chemical reactions are important for both waste disposal and remediation of contaminated sites in fractured media or soils. The feedback of chemical reactions on fluid flow and reactive chemical transport is particularly important because changes in hydraulic properties due to precipitation and dissolution along fractures and rock matrix might be pronouncing. This chapter presents the fundamental physical, chemical, and biological processes that control the movement of fluids, the migration of chemicals, and their interactions with the media. It describes the development and verification of a mechanistic-based numerical model for simulation of coupled fluid flow and reactive chemical transport, including both fast and slow reactions, in variably saturated media. Mathematical formulations, numerical and computer implementations, and numerical experiments are described. Six example problems are employed to illustrate the applications of the model. The first three problems are used to demonstrate the versatility and flexibility in modeling various combinations of fluid flows and reactive chemical transport. The fourth and fifth problems are used to illustrate the interplay and effects of advective/dispersive transport, chemical reactions, and fluid flows. The final problem is used to show that species switching algorithms are necessary to deal with highly nonlinear, stiff geochemistry problems.

[1]Professor, Department of Civil and Environmental Engineering, Penn State University. Email gty2@pau.edu. Presently, Provost Professor, Department of Civil and Environmental Engineering, University of Central Florida, Orlando, FL 32816-2450, USA, gyeh@mail.ucf.edu

[2]Post-doctor Researcher, Division of Flood Mitigation, National Science & Technology Program for Hazards Mitigation. Presently, Assistant Professor, Institute of Hydrology, National Central University, Chunli 32054, Taiwan, mli@ns.naphm.ntu.edu

[3]Principle Member Technical Staff, P.O. Box 5800, Sandia National Laboratories, Albuquerque, NM 87185, msiegel@sandia.gov

1. INTRODUCTION

The volume, extent, distribution, and complexity of contaminated soils and groundwater in the world are pervasive, with pollutants ranging from hazardous chemicals, organic, metals, radionuclides, and mixtures of these classes. Trillions of dollars are estimated to remediate these sites worldwide. Disposal of huge volumes of daily wastes pose a unique and formidable challenge to develop sound characterization, performance assessment, and long-term monitoring technologies that are cost-effective and result in acceptable levels of risk to human health and the environment. To meet the challenge of remediating and assessing both the existing sites and the proposed disposal of wastes, accurate tools to predict contaminant migration and transformation are necessary.

Contaminants may undergo geochemical reactions and transformations and the resultant speciation can enhance or hinder contaminant mobility. Some of these geochemical processes are fast and reversible, others are slow in comparison with transport phenomena; many are dependant upon and cause changes in the acidity (pH) and reduction-oxidation (redox) condition of the subsurface environment. Aqueous complexation including acid-base reactions tend to be rapid, while adsorption-desorption, precipitation-dissolution, and redox processes are often kinetic in nature, with system conditions far from equilibrium.

Experiments have shown that geochemical reactions have profound effects on the mechanics of fluid flows (McGrail et al., 1996). Rapid, dramatic changes in hydraulic conductivity and fluid pore velocity were observed in the unsaturated flow-through testing in columns filled with glass beads. These changes were attributed to secondary phase precipitation, which occurred under reduced saturation conditions. Clearly the removal of water by precipitation would manifest itself if the water is not supplied via hydrologic transport. It is not difficult to see that the solid phase reactions including precipitation and adsorption can plug pores reducing both the porosity and hydraulic conductivity, which in turn would hinder the transport of water into the affected regions. As a result, moisture content would be reduced and precipitation would occur. The onset of further precipitation would remove water from the liquid phase. These complicated interactions between flow dynamics and geochemical reactions must be included in modeling reactive chemical transport, especially when the degree of saturation is low. By coupling unsaturated flow with reactive transport, an effective strategy for investigating contaminant transport and for conducting performance assessment of various waste management and remediation technologies can be developed.

Consideration of the equilibrium chemistry, kinetic chemistry, and hydrologic transport and the interaction between fluid flow and reactive transport is necessary to capture the complexity of many real systems. Analytical solutions of realistically complex reactive transport problems are generally not feasible. Numerical simulations of fluid flow and reactive contaminant transport provide a viable means of analyzing contamination problems and assessing the geochemical mechanisms affecting transformation and therefore transport. Conventional contaminant transport models often take simplistic approaches in dealing with geochemical reactions, which make the upscaling of these models and their applications to real systems problematic. The development of mechanistic-based reactive chemical transport models has exploded in the last two decades. These reactive transport models have had varied scope. Many couple simulation of transport with equilibrium geochemistry (e.g., Valocchi et al., 1981; Jennings et al., 1982; Rubin, 1983; Miller and Benson, 1983; Kirkner et al., 1984; Kirkner et al., 1985; Cederberg, 1985; Hostetler and Erikson, 1989; Liu and Narasimhan, 1989; Yeh and Tripathi, 1989, 1990, 1991; Griffioen, 1993; Simunek and Suarez, 1993; Cheng, 1995; Parkhurst, 1995). Some models couple transport with kinetic geochemistry for specific geochemical processes like precipitation-dissolution (e.g. Lichtner and Setch, 1996; Steefel and Yabusaki, 1996; Suarez and Simunek, 1996) or adsorption (e.g. Theis et al., 1982, Lensing et al., 1994). A few couple a complete suite of geochemical reaction processes (aqueous complexation, adsorption, precipitation dissolution, acid-base, and reduction-oxidation phenomena) and also allow any individual reaction for any of these geochemical processes to be handled as either equilibrium or kinetic as appropriate for the system being considered (e.g. Yeh et al., 1996). Chilakapathi (1995) provides significant flexibility in handling any type of geochemical kinetic reactions coupled with transport, representing equilibrium processes as fast reversible kinetic reactions, but requires problem specific modifications to the code for new simulations to be run. Models that can handle the interaction between unsaturated fluid flows and reactive chemical transport are practically non-existent.

This chapter describes the development and verification of a mechanistic-based numerical model for simulating coupled fluid flow and reactive chemical transport with a full suite of geochemical reactions in variably saturated media. The ability to simulate the full suite of geochemical processes simultaneously, to represent the geochemical reactions mechanistically, and to allow any combination of these reactions and processes to be represented as either equilibrium or kinetic gives this numerical model an extremely flexible scope for handling geochemical reactions. Coupling this capability with simultaneous modeling of fluid flow and their interaction with reactive transport gives the model a broad ability to simulate the complex geochemistry in the subsurface environment. The objective of this chapter is to present the scope, governing equations, and numerical solution technique used to implement the governing equations, and to demonstrate its applications with six example problems. The first three problems illustrate the capability of the model to deal with various combinations of fluid flow and reactive transport problems. The fourth and fifth problems are designed to simulate the interactions and feedback among the fluid flow, advective/dispersive transport, diffusion, and precipitation/dissolution reactions. The final problem is used to illustrate the need of species switching to solve geochemical problems with low concentrations of component species.

2. MATHEMATICAL MODEL

The complete mathematical model that describes both fluid flow and reactive chemical transport in variably saturated subsurface media includes governing equations, initial and boundary conditions, and constituent relationships. These are stated below, detailed derivations can be found elsewhere (Yeh, 1999; Yeh, 2000).

2.1 Fluid Flow

2.1.1 Governing Equations

Based on the (1) mass balance of fluid, (2) mass balance of solids, (3) motion of fluid (Darcy's law), (4) consolidation of the media, and (5) compressibility of water, one can derive the basic equation for density dependent fluid flow as (Cheng et al., 1998)

$$\frac{\rho}{\rho_o} F \frac{\partial h}{\partial t} = \Delta \cdot \left[K \cdot \left(\Delta h + \frac{\rho}{\rho_o} \Delta z \right) \right] + \frac{\rho^*}{\rho_o} q \tag{1}$$

where F is the storage coefficient [1/L] defined as

$$F = \alpha' \frac{\theta}{n_e} + \beta' \theta + n_e \frac{dS}{dh} \tag{2}$$

K is the hydraulic conductivity tensor [L/T]

$$K = \frac{\rho g}{\mu} k = \frac{(\rho / \rho_o)}{(\mu / \mu_o)} \frac{\rho_o g}{\mu_o} k_s k_r = \frac{(\rho / \rho_o)}{(\mu / \mu_o)} K_{so} k_r \tag{3}$$

and the Darcy velocity vector **V** [L/T] can be calculated as follows

$$V = -K \cdot \left(\frac{\rho_o}{\rho} \Delta h + \Delta z \right) \tag{4}$$

where θ is the effective moisture content $[L^3/L^3]$, h is the pressure head [L], t is time [T], z is the potential head [L], q is the source or sink representing the artificial injection or withdrawal of fluid $[(L^3/L^3)/T]$, ρ_o is the referenced fluid density at zero chemical concentration $[M/L^3]$, ρ is the fluid density with dissolved chemical concentrations $[M/L^3]$, ρ^* is the fluid density of either injection ($= \rho^{injection}$) or withdraw ($= \rho$) $[M/L^3]$, μ_o is the fluid dynamic viscosity at zero chemical concentration [M/L/T], μ is the fluid dynamic viscosity with dissolved chemical concentrations [M/L/T], α' is the modified compressibility of the soil matrix [1/L], $\beta'$is the modified compressibility of the liquid [1/L], n_e is the effective porosity [L^3/L^3], S is the degree of saturation of water [dimensionless], g is the gravity [L/T^2], **k** is the permeability tensor [L^2], $\mathbf{k}_s$ is the saturated permeability tensor [L^2], $\mathbf{K}_{so}$ is the referenced saturated hydraulic conductivity tensor [L/T], k_r is the relative permeability or relative hydraulic conductivity [dimensionless].

Equations (1) through (3) and the constitutive relationships among the pressure head, degree of saturation, and hydraulic conductivity tensor, together with associated and appropriate initial conditions and boundary conditions, can be used to compute the temporal-spatial distributions of the hydrologic variables, including pressure head, total head, effective moisture content, and Darcy's velocity.

2.1.2. Initial and Boundary Conditions for Flow Simulations

To complete the mathematical formulation of the saturated-unsaturated problems, Eq. (1) must be supplemented with initial and boundary conditions. The initial condition is stated mathematically as

$$h = h_i(\mathbf{x}) \qquad \text{at} \qquad t = 0 \text{ in } R \tag{5}$$

where R is the region of interest and $h_i(\mathbf{x})$ is the prescribed initial condition function of the spatial coordinate **x**, which can be obtained by either field measurements or by solving the steady-state version of Eq. (1) with time-invariant boundary conditions.

Three basic types of boundary conditions and a river boundary condition can be specified for variably saturated flows. In addition, a variable boundary condition normally at the air-media interface can be specified. These varieties of boundary conditions are considered to handle a variety of physical phenomena that can happen on the boundary. These boundary conditions are stated mathematically as follows:

Dirichlet Conditions. On a Dirichlet boundary, the pressure head is prescribed as

$$h = h_d(\mathbf{x}_b, t) \qquad \text{on } \mathbf{B}_d \tag{6}$$

where $\mathbf{x}_b$ is the spatial coordinate on the boundary, $h_d(\mathbf{x}_b,t)$ is the prescribed head function, and B_d is the Dirichlet boundary segment.

Neumann Conditions. On a Neumann boundary, which normally is the drainage boundary, the gradient flux is specified as

$$-\mathbf{n}\cdot\mathbf{K}\cdot\frac{\rho_o}{\rho}\Delta h = q_n(\mathbf{x}_b, t) \qquad \text{on } B_n \tag{7}$$

where **n** is an outward unit vector normal to the boundary, $q_n(\mathbf{x}_b,t)$ is the prescribed Neumann flux function, and B_n is the Neumann boundary segment.

<u>Cauchy Conditions</u>. On a Cauchy boundary, which normally is an infiltration boundary, the volume flux is prescribed as

$$-\mathbf{n}\cdot\mathbf{K}\cdot\left(\frac{\rho_o}{\rho}\Delta h + \Delta z\right) = q_c(\mathbf{x}_b, t) \qquad \text{on } B_c \tag{8}$$

where $q_c(\mathbf{x}_b,t)$ is the prescribed Cauchy flux function and B_c is the Cauchy boundary segment.

<u>Variable Conditions - During Precipitation Period</u>. On a variable boundary, which is normally the air-media interface, either infiltration, ponding, or seepage can occur during precipitation periods. If infiltration occurs, the maximum infiltration rate is equal to the excess precipitation rate. As to which condition is prevalent, it cannot be determined *a priori*. Rather it must be determined in a cyclic iterative procedure. These physical conditions lead to the mathematical representations

$$h = h_p(\mathbf{x}_b, t) \qquad \text{iff} \quad -\mathbf{n}\cdot\mathbf{K}\cdot\left(\frac{\rho_o}{\rho}\Delta h + \Delta z\right) \geq q_p(\mathbf{x}_b, t) \quad \text{on } B_v \tag{9}$$

$$-\mathbf{n}\cdot\mathbf{K}\cdot\left(\frac{\rho_o}{\rho}\Delta h + \Delta z\right) = q_p(\mathbf{x}_b, t) \quad \text{iff } h \leq h_p(\mathbf{x}_b, t) \quad \text{on } B_v \tag{10}$$

where B_v is the variable boundary segment, $h_p(\mathbf{x}_b,t)$ is the ponding depth function, and $q_p(\mathbf{x}_b,t)$ (numerically negative) is the excess precipitation function. Either Eq. (9) or (10), but not both, is used at any point on the variable boundary at any time during precipitation periods.

<u>Variable Conditions - During Nonprecipitation Period</u>. During non-precipitation periods, either seepage or evaporation can occur on a variable boundary. If evaporation occurs, the maximum amount of evaporation is the potential evaporation. These physical considerations lead to the following mathematical statements:

$$h = h_p(\mathbf{x}_b, t) \qquad \text{iff} \quad -\mathbf{n}\cdot\mathbf{K}\cdot\left(\frac{\rho_o}{\rho}\Delta h + \Delta z\right) \geq q_p(\mathbf{x}_b, t) \quad \text{on } B_v \tag{11}$$

$$h = h_m(\mathbf{x_b}, t) \qquad \text{iff} \quad -\mathbf{n} \cdot \mathbf{K} \cdot (\frac{\rho_o}{\rho} \Delta h + \Delta z) \le q_e(\mathbf{x_b}, t) \quad \text{on } B_v \tag{12}$$

$$-\mathbf{n} \cdot \mathbf{K} \cdot (\frac{\rho_o}{\rho} \Delta h + \Delta z) = q_e(\mathbf{x_b}, t) \quad \text{iff } h \ge h_m(\mathbf{x_b}, t) \quad \text{on } B_v \tag{13}$$

where $h_m(\mathbf{x}_b,t)$ is the allowed minimum pressure function on the variable boundary and $q_e(\mathbf{x}_b,t)$ is the potential evaporation rate function on the variable boundary. Only one of Eq. (11) through (13) is used at any point on the variable boundary at any time during non-precipitation periods.

River Boundary Conditions. At river-media interfaces, two types of boundary conditions can be specified depending on physical conditions. If sediment layers are absent around the wet perimeter, a Dirichlet boundary condition with the pressure head equal to the river depth can be specified. If there are sediment layers around the wet perimeter of the river-media interfaces, then a radiation type of river boundary condition can be imposed:

$$-\mathbf{n} \cdot \mathbf{K} \cdot (\frac{\rho_o}{\rho} \Delta h + \Delta z) = -\frac{K_R}{b_R}(h_R - h) \quad \text{on } B_r \tag{14}$$

where K_R is the hydraulic conductivity of the river bottom sediment layer, b_R is the thickness of the river bottom sediment layer, h_R is the depth of the river bottom measured from the river surface to the top of the sediment layer, and B_r is the segment of river-media interface.

2.2 Reactive Chemical Transport

Two of the most frequently mentioned terms in reactive chemical transport are *components* and *species*. Definitions of these terms loosely follow those of Westall et al. (1976). Components are a set of linearly independent "basis" chemical entities through which every species can be uniquely represented as a linear combination. A component cannot be represented as a linear combination of components other than itself. For example, in a simple system containing water, carbon, and calcium, we can choose Ca^{2+}, CO_3^{2-}, and H^+ as the components, and all other species can be considered the products of these three chemical components, because Ca^{2+} cannot be represented as a product of CO_3^{2} and H^+, CO_3^{2-} cannot be represented as a combination of Ca^{2+} and H^+, and H^+ cannot be represented as a product of Ca^{2+} and CO_3^{2-}. In addition, we require that the global mass of a component be reaction invariant (Rubin, 1983). A species is the product of a chemical reaction involving the components as reactants (Westall et al., 1976). For example, in the above simple system the species HCO$\underline{3}$ is the product of H^+ and CO_3^{2-}.

Let us consider a system of chemical transport, which includes N_a aqueous components (mobile components), N_s adsorbent components (immobile adsorbent sites), and N_e immobile ion-exchange sites. The N_a aqueous components will react with each other to form M_x complexed species and M_p precipitated species. In addition, each aqueous component has a species that is not bound with other components. This species is termed the aqueous component species. The total number of aqueous species, M_a, is the sum of N_a aqueous component species and M_x complexed species. The N_a aqueous components and N_s adsorbent components will react to form M_y adsorbed species for the case of sorption via surface complexation. Each adsorbent component has a species that is not bound with other components. This species is termed the adsorbent component species. In addition, M_z of N_a aqueous component species and M_x complexed species may compete with each other for the ion-exchange sites. The total number of sorbent species, M_s, is the sum of N_s adsorbent component species, M_y adsorbed species, and M_z ion-exchanged species. The total number of chemical species, M, is equal to the sum of M_a, M_s, and M_p. We shall assume that the aqueous component species and complexed species are subject to hydrological transport, whereas the precipitated species, adsorbent component species, adsorbed species, and ion-exchanged species are not subject to hydrologic transport.

2.2.1 Governing Equations

The governing equations for a reactive system are mass balance equations for aqueous components, adsorbent components, equivalent-balance equations for ion exchange sites, mass action equations for equilibrium species, and reaction rate expressions for kinetic species. These equations can be derived based on the principles of mass balance and chemical reactions. They were derived elsewhere (Yeh et al., 1999) and are stated as follows.

Hydrologic transport equations for aqueous components –

$$\theta \frac{\partial T_j}{\partial t} + \frac{\partial \theta}{\partial t}(S_j + P_j) = L(C_j) - \theta\,\Lambda_j^a + M_j^a - QC_j \;,\; j \in N_a \tag{15}$$

in which

$$T_j = c_j + \sum_{i=1}^{M_x} a^x x_i + \sum_{i=1}^{M_y} a_{ij}^y y_i + \sum_{i=1}^{M_z} a_{ij}^z z_i + \sum_{i=1}^{M_p} a_{ij}^p p_i \;,\; j \in N_a \tag{16}$$

$$S_j = \sum_{i=1}^{M_y} a_{ij}^y y_i + \sum_{i=1}^{M_z} a_{ij}^z z_i \,,\; P_j = \sum_{i=1}^{M_p} a_{ij}^p p_i \;,\; C_j = c_j + \sum_{i=1}^{M_x} a_{ij}^x x_i \;,\; j \in N_a \tag{17}$$

$$L(C_j) = -V \cdot \Delta C_j + \Delta \cdot (\theta D \cdot \Delta C_j) \tag{18}$$

$$\mathbf{Q} = \frac{\rho^*}{\rho} q - F\frac{\partial h}{\partial t} + \frac{\partial \theta}{\partial t} - \frac{\rho_o}{\rho} \mathbf{V} \cdot \Delta \frac{\rho}{\rho_o} \tag{19}$$

where T_j is the total analytical concentration of the j-th aqueous component [M/L^3]; S_j is the total sorbed concentration of the j-th aqueous component [M/L^3]; P_j is the total precipitated concentration of the j-th aqueous component [M/L^3]; C_j is the total dissolved concentration of the j-th aqueous component [M/L^3]; Λ_j^a is the total decay rate of the j-th aqueous component [M/L^3/T]; M_j^a is the total rate of source/sink of the j-th aqueous component [M/L^3/T]; N_a is the number of aqueous components; M_x is the number of complexed species; M_y is the number of adsorbed species; M_z is the number of ion exchanged species; M_p is the number of precipitated species; c_j is the concentration of the j-th aqueous component species [M/L^3]; x_i is the concentration of the i-th complexed species [M/L^3]; y_i is the concentration of the i-th adsorbed species [M/L^3]; z_i is the concentration of the i-th ion exchanged species [M/L^3]; p_i is the concentration of the i-th precipitated species [M/L^3]; a_{ij}^x, a_{ij}^y, a_{ij}^z and a_{ij}^p are the stoichiometric coefficients of the j-th aqueous component in the i-th complexed, adsorbed, ion-exchanged, and precipitated species, respectively; L is the advection-dispersion operator; and **D** is the dispersion coefficient [L^2/T].

Mass balance equations for adsorbent components –

$$\theta \frac{\partial W_j}{\partial t} + \frac{\partial \theta}{\partial t} W_j = -\theta \Lambda_j^s + M_j^s \ , \ j \in N_s \tag{20}$$

in which

$$W_j = s_j + \sum_{i=1}^{M_j} b_{ij}^y y_i \, , \qquad j \in N_s \tag{21}$$

where W_j is the total analytical concentration of the j-th adsorbent component [M/L^3], Λ_j^s is the total decay rate of the j-th adsorbent component [M/L^3/T], M_j^s is the total rate of source/sink of the j-th adsorbent component [M/L^3/T], N_s is the number of adsorbent components, s_j is the concentration of the j-th adsorbent component species, and b_{ij}^y is the stoichiometric coefficient of the j-th adsorbent component in adsorbed species i.

Equivalent balance equations for ion exchange sites –

$$\theta \frac{\partial N_k^{eq}}{\partial t} + \frac{\partial \theta}{\partial t} N_k^{eq} = -\theta \Lambda_k^{eq} + M_k^{eq} \ , \ k \in N_z \tag{22}$$

in which

$$N_k^{eq} = \sum_{i \in Mz(k)} v_i z_i \ , \ k \in N_z \tag{23}$$

where N_k^{eq} is the number of equivalents of the k-th ion-exchange sites per liter of solution [M/L^3], □$_k^{eq}$ is the total decay rate of the k-th ion-exchange site [M/L^3/T], M_k^{eq} is total rate of source/sink of the k-th ion-exchange site [M/L^3/T], N_z is the number of ion-exchange sites, $M_{z(k)}$ is the number of ion-exchanged species in the k-th ion-exchanged site, and v_i is the electric charge of the i-th ion exchanged species.

Governing equations for complexed species –

$$xi = \alpha_i^x \prod_{j=1}^{Na} c_j^{a_{ij}^x} \quad , i \in (M_x - K_x) \tag{24}$$

$$\theta \frac{\partial x_i}{\partial t} = L(x_i) - \theta \mathbf{l}_i^x + m_i^x - Qx_i - k_i^{bx} x_i + \alpha_i^{fx} \prod_{j=1}^{Na} c_j^{v_{ij}^x} + \frac{1}{\gamma_i^x} \sum_{k=1}^{Nk} (v_{ik}^x - \mu_{ik}^x) R_k \ , i \in K_x \tag{25}$$

where K_x is the number of kinetically complexed species, α_i^x is the modified stability constant of the i-th equilibrium complex species, k_i^{bx} is the backward rate constant of the i-th basic, kinetic complexation reaction [1/T], α_i^{fx} is the modified forward rate constant of the i-th basic, complexed reaction, γ_i^x is the activity coefficient of the i-th complexed species, N_k is the number of parallel kinetic reactions, $v_{ik}$is the stoichiometric coefficient of the i-th complexed species in the k-th parallel kinetic reaction associated with the products, μ_{ik}^x is the stoichiometric coefficient of the i-th complexed species in the k-th parallel kinetic reaction associated with the reactants, and R_k is the reaction rate of the k-th parallel reaction.

Governing equations for adsorbed species -

$$y_i = \alpha_i^y \left(\prod_{j=1}^{Na} c_j^{a_{ij}^y} \right) \left(\prod_{j=1}^{Na} s_j^{b_{ij}^y} \right) \quad , i \in (M_y - K_y) \tag{26}$$

$$\theta \frac{\partial y_i}{\partial t} + \frac{\partial \theta}{\partial t} y_i = -\theta \mathbf{l}_i^y + m_i^y - k_i^{by} y_i + \alpha_i^{fy} \prod_{j=1}^{Na} c_j^{a_{ij}^y} \prod_{j=1}^{Ns} s_j^{b_{ij}^y} + \frac{1}{\gamma_i^y} \sum_{k=1}^{Nk} (v_{ik}^y - \mu_{ik}^y) R_k \ , i \in K_y \tag{27}$$

where K_y is the number of kinetic adsorption reactions, α_i^y is the modified stability constant of the i-th equilibrium adsorbed species, k_i^{by} is the backward rate constant of the i-th basic, kinetic adsorption reaction [1/T], α_i^{fy} is the modified forward rate constant of the i-th basic, kinetic adsorption reaction, γ_i^y is the activity coefficient of the i-th adsorbed species, v_{ik}^y is the stoichiometric coefficient of the i-th adsorbed species in the k-th parallel kinetic reaction associated with the products, and μ_{ik}^y is the stoichiometric coefficient of the i-th adsorbed species in the k-th parallel kinetic reaction associated with the reactants.

Governing equations for ion exchanged species –

$$\kappa_{iJ(k)} = \frac{\left(z_i/s_{T(k)}\right)^{v_{J(k)}}}{\left(z_{J(k)}/s_{T(k)}\right)^{v_i}} \frac{a_{J(k)}^{v_i}}{a_i^{v_{J(k)}}} \quad , i \in (M_z - K_z), i \neq J(k), i \in M_{z(k)}, k \in N_z \tag{28}$$

$$\theta \frac{\partial z_i}{\partial t} + \frac{\partial \theta}{\partial t} z_i = -\theta I_i^z + m_i^z - \alpha_i^{bz} s_{T(k)} \left(z_i/s_{T(k)}\right)^{v_{J(k)}} a_{J(k)}^{v_i}$$
$$+ \alpha_i^{fz} s_{T(k)} \left(z_{J(k)}/s_{T(k)}\right)^{v_i} a_i^{v_{J(k)}} + \frac{1}{\gamma_i^z} \sum_{k=1}^{Nk} (v_{ik}^z - \mu_{ik}^z) R_k \,, \tag{29}$$
$$i \in K_z,\ i \neq J(k), i \in M_{z(k)}, k \in N_z$$

where K_z is the number of kinetic ion exchanged species; $\kappa_{iJ(k)}$ is the modified selectivity coefficient of the i-th ion exchanged species with respect to the J(k)-th ion exchanged species, or the effective equilibrium constant of the i-th ion-exchanged species; J(k) is the identification number of the referenced ion exchanged species of the k-th exchange site; a_i is the concentration of the i-th aqueous species denoting either c_j or x_i [M/L^3]; z_i is the concentration of the i-th ion-exchanged species [M/L^3]; $s_{T(k)}$ is the total concentration of all ion-exchanged species on the k-th site [M/L^3]; α_i^{bz} is the modified backward rate constant of the i-th basic, kinetic ion exchanged reaction [1/T], α_i^{fz} is the modified forward rate constant of the i-th basic, kinetic ion exchanged reaction, γ_i^z is the activity of the i-th adsorbed species, v_{ik}^z is the stoichiometric coefficient of the i-th ion exchanged species in the k-th parallel kinetic reaction associated with the products, and μ_{ik}^z is the stoichiometric coefficient of the i-th ion exchanged species in the k-th parallel kinetic reaction associated with the reactants.

Governing equations for precipitated species –

$$1 = \alpha_i^p \left(\prod_{j=1}^{Na} c_j^{a_{ij}^p} \right), \ i \in (M_p - K_p) \tag{30}$$

$$\theta \frac{\partial p_i}{\partial t} + \frac{\partial \theta}{\partial t} p_i = -\theta I_i^p + m_i^p + A_i \left(-\alpha_i^{bp} + \alpha_i^{fp} \prod_{k=1}^{Na} c_k^{v_{ik}^p} \right) +$$
$$\frac{A_k}{\gamma_i^p} \sum_{k=1}^{Nk} (v_{ik}^p - \mu_{ik}^p) R_k \ , i \in K_p \tag{31}$$

where K_p is the number of kinetic precipitated species, α_i^p is the modified stability constant of the i-th equilibrium precipitated, α_i^{bp} is the modified backward rate constant of the i-th basic, kinetic precipitation/dissolution reaction [1/T], α_i^{fy} is the modified forward rate constant of the i-th basic, kinetic precipitation/dissolution reaction, γ_i^p is the activity coefficient of the i-th precipitated species, v_{ik}^p is the stoichiometric coefficient of the i-th precipitated species in the k-th parallel kinetic reaction associated with the products, μ_{ik}^p is the stoichiometric coefficient of the i-th precipitated species in the k-th parallel kinetic reaction associated with the reactants, $A_i = 0$ if the i-th basic, kinetic precipitation/dissolution reaction is not to occur or $A_i = 1$ if the reaction is to occur, and $A_k = 0$ if the k-th parallel kinetic precipitation/dissolution reaction is not to occur or $A_k = 1$ if the reaction is to occur.

2.2.2. Initial and Boundary Conditions for Reactive Transport Simulations

To complete the description of the hydrologic transport and mass balance as given by Eq. (15), (20), (22), (25), (27), (29), and (31) initial and boundary conditions must be specified in accordance with dynamic and physical considerations. It will be assumed initially that the total analytical concentrations for all components and all kinetically controlled product species and the number of equivalents of the ion-exchange site are known throughout the region of interest as

$$T_j = T_{jo}(x) \quad \text{at } t=0 \ , j \in N_a \tag{32}$$

$$W_j = W_{jo}(x) \quad \text{at } t=0 \ , j \in N_s \tag{33}$$

$$x_i = x_{io}(x) \quad \text{at } t=0 \ , i \in K_x \tag{34}$$

$$y_i = y_{io}(x) \quad \text{at } t=0 \ , i \in K_y \tag{35}$$

$$z_i = z_{io}(x) \quad \text{at } t=0 \ , i \in K_z \tag{36}$$

$$p_i = p_{io}(x) \quad \text{at } t=0 \ , i \in K_p \tag{37}$$

$$N_k^{eq}(x) = N_{ko}^{eq} \quad \text{at } t=0 \ , i \in N_z \tag{38}$$

where $T_{jo}(\mathbf{x})$ is the initial total analytical concentration of the j-th aqueous component [M/L^3], $W_{jo}(\mathbf{x})$ is the initial total analytical concentration of the j-th adsorbent component [M/L^3], $x_{io}(\mathbf{x})$ is the initial total analytical concentration of the i-th kinetically controlled aqueous complexed species [M/L^3], $y_{io}(\mathbf{x})$ is the initial total analytical concentration of the i-th kinetically controlled adsorbed species [M/L^3], $z_{io}(\mathbf{x})$ is the initial total analytical concentration of the i-th kinetically controlled ion-exchanged species [M/L^3], $p_{io}(\mathbf{x})$ is the initial total analytical concentration of the i-th kinetically controlled precipitated species [M/L^3], and $N_{ko}^{eq}(x)$ is the initial number of equivalents for the k-th ion-exchange site [M/L^3].

Equilibrium product species concentrations are obtained from mass action equations involving the components, and hence do not have to be independently specified. Initial concentrations for aqueous components and kinetically controlled complexed species may be obtained from field measurements or by solving the steady-state version of Equations (15) and (25) with time-invariant boundary conditions.

The mathematical specification of boundary conditions is the one of the most difficult and intricate tasks in multicomponent transport modeling. From the dynamic point of view, a boundary segment may be classified as either flow-through or impervious. From a physical point of view, it is a soil-air interface, a soil-soil interface, or a soil-water interface. From the mathematical point of view, it may be treated as a Dirichlet boundary, for which the total analytical concentration is prescribed; a Neumann boundary, for which the flux due to the gradient of total analytical concentration is known; or a Cauchy boundary, for which the total flux is given. An even more difficult mathematical boundary is the variable condition, in which the boundary conditions are not known *a priori* but are themselves part of the solution. In other words, on the mathematically variable boundary, either Neumann or Cauchy conditions may prevail and change with time. Which condition prevails at a particular time can be determined only in the cyclic process of solving the governing equations (Freeze 1972a, 1972b; Yeh and Ward 1980, 1981).

Whatever point of view is chosen, all boundary conditions eventually must be transformed into mathematical equations for quantitative simulations. Thus, we can specify the boundary conditions from the mathematical point of view in concert with dynamic and physical considerations. Boundary conditions must be specified for all chemical entities that are subject to hydrologic transport, that is all aqueous components and all kinetically controlled aqueous complexed species. Equilibrium aqueous complexed species concentrations are obtained from mass action equations involving the aqueous components, and hence do not have to be independently specified. The boundary conditions imposed on any segment of the boundary can be either Dirichlet, Neumann, Cauchy, or variable for T_j with $j = 1, 2, \ldots, N_a$ and for x_i with $i = 1, 2, \ldots, K_x$ independently of each other. Thus, for any T_j or x_i, the global boundary may be split into four parts: B_D, B_N, B_C, and B_V, denoting Dirichlet, Neumann, Cauchy, and variable boundaries, respectively. The conditions imposed on the first three types of boundaries are given as

$$T_j = T_{jd}(x_b, t) \quad \text{on } B_d \ , \ j \in N_a \tag{39}$$

$$x_i = x_{id}(x_b, t) \quad \text{on } B_d \ , \ i \in K_x \tag{40}$$

$$-\mathbf{n} \cdot \left(\theta \mathbf{D} \cdot \Delta C_j\right) = q_{jn}(x_b, t) \quad \text{on } B_n \ , j \in N_a \tag{41}$$

$$-\mathbf{n} \cdot \left(\theta \mathbf{D} \cdot \Delta x_i\right) = q_{in}(x_b, t) \quad \text{on } B_n \ , i \in K_x \tag{42}$$

$$-\mathbf{n} \cdot \left(\mathbf{V} C_j - \theta \mathbf{D} \cdot \Delta C_j\right) = q_{jc}(x_b, t) \quad \text{on } B_c \ , j \in N_a \tag{43}$$

and

$$-\mathbf{n} \cdot \left(\mathbf{V} x_i - \theta \mathbf{D} \cdot \Delta x_i\right) = q_{ic}(x_b, t) \quad \text{on } B_c \ , j \in K_x \tag{44}$$

where $T_{jd}(\mathbf{x}_b,t)$ is the prescribed Dirichlet total analytical concentration for aqueous component j [M/L^3], $x_{id}(\mathbf{x}_b,t)$ is the prescribed Dirichlet total analytical concentration for kinetic complexed species i [M/L^3], $q_{jn}(\mathbf{x}_b,t)$ is the normal Neumann flux of the j-th aqueous component [$M/L^2/T$], $q_{in}(\mathbf{x}_b,t)$ is the normal Neumann flux of the i-th kinetic complex species [$M/L^2/T$], $q_{jc}(\mathbf{x}_b,t)$ is the normal Cauchy flux of the j-th aqueous component [$M/L^2/T$], $q_{ic}(\mathbf{x}_b,t)$ is the normal Cauchy flux of the i-th kinetic complex species [$M/L^2/T$], and $\mathbf{n}$ is an outward unit vector normal to the boundary.

The conditions imposed on the variable-type boundary, which is normally the soil-air interface or soil-water interface, are either the Neumann with zero gradient flux or the Cauchy with known total flux. The former is specified when the water flow is directed out of the region from the far away boundary, whereas the latter is specified when the water flow is directed into the region. This type of variable condition would normally occur at flow-through boundaries. Written mathematically, the variable boundary condition is given by

$$-\mathbf{n}\cdot\left(\theta\mathbf{D}\cdot\Delta C_j\right)=0 \text{ if } \mathbf{V}\cdot\mathbf{n}>0 \text{ on } B_v\ ,\ j\in N_a \tag{45}$$

$$-\mathbf{n}\cdot\left(\theta\mathbf{D}\cdot\Delta x_i\right)=0 \text{ if } \mathbf{V}\cdot\mathbf{n}>0 \text{ on } B_v\ ,\ i\in K_x \tag{46}$$

and

$$-\mathbf{n}\cdot\left(\mathbf{V}C_j-\theta\mathbf{D}\cdot\Delta C_j\right)=q_{jv}(x_b,t) \text{ if } \mathbf{V}\cdot\mathbf{n}<0 \text{ on } B_v\ ,\ j\in N_a \tag{47}$$

$$-\mathbf{n}\cdot\left(\mathbf{V}x_i-\theta\mathbf{D}\cdot\Delta x_i\right)=q_{iv}(x_b,t) \text{ if } \mathbf{V}\cdot\mathbf{n}<0 \text{ on } B_v\ ,\ i\in K_x \tag{48}$$

where $q_{jv}(\mathbf{x}_b,t)$ is the normal variable flux of the j-th aqueous component [$M/L^2/T$] and $q_{iv}(\mathbf{x}_b,t)$ is the normal variable flux of the i-th kinetic complex species [$M/L^2/T$]. The specification of Dirichlet, Neumann, Cauchy, and variable-type boundary conditions allows a wide range of physical boundaries needed for subsurface transport problems.

2.3 Coupling Between Fluid Flows and Reactive Chemical Transport

The density-dependence of fluid flows are described via the following constitutive relationships (Cheng, 1995)

$$\rho=\rho_o+\sum_{j=1}^{Na}C_jM_j\ ,\ \mu=\mu_o+\sum_{j=1}^{Na}a_jC_jM_j \tag{49}$$

where M_j is the molecular weight of the j-th aqueous component. The interactions between reactive transport and fluid flows are described by the following dependence of moisture content, hydraulic conductivity, and dispersion coefficient on concentrations of minerals (Yeh, et al., 1999)

$$\theta_s=\frac{\theta_{so}}{1+S\varphi_p}\ ,\ \theta=\frac{S\theta_{so}}{1+S\varphi_p},\ K_{so}=K_{soo}\left(\frac{1}{1+S\varphi_p}\right)^n \tag{50}$$

$$\mathbf{D} = \left[a_T \left|\mathbf{V}\right| \delta + \left(a_L - a_T\right)\mathbf{V}\mathbf{V}/\left|\mathbf{V}\right| + \theta D_w \tau \delta\right]\left[\theta^{m-1}\left(1-\varphi_P\right)^m\right] \tag{51}$$

where θ_s is the effective saturated moisture content including the effect of solid or surface species; θ_{so} is the effective saturated moisture content without including the effect of solid or surface chemical species; θ is the effective moisture content, S is the degree of water saturation which is a function of the pressure head; $\phi_p = \Sigma p_i \mathcal{V}_i$, where p_i is the precipitated concentration of the i-th mineral [mole/dm^3 of water], $\mathcal{V}_i$ is the mole volume of the i-th mineral [dm^3 of solid/mole]; K_{soo} is the referenced saturated hydraulic conductivity without density effect nor the effect of solid/surface species; n is the fractal exponent for estimating hydraulic conductivity based on particle size and packing structure; a_T is the transverse diffusivity [L]; |**V**| is the magnitude of the Darcy velocity **V** [L/T]; δ is the Kronecker delta tensor, a_L is the longitudinal diffusivity [L]; D_w is the molecular diffusion coefficient of water without the presence of precipitated species [L^2/T]; τ is the tortuosity; and m is the cementation exponent. Dullien (1979) reports values between 1.3 and 2.5 for the cementation exponent.

2.4 Generalized Reactive Chemical Transport Equations

In Section 2.2 we have presumed that one can easily identify the chemical components so all equilibrium reactions and some kinetic reactions can be written in basic forms. In a complicated reactive system with both kinetic and equilibrium reactions, the designation of chemical components may not be obvious when there are many parallel kinetic reactions. Therefore, to model a reactive transport system under mixed kinetic and equilibrium reactions, we will start from the fundamental and employ matrix methods to derive the governing equations.

A reactive system is completely defined by specifying chemical reactions and counting total number of chemical species involved in the reactions. Let N_R be the number of chemical reactions and M the number of chemical species in the system. The transport of M chemical species under N_R chemical reactions can be described by a system of M partial differential equations from a mathematical point of view as

$$\frac{\partial \theta G_i}{\partial t} = L(G_i) - \theta \Lambda_i + M_i - qG_i + \theta r_i\big|_{N_R}, \quad i \in M \tag{52}$$

in which

$$q = \frac{p^*}{p} q - F\frac{\partial h}{\partial t} - \frac{\rho_o}{\rho}\mathbf{V}\cdot\Delta\frac{\rho}{\rho_o} \tag{53}$$

where G_i (G_i may denote c_j, s_j, x_i, y_i, z_i, or p_i in Section 2.2) is the concentration of the i-th species, t is time, $r_i|_{N_R}$ is the production/consumption rate of the i-th species due to N_R reactions, 7_i is the decay rate of the i-th species, M_i is the source/sink rate of the i-th species, and M is the number of species present in the system. Eq. (52) is a statement of material balance of any species in a reactive system. It simply states that the rate of change of mass of any species in a bulk volume is due to hydrologic transport, decay, artificial source/sinks, and the production/consumption rate due to all chemical reactions that this species is involved.

The central issue in modeling reactive chemical transport is the formulation of $r_i|_{N_R}$ and the determination of parameters associated with it. There may be two general means of formulating $r_i|_{N_R}$: *ad hoc* and reaction-based formulations. *Ad hoc* formulations have been employed to model the production/transformation rate $r_i|_{N_R}$ in practically all kinetic experiments of subsurface geochemical and biochemical processes. In an *ad hoc* formulation, the production/transformation rate is simply obtained with empirical functions as

$$r_i\Big|_{N_R} = f_i(G_1, G_2, \ldots\ldots, G_M; p_1, p_2, \ldots.) \tag{54}$$

where f_i is the empirical function for the i-th species and p_1, p_2, ... are rate parameters used to fit experimental data. This formulation is purely empirical, it does not consider the contribution of individual reactions. The problem with an *ad hoc* approach is that it will result in an inadequate parametrization in which the rate parameters are not true constants; thus they are only applicable to the exact experimental conditions tested. *Ad hoc* approaches litter current literature.

Reaction-based formulations should be taken if we are to obtain true constants broadly applicable to diverse geochemical and biochemical conditions. In a reaction-based formulation, the production/consumption rates per unit bulk volume of M chemical species is described by

$$\theta r_i\Big|_{N_R} = \sum_{k=1}^{N_R} (v_{ik} - \mu_{ik}) R_k, \ i \in M; \text{or } \mathbf{U\theta r}\Big|_{N_R} = \mathbf{vR} \tag{55}$$

where R_k is the reaction rate of the k-th reaction, v_{ik} is the reaction stoichiometry of the i-th species in the k-th reaction associated with the products, μ_{ik} is the reaction stoichiometry of the i-th species in the k-th reaction associated with the reactants, $\theta r\Big|_{N_R}$ is the rate vector of production/consumption with $\theta r_i\Big|_{N_R}$ as its components, **U** is a unit matrix, **v** is the reaction stoichiometry matrix with (v_{ik} - μ_{ik}) as its entries, and **R** is the reaction-rate vector with N_R reaction rates as its components. Once the rate of all reactions R_k are specified, the spatial-temporal distribution of M species can be obtained by integrating Eq. (52) subject to appropriate initial and boundary conditions as described in Section 2.3.

An analytical solution of Eq. (52) is, in general, not possible. Numerical approximations offer the best hope to solve Eq. (52). Numerical integration of Eq.

(52) in its primitive form will encounter two major difficulties. First, the reaction rates of N_R reactions can, in general, range over several orders of magnitude. The time-step size used in numerical integration is dictated by the largest reaction rate among N_R reactions. If at least one of the reactions is infinitely large (i.e., the reaction can reach equilibrium instantaneously), then the time-step size must be infinitely small, which makes integration impractical. Second, the number of independent reactions is less than the number of species (i.e., the rank of the reaction stoichiometry matrix, **v**, is less than M) for almost all problems of practical interest. This implies that there are one or more chemical components whose mass must be conserved during the reactions. Chemical components are distinct from chemical species in that the mass of any chemical component must be conserved with respect to reactions (Rubin, 1983). Under such circumstances, the integration of M simultaneous second-order partial differential equations of species-transport in Eq. (52) may not guarantee mass conservation of chemical components due to numerical errors. Thus, numerical integration of Eq. (52) will make sense only when the following two conditions are met: (1) all reactions are slow and their rates are comparable within a narrow range, and (2) the rank of **v** is equal to M. For practical problems, it is rare that these two conditions can be simultaneously met. Therefore, Eq. (52) must be manipulated to decouple fast equilibrium reactions from slow kinetic reactions and to explicitly enforce mass conservation of chemical components. This can be done via Gauss-Jordian elimination or QR decomposition of **v** (Chilakapati, 1995).

Eq. (52) written in the matrix form can be decomposed based on the type of chemical reactions. Let us assume that among N_R reactions, there are N_E equilibrium (i.e., fast) reactions and N_K kinetic (i.e., slow) reactions: i.e, $N_R = N_E + N_K$. A reaction is fast if it can instantaneously reach equilibrium among all species involved in the reaction. In other words, the rate of an equilibrium reaction is infinity. An infinite rate is mathematically represented by a mass action equation. For equilibrium reactions, we only need to consider the case when N_E reactions are linearly independent since any equilibrium reaction that is linearly dependent on other equilibrium reactions is a redundant reaction. This redundancy results from the representation of equilibrium reactions by mass action. On the other hand, the rate of a slow reaction is finite.

If the rank of the reaction stoichiometry matrix **v** is N_I, then there are N_I linearly independent reactions among N_R chemical reactions. Obviously, N_I must be less than or equal to M based on Eq. (52). Let us denote $N_C = M - N_I$, where N_C represents the number of chemical components, and $N_D = N_R - N_I$, where N_D represents the number of dependent reactions. Furthermore, it is clear that the N_I linearly independent reactions are made of N_E equilibrium reactions and a subset of N_K kinetic reactions. With these definitions, Eq. (52) is decomposed or diagonalized into the following equation via the Gaussian-Jordan elimination or the QR decomposition (Chilakapati et al., 1998)

$$\begin{bmatrix} \mathbf{A}_{11} & \mathbf{A}_{12} \\ \mathbf{A}_{21} & \mathbf{A}_{22} \end{bmatrix} \left(\frac{\partial \theta \mathbf{G}}{\partial t} + \theta \Lambda - \mathbf{M} + q\mathbf{G} \right) = \begin{bmatrix} \mathbf{B}_{11} & \mathbf{B}_{12} \\ \mathbf{B}_{21} & \mathbf{B}_{22} \end{bmatrix} L(\mathbf{G}) + \begin{bmatrix} \mathbf{D} & \mathbf{K} \\ \mathbf{0}_1 & \mathbf{0}_2 \end{bmatrix} \mathbf{R} \tag{56}$$

where **G** is the concentration vector of size M with G_i as its components, □ is the decay rate vector of size M with $□_i$ as its components, **M** is the source/sink vector of size M with M_i as its components, $\mathbf{A}_{11}$ is a submatrix of the reduced **U** with size of N_I x N_I, $\mathbf{A}_{12}$ is a submatrix of the reduced **U** with size of N_I x N_C, $\mathbf{A}_{21}$ is a submatrix of the reduced **U** with size of N_C x N_I, $\mathbf{A}_{22}$ is a submatrix of the reduced **U** with size of N_C x N_C, **D** is the diagonal matrix representing a submatrix of the reduced □ with size of N_I x N_I reflecting the effects of N_I linearly independent reactions on the production rate of all kinetic variables (a kinetic variable is a combination of chemical species; the number of kinetic variables is equal to the number of linearly independent reactions, N_I, minus the number of equilibrium reactions, N_E), **K** is a submatrix of the reduced **v** with size of N_I x N_D reflecting the effects of N_D dependent reactions, $\mathbf{0}_1$ is a zero matrix representing a submatrix of the reduced **v** with size N_C x N_I, $\mathbf{0}_2$ is a zero matrix representing a submatrix of the reduced **v** with size N_C x N_D. $\mathbf{B}_{11}$, $\mathbf{B}_{12}$, $\mathbf{B}_{21}$, and $\mathbf{B}_{22}$ are defined similar to $\mathbf{A}_{11}$, $\mathbf{A}_{12}$, $\mathbf{A}_{21}$, and $\mathbf{A}_{22}$, respectively, but with entries of zeros associated with immobile species.

The reduction of Eq. (52) to Eq. (56) is effectively reducing a set of M simultaneous partial differential equations (PDEs) to three subsets of equations: the first contains N_E nonlinear algebraic equations representing the law of mass action for N_E equilibrium reactions, the second contains (N_I-N_E) simultaneous partial differential equations representing the transport of (N_I-N_E) kinetic variables, and the third contains N_C component transport equations representing mass balance of N_C chemical components. The reduction enables us to make the following inductions. First, from the first N_E PDEs in Eq. (56), any dependent, kinetic reaction that is linearly dependent on only N_E equilibrium reactions is irrelevant to the system. Second, if all N reactions are linearly independent, then the **KR** on the second (N_I-N_E) PDEs is not present. This implies that each kinetic reaction corresponds to a kinetic variable, which is also called the extent of reaction under the circumstance (Lightner and Setch, 1996). Third, the final N_C PDEs reveal that in a reactive system, the mass of each of the N_C chemical components must remain invariant with respect to chemical reactions. Furthermore, the reduction of Eq. (52) to Eq. (56) is not unique. This non-uniqueness will allow an adaption of governing equations for robust numerical solutions.

Rate Formulation

A general, chemical reaction (be it a basic reaction or a parallel reaction) can be written

$$\sum_{i=1}^{M} \mu_{ik} \hat{G}_i \Leftrightarrow \sum_{i=1}^{M} \nu_{ik} \hat{G}_i \ , k \in N_R \tag{57}$$

where G_i is the chemical formula of the i-th species involved in k reactions. The key in modeling a reactive system is the formulation of rate laws for all N_R reactions specified by Eq. (57).

If a reaction is fast that it can reach equilibrium instantaneously, its rate is infinity resulting in the law of mass action as

$$R_K = : \Rightarrow \ K_k^e = \left(\prod_{i=1}^{M} (A_i)^{\nu_{ik}} \right) \Big/ \left(\prod_{i=1}^{M} (A_i)^{\mu_{ik}} \right), \ k \in N_E \tag{58}$$

where R_k is the reaction rate of the k-th reaction, K_k^e is the equilibrium constant of the k-th reaction, A_i is the activity of the i-th species, and N_E is the number of linearly independent equilibrium reactions. The equilibrium constants can be determined one by one with the measurement of the activities of all species using Eq. (58). Furthermore, only linearly independent equilibrium reactions need to be considered because the mass action equation for any linearly dependent reaction can be obtained by the combination of the N_E equations of Eq. (58).

For an elementary kinetic reaction, the finite rate based on the collision theory is used to describe the process as

$$R_K = \left(K_k^f \prod_{i=1}^{M} (A_i)^{\mu_{ik}} - K_k^b \prod_{i=1}^{M} (A_i)^{\nu_{ik}} \right) = \left(k_k^f \prod_{i=1}^{M} (C_i)^{\mu_{ik}} - k_k^b \prod_{i=1}^{M} (C_i)^{\nu_{ik}} \right),$$
$$k \in N_K \text{ with } k_k^f = K_k^f \prod_{i=1}^{M} (\gamma_i)^{\mu_{ik}} \text{ and } k_k^b = K_k^b \prod_{i=1}^{M} (\gamma_i)^{\nu_{ik}} \tag{59}$$

where K_k^f is the forward rate constant of the k-th kinetic reaction, K_k^b is the backward rate constant of the k-th kinetic reaction, k_k^f is the modified forward rate constant of the k-th kinetic reaction (which is numerically equal to K_k^f when the activity coefficients of all reactant species are equal to 1.0), k_k^b is the modified backward rate constant of the k-th kinetic reaction (which is numerical equal to K_k^b if the activity coefficients of all product species are equal to 1.0), and γ_i is the activity coefficient of the i-th species, and N_K is the number of kinetic reactions. It is seen that the forward and backward rate constants cannot be determined one by one anymore with the measurement of concentration-vs-time curves of all species because N_K equations in Eq. (59) are coupled with respect to the forward and backward rate constants.

When a reaction is not of the elementary type, its rate law can be formulated empirically or based on reaction pathways. Although the approaches of empirically formulating reaction rates are one step ahead of *ad hoc* approaches, they are still not satisfactory. For scientific research, not for engineering solutions, one should always try to identify a mechanism or a pathway to represent a non-elementary reaction, even though we realize that identifying the pathway is extremely difficult and each identification may be a Nobel prized-grade research. Without the mechanistic approaches, the science of modeling geochemical kinetics will go nowhere but will continue to be littered with *ad hoc* approaches or with empirical reaction rates, a necessary evil for the immediate term. For long-term research goals, we advocate the use of mechanistic-based reaction rates and discourage the use of empirical reaction rates or *ad hoc* approaches.

A mechanism (pathway) is the sequence of elementary steps (an elementary step is a reaction that can be described by the collision theory or mass action) that describes how the final products are formed from the initial reactants. A mechanism determines the overall reaction. Consider, for example, the decomposition of N_2O_5

$$2N_2O_5 \rightarrow 4NO_2 + O_2 \tag{60}$$

If the above reaction were an elementary step, its rate equation would be given by the collision theory as

$$r = k^f (N_2O_5)^2 - k^b (NO_2)^4 (O_2) \tag{61}$$

where r is the reaction rate; k^f is the modified forward rate constants; k^b is the modified backward rate constant; and (N_2O_5), (NO_2) and (O_2) denote the concentrations of N_2O_5, NO_2, and O_2, respectively. Experimental evidence shows that the reaction rate does not follow Eq. (61) (Ogg, Jr., 1947). Therefore, the reaction was hypothesized to follow a three-step mechanism (Smith, 1981)

$$\begin{gathered} N_2O_5 \underset{k_1}{\overset{k_{-1}}{\Leftrightarrow}} NO_2 + NO_3 ,\; NO_2 + NO_3 \xrightarrow{k_2} NO + O_2 + NO_2 \;\; \text{(slow)} , \\ \text{and } NO + NO_3 \xrightarrow{k_3} 2\,NO_2 \end{gathered} \tag{62}$$

The above three-step mechanism yielded the following rate equations

$$r = \frac{d(O_2)}{dt} = r_2 = \frac{k_1 k_2}{k_{-1}} (N_2O_5) \tag{63}$$

which has proved to describe the decomposition of N_2O_5 quite well (Ogg, Jr., 1947). Equation (60) is not elementary because the N_2O_5 molecule is too complex, contains too many bonds, to decompose all the way to the simple molecules, O_2, and NO_2, in one step. The intrinsic reaction parameters representing the reaction described by Eq. (60) are k_{-1}, k_1, and k_2, not K^b and K^f. To mechanistically model reactive chemical transport, reaction pathways must be identified so that the reaction parameters are true constants. Only when the reaction parameters are true constants, their scale-up from experiments to applications can be adequately performed.

Another example is the dissimilatory iron-reducing bacteria (DIRB)-mediated reduction of ferric oxide with hydrogen as the electron donor

$$2 = FeOH + H_2 + 4H^+ \rightarrow 2Fe^{2+} + P_i's \qquad (64)$$

which represents an oxidation-reduction, where $=Fe(OH)$ is the electron acceptor, H_2 is the electron donor, and $P_i's$ are the final products. No carbon source is shown in the above reaction to imply non-growth conditions for the DIRB. It is very unlikely that the reaction in Eq. (64) is an elementary reaction, i.e., its rate equation is probably not described by the collision theory. Either an empirical rate equation is used or a mechanism must be conjectured. We believe that a conjectured mechanism is better than an empirical approach. Emulating Henri-Michaelis-Benton-Monod's postulation of biodegradation involving both electron donor/carbon sources and an electron acceptor, we may hypothesize the following mechanism (Segel, 1993):

$$E + H_2 \underset{}{\overset{K_D}{\rightleftharpoons}} EH_2$$

$$+ \qquad\qquad +$$

$$2 =FeOH \qquad\qquad 2 =FeOH$$

$$K_A \updownarrow \qquad\qquad \alpha K_A \updownarrow$$

$$E(=FeOH)_2 + H_2 \underset{}{\overset{\alpha K_D}{\rightleftharpoons}} EH_2(=FeOH)_2 \xrightarrow{k_p} E + P_i's$$

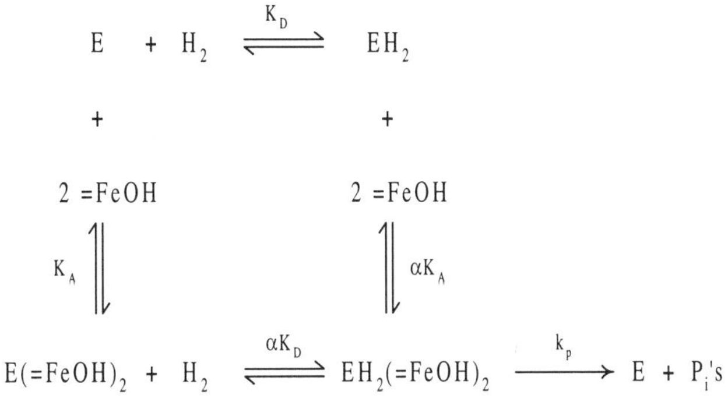

where E is an enzyme, EH_2 is the enzyme-electron donor complex, $E(=FeOH)_2$ is the enzyme-electron acceptor complex, and $EH_2(=FeOH)_2$ is the enzyme-electron donor-electron acceptor complex. These complexes are the intermediate species leading to the final products $P_i's$. In one mechanism, the two electrons donated by hydrogen (represented by the reaction on the top line) are accepted by $=FeOH$ (represented by the reaction on the right column) to form the complex $EH_2(=FeOH)_2$. Since, the order of donation and acceptance is random, there may be another way to state this donation-acceptance reaction. The two electrons accepted by $=FeOH$ (represented by the reaction on the left column) are from the donation of hydrogen to form the complex $EH_2(=FeOH)_2$ (represented by the reaction on left portion of the bottom line). Finally, the complex $EH_2(=FeOH)_2$ is transformed by the DIRB to the final products, $P_i's$ (represented by the reaction on the right portion of the bottom line).

With an appropriate, algebraic manipulation of the above pathway, we obtain the overall rate equation for the reaction in Eq. (64) as

$$r = \frac{d(P_i's)}{dt} = \frac{TOT_E \cdot \mathbf{k}_p(H_2)(=FeOH)^2}{\alpha K_D K_A + \alpha K_A(H_2) + \alpha K_D(=FeOH)^2 + (H_2)(=FeOH)^2} \tag{65}$$

in which

$$TOT_E = (E) + (EH_2) + (E=(Fe(OH)_2) + (EH_2(=FeOH)_2) \tag{66}$$

where r is the overall reaction rate, TOT_E is the total enzymatic site concentration, k_p is the rate constant, α is a characteristic parameter of the enzymatic site, K_D is the half saturation constant for the electron donor, and K_A is the half saturation constant for the electron acceptor (note: $k_p.TOT_E.Y$ is known as the maximum reaction rate, μ_{max}, to microbiologists and environmental chemists, with Y being the specific yield). To avoid the maze of algebraic manipulation to come up with Eq. (66), we can simply conduct a direct simulation of the following four simultaneous reactions

$$EH_2 \Leftrightarrow E + H_2 \ \text{(fast reaction : } K_D\text{)} \tag{67}$$

$$E(=FeOH)_2 \Leftrightarrow E + 2 = FeOH \ \text{(fast reaction : } K_A\text{)} \tag{68}$$

$$EH_2(=FeOH)_2 \Leftrightarrow EH_2 + 2 = FeOH \ \text{(fast reaction : } \alpha K_A\text{)} \tag{69}$$

$$E(=FeOH)_2 \rightarrow E + P_i's \ \text{(irreversible slow reaction : } k_p\text{)} \tag{70}$$

The above four equations describe the proposed pathway. If fact, the overall rate equation was derived based on these four elementary reactions.

3. NUMERICAL SOLUTIONS

The governing equations of the flow and reactive transport with appropriate initial and boundary conditions cannot, in general, be solved analytically using current applied mathematics. Hence, numerical solutions are required. The flow equations are discretized with the Galerkin finite element method. The transport equations are discretized with either the conventional finite element methods or the hybrid Lagrangian-Eulerian finite element methods. The nonlinearities in the fluid flow, reactive transport, and interaction between the two are dealt with the Picard linearization. However, the nonlinearity in geochemistry was dealt with the Newton-Raphson linearization.

3.1 Strategies to Solve Coupled Transport and Geochemical Reaction Problems

Equations (15) through (17) and (20) through (31) constitute a system of equations describing reactive chemical transport for the set of unknowns: T_j, c_j, C_j, S_j, P_j, W_j, s_j, N_k^{eq}, x_i, y_i, z_i, and p_i. Because the number of equations in the system is large, on the order of hundreds for most practical applications, the system is divided into two subsystems: hydrologic transport and geochemical reactions. The subsystem of hydrologic transport includes Eq. (15) for T_j, Eq. (20) for W_j, Eq. (22) for N_k^{eq}, Eq. (25) for x_i where $i \in K_x$. The remaining equations make up the geochemical reaction subsystem.

The strategy of solving coupled hydrologic transport and mixed geochemical equilibrium/kinetic reaction problems is to solve the two subsystems of equations iteratively. Three different approaches may be used to reach a convergent solution. The first approach is a complete iteration between the two subsystems. The second approach is the use of operator splitting, and the third approach is the employment of the predictor-corrector method.

In all three approaches, the total analytical concentrations of all aqueous components are solved in the first subsystem with the total dissolved concentrations of all aqueous components given from the previous iteration. The differences among the three approaches lay in how the kinetically complexed species are solved between the two subsystems. For the first approach, the kinetically controlled complexed species are solved in the first subsystem with the concentrations of all component species given from the previous iteration, and they will not be solved again in the second subsystem. Thus, this approach is applicable only when all the kinetic reactions are "basic kinetic" reactions (a basic reaction refers to a reaction that involves only component species as the reactants and only one product species). Extension of this approach to general "mixed kinetic" reactions (mixed kinetic reactions refer to the presence of both basic kinetic and parallel reactions) would involve large CPU memory to store huge arrays required for the computation of forward and backward rates at every grid point. Thus, when "mixed kinetic" reactions are involved, the second and third approaches are employed.

For the second and third approaches, the concentrations of kinetically complexed species are solved in both the first and second subsystem and there is no need to iterate the concentrations of kinetic x_i between two iterations. The difference between the second and third approaches is how the kinetic x_i are computed in both subsystems (Yeh et al., 1999).

The solution procedure for the first approach, the direct iteration method, for every time step is outlined below:

1. Solve Eq. (20) and (22) for W_j and N_k^{eq}.
2. Solve Eq. (15) for conservative aqueous components T_j.
3. Make an initial guess for nonconservative aqueous components TW_j.
4. Solve Eq. (25) for kinetic complexed species x_i.

5. With known W_j, N_k^{eq}, kinetic x_i, conservative T_j, and guessed nonconservative TW_j, solve Eqs. (16), (21), (23), (24), and (26) through (31) for c_j, s_j, z_i, p_i, equilibrium x_i, and y_i.
6. Compute C_j, S_j, and P_j using Eq. (17).
7. Solve Eq. (15) for nonconservative aqueous components T_j (and $TW_j = T_j$ for conservative aqueous components) using the newly computed C_j, S_j, and P_j. Solve Eq. (25) for kinetic complexed species x_i using all species concentrations from step 5.
8. Revise the guess for nonconservative aqueous components TW_j based on T_j and old TW_j. Revise the guess for kinetic, complexed species based on x_i and xw_i.
9. With known W_j, N_k^{eq}, K_x x_i, conservative T_j, and newly guessed nonconservative T_j, solve Eqs. (16), (21), (23), (24), and (26) through (31) for c_j, s_j, z_i, p_i, equilibrium x_i, and y_i.
10. Compute C_j, S_j, and P_j using Eq. (17).
11. Solve Eq. (15) for nonconservative aqueous components T_j using the newly computed C_j, S_j, and P_j. Solve Eq. (25) for kinetic, complexed species x_i using all species concentrations from step 9.
12. Check convergence by comparing the newly obtained T_j and the revised TW_j in step 8, and comparing the newly obtained x_i and the revised xw_i in step 8.
13. If a convergent solution is obtained, then proceed to the next time-step computation. If the solution is not convergent, repeat Steps 8 through 11.

Let us assume that there are N_c conservative aqueous components and N_r nonconservative aqueous components (i.e., $N_a = N_c + N_r$). The solution of N_s immobile adsorbent components W_j, N_z sites of the cation ion-exchange capacity N_k^{eq}, and N_c conservative aqueous components among the T_j is independent of all other components. However, the solution of N_r nonconservative aqueous components among the T_j is not independent of all other components because the N_r nonconservative aqueous components are coupled. The N_r nonconservative aqueous components must be solved either simultaneously or iteratively. The simultaneous solution of N_r partial differential equations governing the transport of nonconservative aqueous components and the solution of the geochemical reaction equations constitute the major effort in terms of computational time and computer storage.

3.2. Strategies to Solve Coupled Fluid Flow and Reactive Chemical Transport

The strategy of solving coupled fluid flow and reactive chemical transport is to solve two systems of equations iteratively. The solution procedure for every time step is outlined below:

1. Guess total dissolved concentration of aqueous components, C_j and concentrations of precipitated species, p_i.
2. Solve density-dependent and reactive transport-affected flows governed by Eq. (1) through (4) and Eq. (49) and (50) for pressure head h, moisture content θ, and the Darcy velocity **V**.
3. Solve reactive transport problems governed by Eq. (15) through (31) and Eq. (51) to obtain total analytical concentrations of aqueous and adsorbent components, T_j and W_j, ion exchange sites, N_k^{eq}, and concentrations of all chemical species, c_j, s_j, z_i, p_i, x_i, and y_i.
4. Compute total dissolved concentrations of aqueous components, C_j.
5. Check convergence of the pressure head, h, total analytical concentrations of aqueous and adsorbent components, T_j and W_j, ion exchange sites, N_k^{eq}, and concentrations of kinetically controlled complex species x_i.
6. If a convergent solution is obtained, then proceed to the next time-step computation.
7. If the solution is not convergent, revise h, T_j, W_j, N_k^{eq}, and kinetic x_i. Then repeat Steps 2 through 5.

4. EXAMPLE PROBLEMS

A computer code was developed to implement the numerical approximations (Section 3) of the mathematical model (Section 2). A total of six example problems are presented in this section. Problems 1 through 3 exemplify the capability of the code to deal with various combinations of the flow and reactive transport problems. Problem 1 simulates steady-state flow and both steady and transient reactive transport. Here the initial conditions for the transient transport are obtained by its steady-state solution and the steady-state flow is used for both steady-state and transient transport. Problem 2 simulates both the steady-state and transient flows and the transient transport. In this example, the simulation of transient reactive transport requires a transient flow field which is obtained with the transient flow simulation using the simulated steady-state flow as the initial conditions. Problem 3 simulates the steady-state flow and transport and the transient flow and transport. Initial conditions for the transient flow are obtained from the steady-state flow simulation. Similarly, the initial conditions for the transient reactive transport are obtained from its steady-state simulation.

Problem 4 is a diffusion-reaction problem. It was designed to indicate the capability of the code to simulate the coupled diffusion and precipitation-dissolution. The effects of precipitation-dissolution on diffusion are demonstrated. Problem 5 is a coupled flow and reactive transport problem. It is intended to illustrate the capability of the code to simulate the interactions and feedback between the precipitation/dissolution and the fluid flows and advection-dispersion. The effects of precipitation/dissolution on both transport and fluid flow are elucidated. Problem 6 involves reactive transport under a suite of geochemical processes. The geochemical system is rather extensive. As a result, the geochemical problem is difficult to solve. Innovative algorithms are required to tackle this difficulty. Species switching provides a powerful tool to deal with geochemical problems which are characterized by low concentrations of component species. The main purpose of Problem 6 is to demonstrate the success of the species switching in the code.

4.1 Problem 1: Flow and Advective-Dispersive-Reactive Transport

This problem involves steady flow and combined steady and transient transport in a rectangular domain. The region of interest is bounded on the left and right by two symmetric drains (20 m apart), on the bottom by an impervious aquifuge, and on the top by an air-soil interface. Because of the mirror symmetry, the region for numerical simulation will be taken as 0 m < x < 10 m and 0 m < z < 10 m (Fig. 1). The problem domain contains two types of materials as shown in Figure 1. Material 1 has a saturated hydraulic conductivity of $K_{xx} = K_{zz} = 0.1$ dm/day and $K_{xz} = 0$ dm /day. Material 2 has a saturated hydraulic conductivity of $K_{xx} = K_{zz} = 0.085$ dm/day and $K_{xz} = 0$ dm /day. The unsaturated characteristic hydraulic properties of both materials are described by

$$\theta = \theta_r + (\theta_s - \theta_r)\frac{A}{A + |h - h_a|^B} \quad \text{and } Kr = \left(\frac{\theta - \theta r}{\theta_s - \theta_r}\right)^n \tag{71}$$

where $h_a = 0$ m, $A = 10$ m^2, $B = 2$, $\theta_s = 0.25$, and $\theta_r = 0.05$ are the parameters used to compute the water content and n = 2 is the parameter to compute the relative hydraulic conductivity for both materials. Material 1 has a longitudinal dispersivity of 6 dm and a lateral dispersivity of 3 dm. Material 2 has a longitudinal dispersivity of 5 dm and a lateral dispersivity of 2.5 dm. Molecular diffusion is considered negligible in both materials.

For flow simulations, the boundary conditions are given as: no flux on the left (x = 0) and bottom (z = 0) sides; pressure head is assumed to vary from zero at the water surface (z = 2 m) to 2 m at the bottom (z = 0) on the right side (x = 0); and rainfall (infiltration)-seepage boundary conditions are imposed elsewhere.

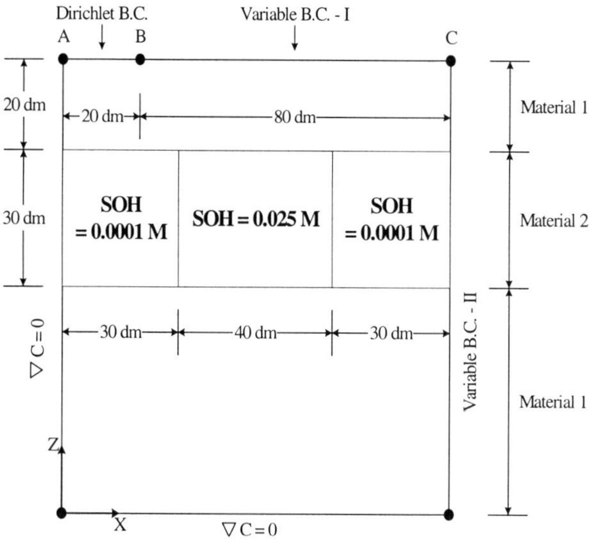

Figure 1. Problem definition for Problem 1

The rainfall rate is 0.006 m/day. The ponding depth is assumed to be 0 m on all variable boundary surfaces. Figure 2 shows the pressure head contours and velocity fields of the steady-state simulation. Driven by this flow field, both steady and transient transport under complexation and adsorbtion/desorption reactions will be simulated in this problem.

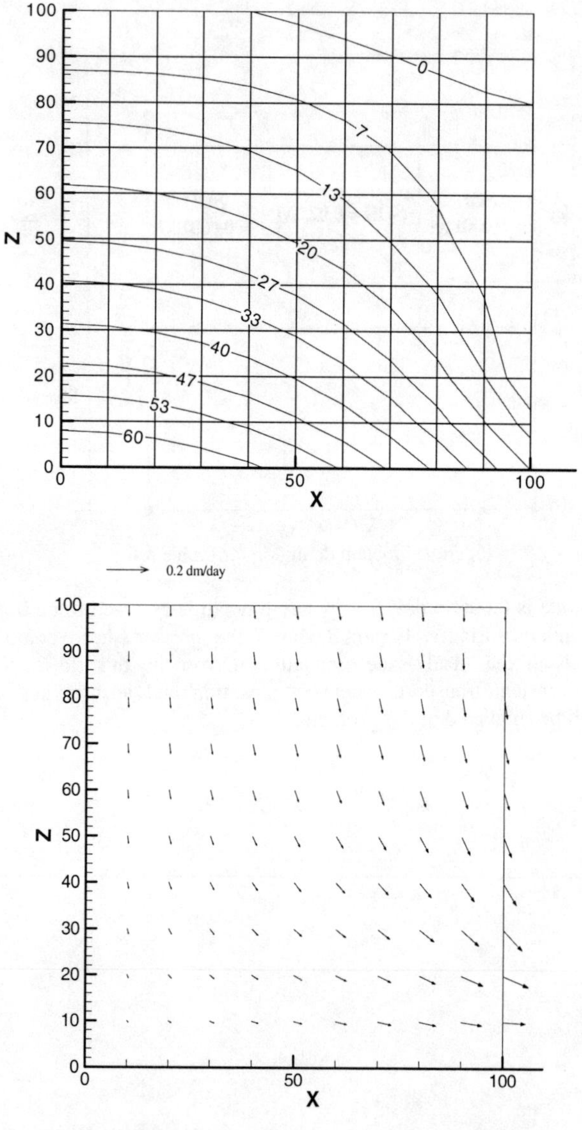

Figure 2. Steady flow results of Problem 1:pressure head contours (top) and velocity fields (bottom)

The reactive system is made of 5 aqueous components (Ca^{2+}, CO_3^{2-}, NpO_2^+, H^+, and H_2O) and one adsorbent component (SOH). The activity of H_2O is assumed 1.0; thus the chemistry of H_2O needs not to be included. The following 12 product species, 9 aqueous species and 3 surface species, are included in the simulation (the number in the parenthesis following each species denotes the Log of equilibrium constants [Log K]):

OH^- (-14.0), $CaCO_3$ (3.22), $CaHCO_3^+$ (11.4), $CaOH^+$ (-12.85), HCO_3^- (10.32), H_2CO_3 (16.67), $NpO_2(OH)$ (-8.85), $NpO_2(CO_3)^-$ (5.60), $NpO_2(CO_3)_2^{3-}$ (7.75), SO^- (-10.3), SOH^+ (5.4), $(NpO_2)(OH)(SOH)$ (-3.5).

For transport simulations, Dirichlet boundary conditions are imposed on the top surface from A to B, variable boundary conditions are specified on the top surface from B to C (Segment I) and on the right side (Segment II), and no flux conditions on the bottom and left sides (Fig. 1). One adsorbing site is present between z = 50 dm and z = 80 dm, and its concentration is 0.001 M for both right and left regions and 0.025 M for the central region (Fig. 1). The initial time step size is 0.25 day, and each subsequent time step size is increased with a multiplier of 1.5, until a maximum time step size of 32.0 days is reached. A total of 100 time steps resulting in a total simulation time of 2880.37 days is performed in this problem. This example is designed to represent a hypothetical clean-up process. The contamination of the domain has reached a steady state via infiltration on the top boundary (Fig. 1) with polluted water. The clean up process (i.e., flush domain with clean infiltrating water) is initiated by removing the pollutant from the infiltrating water. The clean up process is initiated by removing the pollutant from the infiltrating water. To investigate this clean-up scenario, the steady transport simulation is performed first to obtain the initial conditions of contaminants distributions based on the constant boundary conditions: total Ca^{2+} = 10^{-3} M, total CO_3^{2-} = 10^{-4} M, total NpO_2^+ = 10^{-5} M, and total H^+ = 10^{-8} M on both the Dirichlet and variable boundary segments. Then, the transient transport simulation is performed using total zero concentrations on the boundary for all four components.

Figures 3 and 4 show the dissolved neptunium distributions and pH values at different time steps. Before recharging the clean water (steady-state results), only the region with the higher adsorbing-site concentration has lower dissolved neptunium concentration due to the adsorption effect. Most of the regions are contaminated with high concentration of neptunium (the top-left plate in Fig. 3). As the clean water continuously recharging through the top surface, the dissolved neptunium concentration becomes lower with time (Fig. 3). On the other hand, the pH value is initially high throughout the domain due to the high concentrations of calcium (the top-left plate in Fig. 4). Only the region with the high adsorbing-site concentration has the low pH value because of the adsorption effect. As the clean water is continuously recharged through the top surface, the pH values become lower with time (Fig. 4).

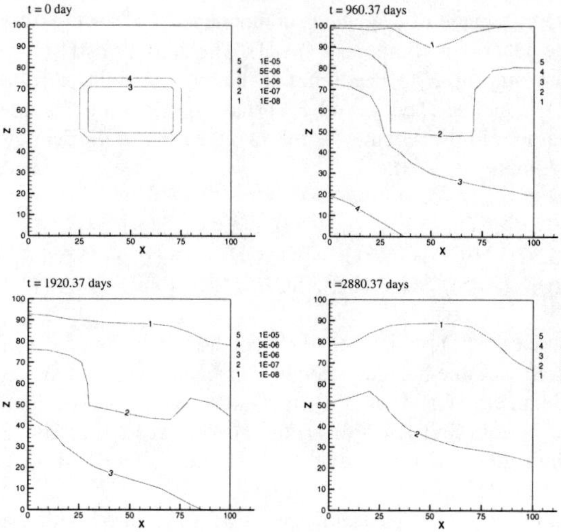

Figure 3. Dissolved neptunium at various time steps for Problem 1

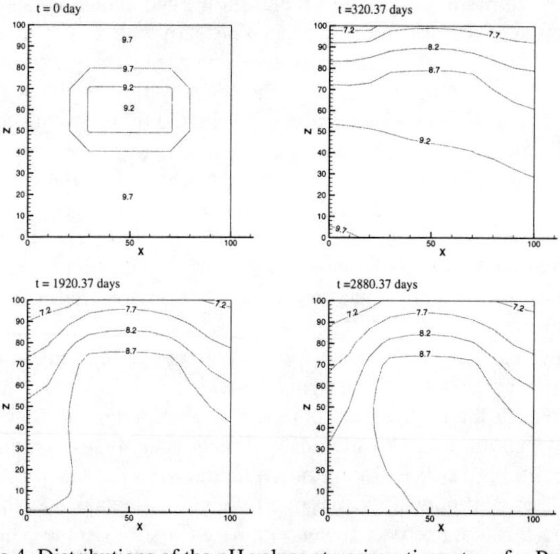

Figure 4. Distributions of the pH values at various time steps for Problem 1

4.2 Problem 2: Flow and Advective-Dispersive-Reactive Transport

This is a modification of the previous example. The major differences of this example from the previous one are: (1) the water table is originally located at z = 20 dm and will change due to recharging on the top surface (2) the domain is initially clean and is contaminated via polluted recharging water. The problem domain, material types, soil property, and transport parameters remain unchanged.

The rainfall rate on the top surface is 0.0 dm/day for steady flow simulations and changes to 0.06 dm/day for the subsequent transient flow computation. The initial time step size is 0.25 day, and each subsequent time step size is increased with a multiplier of 2.0 until a maximum time step size of 128.0 days is reached. A total of 200 time steps resulting in a total simulation time of 24575.75 days is performed in this example.

Figure 5 depicts the simulated water table at various times. While the water table is changing, transient transport is performed. The chemical reactions are identical to those in the previous example. The total concentrations on both the Dirichlet and variable boundary segments are the same as those used for steady- state transport simulation in the previous example. The initial conditions for each component specie are $Ca^{2+} = 10^{-3}$ M, $CO_3^{2-} = 10^{-4}$ M, $NpO_2^{+} = 10^{-8}$ M, and $H^{+} = 0$ M through the entire domain.

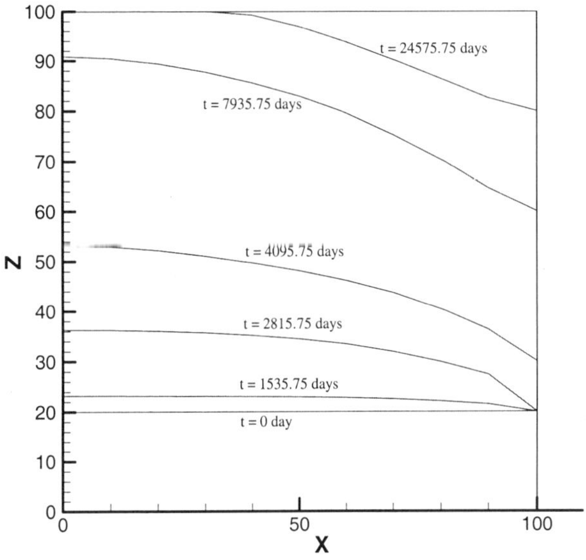

Figure 5. Variations of water table at various times for Problem 2

This example is designed to present an originally clean domain that is contaminated due to the continuous recharge of contaminated water through the top surface. Initially water table is located at z = 20 dm. Transient flow simulations show the gradual rising of the water table (Fig. 5). Meanwhile, neptunium, calcium, and carbonate enter the domain along with recharge through the top surface. Figures 6 and 7 show the dissolved neptunium distributions and pH values at different times, respectively. The dissolved neptunium concentration increases with time. The adsorbing site provides the buffer to adsorb neptunium, which makes the central region (higher SOH) to have the lowest value of neptunium concentration. Due to the existence of higher calcium concentration, the high pH values present initially throughout the domain, and only the region with the higher adsorbing-site concentration has lower pH value due to the adsorption effect. Since calcium is continuously recharged into the system, pH becomes higher as shown in Figure 7.

4.3. Problem 3: Flow and Advective-Dispersive-Reactive Transport

This problem is a modified one from Problem 1 to illustrate both steady and transient simulations for coupled flow and transport. The computed results of steady flow and transport will be used as initial conditions for the subsequent transient flow and transport, respectively. The problem domain, material types, soil properties, and transport parameters are the same as those for Problem 1.

The rainfall rate on the top surface is 0.06 dm/day for steady flow and changes to 0.0 dm/day, no recharging of water, for the subsequent transient flow simulations. Therefore, the transient flow simulation performed in this example is to highlight the drop in the water table from the cessation of rainfall water. A constant time step of 0.25 day is used for 10000 time steps resulting in a total simulation time of 2500 days. As shown in Figure 8, the water table falls more rapidly in early times, such as from 0 to 250 days, than at later times, such as from 750 to 1000 days. The hydraulic gradient is reduced as the water table declines over the period of the simulation, which slows the movement of water.

The chemical reactions considered in this example are identical to those presented for Problem 1. The steady-state transport results are used as the initial distributions of contaminants and are obtained based on the constant boundary conditions presented in Problem 1 under steady-flow conditions. Without the infiltration of clean water on the top surface, transient transport calculations are performed assuming zero incoming contaminant concentration on every specified boundary.

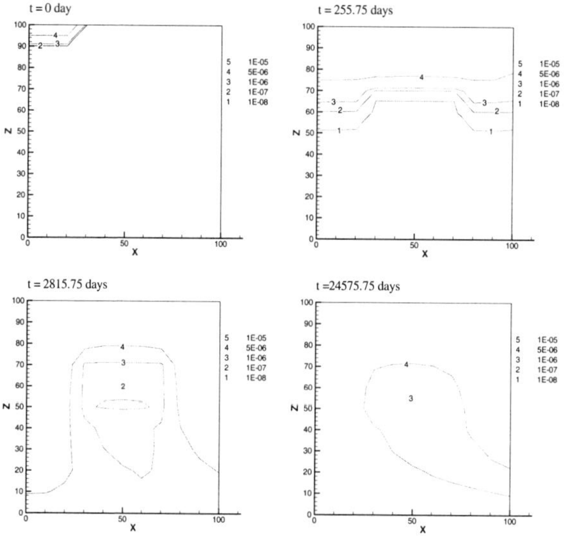

Figure 6. Dissolved neptunium at various time steps for Problem 2

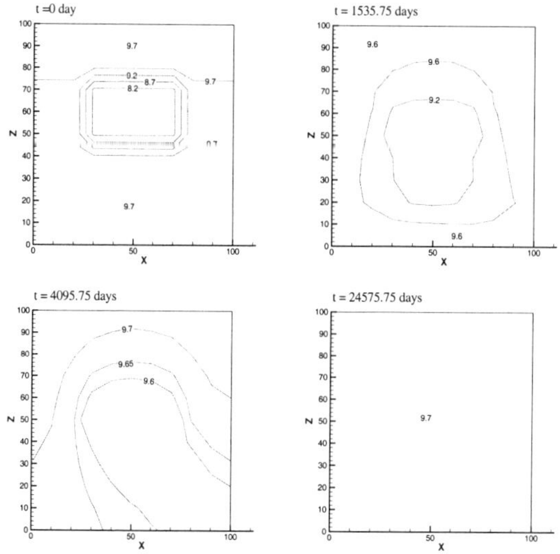

Figure 7. Distributions of the pH values at various time steps for Problem 2

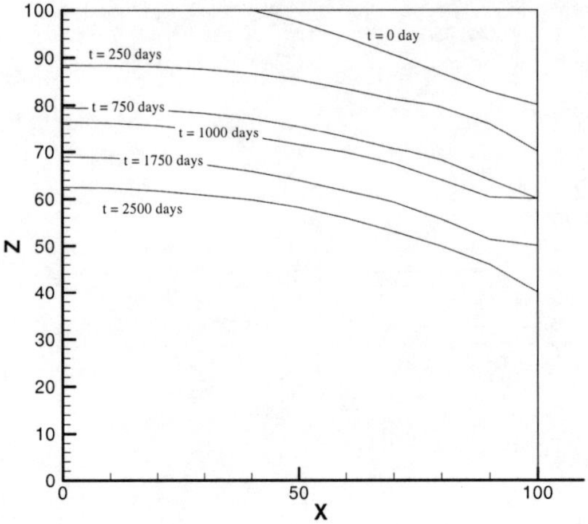

Figure 8. Variations of the water table at various times for Problem 3

Figures 9 and 10 show the dissolved neptunium distributions and pH values at different times. Initially, most regions are contaminated with the high concentration of neptunium (10^{-5} M) due to the continuous recharging of contaminated water, and only the region containing the adsorbing site has a low value of neptunium because of the adsorption effect. When infiltration is eliminated, the concentration of neptunium is reduced with increasing time (Fig. 9) because the contaminated water is moving out of the domain associated with the dropping of water table (Fig. 8). As one can imagine, without the recharging of clean water into the system, the clean-up process is much slower than that shown in Problem 1 with recharging clean water. Initially, the high pH values were found throughout the domain due to the high calcium concentration. As shown in Figure 9, only the region with the higher adsorbing-site concentration (SOH = 0.025 M) has a lower pH value due to the adsorption effect. Without the recharging of clean water, the processing time needed to neutralize a system is much longer than that shown in Problem 1 (compared pH values at t = 1920 days in Fig. 4 versus at t= 2500 days in Fig. 10).

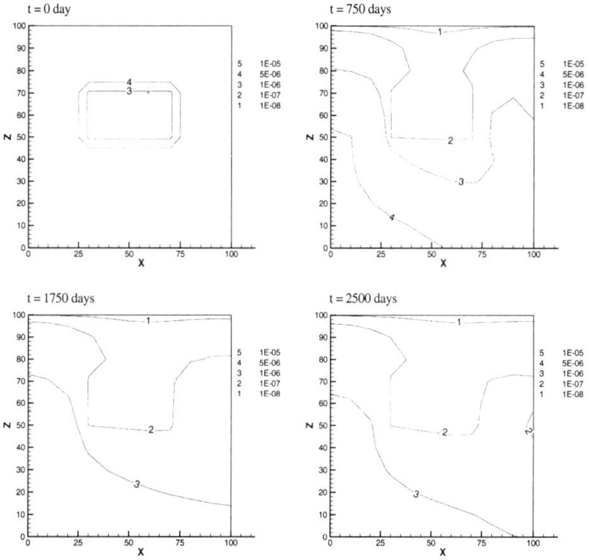

Figure 9. Concentration contours of dissolved neptuniumat various time steps for Problem 3.

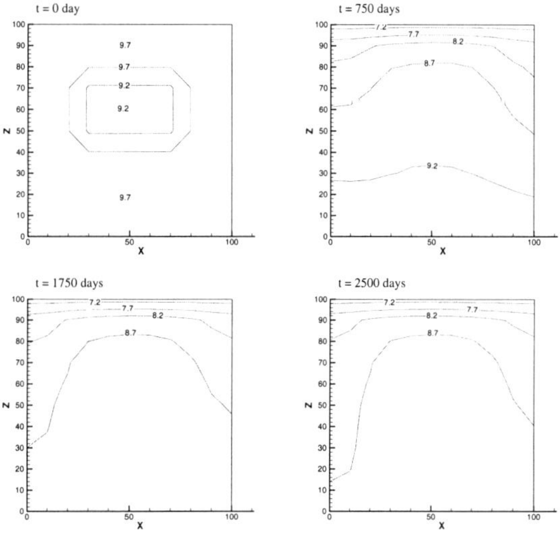

Figure 10. Distributions of the pH values at various time steps for Problem 3

4.4. Problem 4: Diffusion and Precipitation/Dissolution Reaction

This example is an initial value problem involving pure diffusion subject to four fast chemical reactions involving 9 species (Ca^{2+}, CO_3^{2-}, H^+, Na^+, Cl^-, $CaCO_{3(aq)}$, OH^-, $CaCO_{3(s)}$, and $Ca(OH)_{2(s)}$) in two reservoirs and 1.43 dm of porous media (Fig. 11). The porous media block is placed between two reservoirs. The domain of interest is a vertical column having the size of 6.1 dm × 27.63 dm (Fig. 11). The computational mesh is made up of 54 quadrilateral elements (110 nodes) with the largest vertical spacing of 1.0 dm in the reservoir regions away from the block of porous media, decreasing to 0.0715 dm adjacent to and within the block of porous media (Fig. 11). The material information, including diffusion coefficient, water content, and initial conditions of each component in each reservoir and the porous media are given in Figure 11. Two cases are included in this problem: the first considers the effect of precipitation/dissolution on diffusion and the second does not. To reflect the effect of mineral precipitation on diffusion for the first case, the value of the cementation power in the Archie's law is set as 2.0 and the molar volumes of portlandite, $Ca(OH)_2$ (s), and calcite, $CaCO_3$ (s), are set as 0.03 dm^3/mol and 0.037 dm^3/mol, respectively.

With no water flow in or out of the domain, the top and bottom boundaries are defined as variable boundaries for transport simulations which indicate that the concentration at these nodes is to be determined as part of the solution procedure. Since only pure diffusion is considered in this example, both dispersivities (longitudinal and transverse directions) and flow velocity are set to zero. An averaged value of 0.5 was used for the tortuosity (Bear, 1972). The time step size is 4.32×10^4 s for 100 time steps, resulting in a total real simulation time of 1200 hours.

Figures 12 and 13 depict the concentration distributions of the precipitate, calcite, and the tracer, sodium or chloride, at time = 1200 hours for the case that the precipitation and dissolution will not affect diffusion. Due to the diffusion effect, the higher concentration of calcium in the top reservoir will propagate toward the bottom reservoir and vice versa for carbonate. Therefore, the highest concentration of precipitated-carbonate (calcite) is developed in the porous medium where calcium and carbonate will first meet each other in this example (Fig. 12). The development of calcite will lower the concentration of calcium and carbonate in the porous media, which means more calcium and carbonate will diffuse into the region of porous media. On the other hand, the concentrations of sodium and chloride will be equal to each other with respect to the block of porous media because sodium and chloride have the same initial concentrations located in the top and bottom reservoirs, respectively, and neither of them involve any chemical reaction (Fig. 13).

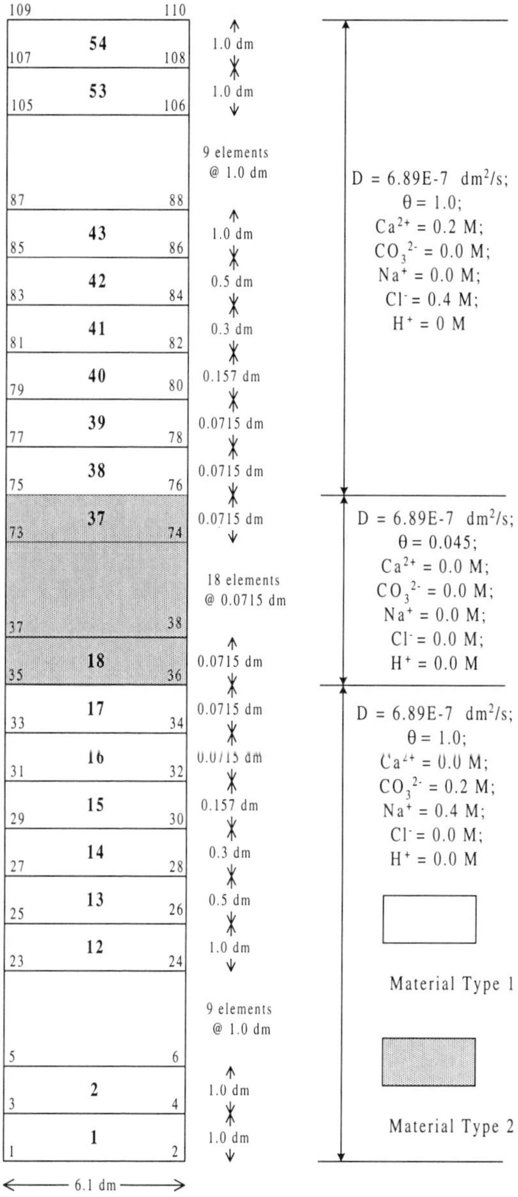

Figure 11. Problem description andfinite element discretization for Problem 4.

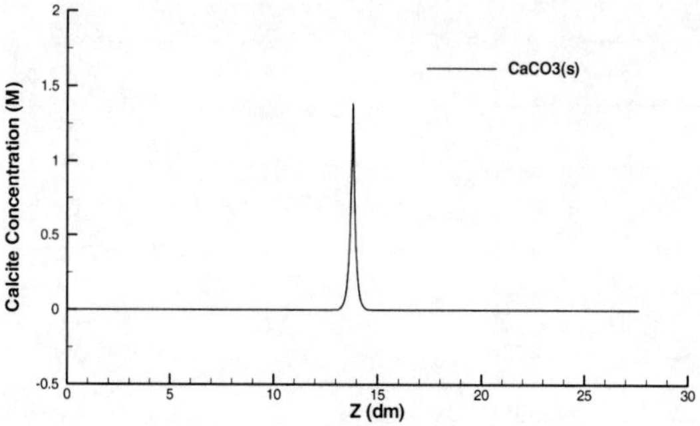

Figure 12. Calcite concentration at t = 1200 hours for Problem 4without precipitation/dissolution effects on diffusion

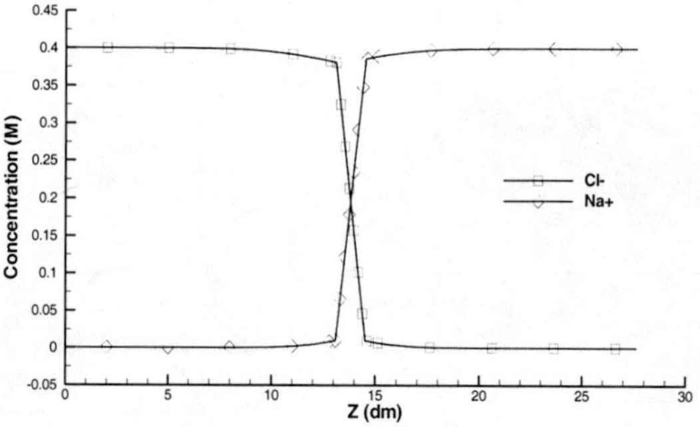

Figure 13. Sodium or chloride concentrations at t = 1200 hoursfor Problem 4 without precipitation/dissolution effects on diffusion

Figures 14 and 15 depict the concentration distributions of the precipitate, calcite, and the tracer, sodium or chloride, at time = 1200 hours for the case that the precipitation and dissolution affect diffusion. With the same vertical scale of Figure 12, Figure 14 shows that the amount of calcite precipitation occurred is much smaller than those of Figure 12. This is because the appearance of mineral precipitations diminished the diffusive transport, which delays the movement of contaminant front. As a result, the development of calcite precipitation is slower than the case of no precipitation/dissolution effects. This slow movement of contaminant front can also be found in the concentration profiles of sodium or chloride. By comparing Figure 13 with Figure 15, the concentration at the location where two profiles cross over in the Figure 15 (about 0.025 M) is smaller than that in Figure 13 (about 0.2M). This demonstrates that the development of mineral precipitations affect the diffusive transport which greatly alter the distribution of both conservative and non-conservative species in this coupled diffusion-reaction system.

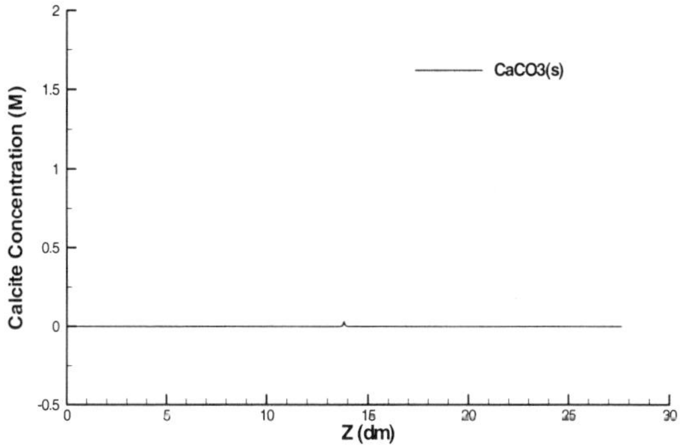

Figure 14. Calcite concentration at t = 1200 hours for Problem 4with precipitation/dissolution effects on diffusion

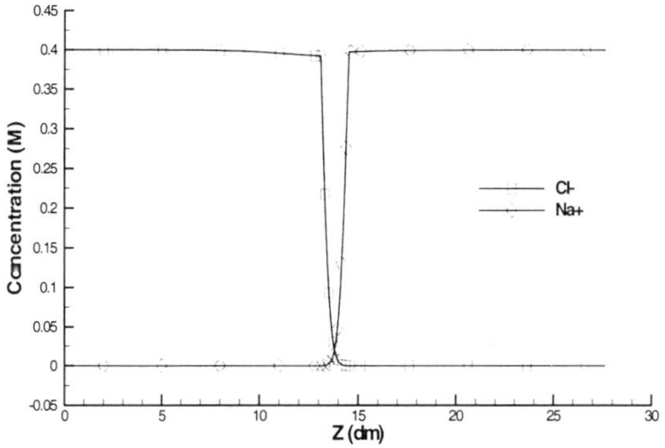

Figure 15. Sodium or chloride concentrations at t = 1200 hours for Problem 4 with precipitation/dissolution effects on diffusion

4.5 Problem 5: Coupled Fracture Flow and Matrix Diffusive-Reactive Transport

This problem simulates the effect of matrix diffusion on solute and reactive transport through fractured media. The model was based on a longitudinal cross-section of a 0.76 meter high column with a 120 : m wide fracture in the center, running the length of the column. Due to symmetry, one-half of the cross-section is used; the resulting domain (top of Fig. 16) is 7.6 dm high by 0.325 dm wide with a 6×10^{-4} dm wide fracture on the left hand side. The fractured media is simulated by using specific elements having a moisture content, 2, of 1.0. The matrix elements have a moisture content (without the presence of precipitated species), tortuosity, and bulk density of 0.35, 0.1, and 1.7 kg/dm^3, respectively. The referenced saturated hydraulic conductivity $\mathbf{K}_{soo}$ in the z-direction for the matrix (material 1) is $K_z = 2.16 \times 10^{-6}$ dm/hr; in the fracture, K_z = 237.0 dm/hr for material 2, K_z = 222.0 dm/hr for material 3, and K_z = 90.0 dm/hr for material 4. The diffusion coefficient, D, is 3.6×10^{-4} dm^2/hr in the fracture cells and 3.6×10^{-5} dm^2/hr in the matrix cells. Longitudinal dispersivity, $_L$ for all elements is set to 7.6 dm while transverse dispersivity, $_T$, is set to 0.0 dm for all elements.

For flow simulations, both top and bottom surfaces are held at a Dirichlet head of 7.615 dm and 7.6 dm, respectively. The matrix is homogeneous and is designated as Material 1. For reactive transport simulations, the upper boundary of the domain is assumed to maintain a constant concentration, Dirichlet boundary. Neumann boundaries are used for the other three boundaries with MC/Mx = 0 for the two vertical boundaries and MC/Mz = 0 for the bottom boundary.

The grid is made up of 21 rows by 17 columns of rectangular elements, with the three left-hand columns of elements representing the fracture (Materials 2, 3, and 4) and the remaining elements representing the matrix (bottom of Fig. 16). The fracture element widths (horizontal direction) are 1.2×10^{-4} dm in the first column and 2.4×10^{-4} dm in the second and third columns. The matrix element widths gradually increased from being equal to the fracture element width, 2.4×10^{-4} dm, at the fracture side to 0.1 dm at the opposite side. The heights (vertical direction) of rows of elements gradually increase from 6×10^{-4} dm to 0.2 dm, from the top to the bottom.

Two cases are simulated in this example; in the first case, flow and reactive transport are coupled through the effect of precipitation on hydraulic conductivity, porosity, and hydrodynamic dispersion. To reflect the effect of mineral precipitation on flow, the fractal exponent for computing the hydraulic conductivity is set equal to 2.5. The cementation exponent in Archie's law is set equal to 2.0 computing dispersion coefficient. Molar volumes (V_i) of 0.03 dm^3/mol and 0.037 dm^3/mol are assumed for portlandite, $Ca(OH)_2$ (s), and calcite, $CaCO_3$(s), respectively. In the second case, although the effects of precipitation/dissolution on flow, conductivity and dispersion are not considered, solute transport and chemical reaction still need to be coupled because heterogeneous reactions (precipitation-dissolution reactions) are involved in the system. A time step size of 0.24 hour was used; a total of six days (144 hours) was simulated using 600 time steps.

The chemistry of the system is relatively simple: the chemical components of the systems are Ca^{2+}, CO_3^{2-}, H^+, and a conservative tracer. Calcite is present in the matrix (8.642×10^{-5} M) in equilibrium with the pore water and fracture water, which contain 10^{-3} M Ca^{2+} and CO_3^{2-} (total concentrations). During the simulation, a solution 10x-richer in both total calcium and total carbonate is injected in the fracture. The complexed species include OH^- (-14.0), $CaCO_3$ (3.0), $CaHCO_3^+$ (11.6), $CaOH^+$ (-12.2), HCO_3^- (10.2), H_2CO_3 (16.5), and $Ca(OH)_{2(s)}$ (-21.9) where the number in the parentheses after each species denotes the log K of the reaction. Calcite ($CaCO_{3(s)}$) is included as solid species subject to kinetic precipitation and dissolution with Log K^b = -5.0 (backward reaction) and Log K^f = 3.5 (forward reaction).

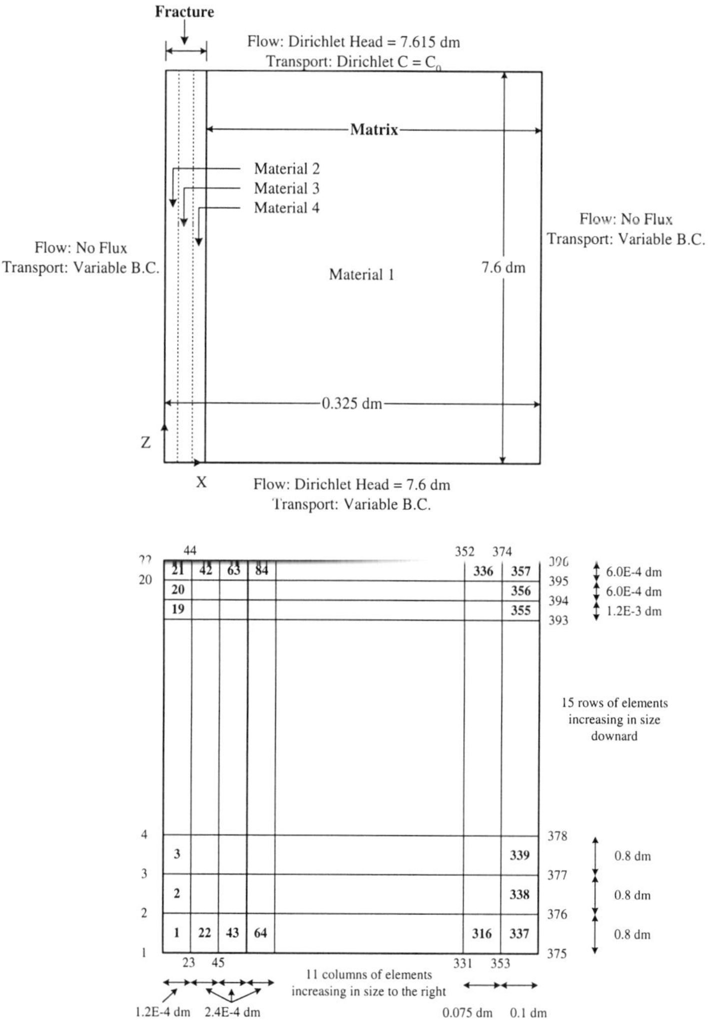

Figure 16. Problem definition (top) and discretization (bottom) for Problem 5

In the first case the precipitation effects are considered, the diffusion transport is diminished due to calcite precipitation which impedes the diffusion of calcium and carbonate into the matrix zone, resulting in less calcite precipitation. This is confirmed by comparing the top and bottom graphs in Figure 17. The concentrations of calcite in the matrix zone with coupling effects are much smaller than those without the coupling effects.

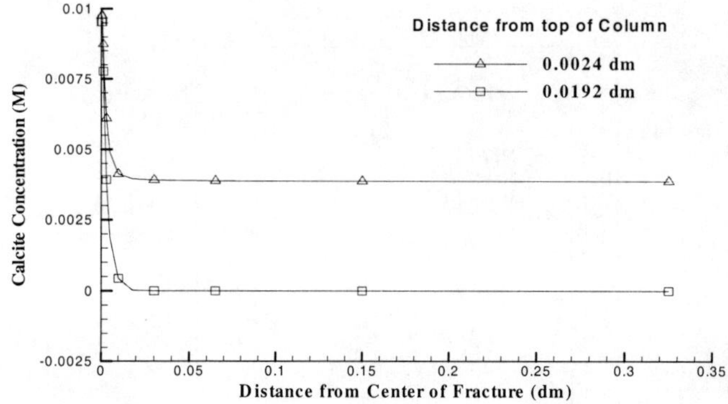

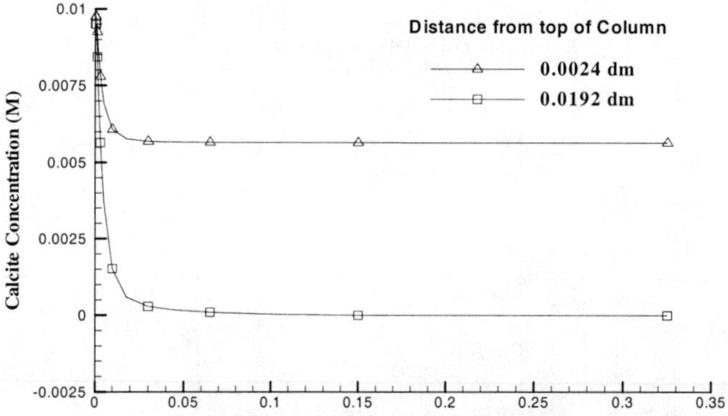

Figure 17. Calcite profiles at time = 144 hours for Problem 5:with (top) and without (bottom) precipitation effects

On the other hand, since less amount of contaminant can diffuse into the matrix zone, a larger amount of contaminant will advect through the fracture zone. This is supported by comparing the top and bottom graphs in Figure 18, where the concentrations of the conservative tracer in the fracture zone (data points on the vertical axis) with the precipitation effect are much higher than those without the effect. As with the calcite distributions, the concentrations of tracer in the matrix zone (data points not on the vertical axis) with the precipitation effect are lower than those without the precipitation effect.

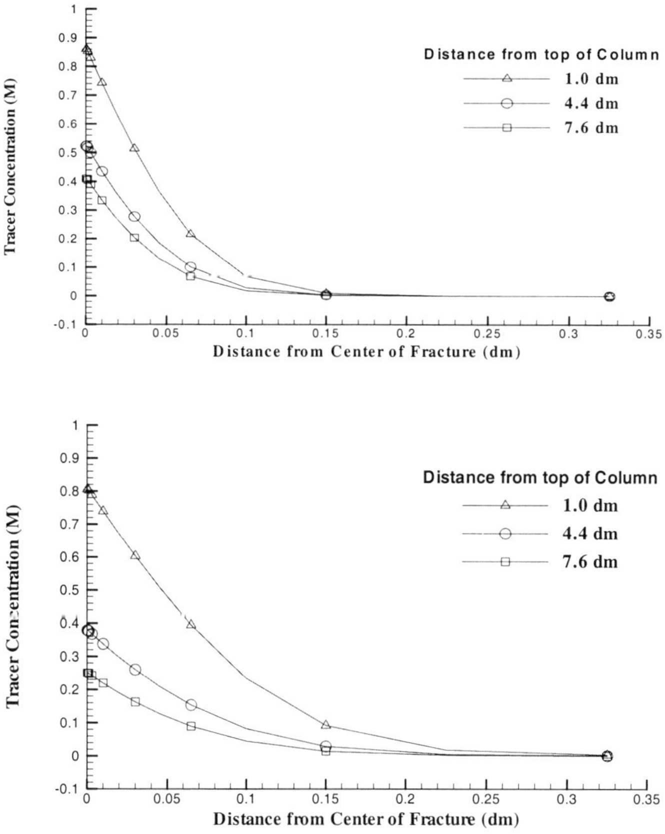

Figure 18. Conservative racer profiles for Problem 5 at time = 144 hours: with (top) and without (bottom) precipitation effects

The change of velocity fields is not significant. The velocity at node 43 (near the top of the fracture) is decreased from 0.44803 dm/hr to 0.44799 dm/hr due to the hydraulic conductivity diminished by this precipitation effect. The reason for the negligible change in flow velocity is that the precipitates developed in the fracture zone constitute only a very small potion of the volume of the fracture.

4.6 Problem 6: Advective-dispersive Reactive Transport

This example involves the transient simulation of advection/dispersion transport and equilibrium complexation/precipitation reactions to demonstrate the successful application of species switching. This technique is required when the calculated concentration of an aqueous component species becomes very low, which results in a mass balance error. In this example, the system domain is a 10 dm × 5000 dm porous medium, with 200 equal size elements, a moisture content of 0.3, a constant flow field of 155.5 dm/day, and a longitudinal dispersivity of 150 dm (Fig. 19). The chemical system includes 7 aqueous components. The initial total concentrations of each component in the domain is given in Figure 19. Three compositional zones are indicated by differences in solution composition and precipitated solids content: Zone 1 contains goethite; Zone 2 contains goethite, gypsum and gibbsite, and Zone 3 contains calcite with minor amounts of gypsum and gibbsite. Variable boundary conditions with the incoming total dissolved concentrations as listed in Table 1 are assigned on the top (nodes 401 and 402) and bottom (nodes 1 and 2) sides. On the right and left sides, a zero flux condition is imposed. A total of 24 chemical reactions is included in this example problem (Table 2)

A constant time step size of 0.025 day was used for 400 time steps resulting in a total real simulation time of 10 days. Without species switching, the simulation stops during the first precipitation cycle of the first time step at node 299 while checking the mass balance of the Al^{3+} component at the first nonlinear iteration (between solving solute transport and chemical reaction). When the species switching option was used to resolve the error in the mass balance, the aqueous complexed species $Al(OH)_4^-$ was selected by the model to replace Al^{3+} as the new component at several nodes. Subsequently, successful simulations for all 400 time steps were carried out.

5. SUMMARY

This chapter communicates the development of a mechanistically-based coupled fluid flow and reactive chemical transport model including both fast and slow reactions in variably saturated media. The theoretical bases were heuristically derived and numerical techniques discussed. Six example problems were presented. Problem 1 through 3 illustrate the capability of the model to deal with various combinations of flow and transport. Problem 4 and 5 demonstrate the capability of the model to simulate coupled fluid flow and reactive chemical transport with both kinetic and equilibrium reactions. Problem 6 exemplifies the need for robust numerical algorithms to solve problems involving complex geochemical reactions with low component species concentrations. Robust mechanistically-based numerical models provide tools to more reliably predict the migration and transformation of chemicals in the subsurface environment. Such capabilities could enhance the ability of environmental scientists, engineers and decision makers to analyze the impact of subsurface contamination and to evaluate the efficacy of alternative remediation techniques prior to incurring expense in the field.

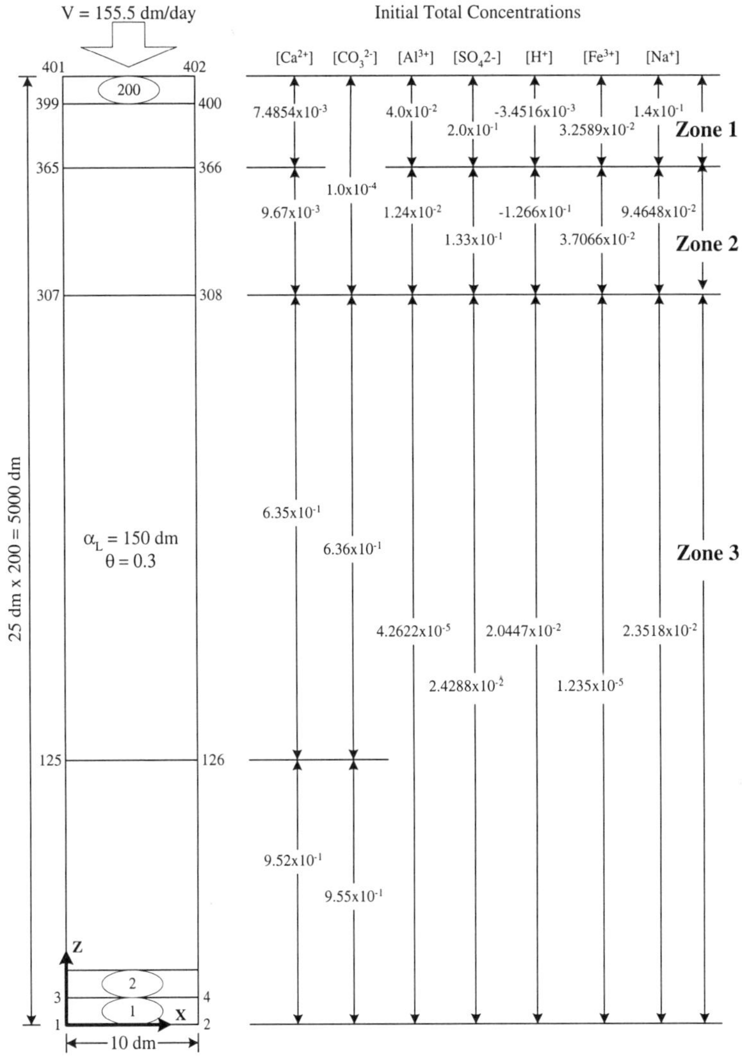

Figure 19. Problem domain, discretization, and initial conditions for Problem 6.

Table 1. Incoming total dissolved concentrations for Problem 6.

Unit: M	t = 0 day	t = 2.5 day	t = 2.6 day	t = 120 day
(Ca^{2+})	7.4854×10^{-3}	7.4854×10^{-3}	3.94×10^{-3}	3.94×10^{-3}
(CO_3^{2-})	1.0×10^{-4}	1.0×10^{-4}	2.5×10^{-3}	2.5×10^{-3}
(Al^{3+})	4.0×10^{-2}	4.0×10^{-2}	9.3077×10^{-9}	9.3077×10^{-9}
(SO_4^{2-})	2.0×10^{-1}	2.0×10^{-1}	4.4424×10^{-3}	4.4424×10^{-3}
(H^+)	1.7071×10^{-3}	1.7071×10^{-3}	2.7628×10^{-3}	2.7628×10^{-3}
(Fe^{3+})	2.093×10^{-3}	2.093×10^{-3}	3.0788×10^{-7}	3.0788×10^{-7}
(Na^+)	1.4×10^{-1}	1.4×10^{-1}	2.3518×10^{-2}	2.3518×10^{-2}

Table 2. List of Chemical Reactions for Problem 6

H^+	+	OH^-	$\Leftrightarrow$	H_2O	Log K = 14.00
Fe^{3+}	+	OH^-	$\Leftrightarrow$	$Fe(OH)^{2+}$	Log K = -2.19
Fe^{3+}	+	$2\ OH^-$	$\Leftrightarrow$	$Fe(OH)_2^+$	Log K = -5.67
Fe^{3+}	+	$3\ OH^-$	$\Leftrightarrow$	Fe(OH)3	Log K = -12.56
Fe^{3+}	+	$4\ OH^-$	$\Leftrightarrow$	$Fe(OH)_4^-$	Log K = -21.60
Al^{3+}	+	OH^-	$\Leftrightarrow$	$Al(OH)^{2+}$	Log K = -5.00
Al^{3+}	+	$2\ OH^-$	$\Leftrightarrow$	$Al(OH)_2^+$	Log K = -10.20
Al^{3+}	+	$3\ OH^-$	$\Leftrightarrow$	$Al(OH)_3$	Log K = -17.20
Al^{3+}	+	$4\ OH^-$	$\Leftrightarrow$	$Al(OH)_4^-$	Log K = -23.00
H^+	+	CO_3^{2-}	$\Leftrightarrow$	HCO_3^+	Log K = 10.33
$2\ H^+$	+	CO_3^{2-}	$\Leftrightarrow$	H_2CO_3	Log K = 16.68
Ca^{2+}	+	CO_3^{2-}	$\Leftrightarrow$	$CaCO_3$	Log K = 3.22
Ca^{2+}	+	HCO_3^{2-}	$\Leftrightarrow$	$CaHCO_3^-$	Log K = 11.44
H^+	+	SO_4^{2-}	$\Leftrightarrow$	HSO_4^+	Log K = 1.99
Ca^{2+}	+	SO_4^{2-}	$\Leftrightarrow$	$CaSO_4$	Log K = 2.30
Al^{3+}	+	SO_4^{2-}	$\Leftrightarrow$	$AlSO_4^+$	Log K = 3.50
Al^{3+}	+	$2\ SO_4^{2-}$	$\Leftrightarrow$	$Al(SO_4)_2^+$	Log K = 5.00
Fe^{3+}	+	SO_4^{2-}	$\Leftrightarrow$	$FeSO_4^+$	Log K = 4.04
Fe^{3+}	+	$2\ SO_4^{2-}$	$\Leftrightarrow$	$Fe(SO_4)_2^+$	Log K = 5.42
Na^+	+	SO_4^{2-}	$\Leftrightarrow$	$NaSO_4^+$	Log K = 0.07
Ca^{2+}	+	CO_3^{2-}	$\Leftrightarrow$	$CaCO_3(s)$	Log K = 8.48

Al^{3+}	+	3 OH^-	⇔	$Al(OH)_3(s)$	Log K = -9.11
Fe^{3+}	+	3 OH^-	⇔	$Fe(OH)_3(s)$	Log K = -4.89
Ca^{2+}	+	SO_4^{2-}	⇔	$CaSO_4(s)$	Log K = 4.58

6. ACKNOWLEDGMENTS

The work of GTY was supported by National Science Foundation under Award No. EAR-0196048 and EAR 9708494 with University of Central Florida and Penn State University, respectively. The work of MDS and MHL was carried out at Sandia National Laboratories and was supported by the United States Department of Energy under contract DE-AC04-94AL85000.

REFERENCES

Bear, J. 1972. *Dynamics of Fluids in Porous Media*. American Elsevier (Also Dover Public. Inc., 1988).

Cederberg, G. A., 1985. A groundwater mass transport and equilibrium chemistry model for multicomponent systems, *Water Resour. Res.*, 21(8), 1095-1104.

Cheng, J. R., 1995. *Numerical Modeling of Three-Dimensional Subsurface Flow, Heat Transfer, and Fate and Transport of Chemicals and Microbes*. Ph.D. Thesis, Department of Civil and Environmental Engineering, The Pennsylvania State University.

Cheng, J. R., Shobl, R. O., Yeh, G. T., Lin, H. C., and Choi, W. H., 1998. Modeling of 2D density-dependent flow and transport in the subsurface, *Journal of Hydrologic Engineering*, ASCE, 3(4), 248-257.

Chilakapati, A., 1995. *RAFT A Simulator for ReActive Flow and Transport of Groundwater Contaminants*. PNL-10636. Pacific Northwest National Laboratory, Richland, Washington.

Chilakapati, A., T. Ginn and J. E. Szecsody, 1998. An analysis of complex reaction networks in groundwater modeling, *Water Resources Res.*, 34, 1767-1780.

Dullien, F. A. L., 1979. Porous Media. Academic Press.

Freeze, R. A., 1972a. Role of subsurface flow in generating surface runoff: 1. Baseflow contributions to channel flow. *Water Resour. Res.* 8:609-623.

Freeze, R. A., 1972b. Role of subsurface flow in generating surface runoff: 2. Upstream Source Areas. *Water Resour. Res.* 8:1272-1283.

Griffioen, J., 1993. Multicomponent Cation Exchange Including Alkalinization/ Acidification Following Flow Through a Sandy Sediment. *Water Resources Res.,* 29(9):3005-3019.

Hostetler, J. C. and R. L. Erickson, 1989. *FASTCHEM Package 5*. Report EA-5870-CCM, Electric Power Research Institute.

Jennings, A. A., D. J. Kirkner, and T. L, Theis, 1982. Multicomponent equilibrium chemistry in groundwater quality models, *Water Resour. Res.*, 18(4), 1089-1096.

Kirkner, D. J., T. L. Theis, and A. A. Jennings, 1984. Multicomponent solute transport with sorption and soluble complexation, *Advances in Water Resources*, 7(3), 120-125.

Kirkner, D. J., A. A. Jennings, and T. L. Theis, 1985. Multicomponent mass transport with chemical interaction kinetics, *J. Hydrol.* 76, 107-117.

Lensing, H. J., M. Voyt, and B. Herrling, 1994. Modeling of Biologically Mediated Redox Processes in the Subsurface. *J. Hydrology*, 159: 125-143.

Lichtner, P. C. and M. S. Setch, 1996. *User's Manual for MULTIFLO: Part II MULTIFLO 1.0 and GEM 1.0, Multicomponent-Multiphase Reactive Transport Model. CNWRA 96-010.* Southwest Research Institute, Center for Nuclear Waste Regulatory Analyses. San Antonio, Texas.

Liu, C. W. and T. Narasimhan, 1989. Redox-controlled Multiple Species Reactive Transport, 1. Model Development. *Water Resources Res.*, 25:869-882.

McGrail, B. P., C. W. Lindenmeier, P. F. Martin, G. R. Holdren, and G. W Gee, 1996. "The Pressurized Unsaturated Flow (PUF) Test: A New Method for Engineered-Barrier Materials Evaluation." In *Proceedings of the American Ceramic Society Conference* held in Indianapolis, Indiana, April 14-18, 1996.

Miller, C. W. and L. V. Benson, 1983. Simulation of solute transport in a chemically reactive heterogeneous system: Model development and application, *Water Resour. Res.*, 19(2), 381-391.

Ogg, R. A., Jr., 1947. *J. Chemical Physics*. 15:337, 613.

Parkhurst, D. L., 1995. *User's Guide to PHREEQC - A Computer Program for Speciation, Reaction-path, Advective Transport, and Inverse Geochemical Calculations*. U. S. Geological Survey, Water Resources Investigations Report 95-4227.

Rubin, J., 1983. Transport of reacting solutes in porous media: Relation between mathematical nature of problem formulation and chemical nature of reactions, *Water Resour. Res.*, 9(5), 1332-1356.

Segel, I. H., 1993. *Enzyme Kinetics*. John Wiley & Sons, Inc., New York. 960 pp.

Simunek J. and D. L. Suarez, 1993. *UNSATCHEM-2D Code for Simulating Two-dimensional Variably Saturated Water Flow, Heat Transport, Carbon Dioxide Production and Transport, and Multicomponent Solute Transport with Major Ion Equilibrium and Kinetic Chemistry*. U. S. Salinity Laboratory Research Report No. 128. U. S. Salinity Laboratory, USDA, Riverside, California.

Smith, J. M., 1981. *Chemical Engineering Kinetics*. R. R. Donnelley & Sons Company. 676 pp.

Steefel, C. I. and S. B. Yabusaki, 1996. *OS3D/GIMRT, Software for Modeling Multi-Component-Multidimensional Reactive Transport, User's Manual and Programmer's Guide*. PNL-11166. Pacific Northwest Laboratory, Richland, WA.

Suarez, D. and J. Šim□nek, 1996. Solute Transport Modeling Under Variably Saturated Water Flow Conditions. In: P. C. Lichtner, C. I. Steefel, and E. H. Oelkers (Editors), *Reactive Transport in Porous Media*, Reviews in Mineralogy, Volume 34. Mineralogical Society of America, Washington, D.C., pp. 229-268.

Theis, T. L., D. J. Kirkner, and A. A. Jennings, 1982. *Multi-Solute Subsurface Transport Modeling for Energy Solid Wastes.* Tech. Progress Rept. C00-10253-3, Department of Civil Engineering, University of Notre Dame, Norte Dame, IN.

Valocchi, A. J., R. L. Street, and P. V. Roberts, 1981. Transport of ion-exchanging solutes in groundwater: Chromatographic theory and field simulation, *Water Resour. Res.*, 17(5), 1517-1527.

Westall, J. C., J. L. Zachary, and F. M. M. Morel, 1976. *MINEQL: A Computer Program for the Calculation of Chemical Equilibrium Composition of Aqueous System.* Technical Note 18, Department of Civil Engineering, Massachusetts Institute of Technology, Cambridge, MA, 91 pp.

Yeh, G. T., and D. S. Ward, 1980. *FEMWATER: A Finite Element Model of WATER Flow through Saturated-Unsaturated Porous Media.* ORNL-5567, Oak Ridge National Laboratory, Oak Ridge, TN.

Yeh, G. T., and D. S. Ward, 1981. *FEMWASTE: A Finite Element Model of WASTE Transport through Saturated-Unsaturated Porous Media.* ORNL-5601, Oak Ridge National Laboratory, Oak Ridge, TN.

Yeh, G. T. and V. S. Tripathi, 1989. A Critical Evaluation of Recent Developments in Hydrogeochemical Transport Models of Reactive Multichemical Components. *Water Resources Res.*, 25 (1): 93-108.

Yeh, G. T. and V. S. Tripathi, 1990. *HYDROGEOCHEM: A Coupled Model of Hydrologic Transport and Geochemical Equilibria in Reactive Multicomponent Systems.* ORNL-6371. Oak Ridge National Laboratory, Environmental Sciences Division.

Yeh G. T. and V. S. Tripathi, 1991. A model for simulating transport of reactive multispecies components: Model development and demonstration. *Water Resour. Res.* 27:3075-3094

Yeh, G. T., K. Salvage, and W. Choi, 1996. Reactive Chemical Transport Controlled by Both Equilibrium and Kinetic Reactions. In: A. A. Aldama, J. Aparicio, C. A. Brebbia, W. G. Gray, I. Herrera, G. F. Pinder (Editors), *Computational Methods in Water Resources XI, Volume 1: Computational Methods in Subsurface Flow and Transport Problems.* Computational Mechanics Publications, Southampton, UK, pp. 585-592.

Yeh, G. T., 1999. *Computational Subsurface Hydrology Fluid Flows.* Kluwer Academic Publishers, Norwell, Massachusetts, 304 pp.

Yeh, G. T., M. H. Li, and M. D. Siegel, 1999. *User's manual for LEHGC: A Lagrangian-Eulerian Finite-Element Model of Coupled Fluid Flows and HydroGeoChemical Transport through Saturated-Unsaturated Media - Version 2.0.* Department of Civil and Environmental Engineering, The Pennsylvania State University, University Park, PA 16802.

Yeh, G. T., 2000. *Computational Subsurface Hydrology Reactions, Transport, and Fate of Chemicals and Microbes.* Kluwer Academic Publishers, Norwell, Massachusetts, 344 pp.

Chapter 8

HEAT AND MASS TRANSPORT IN POROUS MEDIA

Chin-Tsau Hsu[1]

ABSTRACT

Flows and heat and mass transfer through porous media have been the subject of investigation for centuries, because of their wide applications in mechanical, chemical and civil engineering. A review of existing literatures however shows that the description of the phenomena remains fragmented. In this chapter, we attempt to formulate a complete set of macroscopic equations to describe these transport phenomena. The macroscopic transport equations were obtained by averaging the microscopic equations over a representative elementary volume (REV). The average procedure leads to the closure problem where the dispersion, the interfacial tortuosity and the interfacial transfer become the new unknowns in the averaged macroscopic momentum, energy and mass transport equations. The closure relations constructed earlier by others for dispersion, tortuosity and interfacial transfer were summarized, reviewed and adapted to close the equation system. However, several coefficients in the closure relations need to be determined experimentally (or numerically) a priori. Experiments conducted earlier for the determination of these coefficients were reviewed. These experimental results generally supported the validity of the closure relations, but were inconclusive. This leads to the conclusion that more experiments are needed for a complete determination of these coefficients. It is noted that closure relations for heat and mass transfer were obtained under the assumption of steady (or quasi-steady) flows. The applicability of such closure relations to unsteady flows remains an open question.

1. INTRODUCTION

Flows and heat and mass transfer through porous media have been the subject of investigation for centuries, because of their wide applications in mechanical, chemical and civil engineering. Traditionally, Darcy's (1856) law has been applied for the flows when the Reynolds number based on the pore size (or

[1] Department of Mechanical Engineering, Hong Kong University of Sciences and Technology, Clear Water Bay, Kowloon, Hong Kong. mecthsu@ust.hk

particle diameter, d_p) is small. Under this circumstance, the momentum equation for flows in isotropic media is given by

$$-\nabla P = \frac{\mu \mathbf{U}}{K} \tag{1}$$

where P is the pore pressure, μ the fluid viscosity, and **U** the Darcy velocity. In Eq. (1), the permeability, K, takes the well-accepted form of

$$K = \frac{\phi^3 d_p^2}{a(1-\phi)^2} \tag{2}$$

where ϕ is the porosity of porous media and a is a constant depending on the microscopic geometry of the porous materials.

Lately, engineering practices require the packed-bed reactors to be operated at high flow rate, i.e., at high Reynolds number. A nonlinear term to correct for the convection inertia effect was adopted by Forchheimer (1901). Thus Eq. (1) was modified into

$$-\nabla P = \frac{\mu \mathbf{U}}{K} + \frac{F\rho|\mathbf{U}|\mathbf{U}}{\sqrt{K}} \tag{3}$$

where ρ is the fluid density. The nonlinear factor F according to Ergun (1952) is given by $F = b/\sqrt{a\phi^3}$ with the constant b dependent on the microscopic geometry of the porous media. Although Eq. (3) has since been used by researchers with some success in predicting flows in porous media, Hsu and Cheng (1990) showed theoretically that there exists an additional term proportional to $|\mathbf{U}|^{1/2}\mathbf{U}$, which accounts for the viscous boundary layer effect at the intermediate Reynolds number. As a result, in term of permeability, Eq. (3) was modified further into

$$-\nabla P = \frac{\mu \mathbf{U}}{K} + \frac{H\sqrt{\rho\mu|\mathbf{U}|}\mathbf{U}}{K^{3/4}} + \frac{F\rho|\mathbf{U}|\mathbf{U}}{\sqrt{K}} \tag{4}$$

where H is a function of porosity and microscopic solid geometry. The need of including this additional term at the intermediate Reynolds number was confirmed experimentally by Hsu et al. (1999).

Unsteady flows in porous media have recently received great attention. One example is the oscillating flow in the regenerators used in a Stirling engine and catalytic converters. Others are the transient processes in the start-up and shutdown of a capillary heat pipe in mechanical engineering, as well as of well-bore pumping in hydraulic and petroleum engineering. Due to the lack of adequate equations to describe the unsteady flows in porous media, Eq. (3) sometimes was used indiscriminately. For coastal engineering, where ocean waves act on sand sea beds or porous breakwaters, the common practice for the momentum equations is to incorporate into Eq. (1) the terms corresponding to inertia and viscous diffusion (Liu et al., 1996), based on the classical works of Biot (1941) and Dagan (1979). Such equations, as pointed out later by Hsu (2001a), are valid only for flows at long time-scale and low Reynolds number. Hsu (2001a) then constructed a model for unsteady flows in porous media, which to the first order approximation is valid over the entire ranges of time scale and Reynolds number.

Heat and mass transfer in porous media were widely studied for more than a century. The simplest problem in heat transfer in porous media is the pure conduction where the fluid is stagnant. Traditionally, mixture models were used for heat conduction in porous media under the assumption of a local thermal equilibrium between fluid and solid phases. Under this assumption, it becomes possible to define the heat capacitance of the mixture, $(\rho c_p)_m$, as

$$(\rho c_p)_m = \phi \rho c_p + (1-\phi)\rho_s c_{ps} \tag{5}$$

where ρc_p and $\rho_s c_{ps}$ are the heat capacities, with ρ and ρ_s being the densities, of fluid and solid, respectively. As a result, the main task becomes the determination of the effective stagnant thermal conductivity k_{st} appearing in the lumped mixture heat conduction equation given by

$$(\rho c_p)_m \frac{\partial \mathrm{T}}{\partial \mathrm{t}} = \nabla \cdot [k_{st} \nabla \mathrm{T}] \tag{6}$$

The determination of effective stagnant thermal conductivity has been a subject of great effort for more than a century, since the work by Maxwell (1873). A large number of experiments had been carried out to measure the effective stagnant thermal conductivity. Kunii and Smith (1960), Krupiczka (1967), and Crane and Vachon (1977) had compiled these early experimental data. The experimental methods for determining k_{st} were also reviewed by Tsotsas and Martin (1987). Most of these measurements were carried out for materials when the solid to fluid thermal conductivity ratio σ is in the range of $1 < \sigma < 10^3$. Effective thermal conductivities of porous materials with higher value of σ were obtained experimentally by Swift (1966) and Nozad et al. (1985), while those with lower σ were found by Prasad et al. (1989). With advances in computer technology, the effective stagnant thermal conductivities were determined numerically. Deissler and Boegli (1958) were the first to obtain k_{st} for cubic-packing spheres on the basis of a finite-difference scheme, followed by Wakao and Kato (1969) and Wakao and Vortmeyer (1971) for media with a periodic orthorhombic structure. More recently, Nozad et al. (1985) and Sahraoui and

Kaviany (1993) had obtained some numerical results. It should be noted that all the numerical investigations assumed periodic porous structures so that the computation domain could be confined to a unit cell. Since Maxwell (1873), several analytical composite-layer models have been proposed for k_{st} (Kunii and Smith, 1960; Zehner and Schlunder, 1970). Recently, Hsu et al. (1994) extended the model of Zehner and Schlunder (1970) by introducing a particle touching parameter. The model of Kunii and Smith (1960) was improved by Hsu and his co-workers (1995, 1996), using the touching and non-touching geometry of Nozad et al. (1985); they found the predicted results of k_{st} agree remarkably well with the experimental data of Nozad et al. (1985). Kaviany (1991) and Cheng and Hsu (1998) have reviewed the existing models of effective thermal conductivity in detail.

The validity of the assumption of local thermal equilibrium remains an open question, especially when the time scale of a transient heat conduction in the porous media is short and when the thermal conductivity ratio between the fluid and solid deviates greatly from unity. Under thermal non-equilibrium condition, we need to consider the heat transfer in the fluid and solid phases separately to form a two-equation model. Closure modeling of the thermal tortuosity and the interfacial heat transfer becomes inevitable. Quintard and his co-workers (1993, 1995) had made considerable progresses on the two-equation model. Hsu (1999) proposed a transient closure model that provided a method to evaluate the thermal tortuosity and extended the model of Quintard and Whitaker (1995) for interfacial heat transfer by taking into account the dependence on the thermal conductivity ratio of solid to fluid. A review of the transient heat conduction in porous media to assess the validity of local thermal equilibrium assumption was given by Hsu (2000).

When the fluid in the porous media is in motion, a new unknown quantity due to thermal dispersion becomes important and requires closure modeling. In addition, the magnitude of the interfacial heat transfer is enhanced greatly by fluid convection. In contrast to mass dispersion, only a limit amount of work exists on the thermal dispersion. Gunn and De Sousa (1974), Gunn and Khalid (1975), and Vortmeyer (1975) represented some of the early works. More recent works were those by Levec and Carbonell (1985) and Hsu and Cheng (1990). Considerable progresses were made on the modeling of the interfacial heat transfer as enhanced by fluid motion. These can be traced back to the earlier works of Kunii and Suzuki (1967), Nelson and Galloway (1975), Martin (1978) and Wakao et al. (1979). Wakao and Kaguei (1982) provided a comprehensive summary on the interfacial heat transfer. They found a great scattering of the experimental data for low Reynolds number flows. Hsu (2001b) extended his earlier work of interfacial heat transfer for pure conduction (Hsu, 1999) to incorporate the effect of forced convection; the results agreed very well with experimental data for both low and high Reynolds number flows.

The diffusion of mass in fluids and solids is a classical problem. The mass transfer process in porous media in principle is very similar to the heat transfer process. However, the mass transfer in solids is usually neglected because mass diffusivity in the solids is much smaller than the diffusivity in fluids. Different from heat transfer, the adsorption of mass at the solid-fluid interface has to be considered. Considerable works were done for the mass transfer in fluid phase of porous media, especially for the packed beds as encountered frequently in chemical engineering applications. Similar to the heat transfer, the main task in the mass transfer in fluid phase is the determination of the mass dispersion, mass tortuosity, and interfacial mass transfer. The works on the mass dispersion can be traced back as early as those of Aris and Amundson (1957) and Saffman (1959, 1960), followed by Harleman and Rumer (1963), Whitaker (1967), Edwards and Richardson (1968), Carbonell (1979, 1980), Koch and Brady (1985, 1987), and Quintard and Whitaker (1994). The main focus was on the determination of the dispersion mass diffusivity caused by the mixing of velocity and concentration deviations. The effect of the mass tortuosity is to modify the fluid mass diffusivity to an effective stagnant mass diffusivity (Whitaker, 1967; Quintard and Whitaker, 1993). The determination of the rate of mass transfer across the fluid-solid interface is one of the main concerns in chemical and environmental engineering. Kunii and Suzuki (1967), Nelson and Galloway (1975), and Martin (1978) represent some of the earlier works on interfacial mass transfer. A comprehensive review of the works on interfacial mass transfer can also be found in Wakao and Kaguei (1982). More recent works include those by Quintard and Whitaker (1994) and Hsu (2001b).

In this chapter, the convective heat and mass transfer in porous media is treated rigorously by the method of volumetric averaging (Whitaker, 1999), incorporated with an areal averaging procedure (Hsu, 2001a) for the region near a macroscopic boundary. The averaging process renders a macroscopic description of the transport phenomena, but also leads to the closure problem where the closure modeling becomes inevitable for the dispersions, the interfacial tortuosity and the interfacial transfer associated with momentum, energy and mass transport. Closure relations as proposed by the earlier works are summarized to form a closed equation system. The limitations on the closure relations are discussed to offer the possibility for further improvements.

2. MACROSCOPIC GOVERNING EQUATIONS

In this section, we shall obtain the macroscopic governing equations for the transport of momentum, energy and mass concentration in porous media. The scaling law for treating the porous media is first introduced, followed by a description of the transport equations at the microscopic scale. The volumetric and areal averaging procedures are then defined and applied to the microscopic transport equations to obtain the macroscopic transport equations.

2.1 Scaling Law

As depicted in Fig. 1, the macroscopic scale is defined by the coordinate system $\bar{x}_i$, i = 1-3, which has the global length scale of L for the problem under consideration. The increments of the macroscopic scale, $d\bar{x}_i$, are equal to the length l of a representative elementary volume (REV) given by $V = d\bar{x}_1 d\bar{x}_2 d\bar{x}_3$. The local microscopic coordinate system, x_i, also has the length scale l, which is assumed to be large compared to the characteristic length of the solid particles, d_p. Therefore, the increments of the microscopic coordinates, dx_i, are much smaller than d_p, but much larger than the molecular scale l_m. Hence, $L >> l >> d_p >> l_m$, $d\bar{x}_i = l = O(x_i)$, and $d_p >> dx_i >> l_m$. With the last inequality, the concept of continuum mechanics can be applied directly at the microscopic scale.

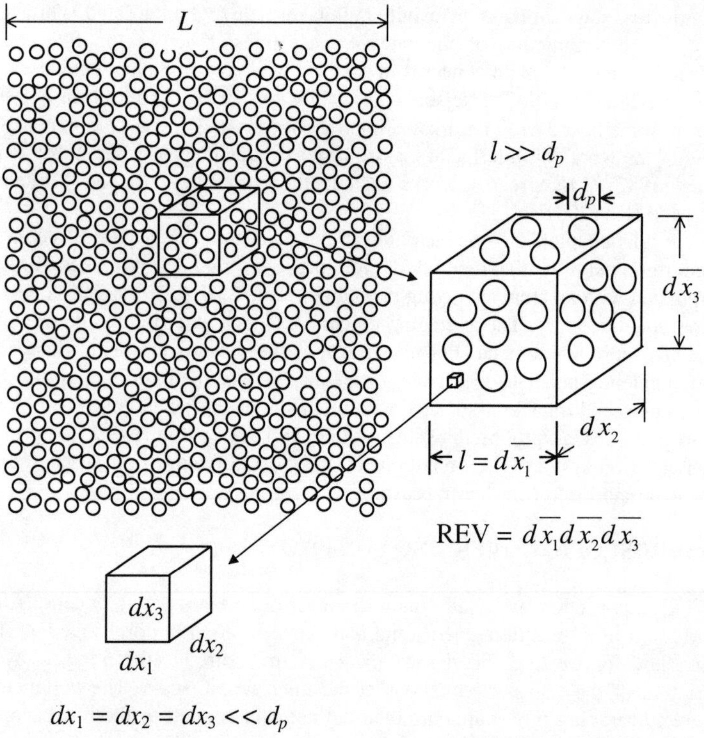

Figure 1. The schematic to illustrate the scaling law and the representative elementary volume (REV) for the volumetric average scheme.

2.2 Microscopic Transport Equations

For simplicity, we shall restrict our discussions to a porous medium with rigid solids. The properties of the fluid such as density, viscosity, thermal conductivity and mass diffusivity are assumed constant. Hence, the fluid is Newtonian and the flow is incompressible. The continuity, momentum, energy and mass-concentration equations of the fluids in pore space at the microscopic scale are given respectively by:

Continuity Equation:

$$\nabla \cdot \mathbf{u} = 0 \tag{7}$$

where **u** is the velocity, ∇ is the microscopic gradient operator.

Momentum Equation:

$$\rho \frac{\partial \mathbf{u}}{\partial t} + \rho \nabla \cdot (\mathbf{uu}) = - \nabla p + \mu \nabla \mathbf{S} \tag{8}$$

where p is the pressure and **S** is the strain tensor given by

$$\mathbf{S} = S_{ij} = \left(\frac{\partial u_i}{\partial x_j} + \frac{\partial u_j}{\partial x_i} \right) \tag{9}$$

Energy Equation in Fluid:

$$\rho c_p \frac{\partial T}{\partial t} + \rho c_p \nabla \cdot (\mathbf{u}T) = k \nabla^2 T \tag{10}$$

where T is the temperature of the fluid, c_p is the specific heat capacity of fluid at constant pressure, k is the thermal conductivity

Mass Concentration Equation in Fluid:

$$\rho \frac{\partial Y}{\partial t} + \rho \nabla \cdot (\mathbf{u}Y) = \rho \eta \nabla^2 Y \tag{11}$$

where Y is the mass concentration and η is the mass diffusivity of the fluid.

While there is no solid motion for rigid media, i.e., $\mathbf{u}_s = 0$, heat conduction and mass diffusion in solids may occur. The energy and mass concentration equations for the solid phase can be obtained from Eqs. (10) and (11) by subscripting the physical quantities with s for the solid phase. They become:

Energy Equation in Solid:

$$\rho_s c_{ps} \frac{\partial \mathrm{T}_s}{\partial \mathrm{t}} = k_s \nabla^2 \mathrm{T}_s \tag{12}$$

where T_s is the temperature, ρ_s the density, c_{ps} the specific heat capacity, and k_s the thermal conductivity, of the solid.

Mass Concentration Equation in Solid:

$$\rho_s \frac{\partial \mathrm{Y}_s}{\partial \mathrm{t}} = \rho_s \eta_s \nabla^2 \mathrm{Y}_s \tag{13}$$

where Y_s is the mass concentration and η_s the mass diffusivity of the solid.

2.3 Volumetric and Areal Averaging Schemes

Volumetric Averaging:

We now follow the procedure of Whitaker (1999) to derive the volumetric average for the divergence of a flux **w**. The divergence theorem reads

$$\int_{\mathrm{V}_f} \nabla \cdot \mathbf{w}\, \mathrm{dV} = \int_{\mathrm{S}_f} \mathbf{w} \cdot \mathbf{ds} \tag{14}$$

where S_f is the surface enclosing the fluid volume V_f and **ds** is the surface areal increment vector which is represented by **n**dA with **n** being the unit vector outward normal to S_f from fluid to solid and dA the scalar areal increment. For the REV depicted in Fig. 1, the surface S_f consists of a fluid/solid interfacial surface A_{fs} and six flat fluid surfaces, two A_{fx_i} each at $\overline{\mathrm{x}}_i \pm \mathrm{d}\overline{\mathrm{x}}_i/2$, i = 1-3. We should assume that the areal porosities in all the three x_i directions are identical to ϕ_a, i.e., $\mathrm{A}_{fx_i}/\mathrm{A}_{x_i} = \phi_a$, and that A_{fx_i} and A_{x_i} are chosen sufficiently large as compared to A_f and A, respectively, to render ϕ_a independent of the size of A. Therefore, Eq. (14) after divided by V becomes

$$\frac{1}{\mathrm{V}} \int_{\mathrm{V}_f} \nabla \cdot \mathbf{w}\, \mathrm{dV} = \overline{\nabla} \cdot \left(\frac{1}{\mathrm{A}} \int_{\mathrm{A}_f} \mathbf{w}\, \mathrm{dA} \right) + \frac{1}{\mathrm{V}} \int_{\mathrm{A}_{fs}} \mathbf{w} \cdot \mathbf{ds} \tag{15}$$

where $\overline{\nabla}$ is the macroscopic gradient operator. The areal and volumetric averages can be taken as the same in the core region of a randomly packed porous medium where both are defined rigorously. As a result, Eq. (15) in terms of the intrinsic average quantity becomes

$$\phi \; \overline{\nabla \cdot \mathbf{w}} \;=\; \overline{\nabla} \cdot (\phi \; \overline{\mathbf{w}}) \; + \frac{1}{V} \int_{A_{fs}} \mathbf{w} \cdot \mathrm{ds} \tag{16}$$

where the overhead bar represents for the intrinsic average over the fluid phase, i.e.,

$$(\overline{\nabla \cdot \mathbf{w}}, \overline{\mathbf{w}}) = \frac{1}{V_f} \int_{V_f} (\nabla \cdot \mathbf{w}, \mathbf{w}) \, \mathrm{dv} \tag{17}$$

By applying the above divergence theorem to w**i** , w**j** and w**k**, respectively, and summing up the resultant expressions into a vector, the averaging procedure for a gradient reads

$$\phi \; \overline{\nabla \mathrm{w}} \;=\; \overline{\nabla}(\phi \; \overline{\mathrm{w}}) \; + \frac{1}{V} \int_{A_{fs}} \mathrm{w} \, \mathbf{ds} \tag{18}$$

Note that the last terms in Eqs. (16) and (18) represent the interfacial transfer terms caused by the interaction between the fluid and solid. For a time dependent quantity, the average over the REV leads to

$$\overline{\phi \frac{\partial \mathrm{w}}{\partial \mathrm{t}}} = \frac{\partial (\phi \, \overline{\mathrm{w}})}{\partial \mathrm{t}} - \frac{1}{V} \int_{A_{fs}} \mathrm{w} \, \mathbf{u} \cdot \mathbf{ds} \tag{19}$$

For rigid solid medium where $\mathbf{u} = 0$ on A_{fs}, last term in Eq. (19) becomes zero.

The above expressions are for the averages in fluid phase. The averages for solid phase can be similarly performed. Then the quantity $\overline{\mathbf{w}}_s$ denoting the solid phase averaged result is defined as

$$\overline{\mathbf{w}}_s = \frac{1}{V_s} \int_{V_s} \mathbf{w} \mathrm{dv} \tag{20}$$

The expressions (16), (18) and (19) then become

$$\phi_s \; \overline{\nabla \cdot \mathbf{w}_s} \;=\; \overline{\nabla} \cdot (\phi_s \; \overline{\mathbf{w}}_s) \; - \frac{1}{V} \int_{A_{fs}} \mathbf{w}_s \cdot \mathrm{ds} \tag{21}$$

$$\phi_s \; \overline{\nabla \mathrm{w}_s} \;=\; \overline{\nabla}(\phi_s \; \overline{\mathrm{w}}_s) \; - \frac{1}{V} \int_{A_{fs}} \mathrm{w}_s \, \mathbf{ds} \tag{22}$$

and

$$\overline{\phi_s \frac{\partial \mathrm{w_s}}{\partial t}} = \frac{\partial (\phi_s \overline{\mathrm{w}}_s)}{\partial t} + \frac{1}{V}\int_{A_{fs}} \mathrm{w}\, \mathbf{u}_s \cdot \mathbf{ds} \qquad (23)$$

for the averages over solid phase, where $\phi_s = 1\text{-}\phi$ is the volume ratio of solid phase. Note that the last terms in Eqs. (21-23) take the different sign from Eqs. (16), (18) and (19) since **ds** is inward into the solid.

Areal Averaging:

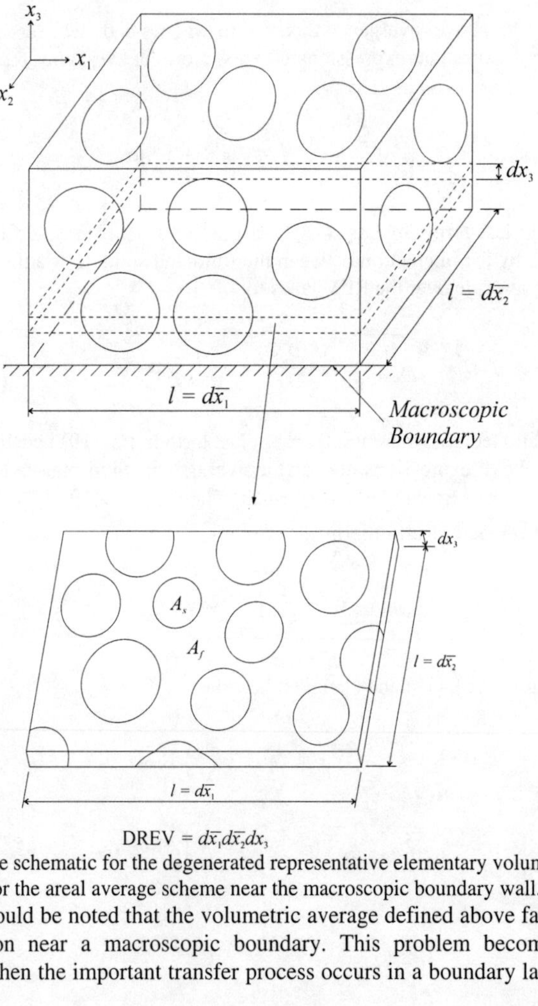

DREV = $d\overline{x}_1 d\overline{x}_2 dx_3$

Figure 2. The schematic for the degenerated representative elementary volume (DREV) for the areal average scheme near the macroscopic boundary wall.

It should be noted that the volumetric average defined above fails to apply in the region near a macroscopic boundary. This problem becomes serious especially when the important transfer process occurs in a boundary layer near an impermeable wall. To circumvent this difficulty, we should follow Hsu (2001a) by degenerating the REV into a thin plate as shown in Fig. 2. Hence, $V = d\overline{x}_1 d\overline{x}_2 dx_3 = A dx_3$ where x_3 is in the normal direction from the macroscopic boundary. Note that x_3 remains at a microscopic scale, i.e., $dx_3 << d_p$. The fluid volume in REV is given by $V_f = A_f dx_3$. Then, the porosity is defined by $V_f/V = A_f/A = \phi_a$. The intrinsic average of a fluid quantity **w** over the fluid phase now becomes

$$\overline{\mathbf{w}} = \frac{1}{V_f}\int_{V_f} \mathbf{w} dV = \frac{1}{A_f}\int_{A_f} \mathbf{w} dA \tag{24}$$

For the degenerated REV, it can be easily shown that the relations given by Eqs. (16), (18) and (19) remain valid, except that the volumetric porosity ϕ is replaced by the areal porosity ϕ_a, that the overhead bar is interpreted as the areal average and that the macroscopic gradient $\overline{\nabla}$ is interpreted as $\overline{\nabla} = (\partial/\partial\overline{x}_1, \partial/\partial\overline{x}_2, \partial/\partial x_3)$. Expressions for the areal average over the solid phase similar to those of Eqs. (21-23) can be obtained, and the corresponding interpretations are presented as follows.

2.4 Macroscopic Transport Equations

By performing volumetric averages to the microscopic equations (7-13), invoking the relations (16-23) and decomposing the velocity, pressure, temperature and mass concentration into $\mathbf{u} = \overline{\mathbf{u}} + \mathbf{u}'$, $p = \overline{p} + p'$, $T = \overline{T} + T'$, $Y = \overline{Y} + Y'$, $T_s = \overline{T}_s + T_s'$ and $Y_s = \overline{Y}_s + Y_s'$, respectively, we find that:

<u>Continuity Equation:</u>

$$\overline{\nabla} \cdot (\phi\, \overline{\mathbf{u}}) = 0 \tag{25}$$

<u>Momentum Equations:</u>

$$\rho\left[\frac{\partial}{\partial t}(\phi\,\overline{\mathbf{u}}) + \overline{\nabla}\cdot(\phi\,\overline{\mathbf{u}}\,\overline{\mathbf{u}})\right] = -\overline{\nabla}(\phi\,\overline{p}) + \rho\left[\nu\overline{\nabla}\cdot(\phi\overline{\mathbf{S}}) + \overline{\nabla}\cdot(-\phi\overline{\mathbf{u}'\mathbf{u}'})\right] + \overline{\mathbf{b}}_{fs} \tag{26}$$

where

$$\overline{\mathbf{S}} = \overline{S}_{ij} = \frac{1}{\phi}\left(\frac{\partial(\phi\,\overline{u}_i)}{\partial x_j} + \frac{\partial(\phi\,\overline{u}_j)}{\partial x_i}\right) \tag{27}$$

In Eq. (26),

$$\overline{\mathbf{b}}_{fs} = \frac{1}{V}\int_{A_{fs}} (\text{-p}\mathbf{I} + \mu\,\mathbf{S}) \cdot \mathrm{ds} = \frac{1}{V}\int_{A_{fs}} (\text{-p}'\,\mathbf{I} + \mu\,\mathbf{S}') \cdot \mathrm{ds} \tag{28}$$

represents the volumetric interfacial force exerted by the solids, with $\mathbf{I} = \delta_{ij}$ being the identity matrix. Note that the volumetric interfacial force is contributed from the pressure and the shear stresses acting on the solid surface. The term, $-\phi\overline{\mathbf{u}'\mathbf{u}'}$, in Eq. (28) represents the additional inertia stresses due to the correlation between the microscopic velocity variations over REV, and is now termed as "momentum dispersion". Its physics may bear some resemblance to the Reynolds stresses in turbulence.

Energy Equations:

Fluid Phase:

$$\rho c_p \frac{\partial(\phi\,\overline{\mathrm{T}})}{\partial\,\mathrm{t}} + \rho c_p \overline{\nabla}\cdot(\phi\,\overline{\mathbf{u}}\,\overline{\mathrm{T}}) = k\,\overline{\nabla}^2(\phi\overline{\mathrm{T}}) + \rho c_p \overline{\nabla}\cdot(-\phi\,\overline{\mathbf{u}'\mathrm{T}'}) + k\,\overline{\nabla}\cdot\mathbf{\Lambda}_{fs} + \mathrm{q}_{fs} \tag{29}$$

where $(-\overline{\mathbf{u}'\mathrm{T}'})$ represents the thermal dispersion,

$$\mathbf{\Lambda}_{fs} = \frac{1}{V}\int_{A_{fs}} \mathrm{T}\cdot\mathrm{ds} = \frac{1}{V}\int_{A_{fs}} \mathrm{T}'\cdot\mathrm{ds} \tag{30}$$

represents the thermal tortuosity and

$$\mathrm{q}_{fs} = \frac{1}{V}\int_{A_{fs}} k\nabla\mathrm{T}\cdot\mathrm{ds} \tag{31}$$

represents the interfacial heat transfer from solid into fluid.

Solid Phase:

$$\rho_s c_{ps} \frac{\partial(\phi_s \overline{\mathrm{T}}_s)}{\partial\,\mathrm{t}} = k_s \overline{\nabla}^2(\phi_s \overline{\mathrm{T}}_s) - k_s \overline{\nabla}\cdot\mathbf{\Lambda}_{fs} - \mathrm{q}_{fs} \tag{32}$$

In Eq. (32), we have invoked the condition of continuity of heat flux across the interface, i.e.,

$$k\nabla\mathrm{T} = k_s \nabla\mathrm{T}_s \qquad \text{on } A_{fs} \tag{33}$$

so that the source term q_{fs} given by Eq. (29) for fluid phase becomes the sink term for solid phase, or vice versa.

Mass Concentration Equations:

Fluid phase:

$$\rho \frac{\partial (\phi\, \overline{\mathrm{Y}})}{\partial \mathrm{t}} + \rho\, \overline{\nabla} \cdot (\phi \overline{\mathbf{u}}\, \overline{\mathrm{Y}}) = \rho\eta \overline{\nabla}^2 (\phi \overline{\mathrm{Y}}) + \rho\, \overline{\nabla} \cdot (-\phi \overline{\mathbf{u}'\mathrm{Y}'}) + \rho\eta \overline{\nabla} \cdot \mathbf{\Omega}_{fs} + \mathrm{m}_{fs} \tag{34}$$

where $(-\overline{\mathbf{u}'\mathrm{Y}'})$ represents the mass dispersion,

$$\mathbf{\Omega}_{fs} = \frac{1}{\mathrm{V}} \int_{\mathrm{A}_{fs}} \mathrm{Y} \cdot \mathbf{ds} = \frac{1}{\mathrm{V}} \int_{\mathrm{A}_{fs}} \mathrm{Y}' \cdot \mathbf{ds} \tag{35}$$

represents the tortuosity of mass diffusion and

$$\mathrm{m}_{fs} = \frac{1}{\mathrm{V}} \int_{\mathrm{A}_{fs}} \rho\eta \nabla \mathrm{Y} \cdot \mathbf{ds} \tag{36}$$

is the interfacial mass transfer from solid to fluid, which is also the rate of mass per unit volume transferred from the solid.

Solid phase:

$$\rho_{\mathrm{s}} \frac{\partial (\phi_{\mathrm{s}} \overline{\mathrm{Y}}_{\mathrm{s}})}{\partial \mathrm{t}} = \rho_{\mathrm{s}}\eta_{\mathrm{s}} \overline{\nabla}^2 (\phi_{\mathrm{s}} \overline{\mathrm{Y}}_{\mathrm{s}}) - \rho_{\mathrm{s}}\eta_{\mathrm{s}} \overline{\nabla} \cdot \mathbf{\Omega}_{fs} - \mathrm{m}_{fs} + \mathrm{m}_{ad} \tag{37}$$

where

$$\mathrm{m}_{ad} = \frac{1}{\mathrm{V}} \int_{\mathrm{A}_{fs}} \mathbf{m}''_{ad} \cdot \mathbf{ds} \tag{38}$$

is the rate of mass adsorbed on the interface per unit volume. In Eq. (37), we have invoked the condition of the mass flux balance across the interface, i.e.,

$$\rho\eta \nabla \mathrm{Y} = \mathbf{m}''_{ad} + \rho_{\mathrm{s}}\eta_{\mathrm{s}} \nabla \mathrm{Y}_{\mathrm{s}} \qquad \text{on } \mathrm{A}_{fs} \tag{39}$$

where $\mathbf{m}''_{ad}$ is the flux of mass adsorption on the solid-fluid interface.

The macroscopic transport equations given above with prescribed boundary conditions cannot be solved directly because there are more unknowns than the number of equations, a so-called *closure problem*. These unknown terms are associated with the momentum, thermal and mass dispersions, the thermal and mass tortuosities and the interfacial momentum, heat and mass transfer. While the dispersions occur in the fluid phase, the tortuosities and interfacial occur at the solid-fluid interface. Constitutive relations have to be sought through closure modeling to relate these new unknowns to the averaged macroscopic quantities.

3. CLOSURE MODELING

The construction of closure relations now becomes the major task for the treatise of the transport phenomena in porous medium. As the closure problem is associated with the dispersions, tortuosity and the interfacial transfer which involve the microscopic deviations, it is logical to examine the behaviors of **u**', p', T' and Y', T_s' and Y_s' for the construction of the closure (constitutive) relations. These can be done by first obtaining the governing equations for the microscopic deviations, then normalizing the resultant equations to reveal the key non-dimensional parameters such as Reynolds number, Peclet number, Schmidt number and Keulegan-Carpenter number that govern the behaviors of the microscopic transports. The closure relations can then be obtained based on the correlation between the microscopic deviations and the macroscopic quantities for different ranges of these key parameters. The composite closure relations valid for all range of the parameters are then constructed. In the following, we shall not engage into the details of the closure modeling, but only summarize the results of these closure relations as obtained earlier by others for the dispersion, tortuosity and interfacial transfer.

3.1 Closure Relations for Momentum Dispersion and Interfacial Force

Hsu (2001a) obtained the closure relations for the hydrodynamic dispersion and interfacial force based on dispersed spheres of uniform diameter of d_p. The resultant composite closure relations that are valid for all Reynolds number, $\mathrm{Re_p} = |\overline{\mathbf{u}}|\, d_p/\nu$, and all range of time scale are expressed as follows.

Momentum Dispersion:

$$\overline{\mathbf{u}'\mathbf{u}'} = c\overline{\mathbf{u}}\,\overline{\mathbf{u}} - \varepsilon\frac{1}{\phi}\overline{\nabla}(\phi\overline{\mathbf{u}}) \tag{40}$$

where ε is the dispersion viscosity given by

$$\varepsilon = c_1\,\bar{l}\,|\overline{\mathbf{u}}| + c_2\bar{l}^2\left|\frac{1}{\phi}\overline{\nabla}\times(\phi\overline{\mathbf{u}})\right| \tag{41}$$

with $\bar{l}$ being the mixing length of momentum dispersion and c_1 and c_2 being the correlation coefficients. The first term on the right hand side of Eq. (41)

resembles the Prandtl's third mixing length theory for turbulent eddy viscosity of a wake behind a sphere and that the second term resembles the eddy viscosity for turbulent boundary layer near a solid wall. However, the dispersion mixing length for flows in porous media is assumed to scale with the particle diameter in the core region far away from an impermeable wall and then to become linearly proportional to the distance $\overline{x}_3$ measured from the impermeable wall. By adapting a van Driest's damping factor A^+, $\bar{l}$ is expressed as

$$\bar{l} = c_3 d_p [1 - \exp(-A^+ \,\overline{x}_3 / d_p)] \tag{42}$$

Interfacial Force:

$$\begin{aligned} \overline{\mathbf{b}}_{fs} = &-\frac{\phi_s \mu c_S}{d_p^2}\overline{\mathbf{u}} - \frac{\phi_s c_B}{d_p^{3/2}}\sqrt{\rho\mu|\overline{\mathbf{u}}|}\,\overline{\mathbf{u}} - \frac{\rho\phi_s c_I}{d_p}|\overline{\mathbf{u}}|\overline{\mathbf{u}} - \frac{\phi_s c_G}{d_p}\sqrt{\frac{\rho\mu}{|\overline{\nabla}\times(\phi\overline{\mathbf{u}})|}}\,\overline{\mathbf{u}}\times[\overline{\nabla}\times(\phi\overline{\mathbf{u}})] \\ &- \rho\phi_s c_L \overline{\mathbf{u}}\times[\overline{\nabla}\times(\phi\overline{\mathbf{u}})] - \rho\phi_s c_V \frac{\overline{D}(\phi\overline{\mathbf{u}})}{\overline{D}t} - \sqrt{\frac{\rho\mu}{\pi}}\frac{\phi_s c_M}{d_p}\int_{-\infty}^{t}\frac{\partial(\phi\overline{\mathbf{u}})}{\partial\tau}\frac{d\tau}{\sqrt{t-\tau}} \end{aligned} \tag{43}$$

where c_S is the drag coefficients due to Stokes flow, c_B and c_I are the viscous and inviscid drag coefficients due to advection, c_G and c_L are the viscous and inviscid lift force coefficients due to mean vorticity, and c_V and c_M are the transient inertia drag coefficient due to the virtual mass and the Basset transient memory effects. Note that the form drag, the inviscid lift force and the virtual mass force are independent of the viscosity, i.e., they are associated mainly with the hydrodynamic pressure. The relation (43) has a considerable resemblance to that of two-phase flows, although there are some differences in detail

Macroscopic Momentum Transport Equations:

By substituting Eq. (40) into Eq. (26), the macroscopic momentum transport equations now become

$$\rho\left[\frac{\partial}{\partial t}(\phi\,\overline{\mathbf{u}}) + \overline{\nabla}\cdot(\phi(1+c)\,\overline{\mathbf{u}}\,\overline{\mathbf{u}})\right] = -\overline{\nabla}(\phi\,\overline{p}) + \rho\overline{\nabla}\cdot\left[(\nu+\varepsilon)\overline{\nabla}(\phi\overline{\mathbf{u}})\right] + \overline{\mathbf{b}}_{fs} \tag{44}$$

where $\overline{\mathbf{b}}_{fs}$ is given by Eq. (43). It is seen that the effects of momentum dispersion are to produce the excess advection velocity and the dispersion viscosity.

3.2 Closure Relations for Thermal Dispersion, Thermal Tortuosity and Interfacial Heat Transfer

The closure for the thermal dispersion, thermal tortuosity and interfacial heat transfer had been studied for decades. Hsu and Cheng (1990) proposed a closure model for the thermal dispersion. The early works on the closure modeling of thermal tortuosity were on the problem of stagnant heat conduction in porous media. Quintard and Whitaker (1993, 1995) gave a first comprehensive account of the thermal tortuosity, followed by the recent work of Hsu (1999). The early works on the interfacial heat transfer were highly experimental (Kunii and Suzuki, 1967; Wakao et al., 1979). A more comprehensive theoretical treatment of the closure model for the interfacial heat transfer was given by Hsu (1999, 2000c). Here we should summarize the closure relations as obtained by Hsu and Cheng (1990), Hsu (1999) and Hsu (2001b) as follows.

Thermal Dispersion:

The thermal dispersion were modeled by Hsu and Cheng (1990), who showed that

$$-\phi\overline{\mathbf{u}'\mathrm{T}'} = \mathbf{A}_{\mathrm{D}} : \overline{\nabla}\,\overline{\mathrm{T}} \tag{45}$$

where $\mathbf{A}_{\mathrm{D}}$ is the dispersion thermal diffusivity tensor, which can be expressed into

$$\mathbf{A}_{\mathrm{D}} = \begin{bmatrix} \alpha_1 & 0 & 0 \\ 0 & \alpha_2 & 0 \\ 0 & 0 & \alpha_3 \end{bmatrix} \tag{46}$$

with α_1, α_2, and α_3 being the dispersion thermal diffusivities in the longitudinal, transverse and lateral directions, respectively. It is anticipated that the transverse and lateral dispersions are of the same order in core region of the porous media, but not in the region near the impermeable wall. According to Hsu and Cheng (1990), the dispersion thermal diffusivities are linearly proportional to the Peclet number, $\mathrm{Pe_p} = |\overline{\mathbf{u}}| d_p/\alpha$, when the Peclet number is large, and becomes proportional to the square of the Peclet number when the Peclet number is low. Therefore, the composite relations for the dispersion thermal diffusivities can be constructed as

$$\alpha_i = (1-\phi)\alpha \frac{a_i \mathrm{Pe_p^2}}{b_i + \mathrm{Pe_p}} \qquad (i = 1, 2 \text{ and } 3) \tag{47}$$

where a_i and b_i are coefficients.

Thermal Tortuosity:

According to the closure modeling of Hsu (1999), the thermal tortuosity is expressed into

$$\mathbf{\Lambda}_{fs} = G(\nabla\overline{\mathrm{T}} - \sigma\nabla\overline{\mathrm{T}}_{s}) \tag{48}$$

where G is the tortuosity parameter and σ the conductivity ratio of solid to fluid. The tortuosity parameter G can be determined from

$$G = \left[\frac{k_{st}}{k} - \phi - \sigma(1-\phi)\right] \Big/ (1-\sigma)^2 \tag{49}$$

where k_{st} is the overall effective stagnant thermal conductivity of the porous media based on a mixture model under the assumption of local thermal equilibrium. A comprehensive review of the effective stagnant thermal conductivity of porous media was given in Cheng and Hsu (1998).

Interfacial Heat Transfer:

From Hsu (1999), the closure relation for the interfacial transfer from solid to fluid is given by

$$q_{fs} = h_{fs} a_{fs} \left(\overline{\mathrm{T}}_{s} - \overline{\mathrm{T}}_{f}\right) \tag{50}$$

where a_{fs} (= A_{fs}/V) is the specific interfacial area and h_{fs} is the interfacial heat transfer coefficient. From Hsu (2001b), the interfacial heat transfer coefficient takes the form of $h_{fs}^*(1 + c_4 \Pr \mathrm{Re}_p)$ at low Reynolds number where h_{fs}^* is the interfacial heat transfer coefficient of stagnant fluid (see Hsu, 1999); however, at large Reynolds number, it becomes proportional $\Pr^m \mathrm{Re}_p^{\,n}$. Therefore, the composite expressions for h_{fs} in terms of the interfacial Nusselt number becomes

$$Nu_{fs} = \frac{h_{fs} d_p}{k} = Nu_{fs}^*\left(1 + \frac{a \Pr \mathrm{Re}_p}{b + Nu_{fs}^* \Pr^{1-m} \mathrm{Re}_p^{1-n}}\right) \tag{51}$$

where Nu_{fs}^* is the stagnant interfacial Nusselt number and a and b are coefficients with $a/b = c_4$. From the theory of convective heat transfer, we have $n > 0.5$ for large Reynolds number. We also have $m = 1/2$ when $\Pr \ll 1$ and $m = 1/3$ when $\Pr \gg 1$. A quasi-steady model for Nu_{fs}^* as proposed by Hsu (1999) on the basis of the parallel conduction layers on fluid and solid sides, respectively, is given by

$$Nu_{fs}^{*} = \frac{h_{fs}^{*} d_p}{k} = \frac{\sigma}{\alpha_A \sigma + \alpha_B} \tag{52}$$

where α_A and α_B represent the dimensionless conduction layer thickness in fluid and solid phases as normalized by the particle diameter, respectively.

Macroscopic Energy Equations:

By substituting the closure relations for the thermal dispersion, thermal tortuosity and interfacial heat transfer into (29) and (32), the macroscopic energy equations become:

Fluid Phase:

$$\rho c_p \frac{\partial(\phi \overline{\mathrm{T}})}{\partial \mathrm{t}} + \rho c_p \overline{\nabla} \cdot (\phi\, \overline{\mathbf{u}}\overline{\mathrm{T}}) = k\, \overline{\nabla}^2(\phi \overline{\mathrm{T}}) + \rho c_p \overline{\nabla} \cdot (\mathbf{A}_{\mathrm{D}} : \overline{\nabla}\, \overline{\mathrm{T}})$$
$$+ k\, \overline{\nabla} \cdot [G(\overline{\nabla}\overline{\mathrm{T}} - \sigma \overline{\nabla}\, \overline{\mathrm{T}}_{\mathrm{s}})] + h_{fs} a_{fs} (\overline{\mathrm{T}}_{\mathrm{s}} - \overline{\mathrm{T}}) \tag{53}$$

Solid Phase:

$$\rho_s c_{ps} \frac{\partial(\phi_s \overline{\mathrm{T}}_{\mathrm{s}})}{\partial \mathrm{t}} = k_s \overline{\nabla}^2 (\phi_s \overline{\mathrm{T}}_{\mathrm{s}}) - k_s \overline{\nabla} \cdot [G(\overline{\nabla}\overline{\mathrm{T}} - \sigma \overline{\nabla}\, \overline{\mathrm{T}}_{\mathrm{s}})] - h_{fs} a_{fs} (\overline{\mathrm{T}}_{\mathrm{s}} - \overline{\mathrm{T}}) \tag{54}$$

Equations (53) and (54) with the dispersion thermal diffusivity, the tortuosity, and the interfacial heat transfer coefficient given by Eqs. (47), (49) and (51), respectively, are the macroscopic governing equations for the unsteady convective heat transfer in porous media. The evaluation of the closure coefficients α_i, G and h_{fs} then becomes one of main tasks.

3.3 Closure Relations for Mass Dispersion, Mass Tortuosity and Interfacial Mass Transfer

As the transfer process of mass concentration in the porous media is very similar to the heat transfer process, it is anticipated that the closure modeling for mass concentration transfer will take the similar form as described in Section 3.2. However, there are some fundamental differences between the two. For example, mass transport is concerned about the movement of contaminant molecules in the carrier fluids and in the solid while the heat transfer with the passing of internal energy due to motion of fluids and inter-molecular interaction. As a result, it is anticipated that the mass diffusion process in the solid is much smaller than that in the fluids, i.e., λ_s ($=\rho_s \eta_s$)<< λ ($=\rho\eta$) and $\eta_s << \eta$. Under this condition, the mass diffusion in solid can be ignored. This inequality may not be true for the conduction in solids and fluids, as conduction in solids can be much larger than that in fluids for the solid materials of metal. The details of the closure relations for the mass transfer were given in Hsu (2001b) and are summarized as follows.

Mass Dispersion:

$$-\phi\overline{\mathbf{u}'\mathrm{Y}'} = \mathbf{M}_{\mathrm{D}} : \overline{\nabla}\,\overline{\mathrm{Y}} \tag{55}$$

where $\mathbf{M}_{\mathrm{D}}$ is the diffusivity tensor of mass dispersion, which can be expressed into

$$\mathbf{M}_{\mathrm{D}} = \begin{bmatrix} \eta_1 & 0 & 0 \\ 0 & \eta_2 & 0 \\ 0 & 0 & \eta_3 \end{bmatrix} \tag{56}$$

with η_1, η_2, and η_3 being the diffusivities of mass dispersion in the longitudinal, transverse and lateral directions, respectively. Again, the transverse and lateral dispersions are of the same order in core region of the porous media, but not in the region near the impermeable wall. According to Koch and Brady (1985), the mass dispersion diffusivities are linearly proportional to mass Peclet number $\mathrm{Pe}_{\mathrm{mp}}$ $(=|\overline{\mathbf{u}}|d_p/\eta)$ when $\mathrm{Pe}_{\mathrm{mp}}$ is large, and becomes proportional to the square of $\mathrm{Pe}_{\mathrm{mp}}$ when $\mathrm{Pe}_{\mathrm{mp}}$ is low. Therefore, the composite relations for the mass dispersion diffusivities can be expressed as

$$\eta_i = (1-\phi)\eta\frac{a_{mi}(\mathrm{Pe}_{\mathrm{mp}})^2}{b_{mi} + \mathrm{Pe}_{\mathrm{mp}}} \qquad (i = 1, 2 \text{ and } 3) \tag{57}$$

where a_{mi} and b_{mi} are coefficients.

Mass Concentration Tortuosity:

The closure relations for the mass concentration tortuosity are very much similar to Eqs. (48) and (49) for heat transfer. However, under the condition of $\lambda_s << \lambda$ the closure relations for mass tortuosity reduce to

$$\mathbf{\Omega}_{fs} = G_m \overline{\nabla \mathrm{Y}} \tag{58}$$

where the mass tortuosity parameter G_m can be determined from

$$G_m = \frac{\lambda_{st}}{\lambda} - \phi \tag{59}$$

where λ_{st} is the effective stagnant mass diffusivity of the porous media under the limit of $\lambda_s << \lambda$, based on a mixture model.

Interfacial Mass Transfer:

When $\lambda_s << \lambda$, the effect of mass diffusion in the solid phase can be neglected and the interfacial transfer is mainly contributed by the surface adsorption. From Hsu (2001b), the closure relation for the interfacial mass transfer can be expressed into

$$\mathrm{m}_{fs} = g_{fs} a_{fs} \left(\mathrm{Y}^* - \overline{\mathrm{Y}}\right) \tag{60}$$

where g_{fs} is the interfacial mass transfer coefficient and Y* is the saturation mass concentration of surface adsorption. The interfacial mass transfer coefficient reduces to a surface adsorption value of g_{ad}* at low Reynolds number. At large Reynolds number, it becomes proportional to $\mathrm{Sc}^m \mathrm{Re_p}^n$ where Sc = ν/η is the Schmidt number. The value of m takes 1/3 since Sc >> 1 for almost all fluids. Therefore, the composite expressions for g_{fs} in terms of the interfacial Sherwood number becomes

$$Sh_{fs} = \frac{g_{fs} d_p}{\lambda} = Sh_{ad}^* + a_m Sc^{1/3}\, \mathrm{Re}_p^n \tag{61}$$

where Sh_{ad}* (= g_{ad}*d_p/λ) is the interfacial Sherwood number due to surface adsorption and the coefficients a_m is a constant. Since Sh_{ad}* is mainly contributed from the surface adsorption, it depends solely on the activation condition on the solid-fluid interface, but not on λ_s when $\lambda_s << \lambda$.

Macroscopic Mass Concentration Equations:

By substituting the closure relations for the mass dispersion, mass tortuosity and interfacial mass transfer into Eq. (34), the macroscopic mass transport equation in fluid phase becomes:

$$\begin{aligned} \rho \frac{\partial (\phi\, \overline{\mathrm{Y}})}{\partial \mathrm{t}} + \rho \overline{\nabla} \cdot (\phi\, \overline{\mathbf{u}}\, \overline{\mathrm{Y}}) &= \lambda\, \overline{\nabla}^2 (\phi \overline{\mathrm{Y}}) + \rho \overline{\nabla} \cdot (\mathbf{M}_\mathrm{D} : \overline{\nabla}\, \overline{\mathrm{Y}}) \\ &+ \lambda\, \overline{\nabla} \cdot (G_m \overline{\nabla \mathrm{Y}}) + g_{fs} a_{fs} (\mathrm{Y}^* - \overline{\mathrm{Y}}) \end{aligned} \tag{62}$$

with the mass diffusion in the solid being neglected. Equation (62) with the coefficients of mass dispersion diffusivity, tortuosity, and interfacial transfer given by Eqs. (57), (59) and (61), respectively, is the macroscopic governing equations for the convective mass transfer in porous media. The closure coefficients in (57), (59) and (61) then have to be evaluated by experiments.

4. MACROSCOPIC EQUATIONS OF SUPERFICIAL FLOWS

The above closure coefficients in the composite macroscopic transport equations and the closure relations were constructed under the limit of $\phi \to 0$ for media of dispersed dilute spheres. They may depend strongly on porosity for media of densely packed particles. It is hard to determine theoretically this porosity dependence; however, evidences from the existing experimental data suggest that the proper scale to account for the contribution due to particle interference to the volumetric interfacial force should be the hydraulic diameter defined by

$$d_h - \frac{\phi}{1-\phi} d_p \tag{63}$$

The flows through the porous media are then postulated as those passing through a series of capillary tubes of diameter d_h. To be in line with classical formulation, we should express the equations in terms of the pore pressure $\mathrm{P} = \overline{\mathrm{p}}$, and the Darcy velocity $\mathbf{U} = \phi \overline{\mathbf{u}}$; however, for the heat and mass transfer, $\overline{\mathrm{T}}, \overline{\mathrm{Y}}$ and $\overline{\mathrm{T}}_s$ will remain to represent the averaged quantities over the respective phases.

4.1. Momentum Equations

In terms of Darcy velocity, pore pressure and the hydraulic diameter, Equations (25) and (44) then become

$$\overline{\nabla} \cdot \mathbf{U} = 0 \tag{64}$$

and

$$\rho\left[\frac{\partial}{\partial \mathrm{t}}(\mathbf{U}) + \overline{\nabla} \cdot ((1+c)\mathbf{UU}/\phi)\right] = -\overline{\nabla}(\phi \mathrm{P}) + \rho\overline{\nabla} \cdot \left[(\nu + \varepsilon)\overline{\nabla}(\mathbf{U})\right] + \overline{\mathbf{b}}_{fs} \tag{65}$$

where the dispersion viscosity in term of the hydraulic diameter is re-written as

$$\varepsilon = c_{\mathrm{h1}}\bar{l}_{\mathrm{h}}|\mathbf{U}| + c_{\mathrm{h2}}\bar{l}_{\mathrm{h}}^{\,2}|\overline{\nabla} \times \mathbf{U}| \tag{66}$$

with the hydraulic dispersion length being modified into

$$\bar{l}_h = d_h[1 - \exp(-\mathrm{A}_h^+ \overline{\mathrm{x}}_3 / d_h)] \tag{67}$$

Note that c_3 in Eq. (42) can be adjusted arbitrarily to render Eq. (67). The volumetric interfacial force now becomes

$$\begin{aligned}\overline{\mathbf{b}}_{fs} = &-\frac{\mu C_S}{d_h^2}\mathbf{U} - \frac{C_B}{d_h^{3/2}}\sqrt{\rho\mu|\mathbf{U}|}\mathbf{U} - \frac{\rho C_I}{d_h}|\mathbf{U}|\mathbf{U} - \frac{C_G}{d_h}\sqrt{\frac{\rho\mu}{|\overline{\nabla}\times\mathbf{U}|}}\mathbf{U}\times(\overline{\nabla}\times\mathbf{U}) \\ &-\rho C_L \mathbf{U}\times(\overline{\nabla}\times\mathbf{U}) - \rho C_V \frac{\overline{D}\mathbf{U}}{\overline{D}t} - \sqrt{\frac{\rho\mu}{\pi}}\frac{C_M}{d_h}\int_{-\infty}^{t}\frac{\partial \mathbf{U}}{\partial \tau}\frac{d\tau}{\sqrt{t-\tau}}\end{aligned} \tag{68}$$

Here, the dependencies of the dispersion viscosity and the volumetric interfacial force on the porosity will appear in the hydraulic diameter and the closure coefficients in Eqs. (66) and (67). Note that these coefficients depend also strongly on the microscopic geometry of the solids; hence, they are to be determined experimentally.

Eqs. (64)-(68) form a closed set of equations that can be solved with proper macroscopic boundary conditions. They are the macroscopic momentum transport equations for unsteady flows in porous media. These equations have taken into account the first-order leading terms over the entire ranges of Reynolds number and time scale. As seen from the right hand side of Eq. (68), the first term of Stokes drag is proportional to μ corresponding to creeping flow at low Reynolds number and the force is contributed from both shear and pressure. The second, fourth and seventh terms are proportional to $\mu^{1/2}$ due to boundary layer effects at intermediate Reynolds number (lower-end of high Reynolds number) and intermediate time scale, where the forces are solely contributed from shear. The third, fifth and sixth terms are independent of μ corresponding to inviscid potential flows at very high Reynolds number and short time scale, and the forces are solely contributed from pressure.

4.2 Energy Equations

The energy transport equations in the fluid phase in terms of Darcy velocity then become:

$$\begin{aligned}\rho c_p \frac{\partial(\phi\overline{T})}{\partial t} + \rho c_p \overline{\nabla}\cdot(\mathbf{U}\overline{T}) = k\,\overline{\nabla}^2(\phi\overline{T}) + \rho c_p \overline{\nabla}\cdot(\mathbf{A}_D : \overline{\nabla}\overline{T}) \\ + k\,\overline{\nabla}\cdot[G(\overline{\nabla}\overline{T} - \sigma\overline{\nabla}\,\overline{T}_s)] + h_{fs}a_{fs}(\overline{T}_s - \overline{T})\end{aligned} \tag{69}$$

where the dispersion thermal diffusivities are given by

$$\alpha_i = \alpha\frac{a_{hi}\mathrm{Pe}_h^2}{b_{hi} + \mathrm{Pe}_h} \qquad (i = 1, 2 \text{ and } 3) \tag{70}$$

with the Peclet number based on hydraulic diameter given by $\mathrm{Pe}_h = |\mathbf{U}|\, d_h/\alpha$, and a_{hi} and b_{hi} being coefficients. The interfacial Nusselt number in term of the hydraulic diameter then becomes

$$Nu_{hfs} = \frac{h_{fs} d_h}{k} = Nu^*_{hfs}\left(1 + \frac{a_h \Pr \mathrm{Re}_h}{b_h + Nu^*_{hfs} \Pr^{1-m} \mathrm{Re}_h^{1-n}}\right) \tag{71}$$

where $\mathrm{Re}_h = |\mathbf{U}|\, d_h/\nu$ is the Reynolds number based on Darcy velocity and hydraulic diameter and $Nu_{hfs}{}^*$ is stagnant Nusselt number given by

$$Nu^*_{hfs} = \frac{h^*_{fs} d_h}{k} = \frac{\sigma}{\alpha_{hA}\sigma + \alpha_{hB}} \tag{72}$$

4.3 Mass Concentration Equations

The mass transport equations in terms of the Darcy velocity becomes:

$$\begin{aligned} \rho \frac{\partial(\phi\, \overline{\mathrm{Y}})}{\partial\, \mathrm{t}} + \rho\overline{\nabla}\cdot(\mathbf{U}\,\overline{\mathrm{Y}}) = \lambda\overline{\nabla}^2(\phi\overline{\mathrm{Y}}\,) + \rho\overline{\nabla}\cdot(\mathbf{M}_{\mathrm{D}} : \overline{\nabla}\,\overline{\mathrm{Y}}) \\ + \lambda\,\overline{\nabla}\cdot(\overline{F\nabla\mathrm{Y}}) + g_{fs} a_{\mathrm{fs}}(\mathrm{Y}^* - \overline{\mathrm{Y}}) \end{aligned} \tag{73}$$

where the dispersion mass diffusivities are given by

$$\eta_i = \eta \frac{a_{mhi}(\mathrm{Pe}_{\mathrm{mh}})^2}{b_{mhi} + \mathrm{Pe}_{\mathrm{mh}}} \qquad (i = 1, 2 \text{ and } 3) \tag{74}$$

with the mass Peclet number based on hydraulic diameter given by $\mathrm{Pe}_{\mathrm{mh}} = |\mathbf{U}|\, d_h/\eta$, and a_{mhi} and b_{mhi} being coefficients. The interfacial Sherwood number in term of the hydraulic diameter then becomes

$$Sh_{hfs} = \frac{g_{fs} d_h}{\lambda} = Sh^*_{had} + a_{mh} Sc^{1/3} \mathrm{Re}_h^n \tag{75}$$

5. EVALUATION OF CLOSURE COEFFICIENTS

In this section, we shall review some of the experiments that are relevant for the determination of the closure coefficients that appear in the closure relations. They are summarized as follows.

5.1 Hydrodynamic Experiments

Most of the early experimental works on the flows in porous media were devoted to the determination of the coefficients in the interfacial force. To the author's knowledge, to date there exist no experimental data for the determination of the coefficients in the closure relation of hydrodynamic dispersion. The main difficulty lays on the fact that the Brinkman layer near an impermeable wall is too thin to be measurable. Even for the interfacial force, most of the early works were conducted on steady flows for determining the permeability to delineate the Darcy's law at very Reynolds number and the Forchheimer inertia effect at very large Reynolds number. To delineate the transient effect, we need to steady the unsteady flows. One of the simplest unsteady flows is the one-dimensional periodically oscillating flow. Recently, Hsu et al. (1999) and Hsu and Fu (2001) measured the velocity and the pressure-drop for both steady and oscillating flows across porous columns packed from wire screens. In the following, we shall present only their experimental results, while leaving the experimental details to the original papers. For better understanding of the experimental results, a brief review of theoretical background is first given. It should be noted that these experiments were only valid for flows in the core region of the porous media.

Theoretical Background:

In the core region of a porous media, Eq. (65) with the substitution of (68) reduces to

$$\rho(1+C_V)\frac{\partial u}{\partial t} = -\frac{\partial(\phi\, p)}{\partial x} - \frac{\mu C_S}{d_h^{\,2}}u - \frac{C_B}{d_h^{\,3/2}}\sqrt{\rho\mu|u|}\,u - \frac{\rho C_I}{d_h}|u|\,u - \sqrt{\frac{\rho\mu}{\pi}}\frac{C_M}{d_h}\int_{-\infty}^{t}\frac{\partial u}{\partial\tau}\frac{d\tau}{\sqrt{t-\tau}} \tag{76}$$

if the flow is one-dimensional, i.e., $\mathbf{U} = (u, 0, 0)$. In Eq. (76), the terms with the coefficients C_S, C_B, C_I, C_V, and C_M are associated respectively with the Stokes drag force, the frictional force due to convection boundary layer, the inviscid form drag, the inviscid virtual mass force and the Basset memory viscous force due to transient boundary layer.

In the limit of low frequency oscillating flows, Eq. (76) reduces further to the quasi-steady form of

$$-\frac{\partial(\phi\, p)}{\partial x} = \frac{\mu C_S}{d_h^{\,2}}u + \frac{C_B}{d_h^{\,3/2}}\sqrt{\rho\mu|u|}\,u + \frac{\rho C_I}{d_h}|u|\,u \tag{77}$$

which was proposed originally by Hsu and Cheng (1990). Equation (77) indicates that the negative pressure gradient and the velocity are in-phase, i.e., maximum pressure drop occurs when the velocity is maximum. Taking the maximums of pressure and velocity oscillations, Eq. (77) becomes

$$f = \frac{C_S}{Re_h} + \frac{C_B}{Re_h^{1/2}} + C_I \tag{78}$$

where $f = \dfrac{\phi \Delta p_{max} d_h}{\rho u_{max}^2 L}$ is the pressure-drop coefficient with Δp_{max} being the maximum pressure-drop across a global length scale L of a porous column, and $Re_h = \dfrac{u_{max} d_h}{\nu}$ the Reynolds number with u_{max} being the amplitude of a sinusoidal velocity, i.e., $u = u_{max} \cos \omega t$.

When the transient inertia force becomes important at high frequency, there will be a phase difference between velocity and pressure-gradient oscillations. A complete description of velocity and pressure-gradient correlation requires both the amplitude correlation and phase difference. If the velocity and pressure gradient are assumed as the real part of the following complex expressions,

$$u = \hat{u}\, e^{i\omega t} \quad \text{and} \quad -\frac{1}{\rho}\frac{\partial P}{\partial x} = \hat{\alpha}\, e^{i\omega t} + \text{harmonics} \tag{79a,b}$$

where $\hat{u}$ (= u_{max}) and $\hat{\alpha}$ represent the complex amplitudes of velocity and pressure gradient, respectively, the substitution of Eq. (79) into Eq. (76), and then collecting the fundamental mode of oscillation, leads to

$$\hat{\alpha} = \frac{1}{\phi}\left[\frac{\nu C_S}{d_h^2} + \frac{C_B}{d_h^{3/2}}\sqrt{\frac{2.64\nu|\hat{u}|}{\pi}} + \frac{2.67 C_I |\hat{u}|}{\pi d_h} + \frac{C_M \sqrt{i\omega\nu}}{d_h} + (1 + C_V)\, i\omega\right]\hat{u} \tag{80}$$

From Eq. (80), it appears that the Basset memory force generates a pressure-gradient component of a 45°-phase difference from the velocity, while the virtual mass force generates a component of a 90°-phase difference. The quasi-steady state then represents the limit case of a 0°-phase difference when $\omega \to 0$. Taking the absolute value to Eq. (80) results in

$$\hat{f} = \left|\frac{C_S}{Re_h} + \frac{C_B}{Re_h^{1/2}}\sqrt{\frac{2.64}{\pi}} + \frac{2.67 C_I}{\pi} + \frac{C_M}{Re_h^{1/2}}\sqrt{\frac{i d_h}{A}} + (1 + C_V)\frac{i d_h}{A}\right| \tag{81}$$

where $\hat{f} = |\hat{\alpha}|\phi\, d_h/|\hat{u}|^2$ is the pressure-gradient coefficient based on the fundamental mode, Re_h = $|\hat{u}|\, d_h/\nu$ is the Reynolds number, and A = $|\hat{u}|/\omega$ is the amplitude of the fluid displacement of the superficial flow. Note that here we have A = $\phi \overline{A}$ with $\overline{A}$ being the intrinsic average of fluid displacement in the pore. The phase angle between the pressure and velocity can be obtained by taking the argument to Eq. (81) to result in

$$\theta = \tan^{-1}\left[\frac{(1+C_V)\dfrac{d_h}{A}+\dfrac{C_M}{\sqrt{2Re_h}}\sqrt{\dfrac{d_h}{A}}}{\dfrac{C_S}{Re_h}+\dfrac{C_B}{\sqrt{Re_h}}\sqrt{\dfrac{2.64}{\pi}}+\dfrac{2.67C_I}{\pi}+\dfrac{C_M}{\sqrt{2Re_h}}\sqrt{\dfrac{d_h}{A}}}\right] \tag{82}$$

Equations (81-82) indicate that the pressure gradient of an oscillating flow in a porous medium depends on two parameters, Re_h and d_h/A. The inverse of d_h/A is the Keulegan-Carpenter number commonly encountered in oscillating flows. In the limit of $d_h/A \to 0$, (i.e., $A \to \infty$ and $\omega \to 0$ while maintaining $|\hat{u}|$ as finite), Eq. (81) reduces to

$$\hat{f} = \frac{C_S}{Re_h}+\frac{C_B}{Re_h^{1/2}}\sqrt{\frac{2.64}{\pi}}+\frac{2.67C_I}{\pi} \tag{83}$$

and the phase angle approaches zero. Note that Eq. (83) is different from Eq. (78) by the factors of $\sqrt{2.64/\pi}$ and $2.67/\pi$ in the last two terms because Eq. (83) uses the amplitude of fundamental mode rather than the maximum of the pressure gradient.

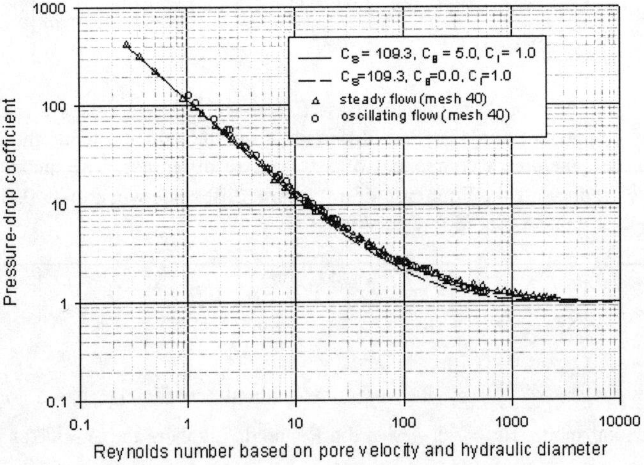

Figure 3. Correlation of pressure-drop coefficient with velocity of steady and oscillating flows through the packed porous column.

<u>Experimental Results</u>:

Figure 3 shows the experimental results of the pressure-drop coefficient varying with the Reynolds number for steady and low frequency oscillating flows across the column packed with wire screens as obtained by Hsu et al. (1999). The most fascinating result is that the oscillating flow data collapse into the steady flow data. This implies that the oscillating flows in porous media in the low frequency limit are indeed quasi-steady. The most important feature in Fig. 3 is that the experimental data covered a wide range of $0.27 < Re_h < 2600$ so that the constants C_S and C_I for the Darcy and the Forchheimer limits at low and high Reynolds numbers, respectively, can be determined with no ambiguity. As a result, C_B can also be determined accurately by fitting the experimental data to Eq. (78). The values of C_S, C_B and C_I as obtained from the best curve-fit are 109.3, 5.0 and 1.0, respectively. For comparison, the curve for $C_B = 0$, which represents the two-term Darcy-Forchheimer correlation commonly used in the porous medium research community, is also plotted in Fig. 3. It is seen that the exclusion of the term with $Re_h^{-1/2}$ on the right hand side of Eq. (78) underestimates the pressure-drop by 20-30% in the intermediate Reynolds number range of $40 < Re_h < 1000$.

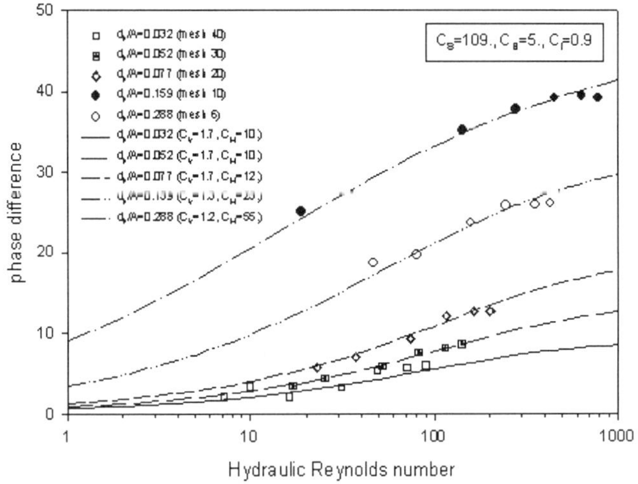

Figure 4. Phase difference between the fundamental mode oscillations of the velocity and the pressure-drop across packed columns made of different sizes of wire-screens.

The experimental results of pressure-drop and velocity correlation for high frequency oscillating flows in the packed column as given by Hsu and Fu (2001) are shown in Figs. 4 and 5 for the phase angle and amplitude, respectively. We note that the amplitude data is not accurate enough to be used for the determination of the coefficients C_M and C_V. Instead, the phase angle data were used. From Fig. 4, it is seen that the phase difference is as much as 40° at the Reynolds number of 780 when d_h/A = 0.288. This implies that the interfacial force due to transient inertia is of the same order in magnitude as that due to convection inertia. With the values of C_S = 109, C_B = 5.0 and C_I = 0.9 from Fig. 3, the curves with the values of C_V and C_M obtained by best fit of the data to Eq. (82) are plotted in Fig. 4. The agreement between the experimental results and the theoretical predictions as shown in Fig. 4 suggests that the inclusion of the transient inertia force into the volumetric interfacial force due to solid resistance is crucial for a complete description of the unsteady flows in porous media. The predictions of amplitude correlation based on Eq. (81) using C_S = 109, C_B = 5.0, C_I = 0.9, and the fitted values of C_M and C_V for different d_h/A are plotted in Fig. 5. For comparison, the steady flow data of Fig. 3 (equivalent to d_h/A = 0) were first converted for Eq. (83) and plotted in Fig. 5. Good agreement is found between the experimental data and the theoretical predictions.

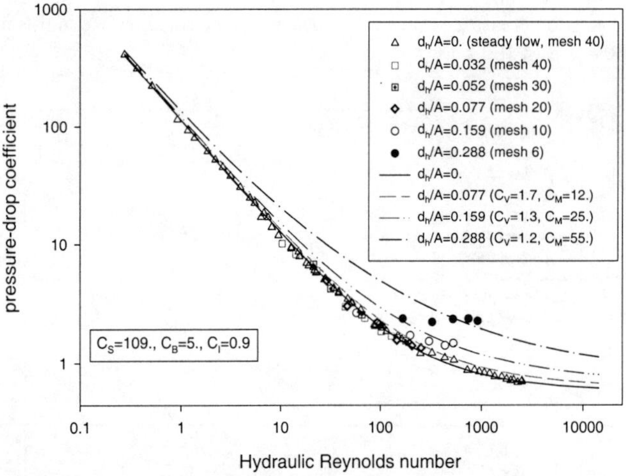

Figure 5. Correlation between the amplitudes of fundamental mode oscillations of velocity and pressure-drop across packed column made of different sizes of wire-screens.

5.2 Heat Transfer Experiments

There exist considerable experiments on the heat transfer in porous media. Wakao and Kaguei (1982) and Kaviany (1991) had comprehensively compiled the previous experimental results. Here we recapture those are relevant to the thermal dispersion, thermal tortuosity and interfacial heat transfer, incorporated with some results from recent experiments by Fu and Hsu (2001).

Thermal Dispersion:

Under the low frequency condition, Fu and Hsu (2001) measured the longitudinal thermal dispersion for oscillating flows through a porous column packed of wire screens. The oscillating flows at such low frequency are quasi-steady as demonstrated in Section 5.1. Figure 6 shows the variation of the effective longitudinal dispersion thermal diffusivity with the hydraulic Peclet number. As seen from Fig. 6, α_l/α increases almost linearly with the Peclet number when $Pe_h >> 10$. As Pe_h decreases toward $Pe_h \approx 10$, the value of α_l/α decreases more rapidly to exhibit the trend of Pe_h^2, although the data range of Pe_h is not low enough to provide a complete picture in the range of low Peclet number. Apparently, the data shown in Fig. 6 are consistent with the quasi-steady closure model for thermal dispersion as given by Hsu and Cheng (1990). The composite expression as given by Eq. (70) were used by Fu and Hsu (2001) to fit the data to obtain a_{h1} =1.94 and b_{h1}=30. The results of the best fit are given in Fig. 6 as the solid curve. It should be noted that the value of b_{h1} may be subject tosome uncertainty because of not enough data in low Peclet number range; however, the value of a_{h1} = 1.94 should give good confidence.

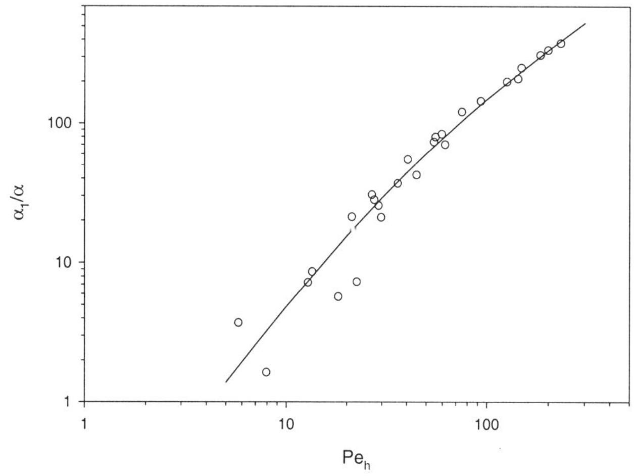

Figure 6. Comparison of the experimental results of dispersion thermal diffusivity (Fu and Hsu, 2001) with the predictions based on the model of Hsu and Cheng (1990).

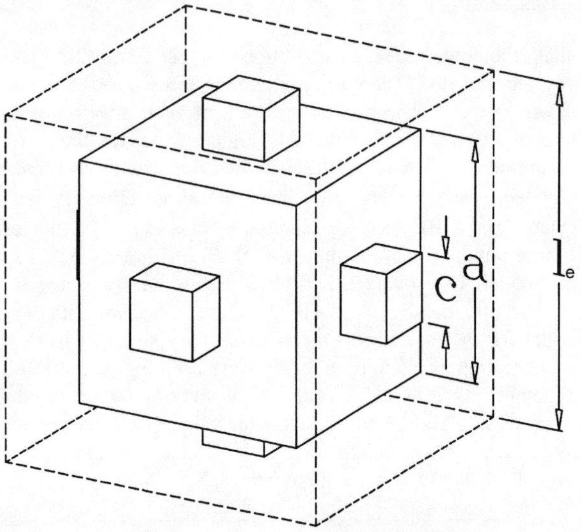

Figure 7. Unit cell of a 3D in-line array of cubes used in Hsu et al. (1995).

Effective Stagnant Thermal Conductivity and Thermal Tortuosity:

As the tortuosity parameter G is related to the effective stagnant thermal conductivity k_{st} by Eq. (48), the main task becomes to experimentally determine k_{st}. A more complete experiment for the determination of k_{st} that covered a wide of solid-to-fluid thermal conductivity ratio was the one conducted by Nozad et al. (1985). More recently, Hsu et al. (1995) proposed the lumped parameter 2D and 3D models to predict the effective stagnant thermal conductivity. For the 3D model of in-line periodic arrays of cubes, the unit cell is shown as in Fig. 7. The expression for the determination of k_{st}/k is then given as:

$$k_{st}/k = 1-\gamma_a^2 - 2\gamma_c\gamma_a + 2\gamma_c\gamma_a^2 + \sigma\gamma_c^2\gamma_a^2 + \frac{\sigma\gamma_a^2(1-\gamma_c^2)}{\sigma+\gamma_a(1-\sigma)} + \frac{2\sigma\gamma_c\gamma_a(1-\gamma_a)}{\sigma+\gamma_c\gamma_a(1-\sigma)} \quad (84)$$

where the particle size parameter γ_a ($=a/l_e$) and the solid-particle contact parameter γ_c ($=c/a$), as shown in Fig. 7, are related to the porosity by

$$1-\phi = (1-3\gamma_c^2)\gamma_a^3 + 3\gamma_c^2\gamma_a^2 \quad (85)$$

For non-touching cubes ($\gamma_c = 0$), $\gamma_a=(1-\phi)^{1/3}$ and (84) reduces to

$$\frac{k_{st}}{k} = \left[1-(1-\phi)^{2/3}\right] + \frac{(1-\phi)^{2/3}\sigma}{[1-(1-\phi)^{1/3}]\sigma + (1-\phi)^{1/3}} \quad (86)$$

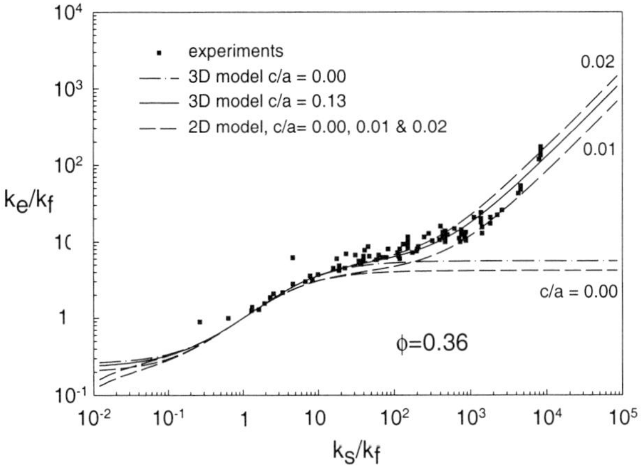

Figure 8. Comparison of the predictions of effective stagnant thermal conductivity based on the 3D cube and 2D square cylinder models of Hsu et al. (1995) with the experimental results of Nozad et al. (1985).

The predictions by the models of Hsu et al. (1995) and the experimental results of Nozad et al. (1985) are given in Fig. 8. The comparison shows excellent agreement. Hsu (1999) then used the above 3D model to calculate k_{st}/k and then evaluated the tortuosity parameter G by Eq. (65). The results of G as a function of σ are given in Fig. 9 for different values of porosity when the solid-particle contact parameter takes a typical value of $\gamma_c = 0.1$.

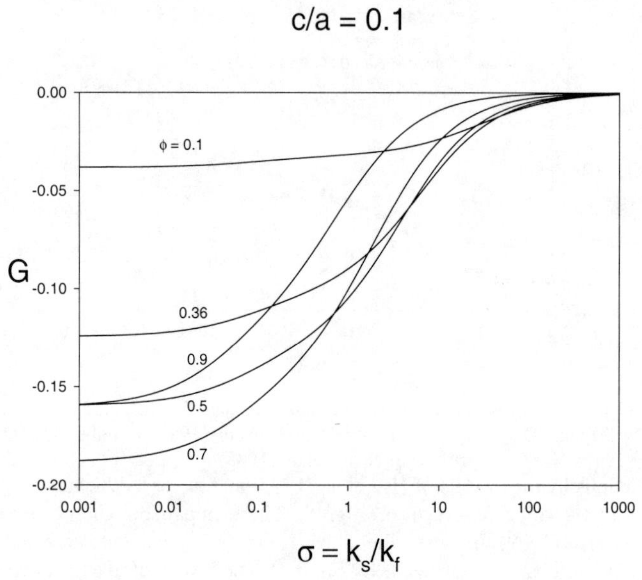

Figure 9. The interfacial thermal tortuosity parameter for different values of porosity with the particle-touching parameter fixed at 0.1.

Interfacial Heat Transfer:

Unlike the thermal dispersion, considerable experimental works on the interfacial heat transfer were done in the past decades, because of its importance in chemical engineering applications. Figure 10 shows the data compiled by Kunii and Suzuki (1967) in the range of low Reynolds number (areas enclosed by solid curves), and by Wakao and Kaguei (1982) in the range of high Reynolds number (areas enclosed by dashed curves). The family of curves for different values of σ in Fig. 10 is the prediction of Eqs. (71-72), with $m = 0.5$, $n = 0.6$, $a_h = 1.29$, and $b_h = 0.001$, $\alpha_{hA} = 0.125$ and $\alpha_{hB} = 0.443$ for air (Pr = 0.7). It appears that the model of Hsu (2000c) predicted the general trend of experimental data with the correct magnitude. However, no solid conclusion can be drawn because of the high scattering in data range (Nu_{hfs} ranges from 10^{-4} to 10^{3}).

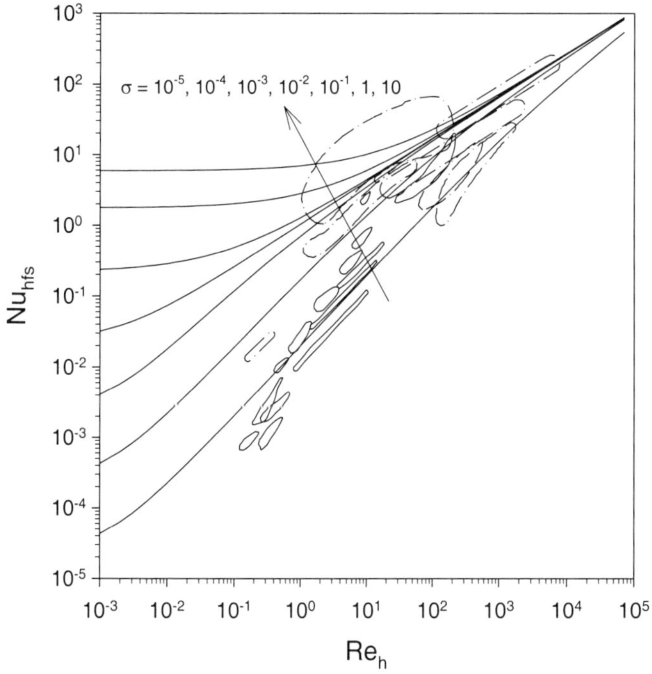

Figure 10. The predictions of the interfacial heat transfer coefficient based on the model of Hsu (2001b) and its comparison with the early experimental data compiled by Kunii and Suzuki (1967) and by Wakao and Kaguei (1982).

5.3 Mass Transfer Experiments

Measurements of the mass diffusion process in porous media to determine the dispersion diffusivity and interfacial mass transfer coefficient were carried out extensively. Eidsath et al. (1983) had compiled the data on the longitudinal mass dispersion diffusivity, while Wakao and Kaguei (1982) compiled those on interfacial heat transfer. On the other hand, there exists little experimental works on the mass tortuosity. From the analog to the heat transfer, the effective stagnant mass diffusivity can be expressed by Eq. (84-86) with k_{st}/k and σ replaced by λ_{st}/λ and σ_m, respectively, where σ_m is the mass diffusivity ratio of solid to fluid. Since $\sigma_m \rightarrow 0$, we find $\lambda_{st}/\lambda = 1\text{-}(1\text{-}\phi)^{2/3}$ for point-touching particles ($\gamma_c = 0$). Therefore, $G_m = (1\text{-}\phi)\text{-}(1\text{-}\phi)^{2/3}$ which depends on the porosity only. Since the results of thermal tortuosity parameter G were presented in the last section, we shall not repeat here for G_m. The experimental data for mass dispersion and interfacial mass transfer are given below.

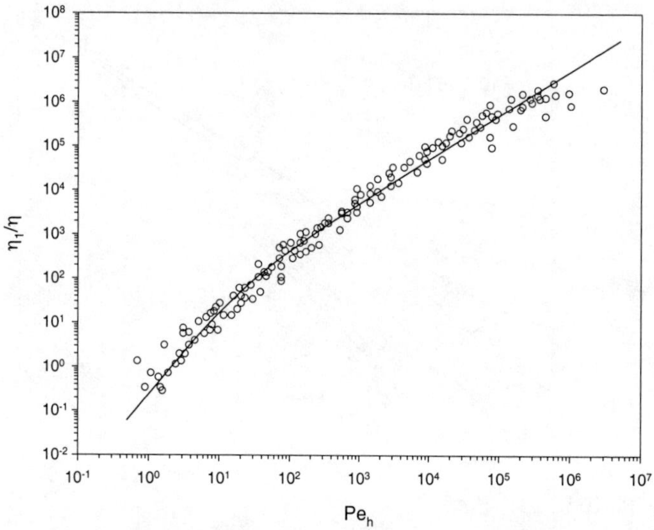

Figure 11 The experimental results of the longitudinal mass dispersion in porous media as compiled by Eidsath et al. (1983) and their fitting to the model equations (Koch and Brady, 1985; Hsu, 2001b).

Mass Dispersion:

The experimental results of longitudinal mass dispersion diffusivity of the early works as compiled by Eidsath et al. (1983) were re-compiled by Hsu (2001b) as shown in Fig. 11 in terms of non-dimensional diffusivity η_1/η and mass Peclet number Pe_{mh}. Apparently, η_1/η increases linearly with Pe_{mh} for large mass Peclet number and becomes proportional to $(\mathrm{Pe}_{mh})^2$ for low mass Peclet number. The solid curve given in Fig. 11 is the result best fit to Eq. (74) with $a_{mh1} = 5.0$ and $b_{mh1} = 20$, which shows excellent agreement.

Interfacial Mass Transfer:

The experimental data on the interfacial mass transfer as obtained earlier by others were re-compiled by Hsu (2001b) using data compiled in Wakao and Kaguei (1982). The re-compiled results are shown in Fig. 12. Also shown in Fig. 12 as the solid curve is the prediction of Eq. (75) with $Sh_{ad}^* = 1.33$, $a_{mh} = 0.94$ and $n = 0.6$. Apparently, the theoretical predictions fit excellently to the experimental data. However, it should be noted that the value of $Sh_{ad}^* = 1.33$ as determined by curve-fitting may subject to considerable uncertainty because of insufficient data appeared in the low Reynolds number range, while the value of

a_{mh} gives good confidence. More experiments in the low Reynolds number range will be needed for a better determination of Sh_{ad}*.

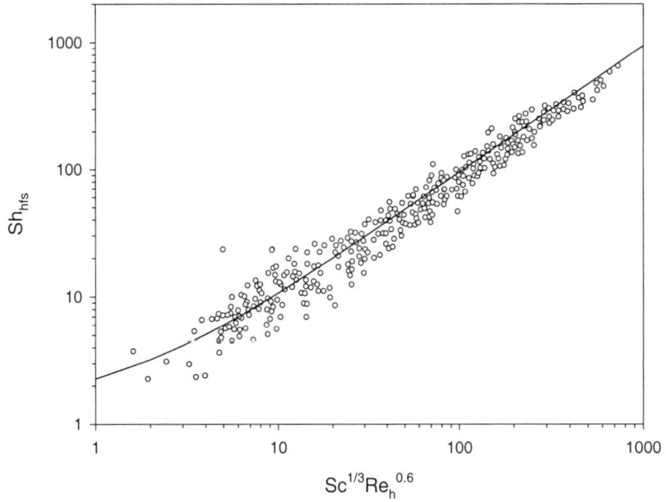

Figure 12. Comparison of the early experimental results on interfacial mass transfer coefficient as compiled by Wakao and Kaguei (1982) with the model predictions of Hsu (2001b).

6. CONCLUDING REMARKS

In this chapter, a complete review of the macroscopic equations that govern the convective heat and mass transfer in porous media is given. The equations were formulated by averaging the microscopic equations over a representative elementary volume (REV). The average procedure leads to the closure problem where the dispersion, the interfacial tortuosity and the interfacial transfer become the new unknowns in the equations. The closure relations as constructed earlier by others for the dispersion, tortuosity and the interfacial transfer were summarized, reviewed and adapted to close the equation system. The equation system plus proper boundary conditions then can be solved either analytically or numerically to predict the convective heat and mass transport phenomena in the porous media. However, there exist several coefficients in the closure relations that have to be determined experimentally (or numerically) a priori. Experiments conducted earlier for the determination of these coefficients were reviewed. These experimental results generally supported the validity of the closure relations, but were not conclusive. This leads to the conclusion that more experiments are needed for the complete determination of these coefficients.

It is noted that the momentum closure relation for the interfacial force as obtained by Hsu (2001a) contains all the components due to drag, lift and transient inertia to the first order approximation. Therefore, the macroscopic momentum equation is expected to be valid for all range of time scale and Reynolds number. On the other hand, the closure relations for interfacial heat and mass transfer as obtained by Hsu (2001b) and Koch and Brady (1985) were derived under the assumption of steady flows. Whether these closure relations are still applicable for unsteady convective heat and mass transfer remains as an open question. This will be the subject of future investigations.

ACKNOWLEDGMENT

This study is supported by the Hong Kong SAR Government under Grant Nos. HKUST575/94E, HKUST815/96E and HKUST6165/98E. The assistance on compiling and drawing the figures by Mr. M. K. Kwan and Mr. J. H. Sun is gratefully appreciated.

REFERENCES

Aris, R. and Amundson, N.R., 1957. Some remarks on longitudinal mixing or diffusion in fixed beds. *AIChE J.*, **3**, 280-282.

Biot, M.A.., 1941. General theory of three dimensional consolidation," *J. Appl. Phy.*, **12**, 155-164.

Carbonell, R.G., 1979. Effect of pore distribution and flow segregation on dispersion in porous media. *Chem. Engng. Sci.*, **34**, 1031-1039.

Carbonell, R.G., 1980. Flow nonuniformities in packed beds, effect on dispersion. Chem. Engng. Sci., **35**, 1347-1356.

Cheng, P. and Hsu, C.T., 1998. Heat conduction. *Transport Phenomena in Porous Media.* Ingham DB, Pop I, eds., Pergamon Press, Elsevier Science, Oxford, 57-76.

Crane, R.A. and Vachon, R.I., 1977. A prediction of the bounds on the effective thermal conductivity of granular materials. *Int. J. Heat Mass Transfer*, **20**, 711-723.

Dagan, G., 1979. The generalization of Darcy's law for non-uniform flows. *Water Resour. Res.*, **15**, 1-7.

Darcy, H., 1856. Les fontaines publiques de la ville de Dijon. *Victor Dalmont*, Paris.

Deissler, R.G. and Boegli, J.S., 1958. An investigation of effective thermal conductivities of powders in various gases. *ASME Trans,* **80**, 1417-1425.

Edwards, M.F. and Richardson, J.E., 1968. Gas dispersion in packed beds. *Chem. Engng. Sci.*, **23**, 109-123.

Eidsath, A., Carbonell, R.G., Whitaker, S. and Herrmann, L.R., 1983. Dispersion in pulsed systems-III. Comparison between theory and experiments for packed beds. *Chem. Engng. Sci.*, **38**, 1803-1816.

Ergun, S., 1952. Fluid flow through packed columns. *Chem. Engg. Prog.*, **48**, 89-94.

Forchheimer, P.H., 1901. *Z. Ver. Dtsch. Ing.,* **45**, 1782-1788.

Fu, H. and Hsu, C.T., 2001. Measurements of longitudinal thermal dispersion of oscillating flows through a packed column. (Submitted)

Gunn, D.J. and De Sousa, J.F.C., 1974. Heat transfer and axial dispersion in packed beds. *Chem. Engng. Sci.*, **29**, 1363-1371.

Gunn, D.J. and Khalid, M., 1975. Thermal dispersion and wall transfer in packed beds. *Chem. Engng. Sci.*, **30**, 261-267.

Harleman, D.R.F. and Rumer, R.R., 1963. Longitudinal and lateral dispersion in packed beds: Effect of column length and particle size distribution. *J. Fluid Mech.*, **16**, 385-394.

Hsu, C.T., 1999. A closure model for transient heat conduction in porous media. *ASME J. Heat Transfer*, **121**, 733-739, 1999.

Hsu, C.T., 2000. Heat conduction in porous media. *Handbook of Porous Media*, ed. K. Vafai, Marcel Dekker, New York, 171-199.

Hsu, C.T., 2001a. A new closure model for unsteady flows in porous media. (Submitted)

Hsu, C.T., 2001b. A unified closure model for convective heat and mass transfer in porous media. (Submitted).

Hsu, C.T. and Cheng, P., 1990. Thermal dispersion in a porous medium. *Int. J. Heat Mass Transfer,* **33**, 1587-1597.

Hsu, C.T., Cheng P. and Wong, K.W., 1994. Modified Zehner-Schlunder models for stagnant thermal conductivity of porous media. *Int. J. Heat Mass Transfer*, **37**, 2751-2759.

Hsu, C.T., Cheng, P., and Wong, K.W., 1995. A lumped parameter model for stagnant thermal conductivity of spatially periodic porous media. *ASME J Heat Transfer*, **117**, 264-269.

Hsu, C.T. and Fu, H., 2001. Measurements of pressure-drop due to high frequency oscillating flows across porous media made of wire screens. (Submitted).

Hsu, C.T., Fu, H. and Cheng, P., 1999. On pressure-velocity correlation of steady and oscillating flows in regenerators made of wire-screens. *ASME J. Fluids Engineering*, **121**, 52-56.

Hsu, C.T., Wong, K.W. and Cheng, P., 1996. Effective thermal conductivity of wire screens. *AIAA J Thermophys Heat Transfer,* **10**, 542-545.

Kaviany, M., 1991. *Principles of Heat Transfer in Porous Media.* Springer-Verlag, New York.

Koch, D.L. and Brady, J.F., 1985. Dispersion in fixed beds. *J. Fluid Mech.*,**154**, 399-427.

Koch, D.L. and Brady, J.F., 1987. A non-local description of advection-diffusion with application to dispersion in porous media. *J. Fluid Mech.*, **180**, 387-403.

Krupiczka, R., 1967. Analysis of thermal conductivity in granular materials. *Int. Chem. Engg.*, **7**, 122-144.

Kunii, D. and Smith, J.M., 1960. Heat transfer characteristics of porous rocks. *AIChE Journal*, **6**,71-78.

Kunii, D. and Suzuki, M., 1967. Particle-to-fluid heat and mass transfer in packed beds of fine particles. *Int. J. Heat Mass Transfer*, **10**, 845-852.

Levec, J. and Carbonell, R.G., 1985. Longitudinal and lateral thermal dispersion in packed beds, I-II. *AIChE. J.*, **31**, 581-590, 591-602.

Liu, L.F., Davis, M.H., and Downing, S., 1996. Wave-induced boundary layer flows above and in a permeable bed. *J. Fluid Mech.*, **325**, 195-218.

Martin, H., 1978. Low Peclet number particle-to-fluid heat and mass transfer in packed beds. *Chem. Engng. Sci.*, **33**, 913-919.

Maxwell, J.C., 1873. *A Treatise on Electricity and Magnetism.* Clarendon Press, Oxford, p 365.

Nelson, P.A. and Galloway, T.R., 1975. Particle-to-fluid heat and mass transfer in dense systems of fine particles. *Chem. Engng. Sci.*, **30**, 1-6.

Nozad, S., Carbonell, R.G. and Whitaker, S., 1985. Heat conduction in multiphase systems. I: Theory and experiments for two-phase systems. *Chem. Engng. Sci.*, **40**, 843-855.

Prasad, V., Kladas, N., Bandyopadhay, A. and Tian, Q., 1989. Evaluation of correlations for stagnant thermal conductivity of liquid-saturated porous beds of spheres. *Int. J. Heat Mass Transfer*, **32**, 1793-1796.

Quintard, M. and Whitaker, S., 1993. One- and two-equation models for transient diffusion processes in two-phase systems. *Adv. Heat Transfer*, **23**, 369-367.

Quintard, M. and Whitaker, S., 1994. Convection, dispersion, and interfacial transport of contaminants: Homogeneous porous media. *Adv. Water Resour.*, **17**, 221-239.

Quintard, M. and Whitaker, S., 1995. Local thermal equilibrium for transient heat conduction: Theory and comparison with numerical experiments. *Int. J. Heat Mass Transfer*, **38**, 2779-2796.

Saffman, P.G., 1959. A theory of dispersion in a porous medium. *J. Fluid Mech.*, **6**, 321-349.

Saffman, P.G., 1960. Dispersion due to molecular diffusion and macroscopic mixing in flow through a network of capillaries. *J. Fluid Mech.*, **7**, 194-208.

Sahraoui, M. and Kaviany, M., 1993. Slip and non-slip temperature boundary conditions at interface of porous, plain media: Conduction. *Int. J. Heat Mass Transfer*, **36**, 1019-1033.

Swift, D.L., 1966. The thermal conductivity of spherical metal powders including the effect of an oxide coating. *Int. J. Heat Mass Transfer*, **9**, 1061-1073.

Tsotsas, E. and Martin, H., 1987. Thermal conductivity of packed beds: A review. *Chem. Eng. Prog.*, **22**, 19-37.

Vortmeyer, D., 1975. Axial heat dispersion in packed beds. *Chem. Engng. Sci.*, **30**, 999-1001.

Wakao, N. and Kaguei, S., 1982. *Heat and Mass Transfer in Packed Beds.* Golden and Breach Science Pub. Inc., New York, p. 294.

Wakao, N. and Kato, K., 1969. Effective thermal conductivity of packed beds. *J Chem. Eng. Japan*, **2**, 24-32.

Wakao, N. and Vortmeyer, D., 1971. Pressure dependency of effective thermal conductivity of packed beds. *Chem. Engng. Sci.*, **26**, 1753-1765.

Wakao, N., Kaguei, S. and Funazkri, T., 1979. Effect of fluid dispersion coefficients on particle-to-fluid heat transfer coefficients in packed beds. *Chem. Engng. Sci.*, **34**, 325-336.

Whitaker, S., 1967. Diffusion and dispersion in porous medium. *A.I.Ch.E. Journal*, **13**, 420-427.

Whitaker, S., 1999. *The Method of Volumetric Averaging*. Kluwer Academic Publishers, Dordrecht, Netherlands.

Zehner, P. and Schlunder, E.U., 1970. Thermal conductivity of granular materials at moderate temperatures. *Chem Ing-Tech* , **42**, 933-941 (in German).

Chapter 9

PARAMETER IDENTIFICATION OF ENVIRONMENTAL SYSTEMS

Ne-Zheng Sun[1] and Alexander Yishan Sun[2]

ABSTRACT

This chapter introduces basic concepts and methods for solving the inverse problem in environmental modeling. The unknown parameters to be identified may include various hydraulic parameters, dispersion coefficients, sorption and release coefficients, and chemical reaction coefficients. Boundary conditions and contaminant sinks and sources can also be identified. A general form of parameterization is used to represent the unknown distributed parameters. It consists of three components: the dimension of parameterization, a set of basis functions, and a set of weighting coefficients. Three kinds of inverse problem are defined. The classical inverse problem (CIP) only identifies the weighting coefficients. The extended inverse problem (EIP) identifies all components of a parameterization. The generalized inverse problem (GIP) combines the parameter identification problem with the reliability of model application and the sufficiency of data. Various criteria of parameter estimation have been introduced in the probability framework. The maximum likelihood estimator (MLE), the weighted least squares (WLS) estimator, and the regularized least squares (RLS) estimator are formulated under certain statistic assumptions. Various numerical optimization methods, including quasi-Newton, Gauss-Newton, genetic algorithm and artificial neural network methods are introduced for solving the CIP. A simple tree-regression procedure is presented for solving the EIP, while a stepwise regression procedure is presented for solving the GIP. In all these algorithms, the sensitivity coefficient matrix, or the Jacobian, plays a very important role. Both the sensitivity equation method (the forward mode of auto differentiation) and the adjoint state method (the reverse mode of auto differentiation) are introduced for calculating the Jacobian. A wise selection between the two methods can significantly save the computation effort. Finally, the observation design problem is discussed briefly.

---

[1]Professor, Department of Civil and Environmental Engineering, UCLA, Los Angeles, CA 90095, nezheng@ucla.edu

[2]Environmental Engineer, R&D Division, Tetra Tech, Inc., Lafayette, CA 94549, aysun45@yahoo.com

1. INTRODUCTION

1.1 Environmental Modeling

In recent years, the use of mathematical models for prediction and management has become very common in all fields of environmental engineering. The transport and fate of a contaminant in environmental media (air, water and soil) is governed by the following *advection-dispersion-reaction equation*:

$$\frac{\partial C}{\partial t} = \frac{\partial}{\partial x_i}(D_{ij}\frac{\partial C}{\partial x_j}) - \frac{\partial}{\partial x_i}(V_i C) - K(C) \quad (1.1)$$

with appropriate initial and boundary conditions. In Eq. (1.1), D_{ij}'s are components of the dispersion coefficient $\mathbf{D}$ (a tensor), V_i's are components of the flow velocity $\mathbf{V}$ (a vector), and $K(C)$ is used to represent all reaction, adsorption and sink/source terms. The dispersion coefficient may have different explanations and magnitudes in different fields. In the groundwater field, it is called the hydrodynamic dispersion coefficient (Bear, 1972) or the macro-dispersion coefficient (Gelhar, 1993). In the surface water field, the dispersion coefficient is also called the turbulent diffusion coefficient and denoted by $\mathbf{E}$ (Schnoor, 1996). In the air pollution field, the turbulent diffusion coefficient is denoted by $\mathbf{K}$ (Zannetti, 1990). In some simple cases, the flow velocity $\mathbf{V}$ can be measured directly in the field, but in most cases it is obtained indirectly from the solution of a flow model.

When the number of chemical compounds considered in a problem is more than one, we must write Eq. (1.1) for each compound and use reaction terms to couple these equations. When the number of flow phases is more than one, we must write Eq. (1.1) for each phase and use mass exchange terms between phases to couple these equations. Therefore, in the cases of multi-phase flow and/or multi-component transport, a distributed parameter model consists of a set of partial differential equations with appropriate initial and boundary conditions. Its general form can be represented by

$$\mathbf{L}(\mathbf{u};\ \mathbf{q};\ \mathbf{p};\ \mathbf{b};\ \mathbf{x}) = \mathbf{0} \quad (1.2)$$

where **L** is a set of partial differential operators, **u** is a set of state variables, **q** is a set of control variables, **p** is a set of model parameters, **b** represents a set of subsidiary conditions, **x** represents a set of spatial and time variables. When **(q; p; b)** are given, the problem of solving the state variables **u** from Eq. (1.2) is called the *forward problem* (FP). Various numerical methods, such as the finite difference method, the finite element method, and relevant software packages have been developed for solving the FP to simulate the mass transport in environmental systems (Wood, 1993; Zlatev, 1995; Sun, 1996; Holzbecher, 1998; among others). The general form of a FP solution can be represented by

$$\mathbf{u} = \mathbf{M}(\mathbf{q};\ \mathbf{p};\ \mathbf{b};\ \mathbf{x}) \tag{1.3}$$

The conventional process of constructing an environmental model consists of the following five steps:

- *Define the problem.* To define the objectives of model development and relevant state and control variables.
- *Collect data.* To collect historical records and measurements on state variables, control variables and environmental parameters, to design and conduct field experiments.
- *Construct a conceptual model.* To select a model structure based on appropriate physical rules, the data available, and some simplifying assumptions.
- *Calibrate the Model.* To adjust the model structure and identify the model parameters such that the model outputs can fit the observed data.
- *Assess the Reliability.* To use a deterministic or a statistic method to estimate the reliability of model predictions and model applications.

1.2 The Inverse Problem

In the above steps of model construction, the key and also the most difficult step is model calibration. With a model given by Eq. (1.3), the observed values of state variables, $\mathbf{u}^{obs}$, at observation locations and times $\mathbf{x}^{obs}$ can be expressed by the following *observation equation*:

$$\mathbf{u}^{obs} = \mathbf{M}(\mathbf{q};\ \mathbf{p};\ \mathbf{b};\ \mathbf{x}^{obs}) + \varepsilon \tag{1.4}$$

where ε consists of both observation and model errors. The objective of model calibration is to determine the unknown part of $\mathbf{q}$ (identification of unknown sinks or sources), $\mathbf{p}$ (identification of unknown physical parameters) and $\mathbf{b}$ (identification of unknown initial or boundary conditions) from (1.4). The problem of model calibration can be seen, in certain sense, as the *inverse problem* (IP) of modeling. The IP seeks the model parameters ($\mathbf{q}$, $\mathbf{p}$, $\mathbf{b}$) when the values of state variables ($\mathbf{u}$) are measured, while the FP predicts the state variables ($\mathbf{u}$) when the model parameters ($\mathbf{q}$, $\mathbf{p}$, $\mathbf{b}$) are given.

For a general discussion on IP, the reader may refer to Tarantola (1989). The IP presented here belongs to the field of identifying a *distributed parameters systems* governed by partial differential equations (Isakov, 1998). The FP of environmental modeling is well-posed, i.e., its solution is always in existence, unique and continuously dependent on data. The IP, however, may be ill-posed, i.e., its solution (the identified parameters) may be non-unique and may be significantly changed when the observed data only change slightly. Examples of ill-posed IP can be found in Sun (1994). Some sufficient conditions have been derived for making the IP *well- posed* (Isakov, 1998). Unfortunately, we cannot use these theoretical results directly in real case studies of environmental modeling because of the following inherent difficulties:

- *High degrees of freedom.* The unknown parameters are usually dependent on location (for example, in the case of identifying the hydraulic conductivity of a heterogeneous aquifer) and may be also on time (for example, in the case of identifying a time variant contaminant source).
- *Very limited data.* The data available for model calibration are usually very limited and incomplete in both spatial and time domains because the measurement of state variables (concentrations, for example) is an expensive operation.
- *Large model error.* The mechanisms of mass transport in environmental media are very complex and the mathematical models used to describe these mechanisms are always based on some idealized assumptions. Values of parameters cannot be defined uniquely as they are scale-dependent (for example, the dispersivity of natural porous media).

As a result, we do not expect that we can find the real values of the unknown parameters through model calibration. A model that fits the observations best may not be the best one for prediction and management. Environmental modelers have recognized that the IP is not simply a "curve fitting" problem. To find a satisfactory inverse solution, we must systematically consider the sufficiency of data, the complexity of model structure, the identifiability of model parameters, and the reliability of model applications.

1.3 Parameterization

It is impossible to identify a distributed parameter with infinite degrees of freedom using a finite set of observation data. *Parameterization* refers to approximately representing a distributed parameter by a function with lower degrees of freedom. The following is a general representation of parameterization

$$\theta(\mathbf{x}) \approx \sum_{j=1}^{m} \theta_j \phi_j(\mathbf{x}, \mathbf{v}) \tag{1.5}$$

where m is called the *dimension of parameterization*, θ_j (j =1,2,...,m) are *weighting coefficients*, $\{\phi_j(\mathbf{x}, \mathbf{v})\}$ is a set of *basis functions* with a set of *shape parameters* $\mathbf{v}$ (vector). Eq. (1.5) is borrowed from the statistical learning theory

(Cherkassky, 1998), in which it is used to represent a probability model, but here we use it to represent a space and/or time distributed parameter.

Various parameterization methods used in modeling a distributed parameter system can be seen as a special case of Eq. (1.5) m) and $\theta(\mathbf{x})$ is represented approximately by a function that has a constant value in each zone (a piecewise constant function). In this case, the dimension of parameterization is equal to the number of zones, θ_j is the value of $\theta(\mathbf{x})$ in zone (Ω_j), and the basis function associated with this zone is defined by $\phi_j(\mathbf{x},\mathbf{v}) = 1$, when $\mathbf{x}$ is in (Ω_j), otherwise, $\phi_j(\mathbf{x},\mathbf{v}) = 0$. The shape parameter $\mathbf{v}$ determines the pattern of zonation. When the *finite element interpolation* method is used for parameterization, m is equal to the number of nodes, θ_j is the nodal value of $\theta(\mathbf{x})$ at the jth node, and $\phi_j(\mathbf{x},\mathbf{v})$ is the basic function associated with the node. In this case, the shape parameter $\mathbf{v}$ is determined by the distribution of the m nodes. The *geostatistical* method (*kriging*) can also be considered as a method of parameterization (see Section 2.5).

In the *classical inverse problem* (CIP), it is assumed that the dimension of parameterization m and all basis functions $\phi_j(\mathbf{x},\mathbf{v})$ are given and only the weighting coefficients need to be identified. The unknown distributed parameter is thus reduced to an unknown vector

$$\theta = (\theta_1, \theta_2, \ldots, \theta_m)^T \tag{1.6}$$

where the upper-script T means transpose. In the *extended inverse problem* (EIP), all components in Eq. (1.5): m, θ_j and $\phi_j(\mathbf{x},\mathbf{v})$ are identified simultaneously based on the available observation data. In the *generalized inverse problem* (GIP), all components in Eq. (1.5) are identified based on not only the available observation data but also the objectives of the model application.

1.4 Organization of This Chapter

Section 2 defines the CIP in the framework of probability. Various criteria of parameter estimation are derived according to the statistical assumptions on observation error and prior information. The geostatistical method (kriging and co-kriging) is also introduced in that section. Section 3 introduces various numerical optimization methods for solving the CIP. Two approaches are given for calculating the sensitivity coefficient matrix (Jacobian), the *sensitivity equation method* (the forward mode of automatic differentiation) and the *adjoint state method* (the reverse mode of automatic differentiation). Materials given in Section 4 are new, in which the EIP is defined and solved by a simple *tree-regression procedure*, and the GIP is defined and solved by a *stepwise regression procedure*. Finally, the observation design problem is briefly discussed. Section 5 summarizes this chapter.

2. STATISTICAL METHODS FOR PARAMETER ESTIMATION

2.1 Parameter Estimation via Information Theory

To emphasize the relationship between the unknown parameter vector θ and the observation vector $\mathbf{u}^{obs}$ of state variables, we rewrite the observation equation (1.4) as

$$\mathbf{u}^{obs} = \mathbf{M}(\theta;\ \mathbf{x}^{obs}) + \varepsilon \tag{2.1}$$

where ε contains both model and observation errors. The dependence of $\mathbf{u}^{obs}$ on known parameters and subsidiary conditions is implied in Eq. (2.1). Due to the random nature of these errors, the identified parameters are always associated with uncertainties and thus can be regarded as *random variables*.

From the point of view of probability, parameter estimation can be considered as a procedure that transfers information from observations ($\mathbf{u}^{obs}$) to unknown parameters (θ) through a model ($\mathbf{u} = \mathbf{M}(\theta; \mathbf{x})$) and thus decreases the uncertainty of the estimated parameters. The best method of parameter estimation should extract information from the observations as much as possible and decrease the uncertainty of the unknown parameters as much as possible. To make this concept clear, we must know:

- How to quantitatively measure information.
- How to transfer information from the observation space to the parameter space.
- How to estimate the uncertainty of the estimated parameters.

Let $p(\theta)$ be the *joint probability density distribution* (pdf) of a parameter vector θ. The uncertainty associated with $p(\theta)$ is measured by its entropy (Bard, 1974):

$$H(p) = -E(\log p) = -\int_{(\Omega)} p(\theta) \log p(\theta) d\theta \tag{2.2}$$

where $E(\log p)$ is the *mathematical expectation* of $\log p(\theta)$ and (Ω) is the whole distribution space. The rationale behind the definition can be seen from the following two examples: (1) For the one-dimensional homogeneous distribution in an interval, we can find $H(p) = \log d$ where d is the length of the interval. This means that the uncertainty of a homogeneous distribution increases along with d. (2) For the one-dimensional normal distribution with variance σ^2, we can find

$H(p) = \log\sigma + \frac{1}{2}(1 + \log 2\pi)$. This means that the uncertainty of a normal distribution increases along with its variance.

The negative value of $H(p)$, i.e., $-H(p) = E(\log p)$, is defined as the *information content* of the distribution $p(\theta)$. The prior information on parameters θ can be described by a pdf, $p_0(\theta)$, which is called the *prior distribution* of θ. After transferring the information from the observed data to the estimated parameters, we will have a new pdf, $p_*(\theta)$, which is called the *posterior distribution* of θ. The information contents contained in the prior and posterior distributions are $-H(p_0)$ and $-H(p_*)$, respectively. The difference between them, i.e.,

$$I(p_0, p_*) = H(p_0) - H(p_*) \tag{2.3}$$

measures the information content transferred from the observed data. The problem is how to find the prior distribution $p_0(\theta)$ and the posterior distribution $p_*(\theta)$.

2.2 Prior and Posterior Distributions

The following two types of prior distribution of multi-variables are often used for parameter estimation:

(1) *Homogeneous distribution* in a given range $\Theta_{ad} = (\theta_L, \theta_U)$. In this case, the upper and lower bounds of the estimated parameters, θ_U and θ_L, are determined by prior information, and the corresponding prior distribution can be expressed by

$$p_0(\theta) = \frac{1}{V}, \text{ when } \theta \in \Theta_{ad}; \quad p_0(\theta) = 0, \text{ otherwise.} \tag{2.4}$$

where V is the volume of the supper box Θ_{ad}.

(2) *Normal distribution* with given mean θ_0 and covariance matrix $\mathbf{V}_\theta$ (an $m \times m$ matrix, where m is the dimension of θ). In this case, θ_0 are the guessed values of the estimated parameters based on the available prior information and $\mathbf{V}_\theta$ measures the reliability of these guessed values. The corresponding prior distribution can be expressed by the following m-dimensional Gaussian distribution

$$p_0(\theta) = (2\pi)^{-m/2}(\det \mathbf{V}_\theta)^{-1/2} \exp\left\{-\frac{1}{2}[\theta - \theta_0]^T \mathbf{V}_\theta^{-1}[\theta - \theta_0]\right\} \tag{2.5}$$

The observation eq. (2.1) defines the relationship between θ and $\mathbf{u}^{obs}$ through a model and an error vector ε. The posterior distribution is the pdf of θ for the given observation data $\mathbf{u}^{obs}$. Thus it is merely the conditional pdf $p(\theta|\mathbf{u}^{obs})$. On the other hand, the conditional pdf $p(\mathbf{u}^{obs}|\theta)$ is the pdf of $\mathbf{u}^{obs}$ for the given estimated parameters θ. If there is no model error, $p(\mathbf{u}^{obs}|\theta)$ is equal to $p(\varepsilon|\theta)$, i.e., the pdf of observation errors when the estimated parameters θ are given. $p(\mathbf{u}^{obs}|\theta)$ is usually called the *likelihood function* of observations and is denoted by $L(\theta)$. According to the *Bayes's theorem*, we have

$$p(\theta|\mathbf{u}^{obs}) = \frac{p(\mathbf{u}^{obs}|\theta)p_0(\theta)}{\int_{(\Omega)} p(\mathbf{u}^{obs}|\theta)p_0(\theta)d\theta} \tag{2.6}$$

or

$$p_*(\theta) = cL(\theta)p_0(\theta) \tag{2.7}$$

where constant $c = \left(\int_{(\Omega)} p(\mathbf{u}^{obs}|\theta)p_0(\theta)d\theta\right)^{-1}$. Eq. (2.7) accomplishes the transfer of information from the observed data to the estimated parameters. Since the prior distribution is given, the posterior distribution $p_*(\theta)$ is completely determined by the likelihood function $L(\theta)$ and a constant factor. When the observation error vector is normally distributed with zero mean and constant covariance matrix $\mathbf{V}_\varepsilon$ (a $n\times n$ matrix, where n is the number of observation data), the likelihood function can be specified as

$$L(\theta) = (2\pi)^{-n/2}(\det \mathbf{V}_\varepsilon)^{-1/2} \times \exp\left\{-\frac{1}{2}[\mathbf{u}^{obs} - \mathbf{M}(\theta;\mathbf{x}^{obs})]^T \mathbf{V}_\varepsilon^{-1}[\mathbf{u}^{obs} - \mathbf{M}(\theta;\mathbf{x}^{obs})]\right\} \tag{2.8}$$

Substituting Eq. (2.4) or Eq. (2.5) and Eq. (2.8) into (2.7), we may have an expression of the posterior distribution that contains both the prior information and the information transferred from the observed data. Thus, we have the *maximum a posterior* (MAP) criterion for parameter estimation as follows.

MAP maximizes the information content to be transferred from the observed data and gives a point estimation of the unknown parameter, *i.e.*,

$$\hat{\theta} = \arg\min_{\theta}\{-p_*(\theta)\} \tag{2.9}$$

Substituting Eq. (2.7) into Eq. (2.9) and using $\ln p_*(\theta)$ to replace $p_*(\theta)$, we have

$$\hat{\theta} = \arg\min_{\theta}\{-\ln L(\theta) - \ln p_0(\theta)\} \tag{2.10}$$

We will use Eq. (2.10) to derive various statistic criteria of parameter estimation.

2.3 Statistical Criteria of Parameter Estimation with Homogeneous Prior Distribution

When $p_0(\theta)$ is given by Eq. (2.4), we have the following *maximum likelihood estimator* (MLE):

$$\hat{\theta} = \arg\min_{\theta}\{-\ln L(\theta)\}, \text{ subject to } \theta \in \Theta_{ad} \tag{2.11}$$

If $L(\theta)$ can be expressed by Eq. (2.8) and the covariance matrix $\mathbf{V}_\varepsilon$ is independent of θ, the above MLE reduces to the following *generalized least squares estimator* (GLS):

$$\hat{\theta} = \arg\min_{\theta} [\mathbf{u}^{obs} - \mathbf{M}(\theta;\mathbf{x}^{obs})]^T \mathbf{V}_\varepsilon^{-1}[\mathbf{u}^{obs} - \mathbf{M}(\theta;\mathbf{x}^{obs})] \tag{2.12}$$
$$\text{subject to } \theta \in \Theta_{ad}$$

Furthermore, if all components of ε are independent of each other, the above GLS reduces to the following *weighted least squares estimator* (WLS):

$$\hat{\theta} = \arg\min_{\theta} [\mathbf{u}^{obs} - \mathbf{M}(\theta;\mathbf{x}^{obs})]^T \mathbf{W}[\mathbf{u}^{obs} - \mathbf{M}(\theta;\mathbf{x}^{obs})] \tag{2.13}$$
$$\text{subject to } \theta \in \Theta_{ad}.$$

where $\mathbf{W}$ is a diagonal matrix with diagonal elements $(1/\sigma_1^2, 1/\sigma_2^2, \ldots, 1/\sigma_n^2)$. Finally, if all variances $\sigma_1^2, \sigma_2^2, \ldots, \sigma_n^2$ are the same, we have the following *ordinary least squares estimator* (OLS):

$$\hat{\theta} = \arg\min_{\theta} [\mathbf{u}^{obs} - \mathbf{M}(\theta;\mathbf{x}^{obs})]^T[\mathbf{u}^{obs} - \mathbf{M}(\theta;\mathbf{x}^{obs})] \tag{2.14}$$
$$\text{subject to } \theta \in \Theta_{ad}.$$

We have clearly shown that under different statistical assumptions, different least squares criteria should be used for parameter estimation.

2.4 Statistical Criteria of Parameter Estimation with Gaussian Prior Distribution

When the prior distribution $p_0(\theta)$ is expressed by the Gaussian distribution (2.5), according to (2.10), we must add a term $-\ln p_0(\theta)$ to the objective function of minimization in all MLE, GLS, WLS and OLS. For example, the GLS estimator becomes

$$\hat{\theta} = \arg\min_{\theta} \left\{ \begin{aligned} & [\mathbf{u}^{obs} - \mathbf{M}(\theta; \mathbf{x}^{obs})]^T \mathbf{V}_\varepsilon^{-1} [\mathbf{u}^{obs} - \mathbf{M}(\theta; \mathbf{x}^{obs})] \\ & + [\theta - \theta_0]^T \mathbf{V}_\theta [\theta - \theta_0] \end{aligned} \right\} \tag{2.15}$$

The second term on the right-hand side of the above equation is called a *regularization term* or a *penalty term*. When $\mathbf{V}_\varepsilon = \sigma^2 \mathbf{I}$ and $\mathbf{V}_\theta = \tau^2 \mathbf{I}$, where $\mathbf{I}$ denotes the unit matrix, (2.15) reduces to

$$\hat{\theta} = \arg\min_{\theta} \left\{ \begin{aligned} & [\mathbf{u}^{obs} - \mathbf{M}(\theta; \mathbf{x}^{obs})]^T [\mathbf{u}^{obs} - \mathbf{M}(\theta; \mathbf{x}^{obs})] \\ & + \lambda [\theta - \theta_0]^T [\theta - \theta_0] \end{aligned} \right\} \tag{2.16}$$

where $\lambda = \tau^2 / \sigma^2$ is called the *regularization factor* and the above criterion is called the *regularized least squares estimator* (RLS). The scalar form of (2.16) is

$$\hat{\theta} = \arg\min_{\theta} \left\{ \sum_{i=1}^{n} [u_i^{obs} - u_i^{cal}(\theta)]^2 + \lambda \sum_{j=1}^{m} (\theta_j - \theta_j^0)^2 \right\} \tag{2.17}$$

where u_i^{obs} is the *i*th observation of the state variables, $u_i^{cal}(\theta)$ is the corresponding model output when θ is used as the unknown model parameter, θ_j is the *j*th component of θ, and θ_j^0 is its corresponding prior estimation. The RLS criterion has been extensively used in the deterministic framework for model parameter identification without checking the necessary assumptions in statistics.

2.5 The Geostatistical Method for Parameter Estimation

We have regarded the unknown parameter as a random variable because it is identified from the observation equation (2.1) and which contains uncontrollable observation and model errors. From the point of view of geostatistics, a distributed physical parameter itself, such as the hydraulic conductivity, may be regarded as a *random field* due to the *variability* of natural formations in different scales. Usually, a random field is described approximately by its first two moments: the *mean* and *covariance* functions. Using the *geostatistical method* (Davis, 1986; Kitanidis, 1997; Wackernagel, 1998), the parameter estimation problem can be solved in two stages. In the first stage, the unknown mean and covariance functions are estimated by the observation data. In the second stage, the estimated random field is conditioned by the same set of observation data for finding the most possible realization. An advantage of the geostatistical method is that it can provide not only the estimated value, $\hat{\theta}(\mathbf{x}_0)$, for any given location $\mathbf{x}_0$, but also the *variance of estimation*. With the variance of estimation, we can measure the reliability of the estimated parameter by calculating, for example, its *confidence interval*. The observation data used for estimating and conditioning a random field include:

(1) A set of direct measurements of the estimated parameter: $\theta_D = \{\theta(\mathbf{x}_1), \theta(\mathbf{x}_2), \ldots, \theta(\mathbf{x}_l)\}$, where $\mathbf{x}_1, \mathbf{x}_2, \ldots, \mathbf{x}_l$ are coordinates of measurement locations.
(2) A set of measurements on a state variable: $\mathbf{u}_D = \{u(\mathbf{x}'_1), u(\mathbf{x}'_2), \ldots, u(\mathbf{x}'_n)\}$, where $\mathbf{x}'_1, \mathbf{x}'_2, \ldots, \mathbf{x}'_n$ are measurement locations (and times if the state variable is time dependent). In general, any measurements can be used for estimating and conditioning a random field provided that they are correlated to it.

In the first stage, the mean and covariance functions must be parameterized by some statistical parameters. A simple example would be that the mean function is a constant (*stationary*) and the covariance function has an *isotropic exponential structure*, i.e.,

$$E[\theta(\mathbf{x})] = \mu, \quad Cov_\theta(\mathbf{x}, \mathbf{x} + \mathbf{h}) = \sigma^2 \exp(-h / r) \tag{2.18}$$

where $\mathbf{h}$ is an increment and h is its length, μ is the constant mean, σ^2 is the variance, and r is called the *correlation length*. When the distance between two points $\mathbf{x}$ and $\mathbf{x} + \mathbf{h}$ are larger than the correlation distance ($h>r$), the correlation between $\theta(\mathbf{x})$ and $\theta(\mathbf{x}+\mathbf{h})$ becomes very small. With (2.18), the estimation of the random field is reduced to the estimation of three statistical parameters μ, σ and r. If the mean function is not constant (*non-stationary*), or the covariance function is anisotropic, we have to estimate more statistical parameters. Generally, we can assume that the mean function is determined by a parameter vector β and the covariance function is determined by a parameter vector ψ. We can use the MLE to estimate these two sets of parameters:

$$(\hat{\beta}, \hat{\psi}) = \underset{(\beta,\psi)}{\arg\min} \left\{ \begin{array}{l} \log\det \mathbf{C}_D(\psi) + \\ [\mathbf{Z}_D - \mathbf{M}_D(\beta)]^T \mathbf{C}_D^{-1}(\psi)[\mathbf{Z}_D - \mathbf{M}_D(\beta)] \end{array} \right\} \tag{2.19}$$

where $\mathbf{Z}_D$ consists of $(l + n)$ measurements θ_D and $\mathbf{u}_D$, $\mathbf{M}_D(\beta)$ consists of the corresponding mean values $\bar{\theta}_D$ (when the mean function is determined by β) and model outputs $\bar{\mathbf{u}}_D$ (when the mean function is used as the model parameter). The $(l + n) \times (l + n)$ covariance matrix $\mathbf{C}_D$ consists of the covariance coefficients between the $(l + n)$ measurements (on both the parameter and the state variable), i.e.,

$$\mathbf{C}_D = \begin{bmatrix} \mathbf{C}_{D,\theta\theta} & \mathbf{C}_{D,\theta u} \\ \mathbf{C}_{D,u\theta} & \mathbf{C}_{D,uu} \end{bmatrix} \tag{2.20}$$

where $\mathbf{C}_{D,\theta\theta} = E\left[(\theta_D - \bar{\theta}_D)(\theta_D - \bar{\theta}_D)^T\right]$ and $\mathbf{C}_{D,uu} = E\left[(\mathbf{u}_D - \bar{\mathbf{u}}_D)(\mathbf{u}_D - \bar{\mathbf{u}}_D)^T\right]$ are $l{\times}l$ and $m{\times}m$ covariance matrices for measurement θ_D and $\mathbf{u}_D$, respectively, $\mathbf{C}_{D,\theta u} = E\left[(\theta_D - \bar{\theta}_D)(\mathbf{u}_D - \bar{\mathbf{u}}_D)^T\right]$ and $\mathbf{C}_{D,u\theta} = \mathbf{C}^T_{D,u\theta}$ are $l{\times}m$ and $m{\times}l$ cross-covariance matrices between θ_D and $\mathbf{u}_D$, respectively. The adjoint state method (see Section 3.5) is a very effective tool for calculating these variance and cross-covariance matrices.

In the second stage, the value $\hat{\theta}(\mathbf{x}_0)$ for any given location $\mathbf{x}_0$ and its variance are estimated by *kriging estimator* (Davis, 1986) when only parameter measurements are available, or by *co-kriging estimator* (Kitanidis, 1997) when measurements of both parameter and state variables are available.

Kriging is the *best linear unbiased estimator* (BLUE) for a random field when only the random field (the unknown parameter) itself is measured. The general form of kriging is:

$$\hat{\theta}(\mathbf{x}_0) = \sum_{i=1}^{l} \lambda_i(\mathbf{x}_0)\theta(\mathbf{x}_i) \tag{2.21}$$

where $\{\theta(\mathbf{x}_i);\ i = 1,2,\ldots,l\}$ is a set of measured values, $\{\lambda_i(\mathbf{x}_0);\ i = 1,2,\ldots,l\}$ is a set of *kriging coefficients* that are dependent on the given point $\mathbf{x}_0$. The minimum variance of estimation given by the kriging estimator is

$$\mathrm{var}[\hat{\theta}(\mathbf{x}_0) - \theta(\mathbf{x}_0)] = \sigma_\theta^2 - \sum_{i=1}^{l} \lambda_i(\mathbf{x}_0) Cov_\theta(\mathbf{x}_0, \mathbf{x}_i) \tag{2.22}$$

where σ_θ^2 is the variance of the estimated parameter before conditioning. After conditioning by the measurements, the variance of estimation decreases an amount as given by the second term on the right-hand side of Eq. (2.22). That is the benefit of conditioning. The kriging coefficients $\{\lambda_i(\mathbf{x}_0);\ i = 1,2,\ldots,l\}$ in above equations are the solution of l *kriging equations*. Under the stationary condition and assuming that the constant mean is filtered out, the kriging equations have the following simple form:

$$\sum_{j=1}^{l} \lambda_j(\mathbf{x}_0) Cov_\theta(\mathbf{x}_i, \mathbf{x}_j) = Cov_\theta(\mathbf{x}_i, \mathbf{x}_0), \; i = 1,2,\ldots,l \tag{2.23}$$

For the non-stationary case, the reader may refer to Sun (1994), Kitanidis (1997), and Wackernagel (1998), among others.

When a set of measurements on the state variable, $\{u(\mathbf{x}'_j); j = 1,2,\ldots,n\}$, is also available, we can use the following co-kriging estimator:

$$\hat{\theta}(\mathbf{x}_0) = \sum_{i=1}^{l} \lambda_i(\mathbf{x}_0)\theta(\mathbf{x}_i) + \sum_{j=1}^{n} \mu_j(\mathbf{x}_0) u(\mathbf{x}'_j) \tag{2.24}$$

with the minimum variance of estimation:

$$\mathrm{var}[\hat{\theta}(\mathbf{x}_0) - \theta(\mathbf{x}_0)] = \sigma_\theta^2 - \sum_{i=1}^{l} \lambda_i(\mathbf{x}_0) Cov_{\theta\theta}(\mathbf{x}_0, \mathbf{x}_i) - \sum_{j=1}^{n} \mu_j(\mathbf{x}_0) Cov_{\theta u}(\mathbf{x}_0, \mathbf{x}'_j) \tag{2.25}$$

The last term on the right-hand-side of the above equation shows the benefit of further conditioning the parameter using the state variable measurements. The *co-kriging coefficients* $\{\lambda_i(\mathbf{x}_0);\; i = 1,2,\ldots,l\}$ and $\{\mu_j(\mathbf{x}_0);\; \mathrm{j} = 1,2,\ldots,n\}$ can be obtained by solving the following *co-kriging equations*:

$$\sum_{k=1}^{l} \lambda_k(\mathbf{x}_0) Cov_{\theta\theta}(\mathbf{x}_i, \mathbf{x}_k) + \sum_{j=1}^{n} \mu_j(\mathbf{x}_0) Cov_{\theta u}(\mathbf{x}_i, \mathbf{x}'_j) = Cov_{\theta\theta}(\mathbf{x}_i, \mathbf{x}_0) \quad (i = 1,2,\ldots,l) \tag{2.26}$$

$$\sum_{i=1}^{l} \lambda_i(\mathbf{x}_0) Cov_{u\theta}(\mathbf{x}_i, \mathbf{x}'_j) + \sum_{k=1}^{n} \mu_k(\mathbf{x}_0) Cov_{uu}(\mathbf{x}'_k, \mathbf{x}'_j) = Cov_{u\theta}(\mathbf{x}'_j, \mathbf{x}_0) \quad (j = 1,2,\ldots,n) \tag{2.27}$$

The geostatistical method was used for groundwater parameter identification by Kitanidis and Vomvoris (1983), Dagan (1985), and Rubin and Dagan (1987) under the condition of steady state flow. Sun and Yeh (1992) extended it to the transient flow condition and used the adjoint state method for efficiently calculating the covariance and cross-covariance matrices. A quasi-linear geostatistical inverse method was given by Kitanidis (1995). In recent years, various methods for incorporating head, concentration, arriving time, well-logs, and seismic measurements into the co-kriging estimation have been presented

(Rubin et al., 1992; Sun et al., 1995; Harvey and Gorelick, 1995; Yeh and Zhang, 1996).

3. ALGORITHMS AND SENSITIVITY ANALYSIS

3.1 Numerical Optimization

In the last section, we have seen that when the prior distribution $p_0(\theta)$ is homogeneous in the admissible region Θ_{ad}, the parameter estimate problem can be transferred into the following *constrained optimization problem*

$$\min_{\theta} E(\theta); \quad \theta \in \Theta_{ad}, \tag{3.1}$$

where the objective function $E(\theta)$ may have different expressions for different estimators. We have also seen that when the prior distribution is Gaussian, the parameter estimation problem can be transferred into an *unconstrained optimization problem* by adding a penalty term to the objective function. Various numerical methods have been developed in mathematics for solving constrained or unconstrained optimization problems (Fletcher, 1987; Kelley, 1999). An iterative numerical method for solving the multivariable optimization problem Eq. (3.1) generally consists of the following three steps:

Step 1. Choose a starting point θ_0.
Step 2. Designate a way to generate a search sequence: $\theta_0, \theta_1, \theta_2, \ldots, \theta_k, \theta_{k+1}, \ldots$
Step 3. Stipulate a termination criterion.

The search sequence has the following general form:

$$\theta_{k+1} = \theta_k + \lambda_k \mathbf{d}_k \tag{3.2}$$

where vector $\mathbf{d}_k$ is called a *displacement direction*, λ_k is a *step size* along the direction. Different optimization methods use different algorithms to generate $\mathbf{d}_k$ and λ_k in each iteration. This iteration process usually leads to or terminates at a local minimum.

In the study of multivariable optimization, the following Taylor expansion plays a very important role:

$$E(\theta_0 + \Delta\theta) = E(\theta_0) + \mathbf{g}^T \Delta\theta + \frac{1}{2} \Delta\theta^T \mathbf{H} \Delta\theta + HOT \tag{3.3}$$

where $g = \left(\frac{\partial E}{\partial \theta_j} \right)_{1 \times m}$ is the *gradient vector*, $\mathbf{H} = \left[\frac{\partial^2 E}{\partial \theta_j \partial \theta_l} \right]_{m \times m}$ is called the *Hessian matrix*, $\Delta\theta$ is an increment, and HOT represents all higher-order terms. All partial derivatives in **g** and **H** are evaluated at θ_0. After defining the gradient vector and the Hessian matrix, we can classify all numerical optimization methods into three categories:

- An optimization algorithm is called a *second order method* if the Hessian matrix **H** needs to be calculated in each iteration.
- An optimization algorithm is called a *gradient method* if only the values of the gradient vector **g** and the objective function need to be calculated in each iteration.
- An optimization algorithm is called a *search method* if only the evaluation of objective function is needed in each iteration.

The *Newton's method* is a typical second-order optimization algorithm, in which

$$\theta_{k+1} = \theta_k - \lambda_k \mathbf{H}_k^{-1} \mathbf{g}_k \tag{3.4}$$

where $\mathbf{H}_k^{-1}$ is the inverse matrix of Hessian $\mathbf{H}(\theta_k)$ and $\mathbf{g}_k = \mathbf{g}(\theta_k)$. Although the Newton's iteration procedure converges fast, it cannot be used for the identification of environmental parameters, as it is impossible to calculate the Hessian matrix by numerical differentiation in each iteration. There are several *quasi-Newton methods*, such as the DFP method and the BFGS method, in which

$$\theta_{k+1} = \theta_k - \lambda_k \widetilde{\mathbf{H}}_k^{-1} \mathbf{g}_k \quad \text{and} \quad \widetilde{\mathbf{H}}_{k+1} = \widetilde{\mathbf{H}}_k + \Delta\widetilde{\mathbf{H}}_k \tag{3.5}$$

where $\widetilde{\mathbf{H}}_k$ is an approximation of the Hessian matrix and is modified by an increment $\Delta\widetilde{\mathbf{H}}_k$ in each iteration. The identity matrix **I** is often used as $\widetilde{\mathbf{H}}_0$, while $\Delta\widetilde{\mathbf{H}}_k$ depends only on gradient vector $\mathbf{g}_k$ and has different definitions in different algorithms. The simplest gradient method is the method of *steepest descent*, in which

$$\theta_{k+1} = \theta_k - \lambda_k \mathbf{g}_k \tag{3.6}$$

Since the objective function $E(\theta)$ does not have an analytical expression, its gradient vector $\mathbf{g}_k$ must be calculated approximately by a numerical method. We will discuss the algorithms of calculating $\mathbf{g}_k$ in Section 3.4 and 3.5. When the objective function is affected by model and observation errors, an *implicit filtering* process is needed in calculating $\mathbf{g}_k$ (Kelley, 1999).

All search methods do not need the calculation of any derivatives of the objective function. The often used search methods include the Nelder-Mead simplex algorithm, the multidirectional search algorithm, and the Hooke-Jeeves algorithm (Kelley, 1999). The use of a search method is strongly suggested when the calculation of $\mathbf{g}_k$ is ineffective and inaccurate.

A guide on useful software packages for numerical optimization can be found in More and Wright (1993).

3.2 Nonlinear Least Squares

The objective function $E(\theta)$ in the WLS criterion has a special form, i.e., a sum of square functions:

$$E(\theta) = \frac{1}{2}\sum_{i=1}^{n} f_i^2(\theta) \tag{3.7}$$

where $f_i(\theta) = w_i[u_i^{obs} - u_i^{cal}(\theta)]$ is called a residual. For such a structure, we can find

$$\frac{\partial E}{\partial \theta_j} = \sum_{i=1}^{n} f_i \frac{\partial f_i}{\partial \theta_j}, \quad (j = 1,2,\ldots,m) \tag{3.8}$$

and

$$\frac{\partial^2 E}{\partial \theta_j \partial \theta_l} = \sum_{i=1}^{n}\left[\frac{\partial f_i}{\partial \theta_j}\frac{\partial f_i}{\partial \theta_l} + f_i \frac{\partial^2 f_i}{\partial \theta_j \partial \theta_l}\right], \quad (j,\ l = 1,2,\ldots,m) \tag{3.9}$$

When θ is close to the minimizer, it is reasonable to assume that the value of residual $f_l(\theta)$ is small and thus the second term on the right-hand side of Eq. (3.9) can be ignored. If we define the Jacobian matrix

$$\mathbf{J} = \left[\frac{\partial f_i}{\partial \theta_j}\right]_{n\times m}, \ (i = 1,2,\ldots,n),\ (j = 1,2,\ldots,m) \tag{3.10}$$

then the gradient vector can be represented by $\mathbf{g} = \mathbf{J}^T\mathbf{f}$, where $\mathbf{f} = (f_1, f_2, \ldots, f_n)^T$, and the Hessian matrix can be represented approximately by $\mathbf{H} \approx \mathbf{J}^T\mathbf{J}$. Substituting them into the Newton's iteration sequence Eq. (3.4), we have the following *Gauss-Newton algorithm*:

$$\theta_{k+1} = \theta_k - \lambda_k (\mathbf{J}_k^T \mathbf{J}_k)^{-1} \mathbf{J}_k^T \mathbf{f}_k \tag{3.11}$$

where $\mathbf{J}_k = \mathbf{J}(\theta_k)$ and $\mathbf{f}_k = \mathbf{f}(\theta_k)$. Note that the Jacobian matrix consists of only first-order derivatives and is much easier to be calculated than the Hessian matrix.

In practice, the Gauss-Newton sequence may not be convergent due to the numerical error associated with the Jacobian matrix, a modified form of Eq. (3.11) given below is called the *Levenberg-Marqardt algorithm*:

$$\theta_{k+1} = \theta_k - (\mathbf{J}_k^T \mathbf{J}_k + \lambda \mathbf{I})^{-1} \mathbf{J}_k^T \mathbf{f}_k \tag{3.12}$$

where $\mathbf{I}$ is the identity matrix and λ is a variable parameter. We start from $\lambda = 0$. In this case, Eq. (3.12) reduces to Eq. (3.11). If the condition $E(\theta_{k+1}) < E(\theta_k)$ is satisfied, we can move to the next iteration, otherwise, we increase the value of λ and try again. When the value of λ is very large, the displacement direction in Eq. (3.12) tends to the steepest descent direction and the step size becomes very small. Therefore, we can expect that the condition $E(\theta_{k+1}) < E(\theta_k)$ should be satisfied and we can move to the next iteration. A code that uses the Levenberg-Marqardt algorithm for solving the inverse problem of groundwater modeling can be found in Sun (1994).

In most real cases of environmental modeling, the number of unknown model parameters is large, the model error is significant, and the quantity and quality of observation data are poor. As a result, we will find that, no matter what optimization method is used, the search sequence may not converge or may be trapped to a *local minimum*, and the value of $E(\theta)$ may stay large and is not sensitive to the change of parameter values. This fact tells us that the parameter estimation problem is not a simple data-fitting problem. We must consider how to control the model error, how to increase the information content of observations, and how to estimate the reliability of the estimated parameters.

3.3 Neural Network and Genetic Algorithm

Genetic algorithm (GA) is a kind of search method that can find the *global optimum* for complex multi-modal functions (Holland, 1992; Goldberg, 1989). It employs the operators of natural population evolution process to gradually optimize the goodness of fit over several generations. The use of a GA for parameter identification consists of the following steps:

1. Define an initial population. According to the prior distribution, we choose a set of possible solutions $\{\theta_1^0, \theta_2^0, \ldots, \theta_s^0\}$ from the admissible set Θ_{ad} as the initial population. For a homogeneous prior distribution, for example, the initial population can be selected randomly.

2. Choose a method to encode the parameters. Usually, the binary string representation method is used. In this case, a unit in the population is represented by a string with elements of either 0 or 1, and the length of the string is fixed.
3. Evaluate the fitness of the initial population. The simulation model is used to calculate the values of $\left\{E(\theta_1^0), E(\theta_2^0), \ldots, E(\theta_s^0)\right\}$.
4. Use genetic operators to create newer and 'fitter' populations. The most common genetic operators are *the selection, crossover* and *mutation*. The selection operator reproduces more units from the population whose fitness values are higher (i.e., the corresponding value of E is smaller). The crossover operation creates two new units (children) from a pair of old units (parents) by exchanging tails after a cutting point (Fig. 3.1). The mutation operator creates a new unit from an old unit by reversing its elements between a segment (Fig. 3.2). This operator forces the algorithm to search new areas.
5. Evaluate the fitness of the new generation and test if a near-optimal solution has been found. If not, return to step 4 to create the next generation.

For a detailed discussion on various genetic algorithms, the reader may refer to Vose (1999). The GA has not been extensively used for environmental parameter identification, because it needs to evaluate the objective function for a large number of possible solutions. The value of $E(\theta)$ is obtained by running the simulation model and that is time consuming. When the number of unknown parameters (m) is large, the GA may become very ineffective and infeasible.

Parent 1;	1 0 0 1 0 | 01 1 0 0 1 1 1 0 0 1
Parent 2:	1 1 0 0 0 | 0 1 0 1 1 0 0 1 0 1 0
Child 1:	1 0 0 1 0 | 0 1 0 1 1 0 0 1 0 1 0
Child 2:	1 1 0 0 0 | 01 1 0 0 1 1 1 0 0 1

Figure 3.1 An example of crossover operation.

Parent:	1 0 0 1 0 | 01 1 | 0 0 1 1 1 0 0 1
Child:	1 0 0 1 0 | 10 0 | 0 0 1 1 1 0 0 1

Figure 3.2 An example of mutation operation.

Artificial neural network (ANN) can be considered as a function (or mapping) approximator (Hecht-Nielson, 1990; Pham and Liu, 1995). Figure 3.3 shows a feed-forward network with three layers: an input layer (θ), an output layer (**u**) and an intermediate or hidden layer (**y**). Each neuron y_j in the hidden layer is a function of all input neurons θ_i $(i = 1,2,\ldots,m)$ defined by

$$y_j = f\left(\sum_{i=1}^{m} w_{ji}\theta_i\right), \quad j = 1,2,\ldots,l \tag{3.13}$$

where w_{ji} is the weight coefficient linking the input neuron i to the hidden neuron j, and the hyperbolic tangent is often used as the transfer function $f(\cdot)$, i.e., let $f(\cdot) = \tanh(\cdot)$. Similarly, a neuron u_k $(k = 1,2,\ldots,n)$ in the output layer is determined by all hidden neurons as

$$u_k = f\left(\sum_{j=1}^{l} v_{kj} y_j\right), \quad k = 1,2,\ldots,n \tag{3.14}$$

Now let us consider how to use this neural network to approximate a mapping from the parameter space to the observation space: $\mathbf{u}^{cal}(\theta) = \mathbf{M}(\theta, \mathbf{x}^{obs})$. We can use the following steps:

1. Randomly generate a set of parameters $\{\theta_1, \theta_2, \ldots, \theta_s\}$ in the admissible region Θ_{ad}.
2. Use the simulation model to generate corresponding model outputs in the observation space, i.e., $\{\mathbf{u}^{cal}(\theta_1), \mathbf{u}^{cal}(\theta_2), \ldots, \mathbf{u}^{cal}(\theta_s)\}$.
3. Train the network by these sampling data. The objective is to determine the two sets of weighting coefficients $\{w_{ji}\}$ and $\{v_{kj}\}$ such that the outputs of the network can best fit the outputs of the simulation model. This can be done by minimize the following least squares criterion:

$$I = \frac{1}{2}\sum_{k=1}^{n}(u_k - u_k^{cal})^2 \tag{3.15}$$

for which the same θ is used to generate u_k (the network output) and u_k^{cal} (the model output). A modified steepest descent iteration procedure is often used in training the network (Adeli, 1995).

After the network is successfully trained and tested, we can use it to generate $\mathbf{u}^{cal}(\theta)$ for any θ without solving the simulation model.

The first way of using ANN for parameter identification is to combine it with a GA. Instead of solving a complicated simulation model, we use a trained network to evaluate all trials generated in the GA and thus make the GA more effective (Rogers et al., 1995; Petridis et al., 1998). Morshed and Kaluarachchi (1998) presented an example of using ANN-GA for identifying the hydraulic parameters of an unsaturated aquifer. Another way of using ANN for parameter identification is to develop a network that can approximate the inverse mapping $\theta = \mathbf{M}^{-1}(\mathbf{u}^{cal})$. In that case, we use the model outputs $\mathbf{u}^{cal}$ as the network inputs, and the model inputs θ as the network outputs in training the network. After training, we can use the observations $\mathbf{u}^{obs}$ as the network inputs and the identified parameters can be obtained directly from the network outputs (Hecht-Nielson, 1990; Morshed and Kaluarachchi, 1998). This technique, however, may not be successful because the inverse mapping may not exist.

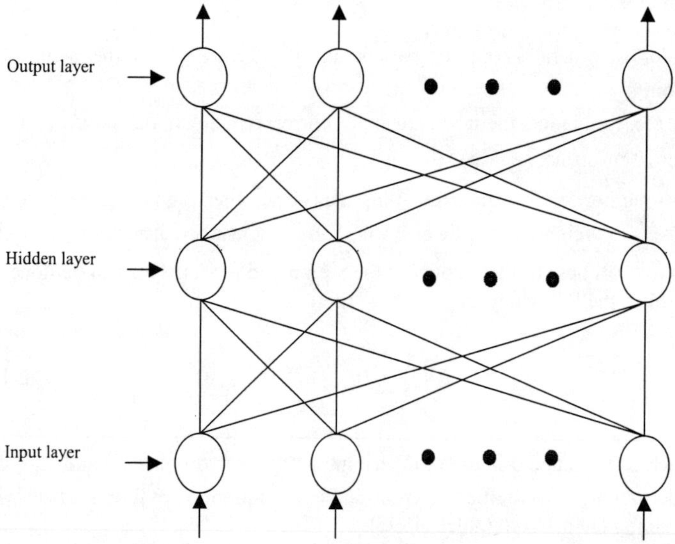

Figure 3.3 A three layer feed-forward neural network.

3.4 Sensitivity Coefficients

A simulation model can only give the values of state variables $\mathbf{u} = \mathbf{M}(\theta, \mathbf{x})$ and thus the value of the objective function $E(\theta)$. When a quasi-Newton method is used for parameter identification, we have to find a way to calculate the gradient $\mathbf{g}$ of $E(\theta)$. The simplest method for this purpose is the use of *first-order finite difference approximation*:

$$\frac{\partial E}{\partial \theta_j} \approx \frac{E(\theta + h_j \mathbf{e}_j) - E(\theta)}{h_j}, \quad j = 1,2,\ldots,m \tag{3.16}$$

where $\mathbf{e}_j$ is an m-dimensional vector whose components are all zero except the jth component, which is equal to one, and h_j is an increment (perturbation). Usually, h_j is taken in proportion to the value of θ_j, i.e., let $h_j = \lambda 0_j$, where $\lambda = 1\% \sim 10\%$. When the implicit filtering technique (Kelley, 1999) is used, the value of λ is decreased gradually in the iteration process. To obtain the gradient $\mathbf{g}$ by (3.16) for an updated θ, we need to run the simulation model m+1 times, in which one run is used for calculating $E(\theta)$.

When the Gauss-Newton method is used for parameter identification, we have to calculate the Jacobian (3.10) in each iteration. This is equal to the calculation of all $\partial u_i / \partial \theta_j (i = 1,2,\ldots,n; j = 1,2,\ldots,m)$. These partial derivatives are called the *sensitivity coefficients* and can be calculated approximately by

$$\frac{\partial u_i}{\partial \theta_j} \approx \frac{u_i(\theta + h_j \mathbf{e}_j) - u_i(0)}{h_j}, \quad i = 1,2,\ldots,n, \ j = 1,2,\ldots,m \tag{3.17}$$

To obtain all sensitivity coefficients for an updated θ, we need to run the simulation model m+1 times and that is independent of the number of observations. The finite difference approximation Eq. (3.17) may give poor results because its value is influenced significantly by the error in u_i and the value of h_j (the approximation Eq. (3.16) has the same disadvantage). Replacing the forward finite difference approximation used in Eq. (3.16) and Eq. (3.17) by the central finite difference approximation can only slightly improve the results, but the simulation model has to be run more times.

Another way for calculating the sensitivity coefficients is the *sensitivity equation method*. Differentiating the governing equation, for example, the advection-dispersion Eq. (1.1) with respect to a parameter θ_k and letting $C'_k = \partial C / \partial \theta_k$, we have the following *sensitivity equation* for the parameter:

$$\frac{\partial C'_k}{\partial t} = \frac{\partial}{\partial x_i}(D_{ij}\frac{\partial C'_k}{\partial x_j}) - \frac{\partial}{\partial x_i}(V_i C'_k) - \frac{\partial K}{\partial C}C'_k + S_k \tag{3.18}$$

where $S_k = \frac{\partial}{\partial x_i}\left(\frac{\partial D_{ij}}{\partial \theta_k}\frac{\partial C}{\partial x_j} - \frac{\partial V_i}{\partial \theta_k}C\right)$. The initial and boundary conditions of Eq. (3.18) can be obtained by differentiating the initial and boundary conditions of the original Eq. (1.1). The sensitivity equation has the same structure as the original governing equation and thus it can be solved by the same routine. To obtain the sensitivity coefficients for all m parameters, we need to run the simulation model m+1 times, in which one run is used for solving the original simulation model. Instead of using the finite difference approximation, we should use the sensitivity equation method to calculate the Jacobian when the Gauss-Newton method is used for parameter identification because the sensitivity equation method can avoid the inherent difficulty of specifying parameter perturbations in Eq. (3.17). From the point of view of *automatic differentiation*, using sensitivity equation method to calculate the Jacobian is merely a "forward" or "bottom-up" algorithm. The code for "forward mode" of automatic differentiation can be obtained directly from the code of the original simulation model (Corliss and Rall, 1996).

3.5 The Adjoint State Method for Sensitivity Analysis

Sensitivity coefficients have many applications in the study of distributed parameter systems. In environmental modeling, we need to calculate sensitivity coefficients not only for classical parameter estimation, but also for parameter structure identification, prediction reliability analysis, and observation design.

Suppose that we want to find how sensitive the concentration $\hat{C}$ is at a specified location and a specified time with respect to a distributed dispersion coefficient D. In this case, we need to calculate N sensitivity coefficients $\partial\hat{C}/\partial D_j$, ($j = 1,2,\ldots,N$), where N is the *number of nodes* and D_j is the nodal value of D at the jth node. For such a problem, the sensitivity equation method is inefficient because it needs to run the simulation model N+1 times and N is usually very large.

In this section, we will introduce a more efficient approach, *the adjoint state method*, for sensitivity analysis. Using the adjoint state method to calculate all $\partial\hat{C}/\partial D_j$ ($j = 1,2,\ldots,N$), we only need two simulation runs, one for solving the original problem and another one for solving the adjoint problem.

Let us consider the concentration distribution $C(x,y,t)$ satisfying the following advection-dispersion equation

$$\frac{\partial C}{\partial t} - \frac{\partial}{\partial x}(D_x\frac{\partial C}{\partial x}) - \frac{\partial}{\partial y}(D_y\frac{\partial C}{\partial y}) + \frac{\partial}{\partial x}(V_x C) + \frac{\partial}{\partial y}(V_y C) + K(C) = 0 \tag{3.19}$$

$$(x,y) \in (\Omega), \quad 0 \le t \le t_f$$

with initial condition $C(x,y,0) = C_0$ and boundary conditions

$$C|_{(\Gamma_1)} = C_B$$

$$\left\{ -(D_x \frac{\partial C}{\partial x} - V_x C) n_x - (D_y \frac{\partial C}{\partial y} - V_y C) n_y \right\}_{(\Gamma_2)} = g \qquad (3.20)$$

where D_x and D_y are distributed longitudinal and transverse dispersion coefficients (functions of *x* and *y*), respectively, C_0, C_B and *g* are given functions, (Ω) is the flow region with boundary sections (Γ_1) and (Γ_2), and (n_x, n_y) is the normal direction of (Γ_2). For the following general performance function

$$E(C, D_x, D_y) = \int_0^{t_f} \int_{(\Omega)} f(C, D_x, D_y; x, y, t) d\Omega dt \qquad (3.21)$$

we can find an *adjoint state* $\psi(x,y,t)$ by solving the *adjoint problem*

$$\frac{\partial \psi}{\partial t} + \frac{\partial}{\partial x}(D_x \frac{\partial \psi}{\partial x}) + \frac{\partial}{\partial y}(D_y \frac{\partial \psi}{\partial y}) + V_x \frac{\partial \psi}{\partial x}$$
$$+ V_y \frac{\partial \psi}{\partial y} + \frac{\partial K}{\partial C} \psi - \frac{\partial f}{\partial C} = 0 \qquad (3.22)$$

with the final condition, $\psi(x,y,t_f) = 0$, and the boundary conditions

$$\psi|_{(\Gamma_1)} = 0 \quad \text{and} \quad \left\{ -D_x \frac{\partial \psi}{\partial x} n_x - D_y \frac{\partial \psi}{\partial y} n_y \right\}_{(\Gamma_2)} = 0 \qquad (3.23)$$

The adjoint problem has the same structure as the original problem and thus both problems can be solved by the same routine. With the adjoint state $\psi(x,y,t)$, we can find the *functional partial derivatives*

$$\frac{\partial E}{\partial D_x} = \int_0^{t_f} \int_{(\Omega)} \left[\frac{\partial f}{\partial D_x} + \frac{\partial \psi}{\partial x} \frac{\partial C}{\partial x} \right] d\Omega dt \qquad (3.24)$$

and

$$\frac{\partial E}{\partial D_y} = \int_0^{t_f} \int_{(\Omega)} \left[\frac{\partial f}{\partial D_y} + \frac{\partial \psi}{\partial y} \frac{\partial C}{\partial y} \right] d\Omega dt \qquad (3.25)$$

For details on deriving Eq. (3.24) and Eq. (3.25), the reader may refer to Sun (1994). Assuming that the unknown longitudinal dispersion coefficient D_x can be parameterized by

$$D_x(x,y) \approx \sum_{j=1}^{m} \theta_j \phi_j(x,y) \tag{3.26}$$

and using Eq. (3.24), we can find all gradient components of E with respect to θ as

$$\frac{\partial E}{\partial \theta_j} = \int_0^{t_f} \int_{(\Omega)} \left[\frac{\partial f}{\partial D_x} + \frac{\partial \psi}{\partial x} \frac{\partial C}{\partial x} \right] \phi_j(x,y) d\Omega dt \; , \; j = 1,2,\ldots,m \tag{3.27}$$

There are similar results for the transverse dispersion coefficient D_y. To obtain all gradient components $\partial E / \partial \theta_j$ by Eq. (3.27), we only need the solution C of the original problem and the solution ψ of the adjoint problem. The total computation effort is equivalent to running the same simulation model twice and that is independent of the number of parameters, m.

By combining the adjoint state method with a quasi-Newton method, we can form a highly efficient algorithm for environmental parameter identification. For the above advection-dispersion problem, if we define

$$f = \sum_{i=1}^{n} w_i^2 [C_i(D_x, D_y) - C_i^{obs}]^2 \delta(\mathbf{x} - \mathbf{x}_i^{obs}) \tag{3.28}$$

The performance function E in Eq. (3.21) becomes the WLS criterion and its gradient $\boldsymbol{g}$ can be calculated by Eq. (3.27) with the computation effort being equal to only two simulation runs. In Eq. (3.28), C_i^{obs} is the observed concentration of the ith observation location and time $\mathbf{x}_i^{obs}$, C_i is the corresponding model output, $\delta(\cdot)$ is the Dirac-δ function. The following are some examples of combining the adjoint state method with a quasi-Newton method for parameter identification: Chavent et al. (1975) used this approach to identify the hydraulic conductivity for petroleum reservoir modeling. Neuman (1980) and Carrera and Neuman (1986) used this approach to identify the hydraulic conductivity for groundwater modeling. Recently, Navon (1998) reviewed the use of this approach for parameter estimation in meteorology and oceanography, Piasecki and Katopodes (1999) used this approach to identify the dispersion coefficient of rivers, Kaminski et al. (1999) used this approach to identify the sinks and sources.

The adjoint state method is also an ideal tool for *sensitivity analysis*. To calculate a sensitivity coefficient $\partial C_i / \partial \theta_j$, we can take

$$f = C\,\delta(\mathbf{x} - \mathbf{x}_i^{obs}) \tag{3.29}$$

The performance function E in Eq. (3.21) now becomes C_i and thus $\partial C_i / \partial \theta_j$ can be calculated by Eq. (3.27) for any θ_j. In fact, we can calculate it for all nodes ($j = 1,2,\ldots,N$) and draw contours to show the sensitivity of a predicted concentration (C_i) with respect to a distributed parameter (D_x or D_y).

Sun and Yeh (1990) presented a set of *adjoint operation rules* that can help derive the adjoint state equations and functional partial derivatives for coupled flow and mass transport problems. The discrete form of the adjoint state method can be derived directly from the discrete form of the original problem. The coefficient matrices for the two problems are the same. From the point of view of automatic differentiation, the use of the adjoint state method to calculate the Jacobian is merely a "reverse" or "top-down" algorithm. The code for "reverse mode" of automatic differentiation can also be obtained directly from the code of the original simulation model (Corliss and Rall, 1996; Giering and Kaminski, 1998).

Note that we should not combine the adjoint state method with the Gauss-Newton method for parameter identification because in the case of parameter identification we always have $m<<n$, where m is the number of unknown parameters and n is the number of observation data. To calculate all sensitivity coefficients, the adjoint state method needs to solve the adjoint problem n times and the original problem once, whereas the sensitivity equation method needs to solve the original problem (m+1) times. In this case, the sensitivity equation method is more efficient. On the other hand, if we want to calculate the Jacobian under the condition $n<<m$, the adjoint state method becomes more efficient. This case can be met, for example, when the geostatistical method is used for parameter identification (Sun, 1994). In the field of atmospheric transport, the adjoint method was used recently by Kaminski et al. (1999) in calculating the sensitivity matrix of CO_2 concentration at n observation locations with respect to m sinks and sources when $n<<m$. In the next section, we will use the adjoint method for parameter structure identification and observation network design.

4. THE EXTENDED AND GENERALIZED INVERSE PROBLEMS

4.1 Parameter Structure Identification

So far, we have only considered the classical inverse problem (CIP), i.e., estimating the weighting coefficients θ_j ($j = 1,2,\ldots,m$) in the parameterization representation (1.5) based on prior information and observation data. In practice, however, the dimension of parameterization m, the basis functions $\{\phi_j(\mathbf{x},\mathbf{v})\}$, and the shape parameter vector $\mathbf{v}$ in the representation cannot be given a priori.

For example, when the unknown hydraulic conductivity of an aquifer is parameterized for identification, first we have to determine which kind of basis function should be selected. If the zonation method is selected, we have to further determine the number of zones and the zone patterns. If the linear interpolation method is selected, we have to further determine the number of basis points and their locations. A different selection on m and $\{\phi_j(\mathbf{x},\mathbf{v})\}$ means a different selection on *parameter structure*. With different parameter structures for parameter estimation, we will obtain different parameter values and different fitting residuals. In other words, without using a correct parameter structure, we cannot find correct parameter values. The *extended inverse problem* (EIP) requires the identification of all components of parameterization based on available prior information and observed data.

Sun and Yeh (1985) presented a methodology that can simultaneously identify the parameter structure and values based on observed data. In Sun and Yeh (1985), the distributed unknown parameter is represented by a vector, θ_N, where N is the number of nodes used in the simulation model. A set of modified distance weighted interpolation functions with m movable basis points is used to define the basis functions of parameterization. With parameterization, the N-dimensional parameter vector θ_N is represented by an m-dimensional parameter vector θ_m as follows

$$\theta_N = \mathbf{G}\theta_m \tag{4.1}$$

where the $N\times m$ matrix $\mathbf{G}$ is called a *structure matrix*. It depends on two shape parameters (α and β) and the coordinates of m basis points. By changing the values of α and β, the structure of the unknown parameter can be changed from a piecewise constant shape to a continuous shape. For example, when $\alpha>2$, the unknown parameter will have a special zonation structure known as the Voronol partition (Figure 4.1).

The modified Gauss-Newton iteration sequence for this case is changed to

$$\theta_{k+1} = \theta_k - (\mathbf{G}_k^T\mathbf{J}_k^T\mathbf{J}_k\mathbf{G} + \lambda\mathbf{I})^{-1}\mathbf{G}_k^T\mathbf{J}_k^T\mathbf{f}_k \tag{4.2}$$

In each iteration, the shape parameters and the coordinates of basis points are systematically varied to get the best fitting. The parameter structure identification problem is thus formulated into a *combinatorial optimization problem*. This procedure, unfortunately, needs too many additional simulation runs and thus becomes inefficient when the dimension of parameterization m is large.

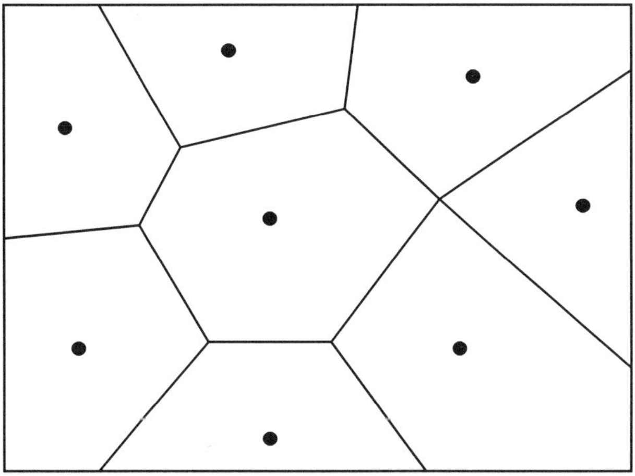

Figure 4.1 A distribution of basis points and its relevant Voronol partition.

Now we present a simple *tree-regression procedure* that can find a nearly optimal parameter structure. In this procedure, the parameterization method is limited to zonation. The zonation pattern is determined by the best fitting of observation data, and the parameter structure is improved gradually by forming a structure sequence:

$$S_1 \subset S_2 \subset S_3 \subset \cdots \subset S_m \subset S_{m+1} \subset \cdots \tag{4.3}$$

where S_1 is a homogeneous structure, S_2 is a structure with two zones, and so forth. Generally, S_{m+1} is obtained from S_m by dividing one zone of S_m into two sub-zones, and thus structure S_m is a special case of structure S_{m+1}. The following steps show how this sequence is formed:

1. Use the homogeneous structure as S_1. The associated optimal parameter value θ_1 is obtained by minimizing a least squares criterion.
2. Divide S_1 into two zones by a linear boundary (a straight line in the 2-D case and a plane in the 3-D case). The least squares criterion is used to determine the optimal location of the linear boundary as well as the optimal parameter values θ_1 and θ_2 of the two zones. The method used in this step is the same as what used in Sun and Yeh (1985) when there are only two basis points. The adjoint state method should be used to update the Jacobian during the search procedure. In this step, a two-zone structure S_2 is formed from S_1.

3. Determine the parameter of which zone is more sensitive to the observation data. This can be done by calculating the absolute values of $\partial E / \partial \theta$ associated with the two zones. The zone with larger $|\partial E / \partial \theta|$ is selected and is subsequently divided into two sub-zones. The optimal location of the linear boundary between the two sub-zones and the parameter values $(\theta_1, \theta_2, \theta_3)$ associated with the total three zones are determined by minimizing the least squares criterion. In this step, a three-zone structure S_3 is formed from S_2.
4. Select one zone from the three zones for which the associated parameter is the most sensitive one to the observation data (i.e., the value of $|\partial E / \partial \theta|$ is the largest one for that zone). The so-selected zone is divided into two sub-zones. The optimal location of the linear boundary between the two sub-zones and the parameter values $(\theta_1, \theta_2, \theta_3, \theta_4)$ associated with the total four zones are then determined by minimizing the least squares criterion. In this step, a four-zone structure S_4 is formed from S_3.
5. Repeat the above procedure to add more and more zones until a stopping criterion is satisfied. There are three criteria to be selected from, (1) $E(\theta_m)$ becomes small (comparing with the observation error); (2) The difference between $E(\theta_m)$ and $E(\theta_{m+1})$ becomes small; and (3) The values of $|\partial E / \partial \theta|$ associated with all zones become small. In this case, no more information can be extracted from the observation data.

The above tree-regression procedure is described in Figures 4.2 and 4.3. It is simple and yet, it has several advantages: (1) the over-parameterization problem can be avoided; (2) the number of structure parameters is minimized; (3) all parameters involved in the optimization procedure are sensitive to the observation data. The so-identified parameter structure, of course, is only nearly optimal, because the boundaries between zones are limited to the linear type. This approximation, however, should be enough for most problems of environmental modeling.

We have developed a methodology for solving the EIP that consists of the solution of a series of CIPs. From the point of view of fitting observation data, the solution of EIP is indeed better than the solution of a specified CIP without optimizing the parameter structure. From the point of view of model application, however, the so-identified parameter structure and the associated parameter values may not be optimal, or may even be unacceptable.

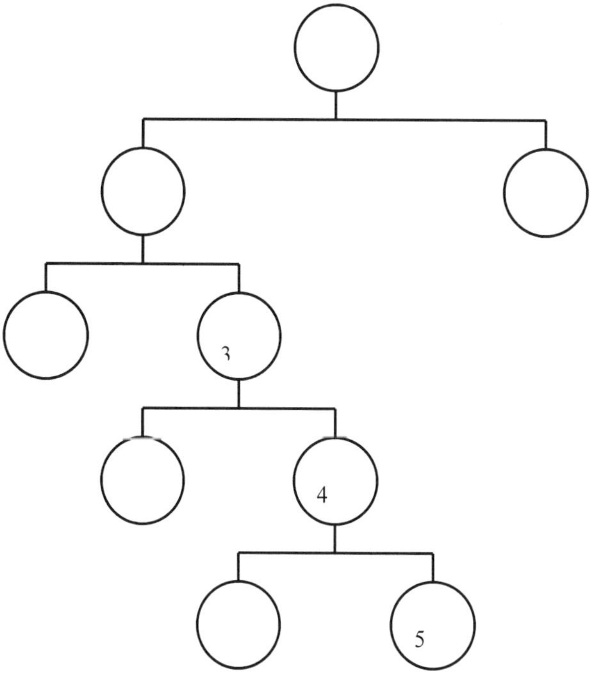

Figure 4.2 A tree-regression procedure.

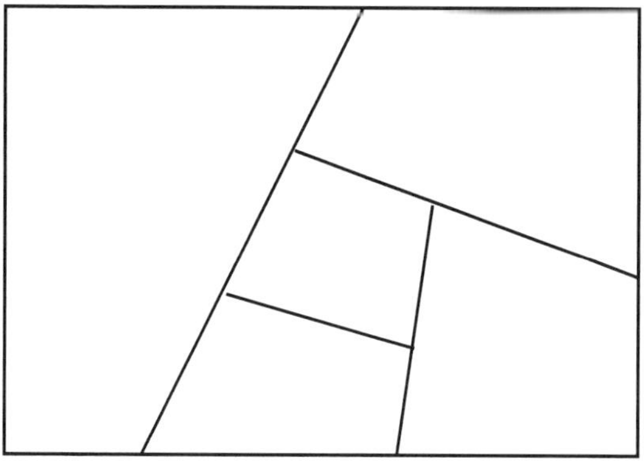

Figure 4.3 The zonation pattern generated by the tree-regression procedure.

This fact is easy to understand. For example, a pumping test was conducted in a well, but we want to predict the head of another well when the pumping rate is increased. The hydraulic conductivity, which is the best parameter to fit the pumping test, may not be the best one for prediction. In the following sections, we will introduce some new concepts in order to find the optimal parameter structure for special purposes of model application.

4.2 Structure Error and Equivalent Structures

A distributed parameter may have infinite degrees of freedom and can never be accurately identified by finite observation data. In order to identify it approximately, we have to simplify its structure, i.e., to decrease its degrees of freedom to a small number by parameterization. Of course, we prefer to select the simplest parameter structure if the accuracy requirement of model application can be satisfied, because a simpler structure needs less observation data to calibrate. But, how do we measure the difference between two parameter structures? Or, how do we calculate the structure error when a parameter structure is replaced by a simplified one?

We will use (S,θ) to denote a *parameterization representation* (PR) of a distributed parameter, where S represents a parameter structure and θ is the parameter vector (values) associated with the structure. The same distributed parameter may have different PRs when it is approximated by different structures. Letting $M_A = (S_A, \theta_A)$ and $M_B = (S_B, \theta_B)$ be two different PRs of a distributed parameter, the distance between them can be measured in parameter space, in observation space as well as in prediction (or management) space, respectively, by

$$d_P(M_A, M_B) = \|\overline{\theta}_A - \overline{\theta}_B\|_P \tag{4.4}$$

$$d_D(M_A, M_B) = \|\mathbf{u}_D(M_A) - \mathbf{u}_D(M_B)\|_D \tag{4.5}$$

$$d_E(M_A, M_B) = \|\mathbf{g}_E(M_A) - \mathbf{g}_E(M_B)\|_E \tag{4.6}$$

where $\overline{\theta}_A$ and $\overline{\theta}_B$ are spans of θ_A and θ_B associated with the parameter space P, $\mathbf{u}_D$ is the model outputs corresponding to an observation design D, $\mathbf{g}_E$ is the model predictions for a set of given prediction objectives E, $\|\cdot\|$ means a norm defined in a space. A detailed explanation on these definitions can be found in Sun et al. (1998). The total distance d between M_A and M_B is defined as the weighted summation of their distances measured in parameter, observation and prediction spaces, i.e.,

$$d(M_A, M_B) = d_E + \mu d_D + \lambda d_P \tag{4.7}$$

When M_A and M_B are given, $d(M_A, M_B)$ is computable. The weighting coefficients μ and λ in (4.7) may be determined by the decision-maker, based on the principle of multiple objective decision making. Assume that we use two PRs, M_1 and M_2, to approximate a PR, M_0. Letting $\lambda = 0$ and $\mu = 0$ means that we care only the difference between the two PR's in the prediction space. If $d_E(M_1,M_0) = d_E(M_2,M_0)$, then we think that the goodnesses of M_1 and M_2 are the same. Letting $\mu > 0$ means that we care also the difference between the two PRs in the observation space. If $d_E(M_1,M_0)=d_E(M_2,M_0)$ but $d_D(M_1,M_0) < d_D(M_2,M_0)$, M_1 is considered to be closer to M_0 and thus is better than M_2.

After defining the distance between two PRs, we can define the distance between a PR and a set of PRs that have the same structure. Let $M_A = (S_A, \theta_A)$ be a PR and $\{M_B\}$ be a set of PRs. All PRs in the set have the same structure S_B and which is different from S_A. In $\{M_B\}$, we can find a PR $M_{BA}(S_B, \theta_{BA})$ that is the closest PR to M_A than all other PRs in the set. The distance between M_A and M_{BA}, $d(M_A, M_{BA})$, is then defined as the distance between M_A and the set $\{M_B\}$, $d(M_A, S_B)$. The parameter θ_{BA} of PR M_{BA} can be found by solving the following optimization problem:

$$\min_{\theta_B} \left\{ \begin{matrix} \|\mathbf{g}_E(M_A) - \mathbf{g}_E(S_B,\theta_B)\|_E + \mu\|\mathbf{u}_\mathbf{D}(M_A) - \mathbf{u}_\mathbf{D}(S_B,\theta_B)\|_D \\ +\lambda\|\overline{\theta}_A - \overline{\theta}_B\|_P \end{matrix} \right\} \tag{4.8}$$

This is, actually, a classical inverse problem: using a fixed parameter structure S_B and changing the parameter values θ_B to best fit the model outputs $\mathbf{u}_\mathbf{D}(M_A)$ and $\mathbf{g}_\mathbf{E}(M_A)$, when both observation and prediction spaces are considered. Values of $\mathbf{u}_\mathbf{D}(M_A)$ and $\mathbf{g}_\mathbf{E}(M_A)$ can be obtained by running the simulation and prediction (or management) models.

After the distance between two PRs is defined, the parameter structure error resulting from replacing the parameter structure S_A using the parameter structure S_B can then be defined by the following max-min problem:

$$SE(S_A,S_B) = \max_{\theta_A} \min_{\theta_B} d(S_A,\theta_A;\ S_B,\theta_B) \tag{4.9}$$

where θ_A and θ_B must be chosen from their admissible regions Θ_A and Θ_B, respectively. Generally, $SE(S_A,S_B) \neq SE(S_B,S_A)$. When S_A is a simplification of S_B, we have $SE(S_A,S_B) = 0$. The max-min problem (4.9) may be decomposed into

$$SE(S_A,S_B) = \max_{\theta_A} F(\theta_A), \quad \theta_A \in \Theta_A \tag{4.10}$$

and

$$F(\theta_A) = \min_{\theta_B} d(M_A,M_B), \quad \theta_B \in \Theta_B \tag{4.11}$$

Problem (4.11) is the same as Problem (4.8). In order to find the solution of Problem (4.10), Problem (4.11) must be solved many times. When S_B is a simplification of S_A, the solution of Problem (4.10), $\hat{\theta}_A \in \Theta_A$, can often be guessed from the structure pattern and the physical problem, as $(S_A, \hat{\theta}_A)$ is such a PR that its structure is the most difficult one to be simplified. In this case, the computation effort of calculating a structure error is equal to solving a typical CIP. One example of calculating the parameter structure error can be found in Sun (1996), in which the hydraulic conductivity with 20 heterogeneous zones is replaced by a simpler structure with only 4 zones. Another example can be found in Sun et al. (1998), in which the hydraulic conductivity generated from a non-stationary random field is replaced by a structure with five zones.

If both $SE(S_A, S_B)$ and $SE(S_B, S_A)$ are within a given error tolerance, the two structures are said to be equivalent. When S_A is a simplification of S_B, to judge their equivalence we only need the value of $SE(S_B, S_A)$. If we take λ=0 and μ=0 in Eq. (4.7), the equivalence of two parameter structures means that we can use one structure to replace another one for specified model applications. For example, a continuously varying 3-D structure may be replaced by a simple 2-D zonation structure.

4.3 A Generalized Inverse Problem and Its Solution

In this section, we will introduce a *generalized inverse problem* (GIP) that identifies the parameter structure and associated parameter values based on not only the observed data but also the objective of model construction and accuracy requirement. We will use J to represent a specified objective. If the number of objectives of interest is more than one, J is considered as a vector. After J is specified, we must also specify an accuracy requirement ε to J. The value of J can be calculated by running a simulation model, $u = u(p;\, q_E;\, \mathbf{x})$, where $q_E(\mathbf{x})$ represents a given control variable and $p(\mathbf{x})$ is a distributed parameter. When $p(\mathbf{x})$ is replaced by a PR, $M=(S,\theta)$, the reliability requirement of model prediction (or management) may be stated as

$$\left\| J(S,\theta; q_E) - J^t(p; q_E) \right\|_E < \varepsilon \tag{4.12}$$

where $\|\cdot\|$ is a norm defined in the objective (prediction or management) space, J^t is the true value of J which , of course, is unknown. PRs satisfying Eq. (4.12) are not unique and may have different parameter structures and different parameter values. Assume that we have the following data for estimating the unknown parameter:

- Prior information on the upper and lower bounds of the true parameter $p(x)$: $p_l(x) \le p(x) \le p_u(x)$ that defines an admissible set P_{ad} of the estimated parameter.
- A set of observations, $\{u_D^{obs}\}$, of the state variable under a certain control q_D defined by an experimental design D.
- An objective of model application, J, and its accuracy requirement ε.

The GIP requires finding the *simplest* parameter structure S^* and its associated model parameter θ^* in P_{ad} such that the condition Eq. (4.12) is satisfied when the PR $M^* = (S^*, \theta^*)$ is used as $M = (S, \theta)$ in that equation.

The so defined GIP has the following advantages. First, the reliability of model application is incorporated into the identification procedure. Second, the parameter identification problem is replaced by a weak requirement, i.e., the equivalent condition (4.13). This condition may be satisfied by such parameters that are not close to the true parameter in the parameter space. Third, the data requirement is minimized because the GIP attempts to find the simplest parameter structure. In GIP, the complexity of parameter structure is determined by the requirement of model application. Once the complexity is determined, the sufficiency of existing data can be judged.

The GIP can be solved by a *stepwise regression method* presented by Sun et al. (1998). The solution procedure is to form a structure sequence:

$$S_1 \subset S_2 \subset S_3 \subset \cdots \subset S_m \subset S_{m+1} \subset \cdots \tag{4.13}$$

where m is the dimension of parameterization. We require that S_m be a simplification of S_{m+1} for all $m = 1, 2, \cdots$, and any structure complexity can be represented by the increase of m. For each m, we use the method presented in Section 4.1 to find the optimal PR of the unknown parameter by fitting the observation data. The following residual of fitting thus can be obtained:

$$RE_m = \min_{S_m, \theta_m} \left\| u_D^{obs} - u_D^{cal}(S_m, \theta_m; q_D) \right\|_D \tag{4.14}$$

At the same time, we calculate the structure error of using S_{m-1} to replace S_m, which is measured in the objective space, i.e.,

$$SE_m = \max_{\theta_m} \min_{\theta_{m-1}} \left\{ \left\| J(S_m, \theta_m; q_E) - J(S_{m-1}, \theta_{m-1}; q_E) \right\|_E \right\} \tag{4.15}$$

The stepwise regression procedure designed for solving the GIP consists of the following steps:

1. Let S_1 be the homogeneous structure. By solving problem (4.14), we can find a PR, $M_1 = (S_1, \theta_1)$, and its associated residual RE_1.
2. Divide S_1 into two zones. The location of the linear boundary between the two zones and associated parameter values can be found by minimizing the *RE*. In this step, we find a PR, $M_2 = (S_2, \theta_2)$, and its associate residual RE_2. We must have $RE_2 < RE_1$ because a homogeneous structure is replaced by a structure with more degrees of freedom to fit the same observations.
3. Calculate the structure error SE_2 of using S_1 to replace S_2 by solving the max-min problem (4.15). If both SE_2 and RE_2 are large compared with the accuracy requirement of prediction (or management) and the upper bound of observation error, respectively, we need to find a more complicated PR, $M_3 = (S_3, \theta_3)$. Based on the identified model M_2, we can calculate the sensitivity coefficients of prediction (or management) to the parameter associated with the two zones. The most sensitive zone is selected to divide into two sub-zones. The optimal location of the linear boundary between the two sub-zones and the associated parameter values for all zones can be found by minimizing *RE*. The best fitting PR, $M_3 = (S_3, \theta_3)$, and corresponding residual RE_3 are thus obtained. Then we can calculate the structure error SE_3 of using S_2 to replace S_3. This step can be repeated. Assume that we have calculated M_m, RE_m and SE_m for a subscript *m*.
4. Now we consider the following four cases:
   (1) If both SE_m and RE_m are large, increase *m* by one and repeat Step 3.
   (2) If both SE_m and RE_m are small, stop and use M_m as the identified parameter.
   (3) If SE_m is small but RE_m is large, either stop or continuously increase the complexity until RE_m becomes small;
   (4) If SE_m is large but RE_m is small, new data need to be collected.

In Case 1, the identified parameter cannot satisfy the accuracy requirement of the given model application, but at the same time, the existing data still have the potential to provide more information. Therefore, we try more complicated parameter structures. In Case 2 and Case 3, there is no significant improvement to the given model application when the complexity of parameter structure is increased. Thus, we can either accept the identified model or continuously increase the model complexity if there is more information contained in the existing data. In Case 4, the information contained in the existing data is not sufficient for identifying a reliable model, and thus, we need to collect more data.

The above stepwise regression procedure can effectively solve the GIP and can determine if existing observation data are sufficient. Note that the structure identified by this procedure is different from what identified by solving the EIP. In other words, the search sequences obtained in Eq. (4.3) and Eq. (4.13) are different even the same observation data are used.

When initial conditions, boundary conditions, and source terms are not completely known, their structures and values can also be identified by solving the GIP. When (1.5) is used to represent the unknown boundary conditions, all basis functions and basis points must be defined on boundary surfaces. To identify a discrete source, the basis functions may be defined by

$$\phi_j(\mathbf{x}) = 1,\ \text{when}\ \mathbf{x} = \mathbf{x}_j;\quad \phi_j(\mathbf{x}) = 0,\ \text{otherwise} \tag{4.16}$$

where $\mathbf{x}_j$ is the location of jth basis point. This approach may find interesting applications in locating the unknown contaminant sources.

In Sun et al. (1998), a remediation design problem was studies by solving the GIP. The real hydraulic conductivity is a realization of a non-stationary random field but unknown. We found that a five- zone structure of hydraulic conductivity can produce the same cleanup schedule as the real one does, and the five-zone structure can be identified by very limited observation data.

4.4 Observation Design

In Section 2, we have learned that the problem of parameter identification actually is a problem of how to transfer information from observation data to the estimated parameter. No identification method is effective if the quantity and quality of observation are insufficient. The objective of observation design is to provide the maximum information with a certain budget, or to provide certain information with the minimum budget. *Observation design* is a special case of experimental design and sampling techniques. For environmental modelling, it includes the determination of locations and frequency of measuring concentration and other state variables (water head, pressure, and temperature) in the field. Sun (1994) presented a detailed discussion on experimental design for groundwater modeling.

In statistics, the theory of experimental design has been well developed for linear models (Pazman, 1986). For a *linear model*, the observation equation can be represented by

$$\mathbf{u}_D = \mathbf{A}_D\theta + \varepsilon_D \tag{4.17}$$

where $\mathbf{A}_D$ is a $n\times m$ matrix. If the observation error ε_D is normally distributed with zero mean and variance σ^2, the covariance matrix of the estimated parameter $\hat{\theta}$ is given by

$$Cov(\hat{\theta}) = \sigma^2(\mathbf{A}_D^T\mathbf{A}_D)^{-1} \tag{4.18}$$

The determinant of the covariance matrix is a measurement of the *uncertainty* of the estimated parameter. The matrix $\mathbf{A}_D^T\mathbf{A}_D$ is called the *information matrix*. To maximize the determinant of the information matrix means to minimize the uncertainty of the estimated parameter. Thus, we have the following *D-optimal design criterion*: finding a design D^* from all admissible designs D_{ad} such that

$$D^* = \arg\max_D \det(\mathbf{A}_D^T\mathbf{A}_D),\ D \in D_{ad} \tag{4.19}$$

For environmental modeling, the simulation model is usually nonlinear and the observation equation is given by (2.1), i.e.,

$$\mathbf{u}_D = \mathbf{M}(\theta;\ \mathbf{x}_D) + \varepsilon_D \tag{4.20}$$

This equation can be linearized by first-order approximation in the neighborhood of a parameter θ_0 (an initial estimation of the unknown parameter). The matrix $\mathbf{A}_D$ for the linear model now is replaced by the Jacobian $\mathbf{J}_D$, in which all partial derivatives are evaluated at θ_0. As a result, the information matrix is approximated by $\mathbf{J}_D^T\mathbf{J}_D$ and the D-optimal design criterion (4.19) becomes

$$D^* = \arg\max_D \det(\mathbf{J}_D^T\mathbf{J}_D),\ D \in D_{ad} \tag{4.21}$$

Since the Jacobian $\mathbf{J}_D$ depends on the initial estimation of the parameter to be estimated, the D-optimal design can only be used in a sequential design procedure.

D-optimal design is not an effective design method in the field of environmental engineering because a design obtained by the D-optimal criterion is neither necessary nor sufficient for the purpose of model application. If the error in parameter structure can be ignored, the reliability of model prediction can be estimated from the reliability of the estimated parameter vector. Using the first-order approximation, we have

$$Cov(\hat{\mathbf{g}}_E) = \mathbf{J}_E Cov(\hat{\theta})\mathbf{J}_E^T = \sigma^2\mathbf{J}_E(\mathbf{J}_D^T\mathbf{J}_D)^{-1}\mathbf{J}_E^T \tag{4.22}$$

where $\hat{\mathbf{g}}_E$ is a set of model predictions, $\mathbf{J}_E = [\partial \mathbf{g}_E / \partial\theta]$ is the Jacobian of the predictions with respect to the parameter vector. A design criterion that minimizes the uncertainty of a set of given model predictions is thus obtained: finding a design D^* from all admissible designs D_{ad} such that the determinant of $\mathbf{J}_E(\mathbf{J}_D^T\mathbf{J}_D)^{-1}\mathbf{J}_E^T$ is minimized, which is known as the *G-optimal design.*

For a distributed parameter, the parameter structure error caused by parameterization can never be avoided. A parameter with more complicated structure needs more data to identify and vice versa. Therefore, the identification of parameter structure must be considered in the design stage. The problem of observation design for parameter structure identification is a new topic. Without knowing the complexity of the estimated parameter, we cannot decide whether or not the data provided by a design is sufficient. Lacking the knowledge of data sufficiency, there is no basis to determine if a design is optimal. The concept of GIP given in Section 4.3 may provide new criteria for observation design because it can determine an appropriate complexity of parameter structure and judge the sufficiency of observation data. The optimal design can be found from all sufficient and admissible designs based on a cost-effective criterion.

5. CONCLUSIONS

We have reviewed the basic concepts, criteria, and methods for the identification of distributed parameters in environmental modeling. We started from the general form (1.5) of parameterization. The dimension of parameterization, *m*, and the *m* basis functions, $\{\phi_j(\mathbf{x},\mathbf{v})\}$, define the structure of the estimated parameter. The classical inverse problem (CIP) assumes that the parameter structure is known and only the weighting coefficients, θ_j ($j=1,2,\ldots,m$), need to be identified, while the extended inverse problem (EIP) requires the identification of all components of parameterization, including both parameter structure and parameter values. The generalized inverse problem (GIP) combines the parameter identification problem with the reliability of model application and the sufficiency of data.

The problem of parameter estimation is defined from the point of view of information transfer. The criterion of maximizing a posterior distribution (MAP) can be used to derive other often used criteria of parameter identification, such as the maximum likelihood estimation (MLE), the general least squares estimation (GLS), the weighted least squares estimation (WLS) and the ordinary least squares estimation (OLS). When the prior distribution is assumed to be normal, a regularization term (or a penalty term) should be added to these criteria. The CIP is thus transferred into a constrained non-linear optimization problem. The later can be solved by various methods of numerical optimization or by a modified Gauss-Newton method. The genetic algorithm (GA) and artificial neural network (ANN) have been used for parameter identification by some authors when the dimension of parameterization is low.

The adjoint state method is very effective for calculating the gradient of objective function when a quasi-Newton method is used for parameter identification. The sensitivity matrix, or the Jacobian, may be calculated by either the sensitivity equation method (the bottom-up mode of auto differentiation) or the adjoint state method (the top-down mode of auto differentiation). The sensitivity equation method should be chosen when the Gauss-Newton method is used for parameter identification. The adjoint state method has found important applications in sensitivity analysis, parameter structure identification and observation design problems.

We have presented a simple tree-regression process to solve the EIP that can find an approximately optimal parameter structure and associated parameter values. In this process, the dimension of parameterization and number of structure parameters are minimized. Since the most sensitive zone to the observation data is selected to be divided into two sub-zones, the sensitivity coefficients associated with all zones will not become too small before stopping the increase of the number of zones. As a result, the over-parameterization problem can be avoided.

The GIP considers not only the parameter and observation spaces, but also the prediction and management spaces. We have presented a step-wise regression method for solving the GIP, in which, the most sensitive zone to the prediction (or management) is selected to divide into two sub-zones. Based on solving the GIP, we can develop a new methodology that allows for a systematic procedure of model construction. Future work is needed to solve the observation design problem for EIP and GIP.

REFERENCES

Adeli, H. and Hung, S.-L. (1995) *Machine Learning: Neural Networks, Genetic Algorithms, and Fuzzy Systems*, John Wiley, New York, 211 pp.

Bard, Y. (1974) *Nonlinear Parameter Estimation*, Academic, San Diego, CA., 341 pp.

Bear, J. (1972) *Dynamics of Fluids in Porous Media*, Elsevier, New York, 764pp.

Carrera, J. and Neuman, S. P. (1986) "Estimation of aquifer parameters under transient and steady state conditions, 1. Maximum likelihood method incorporating prior information," *Water Resour. Res.*, 22(2), pp. 199-210.

Cherkassky, V. S. (1998) *Learning From Data: Concepts, Theory, and Methods*, John Wiley & Sons, New York, 441pp.

Chavent, G., Dupuy, M. and Lemonnier, P. (1975) "History matching by use of optimal control theory," *Soc. Pet. Eng. J.*, 15(1), pp. 74-86.

Corliss, G. and Rall, L. B. (1996) "An introduction to automatic differentiation," in *Computational Differentiation: Techniques, Applications, and Tools*, edited by M. Berz, C. Bischof, G. Corliss, and A. Griewank, pp.1-18, SIAM, Philadelphia, PA.

Dagan, G. (1985) "Stochastic modeling of groundwater flow by unconditional and conditional probabilities: The inverse problem," *Water Resour. Res.*, 21(1), pp. 65-72.

Davis, J. C. (1986) *Statistics and Data Analysis in Geology*, John Wiley, 646 pp.

Fletcher, R. (1987) *Practical Methods of Optimization*, John Wiley, New York, 436 pp.

Gelhar, L. W. (1993) *Stochastic Subsurface Hydrology*, Prentice-Hall, N. J., 390 pp.

Giering, R. and Kaminski, T. (1998) "Recipes for adjoint code construction," *ACM Transactions on Mathematical Software*, 24(4), pp. 437-474.

Goldberg, D. E. (1989) *Genetic algorithms in search, optimization, and machine learning*, Reading, Mass., Addison-Wesley, 412 pp.

Harvey C. F. and Gorelick, S. M. (1995) "Mapping hydraulic conductivity - sequential conditioning with measurements of solute arrival time, hydraulic-head, and local conductivity," *Water Resour. Res.*, 31(7), pp.1615-1626.

Hecht-Nielsen, R. (1990) *Neurocomputing*, Addison-Wesley, 433 pp.

Holland, J. H. (1992) *Adaptation in Natural and Artificial Systems: An Introductory Analysis with Applications to Biology, Control, and Artificial Intelligence*, MIT Press, Cambridge, 211 pp.

Holzbecher, E. O. (1998) *Modeling Density-Driven Flow in Porous Media: Principles, Numerics, Software*, Springer –Verlag, Berlin, 286 pp.

Isakov, V. (1998) *Inverse Problems for Partial Differential Equations*, Springer – Verlag, New York, 284 pp.

Kaminski, T., Heimann, M. and Giering, R. (1999) "A coarse grid three-dimensional global inverse model of the atmospheric transport - 1. Adjoint model and Jacobian matrix," *J. Geophys. Res. Atmos.*,104(D15), pp.18535-18553.

Kelley, C. T. (1999) *Iterative methods for optimization*, SIAM, Philadelphia, 180pp.

Kitanidis, P. K. and Vomvoris, E. G. (1983) "A geostatistical approach to the inverse problem in groundwater modeling (steady state) and one-dimensional simulations," *Water Resour. Res.*, 19(3), pp. 677-690.

Kitanidis, P. K. (1995) "Quasi-linear geostatistical theory for inversing," *Water Resour. Res.*, 31(10), pp. 2411-2420.

Kitanidis, P. K. (1997) *Introduction to Geostatistics: Applications to Hydrogeology*, Cambridge University Press, New York , 249 pp.

More J. J. and Wright, S. J. (1993) *Optimization Software Guide*, SIAM, Philadelphia.

Morshed, J. and Kaluarachchi, J. J. (1998) "Parameter estimation using artificial neural Network and genetic algorithm for free-product migration and recovery," *Water Resour. Res.*, 34(5). Pp.1101-1113.

Navon, I. M. (1998) "Practical and theoretical aspects of adjoint parameter estimation and identifiability in meteorology and oceanography," *Dynam. Atmos. Oceans.*, 27(1-4), pp. 55-79.

Neuman, S. P. (1980) "A statistical approach to the inverse problem of aquifer hydrology, 3, Improved solution method and added perspective," *Water Resour. Res.*, 16(2), pp. 331-346.

Pazman, A. (1986) *Foundations of Optimum Experimental Design*, D. Reidel, Kluwer Academic Publishers, Dordrecht, Holland, 228 pp.

Petridis, V., Paterakis, E. and Kehagias, A. (1998) "Hybrid neural-genetic multimodal parameter estimation algorithm," *IEEE, Neural Network*, 9(5), pp. 862-876.

Pham, D. T. and Liu, X. (1995) *Neural Networks for Identification, Prediction and Control*, Springer-Verlag, New York, 238 pp.

Piasecki, M. and Katopodes, N. D. (1999) "Identification of stream dispersion coefficients by adjoint sensitivity method," *J. Hydraul. Eng., ASCE*, 125(7), pp. 714-724.

Rogers, L. L., Dowla, F. U. and Johnson, V. M. (1995) "Optimal field-scale groundwater remediation using neural networks and the genetic algorithm," *Environ. Sci. Technol.*, 29(5), pp.1145-1155.

Rubin, Y., Mavko, G. and Harris, J. (1992) "Mapping permeability in heterogeneous aquifers using hydrological and seismic data," *Water Resour. Res.*, 28(7), pp. 1792-18.

Schnoor, J. L. (1996) *Environmental Modeling: Fate and Transport of Pollutants in Water, Air, and Soil*, John Wiley, New York, 682 pp.

Sun, N.-Z. and Yeh, W. W-G. (1985) "Identification of parameter structure in groundwater inverse problem," *Water Resour. Res.*, 21(6), pp. 869-883.

Sun, N.-Z. and Yeh, W. W-G. (1990) "Coupled inverse problem in groundwater modeling, 1, Sensitivity analysis and parameter identification," *Water Resour. Res.*, 26(10), pp. 2507-2525.

Sun, N.-Z. and Yeh, W. W-G. (1992) "A stochastic inverse solution for transient groundwater flow: Parameter identification and reliability analysis," *Water Resour. Res.*, 28(12), pp. 3269-3280.

Sun, N.-Z. (1994) *Inverse Problems in Groundwater Modeling*, Kluwer Academic Publishers, 337 pp.

Sun, N.-Z., Jeng, M-C. and Yeh, W. W-G. (1995) "A proposed geological parameterization method for parameter identification in three-dimensional groundwater modeling," *Water Resour. Res.*, 31 (1), pp. 89-102.

Sun, N.-Z. (1996) *Mathematical Modeling of Groundwater Pollution*, Springer-Verlag, New York, 377 pp.

Sun, N.-Z. (1996), "Identification and reduction of model structure for modeling distributed parameter systems," in *Parameter Identification and Inverse Problems in Hydrology, Geology anf Ecology*, edited by J. Gottlieb and P. DuChateau, Kluwer Academic Publishers.

Sun, N-Z., Yang, S-L. and Yeh, W. W-G. (1998) "A proposed stepwise regression method for model structure identification," *Water Resour. Res.*, 34(10), pp. 2561-2572.

Tarantola, A. (1987) *Inverse Problem Theory: Methods for Data Fitting and Model Parameter Estimation*, Elsevier, New York, 613 pp.

Vose, M. D. (1999) *The Simple Genetic Algorithm: Foundations and Theory,* MIT Press, 251pp.

Wackernagel, H. (1998) *Multivariate Geostatistics: An Introduction with Applications*, Springer-Verlag, New York, 291 pp.

Wood, W. L. (1993) *Introduction to Numerical Methods for Water Resources,* Oxford University Press, New York, 255 pp.

Yeh, J. T.-C. and Zhang, J. (1996) "A geostatistical inverse method for variably saturated flow in the vadose zone," *Water Resour. Res.*, 32(9), pp. 2757-2766.

Zannetti, P. (1990) *Air Pollution Modeling: Theories, Computational Methods, and Available Software,* Computational Mechanics Publications, New York, 444 pp.

Zlatev, Z. (1995) *Computer Treatment of Large Air Pollution Models*, Kluwer Academic Publishers, 358 pp.

Chapter 10

FINITE ELEMENT ANALYSIS OF STRATIFIED LAKE HYDRODYNAMICS

Der-Liang Frank Young[1]

ABSTRACT

The detailed formulation of laterally averaged two-dimensional stratified lake modeling with free surfaces by the finite element analysis is given in this chapter. The Galerkin finite element method with arbitrary Lagrangian-Eulerian (ALE) computing mesh enables us to treat the Navier-Stokes equations of stratified flows with free surfaces under the assumptions of fluid incompressibility, hydrostatic pressure distribution, and the Boussinesq approximation. The model is first tested for the laboratory data of the gravitational underflows of the Generalized Reservoir Hydrodynamics (GRH) flume model taken at the Waterways Experiment Station, U.S. Army Corps of Engineers. After the successful verification with the laboratory scale, the model is then applied to simulate the field scale of two-year annual cycles of the stratified man-made lake of the Te-Chi reservoir located at central Taiwan.

1. INTRODUCTION

A lake can be considered as an open system with an upper free surface boundary exposed to the atmosphere and a solid boundary, called the basin, made of minerals. Various geological processes such as weathering, mass wasting, erosion and deposition by streams are responsible for the existence of these basins of natural lakes. Artificial lakes are also created for storing water obtained from rain and other natural streams. The stream-eroded valleys serve as basins for most of the reservoirs that are built behind dams. Apart from serving as reservoirs for storing water, lakes and lake sediments provide useful information about environmental changes. The analysis of fossil pollen provides information about the changes that have taken place in the nearby vegetation. The changes in the shoreline elevations can be used along with the pollen records to study the recent changes in the environment. The variations in lake levels also serve as an indication on the amount of rainfall or evaporation. Some geologists have

[1]Professor, Department of Civil Engineering and Hydrotech Research Institute, National Taiwan University, Taipei, Taiwan-10617, Republic of China, **dlyoung@hy.ntu.edu.tw**

discovered that lakebeds are a common source of petroleum. Hence the study of old lake deposits has become a major research activity.

Lakes serve as a major source of water supply for human activities. The demand for water is continuously increasing with growing population and civilization. In recent years there is much concern about the quality of lake water, since the lake water is contaminated due to different kinds of interactions of a lake with its surroundings. These interactions vary widely as listed below:

(i) Lake water absorbs heat from solar radiation and also from power plant cooling water while released to the lake; it also loses heat to the atmosphere.
(ii) Lake receives water as inflow from streams and also in the form of precipitation; water leaves the lake as outflow for various purposes.
(iii) Lake receives contaminated water and liquids in varying densities.
(iv) Wind blowing over the lake surface changes the wave pattern of water surface and thus the surface roughness thereby increases the momentum and energy transfer.
(v) Lake interacts with the ground water table; this may affect the mass and momentum balances in the lake.

Many of these interactions result in coupled activities, affecting the balances of mass, momentum, heat and species concentration in the lake water. Furthermore, the flows associated with lakes are always turbulent and the turbulence varies with time and space at different scales. In order to have control over the quality of lake water, knowledge about the dynamics of lake water is essential. The study of variation of velocity, temperature and species concentration in lake water due to various kinds of interactions of a lake with its surroundings is called 'lake hydrodynamics', also called the 'physical limnology'.

2. MASS, MOMENTUM, HEAT AND CONCENTRATION TRANSPORTS IN LAKES

One important characteristic of a lake is its density stratification (Harleman, 1961, and Yih, 1980). The density stratification in lakes takes place both by means of thermal effects as well as by the inflow of liquids with varying densities. The thermal stratification in a lake is formed primarily by solar radiation. The heat transfer mechanism at the water surface together with solar radiation plays a major role in deciding the surface water temperature (Verevochkin and Startsev, 2000). Consequently, the temperature of water in a lake varies seasonally. In spring the water temperature increases due to heating of surface water by growing solar energy. The resulting thermal stratification becomes strong in the late summer. The various thermal regimes developed inside a lake due to thermal stratification are shown in Fig.1. The warm upper layer is called the 'epilimnion' and the cold lower layer is referred to as the 'hypolimnion'. The region of the largest temperature gradient is often denoted as the 'metalimnion'. When the continuous stratification is idealized by a two-layer model with a light upper layer and a heavier lower layer, the plane of separation is called the 'thermocline'. This is usually taken to be the position of the strongest vertical temperature gradient. The stability of thermal stratification is characterized by the Richardson number, defined as

$$Ri = -\frac{g}{\rho}\frac{\partial \rho}{\partial z} \bigg/ \left(\frac{\partial u}{\partial z}\right)^2$$

The stability of thermal stratification increases when Ri increases.

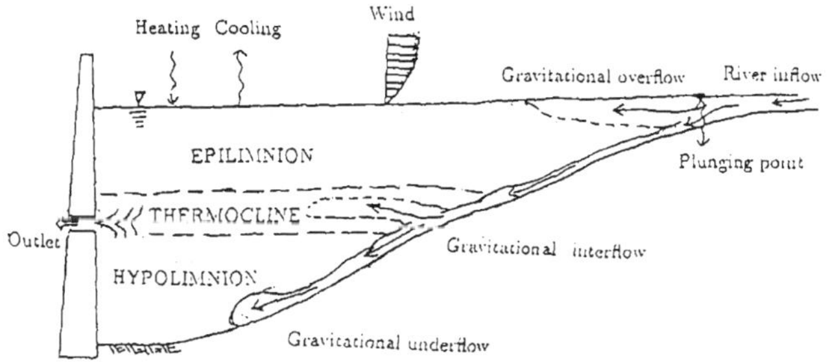

Fig. 1. Schematic diagram of a stratified lake or reservoir showing the transport processes of mass, momentum, heat, and sediment.

Apart from thermal effects the density of the incoming flow into the lake also affects the stratification. The inflow of liquids with varying densities results in different types of flow within the lake. When the density of the incoming flow is lower than that of the water in the lake, a layer of flow will move along the free surface. This type of flow is called 'overflow'. When the density of the incoming flow is greater than that of the water in the lake, then a layer of flow will move along the reservoir bed and this flow is called 'underflow'. When the density of water in a lake is not homogeneous due to thermal stratification, the incoming flow will move towards an intermediate layer and this is called 'interflow'. When a flow with sediment reaches the backwater zone of a reservoir, the suspended sediments start to deposit at the bottom of the lake. When the sediment-laden flow plunges beneath the free surface under certain circumstances, the layer of underflow formed is called 'turbidity current' (Lee and Yu, 1997). The nature of density current also depends on whether the flow is steady or unsteady and continuous or instantaneous.

Thermal stratification plays a vital role in water quality modeling. When thermal stratification exists for a longer period, the water in the hypolimnion layer is cut off from contact with the atmosphere with respect to oxygen supply and this will have an adverse effect on the quality of water. A lake can be destratified by artificial means to overcome this problem. However, sometimes nature also helps us to weaken the stratification by turbulent mixing. The flow into a lake is always turbulent. When such flow enters the lake, there will be turbulent exchange of momentum across the interface of the density current. The low momentum liquid easily enters into the density current by interfacial mixing. The interfacial mixing is an important process for destratification to maintain water quality. This depends on the shear stress and hydrostatic pressure gradient at the interface. Hence knowledge about density stratification is essential for water quality modeling. The study of lake hydrodynamics mainly concentrates on the determination of various types of density currents developed for the imposed conditions on a lake.

3. MATHEMATICAL MODELING OF LAKE HYDRODYNAMICS

The hydrodynamics of lakes has been studied by collecting field data and by conducting experiments on laboratory scale models. Since the hydrodynamic processes are in general transient, coupled and non-linear in nature, pure experimental approach will demand an enormous amount of tests. The physics of the phenomena can be well understood by representing the various physical processes and interactions by suitable mathematical model and making use of the field data (Hutter, 1984, Hutter, 1987, and Mortimer, 1974). A mathematical model can be built to represent various interactions such as (i) heating of water in lake due to solar radiation, (ii) heat exchange at the air-water interface, (iii) vertical mixing of water due to wind action, (iv) convective water movement due to density currents, (v) turbulent nature of flow within the lake and (vi) interaction of lake bottom with the ground water movement. Developing a mathematical model including all the above effects is very complex and difficult. Moreover, the study of lake hydrodynamics may vary depending upon the applications. Hence a suitable mathematical model for lake hydrodynamics is problem dependent.

In all the mathematical models it is assumed that the principles of conservations of mass, momentum, energy and concentration provide the basic equations for describing the hydrodynamics of lakes. With respect to the size of the lakes studied, in general, the Coriolis parameter is neglected in the study of lake hydrodynamics except for the Great Lakes. Most numerical models assume hydrostatic approximation (due to small value of depth to length ratio, i.e. $H/L \ll 1$), Boussinesq approximation and incompressible flows. The hydrostatic approximation implies that any change of pressure due to the movement of air-water interface, is transmitted through the entire water column instantly. The Boussinesq approximation (Gray and Girgini, 1976) treats density as constant throughout the governing equations except those associated with gravitational terms (Turner, 1973). With the above approximations, the variation of density is only explicitly included in a limited way. The additional influences of density gradients will appear in the turbulent mixing coefficients (Cheng, 1977). With the above assumptions a mathematical model can be developed to study the hydrodynamic behavior of a lake. Depending upon the topography, meteorological conditions and the nature of inflow and outflow into a lake, the hydrodynamic model may be used to focus on specific phenomenon such as thermal stratification, density circulation, wind-induced circulation etc. With an aim to make the reader appreciate the on-going research in various fields of lake hydrodynamics, an overview of some important literatures are highlighted in the following sections.

3.1 Thermal stratification

The water temperature in a lake is affected by the phenomenon of stratification. The development of thermocline and erosion of thermocline have been studied using mathematical models (see Csanady, 1975, Graf and Mortimer, 1979, Hutchinson, 1957, Kawahara, 1977, Spalding and Svensson, 1977, and Street et al., 1977). Herleman (1982) gives a review of vertical temperature structure of lakes using different models such as diffusion models and mixed-layer models. The thermal stratification can be studied using simple models like cavity flow models (Young et al., 1976a and 1976b). Sundermann (1978) discusses circulation in lakes using barotropic one-layer model, barotropic Ekman type model and baroclinic multi-layer model. Three-dimensional transient models also have been used to study lake hydrodynamics (Bauer, 1978, and Wittmiss, 1978).

3.2 Density-driven circulation

Mass, in the form of suspended sediments, algae or nutrients such as phosphorus, is distributed in lake water by the action of hydrothermal processes. They also undergo biogeological transformations simultaneously. Density currents may be created either by dense or lighter inflows into stratified or unstratified lakes and reservoirs (Alavian et al., 1992). Turbidity currents are generated by sediment-laden flow into the reservoir (Lee and Yu, 1997). The contaminant dispersion in lakes (Salmon et al., 1980) also creates density currents giving rise to different flow regimes such as plunging flow, underflow and interflow (Chung and Gu, 1998). The migration of plunge point in reservoir affects the hydraulic characteristic of the reservoir.

3.3 Wind-driven circulation

The wind-driven circulation in lakes has been studied using different types of numerical models (Cheng, 1977). Commonly used models include layered models, Ekman type models and three-dimensional models. Wind acting on the surface of a stratified lake gives rise to large changes in the distribution of pressure. These pressure changes cause circulation gyre and modify the position of the interface. The shear stress exerted by the earth on the atmosphere slows down the air currents. In response to this resistance, the water surface generates a drift current and a wave pattern (Simpson, 1995, and Stoker, 1957) as a result of boundary layer effects. The drift velocity profile and wave pattern evolve with space and time (Cheng, 1977, Hansen, 1975, Hansen, 1978, Plate and Wengefeld, 1978, and Ramming, 1978) and this causes changes in the airflow pattern. Due to the drift current, a "slip" condition prevails at the lower boundary of the airflow.

Finally the shear stress is altered resulting from the above phenomena. Another effect due to wind is the deepening of upper layer caused by the entrainment of water from the lower denser layer (Hansen, 1975). The earth rotation also affects the changes that are taking place in a stratified lake in addition to wind effects (Mortimer, 1984).

3.4 Interaction between lake and ground water systems

Surface water bodies such as lakes, ponds, streams etc. are all integral parts of groundwater flow systems. The interaction of a surface water body such as a lake with ground water flow systems, affect the distribution pattern of hydraulic head, specific discharge and stream lines. This interaction depends on the positions of the lake with respect to groundwater flow system, geologic characteristic of their beds and their climatic settings (Winter, 1999). Kacimov (2000) developed an analytical model for a regional 3-D groundwater flow interacting with a lake. He found that any lake can change its regime from gaining to losing or flow-through, depending on the varying incident ground water flow velocity, evaporation-precipitation-runoff rates in the lake, lake sizes, thickness of silt layer at the bottom of the lake and other hydrological characteristics. The information on gaining or losing of lakes is important for water budget calculation. Water quality also can be assessed by this information because it is affected by the contaminants present in the ground water flow systems.

3.5 Water quality

Lake is a major source of fresh water supply. The quality of lake water is affected by the biological and biochemical reactions taking place within the water body. In addition, inflow water also carries a number of contaminants. The mixing of these contaminants with the lake water causes degrade of water. The degree of degrade depends upon how well these contaminants are mixed with the water. The hydrodynamic transport mechanisms such as convection and advection and thermal stratification influence the mixing of these contaminants. Furthermore, these mixing take place in different time scales. A numerical model developed by Imberger et al. (1978) incorporated all the dominant physical processes taking place at different time scales. Imberger and Patterson (1981) used a one-dimensional model for daily prediction of salinity and temperature for small lakes. The hydrodynamic processes within a lake are also influenced by the thermal stratification. Markofsky (1978) suggested that a coupled thermal stratification-water quality model has to be used in order to correctly predict the quality of water.

3.6 Numerical methods

The governing equations used to study the lake hydrodynamics are transient, coupled and non-linear and hence they are solved only by numerical methods. Finite difference method (FDM) has been widely used to solve such complex equations. In the case of FDM, the governing equations are reduced to difference equations using Taylor's series expansion. Hence the governing equations are modified by representing the differential terms in terms of finite values of the variables at fixed nodes. The approximation can be made either by using forward difference, backward difference or central difference schemes. The difference equations are obtained for all nodes in a computational domain, resulting in a number of simultaneous algebraic equations.

In the case of finite element method (FEM), an approximate solution for the nodal variables of the finite element is assumed. In addition, the piecewise variation of the given variables between the nodes is represented by means of a function called the shape function. When the approximate solution is substituted into the governing equation, a residue is obtained in the case of weighted residual method. The residue is minimized with the help of a weighting function. The shape function itself is taken as the weighting function in the case of Galerkin's weighted residual method. The residue is minimized in the entire computational domain. This procedure also results in a number of simultaneous nodal equations.

It is very difficult to represent irregular geometries by FDM. This is a major disadvantage of these models (Bauer, 1978). As it is well known, the boundaries of the basins of lakes or reservoirs are highly irregular in geometry. The problem of representing irregular geometry is overcome by FEM. Thacker (1977) proposed an irregular grid FDM by combining the advantage of a smooth geometrical representation using FEM and the advantage of fast execution using FDM.

3.7 Application of finite element method for lake hydrodynamic model studies

With the advancement of high-speed computers, FEM has become the most powerful method to treat irregular boundaries. FEM provides the user with flexibilities with respect to the computational effort and accuracy by appropriate discretization of the computational domain. In addition the spatial variations of physical properties can easily be taken into account by FEM. Moreover, isoparametric elements can be used to effectively map curvilinear boundaries. This is another attractive feature of FEM compared to other numerical methods. Since the boundary of a lake is of irregular geometry, FEM is found to be an efficient numerical tool to handle such situations. FEM has been used successfully by a number of researchers to solve the governing equations of lake hydrodynamics. Detailed study on the application of finite element method for lake hydrodynamics using different types of models may be found in the literature (Gallagher et al., 1973, Gallagher and Chan, 1973, Young, 1980, Young et al., 1976(a) & 1976(b), and Young and Liggett, 1977).

4. DERIVATION OF GOVERNING EQUATIONS

The governing equations for lake hydrodynamics are obtained by applying the conservation laws of mass, momentum, energy and species concentration to an elemental control volume. These equations differ from the general conservation equations in certain aspects: the general conservation equations are derived based on the assumption that the control volume is uniform. In the case of lakes, the width varies both in vertical and in longitudinal directions. In order to make the reader appreciate the application of conservation laws to a real system such as a lake, the important steps in the derivation are discussed in the following section. The coordinate system used to represent different dimensions of the lake is shown in Fig. 2a.

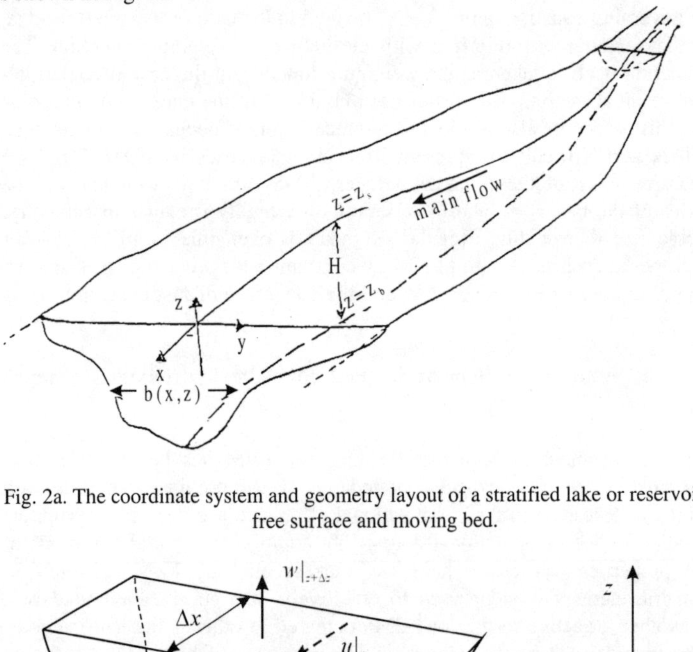

Fig. 2a. The coordinate system and geometry layout of a stratified lake or reservoir with free surface and moving bed.

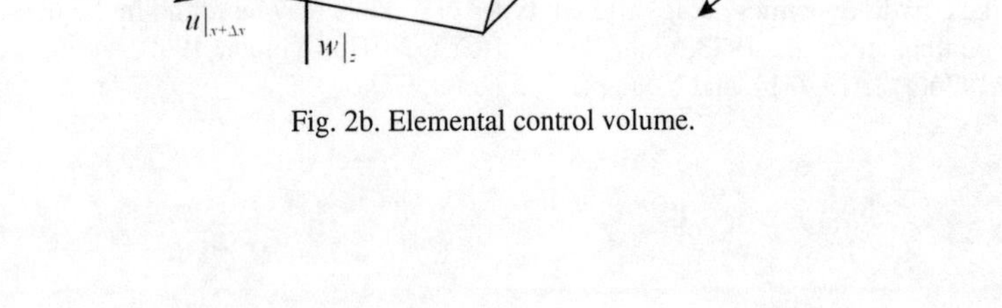

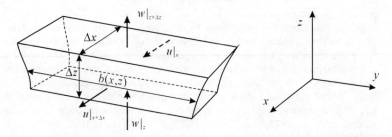

Fig. 2b. Elemental control volume.

4.1 Assumptions

The following assumptions are made in the derivation of conservation equations.

(i) The fluid is incompressible.
(ii) Beoussinesq approximation applies.
(iii) The flow in the lake satisfies hydrostatic approximation.
(iv) Surfaces perperndicular to the y-direction are impermeable for any mode of transport (laterally averaged).

4.2 Continuity equation

The continuity equation is derived based on the conservation of mass principle applied to a small elemental control volume as shown in Fig.2b.
We know from the conservation of mass principle,

$$\begin{pmatrix}\text{rate of mass}\\ \text{increase in the}\\ \text{elemental volume}\end{pmatrix} = \begin{pmatrix}\text{rate of mass}\\ \text{entering the}\\ \text{elemental volume}\end{pmatrix} - \begin{pmatrix}\text{rate of mass}\\ \text{leaving the}\\ \text{elemental volume}\end{pmatrix} + \begin{pmatrix}\text{rate of mass}\\ \text{addition by}\\ \text{lateral flows}\end{pmatrix}$$

x-direction:
Rate of mass flow into the elemental volume is given by

$$(\rho u)\Big|_x b(x,z)\Delta z = (\rho ub)\Big|_x \Delta z$$

Rate of mass flow out of the elemental volume is given by

$$(\rho u)\Big|_{x+\Delta x} b(x,z)\Delta z = (\rho ub)\Big|_{x+\Delta x} \Delta z$$

z-direction:
Rate of mass flow into the elemental volume is given by

$$(\rho w)\Big|_z b(x,z)\Delta x = (\rho wb)\Big|_z \Delta x$$

Rate of mass flow out of the elemental volume is given by

$$(\rho w)\Big|_{z+\Delta z} b(x,z)\Delta x = (\rho wb)\Big|_{z+\Delta z} \Delta x$$

Rate of mass increase in the elemental volume is given by

$$\left(\frac{\partial \rho}{\partial t}\right) b(x,z)\Delta x\Delta z$$

Rate of mass addition to the elemental volume due to lateral flow is

$$\rho q_t \Delta x \Delta z$$

Hence the application of mass conservation principle gives

$$\left(\rho ub\right)\Big|_{x}\Delta z-\left(\rho ub\right)\Big|_{x+\Delta x}\Delta z+\left(\rho wb\right)\Big|_{z}\Delta x-\left(\rho wb\right)\Big|_{z+\Delta z}\Delta x+\rho q_t\Delta x\Delta z=\left(\frac{\partial\rho}{\partial t}\right)b(x,z)\Delta x\Delta z$$

Dividing throughout by $\Delta x\Delta z$ and applying the principle of differential equation we get

$$-\frac{\partial}{\partial x}(\rho ub)-\frac{\partial}{\partial z}(\rho wb)+\rho q_t=b(x,z)\frac{\partial\rho}{\partial t}$$

Since the fluid density is assumed to be constant except when associated with a gravitational term, the term on the right hand side of the above equation becomes zero. The final form of the continuity equation can be written as follows:

$$\frac{\partial(ub)}{\partial x}+\frac{\partial(wb)}{\partial z}=q_t \tag{1}$$

4.3 Momentum equation

The momentum equation is also derived based on the principle used in any fluid mechanics texts. The only difference in the present formulation is the variation of lateral width of the lake b(x, z). Since this term is used frequently in the derivations, henceforth only 'b' will be used in place of 'b(x, z)'. Only the x-momentum equation is derived. The z-momentum equation is represented by the hydrostatic equation and the gravity term. The shear stress on the x-z plane is neglected.

x-momentum equation:

From Newton's second law of motion we can write the force balance for the elemental control volume shown in Fig.2b as follows:

$$\begin{pmatrix}\text{rate of momentum}\\\text{increase within}\\\text{the elemental volume}\end{pmatrix}+\begin{pmatrix}\text{net efflux of}\\\text{momentum out of}\\\text{the elemental volume}\end{pmatrix}-$$

$$\begin{pmatrix}\text{rate of momentum}\\\text{entering the elemental volume}\\\text{due to lateral flow velocity}\end{pmatrix}=\begin{pmatrix}\text{rate of external}\\\text{forces acting on}\\\text{the elemental volume}\end{pmatrix}$$

Net efflux of momentum of the elemental volume=rate of momentum leaving the elemental volume – rate of momentum entering the elemental volume

Rate of accumulation of momentum within the elemental volume is given by

$$\frac{\partial(\rho ub)}{\partial t}\Delta x\Delta z=b\rho\frac{\partial u}{\partial t}$$

Rate at which x-momentum enters the element at x $=\left(\rho uub\right)\Big|_{x}\Delta z$

Rate at which x-momentum leaves the element at x+ $\Delta x=\left(\rho uub\right)\Big|_{x+\Delta x}\Delta z$

Rate at which x-component of momentum enters the element at z= $\left(\rho uwb\right)\big|_{z} \Delta x$

Rate at which x-component of momentum leaves the element at

z+ $\Delta z = \left(\rho uwb\right)\big|_{z+\Delta z} \Delta x$

After dividing by $\Delta x \Delta z$, the net efflux of momentum from the elemental volume in the x-direction can be written as follows:

$$\frac{\partial(\rho ubu)}{\partial x}+\frac{\partial(\rho ubw)}{\partial z}=\rho u\frac{\partial(ub)}{\partial x}+\rho ub\frac{\partial(u)}{\partial x}+\rho u\frac{\partial(wb)}{\partial z}+\rho wb\frac{\partial(u)}{\partial z}$$

Since the density is constant, the above equation can be rewritten as

$$\rho u\left[\frac{\partial(ub)}{\partial x}+\frac{\partial(wb)}{\partial z}\right]+\rho b\left[u\frac{\partial u}{\partial x}+w\frac{\partial u}{\partial z}\right]=\rho uq_t+\rho b\left[u\frac{\partial u}{\partial x}+w\frac{\partial u}{\partial z}\right]$$

Because the total external forces acting on the element is the sum of forces due to liquid pressure and due to shear stresses, we next derive all external forces as follows:

Forces due to liquid pressure:

The net pressure force acting on the element in the x-direction is

$$\left(bp\big|_{x}-bp\big|_{x+\Delta x}\right)\Delta z$$

After dividing by $\Delta x \Delta z$ we get the differential form of the equation as

$$-\frac{\partial(bp)}{\partial x}$$

Forces due to shear stresses:

The forces due to the shear stresses in the x-direction are:

$$\left(\tau_{xx}b\big|_{x+\Delta x}-\tau_{xx}b\big|_{x}\right)\Delta z+\left(\tau_{zx}b\big|_{z+\Delta z}-\tau_{zx}b\big|_{z}\right)\Delta x$$

After dividing by $\Delta x \Delta z$ we get the final form of the equation as

$$\frac{\partial(b\tau_{xx})}{\partial x}+\frac{\partial(b\tau_{zx})}{\partial z}$$

For turbulent flow the shear stresses can be written in terms of eddy viscosities and velocity gradients as

$$\tau_{xx}=\rho K_x^M\left(\frac{\partial u}{\partial x}\right)\text{and}\,\tau_{zx}=\rho K_z^M\left(\frac{\partial u}{\partial z}\right)$$

where K_x^M, K_z^M - longitudinal and vertical eddy viscosities respectively.

The final form of the equation for forces due to shear stresses is written as

$$\frac{\partial}{\partial x}\left(\rho bK_x^M\frac{\partial u}{\partial x}\right)+\frac{\partial}{\partial z}\left(\rho bK_z^M\frac{\partial u}{\partial z}\right)$$

Now we derive the inertial terms as follows.

Rate of momentum entering into the elemental volume due to the lateral flow:

Rate of x-component of momentum entering at $x = \rho ub\big|_x \Delta z u_t$

Rate of x-component of momentum leaving the elemental volume at $x+\Delta x = \rho ub\big|_{x+\Delta x} \Delta z u_t$

Rate of x-component of momentum entering the elemental volume at $z= \rho wb\big|_z \Delta x u_t$

Rate of x-component of momentum leaving the elemental volume at $z+\Delta z = \rho wb\big|_{z+\Delta z} \Delta x u_t$

where u_t is the lateral inflow velocity component parallel to the longitudinal direction.

Following similar steps used in the derivation of the x-momentum equation, the differential form of the rate of change of momentum due to lateral flow velocity is given by

$$\left[\frac{\partial}{\partial x}(ub)+\frac{\partial}{\partial z}(wb)\right]\rho u_t = \rho u_t q_t$$

By substituting various force terms in the momentum conservation equation, we get finally

$$\rho b\frac{\partial u}{\partial t}+\rho b\left(u\frac{\partial u}{\partial x}+w\frac{\partial u}{\partial z}\right)+\rho u q_t-\rho u_t q_t=-\frac{\partial(bp)}{\partial x}+\frac{\partial}{\partial x}\left(\rho b K_x^M\frac{\partial u}{\partial x}\right)+\frac{\partial}{\partial z}\left(\rho b K_z^M\frac{\partial u}{\partial z}\right)$$

Dividing the above equation throughout by ρb the final form of the x-momentum equation is obtained as follows:

$$\frac{\partial u}{\partial t}+u\frac{\partial u}{\partial x}+w\frac{\partial u}{\partial z}=-\frac{1}{\rho b}\frac{\partial(bp)}{\partial x}+\frac{1}{b}\frac{\partial}{\partial x}\left(bK_x^M\frac{\partial u}{\partial x}\right)+\frac{1}{b}\frac{\partial}{\partial z}\left(bK_z^M\frac{\partial u}{\partial z}\right)+\frac{q_t}{b}(u_t-u) \quad (2)$$

4.4 Energy equation

The energy equation is derived based on the conservation of energy principle as postulated by the First law of Thermodynamics. The First law for an open system like the one shown in Fig.2b states that

$$\frac{dE_{cv}}{dt}=\frac{dQ_{cv}}{dt}-\frac{dW_{cv}}{dt}+\dot{m}e_{out}-\dot{m}e_{in}$$

which means that

$$\begin{pmatrix}\text{rate of storage of}\\ \text{internal energy in}\\ \text{the elemental volume}\end{pmatrix}=\begin{pmatrix}\text{rate of heat}\\ \text{transfer into the}\\ \text{elemental volume}\end{pmatrix}-\begin{pmatrix}\text{rate of work}\\ \text{done within}\\ \text{the elemental volume}\end{pmatrix}$$
$$+\begin{pmatrix}\text{net efflux of}\\ \text{energy from the}\\ \text{elemental volume}\end{pmatrix}+\begin{pmatrix}\text{rate of energy}\\ \text{input due to}\\ \text{lateral flow}\end{pmatrix}$$

Rate of internal energy storage within the elemental volume= $\frac{\partial}{\partial t}(\rho e)b\Delta x\Delta z$

Total internal energy leaving the elemental volume is given by

$$(\rho eub)\Big|_{x+\Delta x}\Delta z+(\rho ewb)\Big|_{z+\Delta z}\Delta x$$

Total internal energy entering the elemental volume is given by

$$(\rho eub)\Big|_{x}\Delta z+(\rho ewb)\Big|_{z}\Delta x$$

Hence the net efflux of energy leaving the elemental volume is obtained by dividing the differences of the above two expressions by $\Delta x\Delta z$. The final equation obtained is

$$\frac{\partial}{\partial x}(\rho eub)+\frac{\partial}{\partial z}(\rho ewb)=\rho e\left[\frac{\partial}{\partial x}(ub)+\frac{\partial}{\partial z}(wb)\right]+\rho b\left[u\frac{\partial e}{\partial x}+w\frac{\partial e}{\partial z}\right]$$

$$=\rho e\,q_t+\rho b\left[u\frac{\partial e}{\partial x}+w\frac{\partial e}{\partial z}\right]$$

(where Eq. (1) is used.)

The net rate of heat flux entering the control volume is given by

$$qb\Big|_{x+\Delta x}\Delta z-qb\Big|_{x}\Delta z+qb\Big|_{z+\Delta z}\Delta x-qb\Big|_{z}\Delta x$$

After dividing by $\Delta x\Delta z$ the differential form of the above expression is

$$\frac{\partial}{\partial x}(q_x b)+\frac{\partial}{\partial z}(q_z b)$$

where $q_x=k_x\frac{\partial T}{\partial x}$ and $q_z=k_z\frac{\partial T}{\partial z}$ and k_x,k_z are the thermal conductivities in the x and z directions respectively.

The energy input to the elemental control volume due to the lateral flow is

$$\left(\rho ub\Big|_{x+\Delta x}-\rho ub\Big|_{x}\right)\Delta z e_t+\left(\rho wb\Big|_{z+\Delta z}-\rho wb\Big|_{z}\right)\Delta x e_t$$

where e_t is energy per unit mass due to the lateral flow.

After dividing by $\Delta x\Delta z$, the differential form of the above expression is

$$\rho e_t\left[\frac{\partial}{\partial x}(ub)+\frac{\partial}{\partial z}(wb)\right]=\rho e_t q_t$$

The internal energy for an incompressible fluid is given by $e=CT$, where C=specific heat capacity and T= temperature.

By substituting the respective expressions in the energy balance equation we get

$$\rho bC\frac{\partial T}{\partial t}+\rho bC\left[u\frac{\partial T}{\partial x}+w\frac{\partial T}{\partial z}\right]+\rho CT\,q_t=\frac{\partial}{\partial x}\left(bk_x\frac{\partial T}{\partial x}\right)+\frac{\partial}{\partial z}\left(bk_z\frac{\partial T}{\partial z}\right)+\rho C\,T_t q_t$$

By dividing the above equation throughout by $b\rho C$, we get

$$\frac{\partial T}{\partial t}+u\frac{\partial T}{\partial x}+w\frac{\partial T}{\partial z}=\frac{1}{b}\frac{\partial}{\partial x}\left(b\frac{k_x}{\rho C}\frac{\partial T}{\partial x}\right)+\frac{1}{b}\frac{\partial}{\partial z}\left(b\frac{k_z}{\rho C}\frac{\partial T}{\partial z}\right)+\frac{q_t}{b}(T_t-T)$$

For turbulent flows, the thermal diffusivity terms are replaced by means of eddy diffusivity coefficients. That is

$\frac{k_x}{\rho C} = K_x^H, \frac{k_z}{\rho C} = K_z^H$ where K_x^H, K_z^H = longitudinal and vertical eddy diffusivity coefficients respectively. After substituting these coefficients, the final form of the energy equation is reduced to the following form

$$\frac{\partial T}{\partial t} + u\frac{\partial T}{\partial x} + w\frac{\partial T}{\partial z} = \frac{1}{b}\frac{\partial}{\partial x}\left(bK_x^H \frac{\partial T}{\partial x}\right) + \frac{1}{b}\frac{\partial}{\partial z}\left(bK_z^H \frac{\partial T}{\partial z}\right) + \frac{q_t}{b}(T_t - T) \qquad (3)$$

4.5 Sediment transport equation

Applying the principle of conservation of mass principle to sediment transport through the elemental control volume, we get

$$\begin{pmatrix}\text{rate of storage of}\\ \text{sediment in the}\\ \text{elemental volume}\end{pmatrix} = \begin{pmatrix}\text{rate of sediment}\\ \text{entering the}\\ \text{elemental volume}\end{pmatrix} - \begin{pmatrix}\text{rate of sediment}\\ \text{leaving the}\\ \text{elemental volume}\end{pmatrix}$$
$$+ \begin{pmatrix}\text{diffusive flux of}\\ \text{sediment entering into}\\ \text{the elemental volume}\end{pmatrix} + \begin{pmatrix}\text{rate of sediment}\\ \text{input due to}\\ \text{lateral flow}\end{pmatrix}$$

Rate of storage of sediment in the elemental volume = $\left(\frac{\partial C}{\partial t}\right) b\Delta x\Delta z$

The net efflux of sediment from the elemental volume is given by

$$\left(ucb\big|_{x+\Delta x}\Delta z - ucb\big|_x \Delta z\right) + \left(wcb\big|_{z+\Delta z}\Delta x - wcb\big|_z \Delta x\right)$$

After dividing by $\Delta x\Delta z$ we get the differential form of the above equation as

$$\frac{\partial}{\partial x}(uCb) + \frac{\partial}{\partial z}(wCb) = \left[ub\frac{\partial C}{\partial x} + wb\frac{\partial C}{\partial z}\right] + C\left[\frac{\partial}{\partial x}(ub) + \frac{\partial}{\partial z}(wb)\right]$$
$$= b\left(u\frac{\partial C}{\partial x} + w\frac{\partial C}{\partial z}\right) + Cq_t$$

(where Eq. (1) is used.)

Net rate of diffusive sediment transport into the elemental volume is given by

$$\dot{m}b\Big|_{x+\Delta x}\Delta z - \dot{m}b\Big|_x \Delta z + \dot{m}b\Big|_{z+\Delta z}\Delta x - \dot{m}b\Big|_z \Delta x$$

After dividing the above by $\Delta x\Delta z$, we get the differential form of the equation as

$$\frac{\partial}{\partial x}\left(\dot{m}_x b\right) + \frac{\partial}{\partial z}\left(\dot{m}_z b\right)$$

Applying the Fick's law of diffusion, the above equation can be rewritten as

$$\frac{\partial}{\partial x}\left(bD_x \frac{\partial C}{\partial x}\right) + \frac{\partial}{\partial z}\left(bD_z \frac{\partial C}{\partial z}\right)$$

where the mass flux $\dot{m}_x = D\dfrac{\partial C}{\partial x}$ and $\dot{m}_z = D\dfrac{\partial C}{\partial z}$.

It can be shown that the sediment inflow due to lateral flow is given by the following expression using similar steps for the derivation of lateral energy flow in the energy equation.

$C_t\left[\dfrac{\partial}{\partial x}(ub)+\dfrac{\partial}{\partial z}(wb)\right] = C_t q_t$ where C_t = the concentration of the sediment due to the lateral flow.

After substituting the respective expressions in the sediment conservation equation, we get the following expression

$$b\frac{\partial C}{\partial t}+b\left(u\frac{\partial C}{\partial x}+w\frac{\partial C}{\partial z}\right)+Cq_t - C_t q_t = \frac{\partial}{\partial x}\left(bD_x\frac{\partial C}{\partial x}\right)+\frac{\partial}{\partial z}\left(bD_z\frac{\partial C}{\partial z}\right)$$

For turbulent flow, the diffusion coefficients are replaced by turbulent eddy diffusion coefficients as was done in the derivation of the momentum and energy equations. With this substitution and dividing the above equation throughout by 'b', the final form of the sediment transport equation is obtained as

$$\frac{\partial C}{\partial t}+u\frac{\partial C}{\partial x}+w\frac{\partial C}{\partial z} = \frac{1}{b}\frac{\partial}{\partial x}\left(bK_x^C\frac{\partial C}{\partial x}\right)+\frac{1}{b}\frac{\partial}{\partial z}\left(bK_z^C\frac{\partial C}{\partial z}\right)+\frac{q_t}{b}(C_t - C) \qquad (4)$$

where C = concentration of suspended sediment; q_t = lateral inflow per unit area; C_t = lateral inflow concentration of suspended sediment; K_x^C, K_z^C = longitudinal and vertical eddy diffusivity for sediment transport, respectively.

5. FINITE ELEMENT FORMULATION

The mass, momentum, energy and sediment transport equations can be solved using Galerkin's weighted residual method which is widely used for field problems. Considering a triangular element, any field variable can be written in terms of its nodal values and the respective shape functions as follows:

$$\phi = N_i\phi_i + N_j\phi_j + N_k\phi_k$$

The above expression represents the form of solution assumed for the variable. When this solution is substituted into the differential equation, a residue will be obtained because the assumed solution is not the exact solution. The principle of weighted residual method is that the given differential field equation multiplied by a weighting function will be integrated (in the case of Galerkin's method, the shape function itself is the weighting function) throughout the computational domain in order to make the residue minimum. The resultant integrated equations will yield a number of simultaneous algebraic equations. The simultaneous equations may be linear or non-linear depending upon the type of governing equations. Details about the application of FEM for the solution of the governing equations pertaining to lake hydrodynamics are discussed by Gallagher et al. (1973), Gallagher and Chan (1973), King (1982, 1982a & 1982b), Lin (1992), Lin and Young (1995), Young (1980 & 1999), Young et al. (1976a & 1976b), Young and Liggett (1977), and Young and Lin (1993). Standard textbooks on finite element method by Gallagher (1975), Segerlind (1984) or Zinkiewicz (1983 & 1977) can also be referred.

6. AN EXAMPLE PROBLEM

The governing equations developed for the study of lake hydrodynamics will be applied to a specific problem in order to understand the physics of the phenomenon. Details about the application of the mathematical model are discussed in the following sections.

6.1 Numerical scheme for moving boundary problems

Simulation of free surface flows is an interesting and challenging subject in the realm of computational hydraulics. The boundary conditions on the free surface or moving surface are not known *a priori* and hence it is difficult to numerically model such flow situations using conventional numerical schemes. In general, there are three approaches to deal with the dynamics of free surface or moving boundary flow problems, namely the Lagrangian, Eulerian, and arbitrary Lagrangian-Eulerian (ALE) methods (Young and Lin, 1993).

In the purely Eulerian description (Harlow and Welch, 1965) the mesh is fixed in space, and is very difficult to treat the free surfaces or moving boundaries with such fixed mesh. On the other hand, a purely Lagrangian coordinate system follows the movement of the fluid particle and hence it is better for dealing with the free surface flows by a moving mesh. However, the mesh can get distorted very easily and hence this scheme is not convenient to model the free surface problems. In contrast, arbitrary Lagrangian-Eulerian (ALE) is based on the arbitrary movement of the reference frame, which is continuously rezoned in order to allow a precise description of the moving interfaces. Noh (1964) and Hirt et al. (1974) developed the ALE scheme for FDM and later Donea et al. (1975 & 1982) and Hughes et al. (1981) developed it for FEM.

6.2 Justification for two-dimensional model

The lake or reservoir hydrodynamics has to be analyzed using three-dimensional models in order to obtain realistic predictions for the natural environments. Further the numerical predictions have to be obtained over a period of a year or more. But the cost of three-dimensional computation is too expensive to be used for multiple year computations. Therefore, to simplify the problem, some appropriate assumptions are employed. For lakes or reservoirs with the characteristics of relatively narrow width and long reach, a laterally averaged two-dimensional model in the longitudinal-vertical profile is a good choice (Street, et al., 1977, Edinger and Buchak, 1979, Buchak and Edinger, 1982, King, 1985a, Karpik and Raithby, 1990). Because the most important features of the stratified lakes or reservoirs, such as the circulation gyres, density currents, as well as the thermal structures like thermocline process, will be retained in the longitudinal-vertical plane (Young et al., 1976a & 1976b, and Meyer, 1986). Other depth-averaged horizontal two-dimensional model can be applied to the shallow homogenous lakes or reservoirs. The laterally averaged two dimensional, longitudinal-vertical model will automatically rule out the consideration of the Coriolis forces for large lakes or reservoirs, since the Coriolis forces are relevant only in three dimensional modeling (Pedlosky, 1982). Furthermore, since there is no vortex stretching in two-dimensional flows, there is no need to include the Coriolis force effect into the governing equations.

6.3 Turbulence closure model

It is visualized that flows occurring in lakes or reservoirs are highly turbulent. One of the most difficulties in modeling the stratified lakes or reservoirs is the choice of turbulence closure model. There are different types of turbulence closure models such as the Direct Numerical Simulation (DNS), Large Eddy Simulation (LES) or higher order turbulence closure models, such as the $k-\varepsilon$ models (Rodi, 1980) available in literature, for developing flows and thermal structures and its associated transport processes. However, for the sake of economical computations and practical engineering applications, in this example, the semi-empirical formulation of turbulence closure models will be used. By the consideration of the influence of stability parameter, such as of the Richardson number, Munk and Anderson (1948) and Mamayev (1957) were able to propose their semi-empirical formulae of the vertical turbulent mixing for stratified flows.

6.4 Modified governing equations for moving boundary problem

The governing equations for the study of stratified lakes or reservoirs are the conservation laws of the fluid mass, linear momentum, and transport processes of heat and suspended sediment derived in section 4 and the equation of state, to link the constitutive relationship between density, temperature and sediment concentration. The equations become complete with the introduction of turbulence closure model. The continuity equation remains unaltered. Since the numerical model has to take care of the moving boundaries, mesh velocities in x and z directions are introduced in the convective terms of the momentum, energy and sediment transport equations from the ALE considerations. Due to the settling velocity of the sediments, the vertical velocity term in the convective part of the sediment transport equation will have an additional velocity term. With these modifications, the governing equations developed in section 4 are modified as follows (Lin, 1992):

a) Continuity equation (local)

$$\frac{\partial}{\partial x}(ub)+\frac{\partial}{\partial z}(wb)=q_t \tag{1}$$

b) Longitudinal momentum equation

$$\frac{\partial u}{\partial t}+(u-\tilde{u})\frac{\partial u}{\partial x}+(w-\tilde{w})\frac{\partial u}{\partial z}=-\frac{1}{b}\frac{1}{\rho}\frac{\partial(bp)}{\partial x}$$
$$+\frac{1}{b}\frac{\partial}{\partial x}(bK_x^M\frac{\partial u}{\partial x})+\frac{1}{b}\frac{\partial}{\partial z}(bK_z^M\frac{\partial u}{\partial z})+\frac{q_t(u_t-u)}{b} \quad (2a)$$

c) Vertical momentum equation

$$0=-\frac{1}{\rho}\frac{\partial p}{\partial z}-g \quad (2b)$$

d) Heat transport equation

$$\frac{\partial T}{\partial t}+(u-\tilde{u})\frac{\partial T}{\partial x}+(w-\tilde{w})\frac{\partial T}{\partial z}=\frac{1}{b}\frac{\partial}{\partial x}(bK_x^H\frac{\partial T}{\partial x})$$
$$+\frac{1}{b}\frac{\partial}{\partial z}(bK_z^H\frac{\partial T}{\partial z})+\frac{q_t(T_t-T)}{b} \quad (3a)$$

e) Suspended sediment transport equation

$$\frac{\partial C}{\partial t}+(u-\tilde{u})\frac{\partial C}{\partial x}+(w-w_s-\tilde{w})\frac{\partial C}{\partial z}=\frac{1}{b}\frac{\partial}{\partial x}(bK_x^C\frac{\partial C}{\partial x})$$
$$+\frac{1}{b}\frac{\partial}{\partial z}(bK_z^C\frac{\partial C}{\partial z})+\frac{q_t(C_t-C)}{b} \quad (4a)$$

f) Equation of state for pure water to determine density in terms of temperature

$$\rho_w=999.841+6.59585*10^{-2}T\text{-}8.45123*10^{-3}T^2+5.29157*10^{-5}T^3 \quad (5)$$

Ambient fluid density

$$\rho=(1\text{-}C)\,\rho_w+C\,\rho_c \quad (6)$$

g) Turbulence closure model (Munk and Anderson, 1948)

$$K_z^M=K_0^M(1+10R_i)^{-0.5} \quad (7a)$$
$$K_z^H=K_0^H(1+3.33R_i)^{-1.5} \quad (7b)$$
$$K_z^C=K_z^H \quad (7c)$$
$$R_i=\frac{-g\dfrac{\partial\rho}{\partial z}}{\rho(\dfrac{\partial u}{\partial x})^2} \quad (7d)$$

h) Mesh velocity

$$\tilde{u}=\frac{dx}{dt}\quad\text{and}\quad\tilde{w}=\frac{dz}{dt} \quad (8)$$

where ρ_w = pure water density; ρ_c = sediment density; K_0^M, K_0^H = eddy viscosity and diffusivity for homogeneous lake calculated using empirical relations, as functions of velocity of water within the eddies, the sizes of the eddies, wind shear and size of lakes; $\tilde{u},\tilde{w}$= mesh velocity in the ALE system for x and z direction respectively; w_s = settling velocity of suspended sediment; Ri is the Richardson number for the consideration of density stratification, Ri=0 in a homogeneous environment.

6.4.1 Velocity Boundary conditions

The above governing equations need appropriate initial and boundary conditions for obtaining the numerical solutions. For the lake hydrodynamic simulation, the initial conditions (IC) are unavailable in general. A more practical and feasible way is to execute a start-up run from any accepted initial conditions, generally from a motionless and homogeneous case or even from a potential flow condition. The associated boundary conditions (BC) are listed as follows:

1. The kinematic free surface BC

$$\frac{\partial z_\infty}{\partial t} + (u_\infty - \tilde{u}_\infty)\frac{\partial z_\infty}{\partial x} - w_\infty = 0 \tag{9}$$

2. The dynamic free surface BC

$$\tau_\infty = \Im \rho_a U_{10}|U_{10}| \tag{10}$$

3. At the bottom of lake or reservoir

$$\frac{\partial z_0}{\partial t} - \tilde{u}_0 \frac{\partial z_0}{\partial x} = 0 \tag{11}$$

4. At the solid boundaries (no-slip BC)

$$u = u_f \ , \quad w = w_f \tag{12}$$

where τ_∞ = wind shear stress on the free surface; ρ_a= air density; $\Im$ = an empirical constant = 0.0026 in this study; U_{10} = wind speed at 10 m above the water surface; u_f, w_f = the specified boundary velocity in x and z direction respectively, such as the inflow/outflow velocity; the subscript '0' stands for the physical properties associated with the bottom of the lake and the subscript '∞' stands for the physical properties with respect to the free surface.

6.4.2 Temperature and sediment boundary conditions

As far as the boundary conditions of temperature and sediment concentration are concerned, the most commonly used boundary conditions are, the Dirichlet BC of specified temperature or concentration, or the Neumann BC of specified heat flux or flux of sediment concentration. On the free surfaces, the heat transfer on the air-water interface is specified by the Newton's law of cooling which represents the net heat flux between the water surface and atmosphere, as

$$q_\infty = h(T_e - T_\infty) \tag{13}$$

where q_∞ = heat flux on the free surface; T_e = the equilibrium temperature; T_∞ = surface water temperature; h = convective heat transfer coefficient.

6.4.3 Calculation of equilibrium temperature

The equilibrium temperature, T_e, is a calculated, hypothetical surface water temperature obtained by assuming a balance between the incoming and outgoing heat flux across the air-water interface. Under normal conditions at which the surface water temperature is higher than the equilibrium temperature, the lake or reservoir dissipates excess heat into the atmosphere and acts as a cooling body of water. However, during calm and hot summer days when the equilibrium temperature is higher than the surface temperature, the lake or reservoir gains heat from the atmosphere. Initially, the water at the surface of the lake will warm up and eventually the thermal stratification is formed in due course.

The heat transfer coefficient h depends on the thermo-physical properties of water at the surface of a lake. Since the properties of water will vary with temperature, h is evaluated by two different empirical formulae, one proposed by Rimsha and Dochenko (1958) for surface water temperature below 41° F (5 ° C) and the other proposed by and Brady et al. (1969), for surface water temperature above 41° F (5° C) and they are as given below:

The formula proposed by Rimsha and Dochenko (1958), applicable for $T_\infty <$ 41° F (5° C):

$$h = 2.32 + \{1.56[8.0 + 0.35(T_\infty - T_a)] + 6.08U_2\}[\beta + 0.61] \tag{14a}$$

where

$\beta = 0.612 - 0.0204T_m + 0.00049T_m^2$ and $T_m = (T_a + T_\infty)/2$

In the above, T_a is the air temperature and U_2 is the wind speed at 2 meter above the water surface. The formula proposed by Brady et al. (1969), applicable for $T_\infty \geq$ 41° F (5° C):

$$h = 15.7 + (\beta + 0.26)(70 + 0.7U_2^2) \tag{14b}$$

where $\beta = 0.255 - 0.085T_m + 0.000204T_m^2$

The equilibrium temperature, T_e, can be obtained using the following expressing.

$$T_e = T_d + \frac{I}{h} \tag{14c}$$

where I = Solar radiation and T_d = dew point temperature.

6.4.4 Hydrostatic pressure distribution

The assumption of hydrostatic pressure distribution enables us to integrate the vertical momentum Eq. (2b) to get the pressure field as:

$$p(x,z,t) = p_\infty(x,t) + \rho g(z_\infty - z) + \hat{p}(x,z,t) \tag{15}$$

$$\hat{p} = \int_z^{z_\infty} [\rho(x,z,t) - \rho_r] g dz$$

where p_∞ = pressure on the water surface and is included for the consideration of the pressure depression on the water surface; ρ_r = the reference fluid density; $\hat{p}$ = perturbed pressure due to the density stratification and becomes equal to zero in a homogeneous environment. It is also related to the index for the measure of the potential energy associated with the stability of density stratification. The first two terms on the right hand side of Eq. (15) are called the barotropic or external pressure, which is governed by the atmospheric pressure and the positions of free surface and fluid. On the other hand, the integral term is named the baroclinic or internal pressure, which is dependent upon the internal density distribution by stratification.

Substituting the pressure gradient of $\partial p/\partial x$ from Eq. (15) into the horizontal momentum Eq. (2a), we get

$$\frac{\partial u}{\partial t}+(u-\tilde{u})\frac{\partial u}{\partial x}+(w-\tilde{w})\frac{\partial u}{\partial z}=-g(\frac{\partial H}{\partial x}+S_0)-\frac{1}{b\rho}\frac{\partial(bp_\infty)}{\partial x}-\frac{1}{b\rho}\frac{\partial(b\hat{p}_\infty)}{\partial x}+\frac{1}{b}\frac{\partial}{\partial x}(bK_x^M\frac{\partial u}{\partial x})$$

$$+\frac{1}{b}\frac{\partial}{\partial z}(bK_z^M\frac{\partial u}{\partial z})+\frac{q_l(u_l-u)}{b} \quad (16)$$

where $H = z_\infty - z_0$ = total water depth; $S_0 = \partial z_0/\partial x$ = bottom slope.

6.4.5 Relation between water depth equation and continuity equation

In order to link the calculation of the water depth H and also to filter out the instability caused by the gravity wave on the water surface, the local continuity Eq. (1) together with the kinematic BC on the free surface of Eq. (9) and lake bottom of Eq. (11), are used to yield the following water depth equation..

$$\frac{\partial H}{\partial t}+(\frac{\overline{ub}}{b_\infty}-\tilde{\bar{u}})\frac{\partial H}{\partial x}+\frac{H}{b_\infty}\frac{\partial(\overline{ub})}{\partial x}=\frac{q_l}{b_\infty} \quad (17)$$

in which, $\overline{ub}=\frac{1}{H}\int_{z_0}^{z_\infty}ubdz$ and $\tilde{\bar{u}}=\frac{1}{H}\int_{z_0}^{z_\infty}\tilde{u}dz$, where b_∞ = surface width; $\overline{ub}$ = average lateral velocity flux; $\tilde{\bar{u}}$ = average mesh velocity. It is noted that Eq. (17) is a one-dimensional equation in *x*, since the *z* direction has already been integrated from the lake bottom to the surface.

Finally, to assure a better imposition of boundary condition for the vertical velocity, *w*, we have to differentiate the local continuity Eq. (1) with respect to *z* and get the following differentiated form for the continuity equation (King, 1985a):

$$\frac{\partial^2(ub)}{\partial x\partial z}+\frac{\partial^2(wb)}{\partial z^2}=\frac{\partial q_l}{\partial z} \quad (18)$$

In summary, Eqs (15), (16), (17), (18), (3a), (4a), (5) and (6) are the final governing equations to be solved for the unknown variables *u, w, H* or z_∞ , ρ, *T, C*, and *p*.

6.5 Numerical scheme

The arbitrary Lagrangian-Eulerian (ALE) finite element method (FEM) is employed to discretize the space domain for obtaining the solution of the nonlinear partial differential governing Eqs (15-18) and (3a-6). To avoid the time step constraint of numerical instability, or the so-called Courant-Friedrichs-Levy

(CFL) condition, an implicit finite difference method (FDM) is used for the time and finite element method (FEM) is used for the space domain, for solving the one-dimensional water depth Eq. (17). On the other hand, to achieve the most economic computation, other governing equations used to solve the longitudinal velocity, temperature, and sediment concentration Eqs (16), (3a) and (4a) are all discretized by the explicit finite difference scheme in time domain and by finite element scheme in space domain. The existence of free surface not only gives complicated boundary condition, but also restricts the severe time step used for the numerical computation. By this implicitness in one dimension and explicitness in two dimensions, we have circumvented the difficulty of the excessive computation. The advantage of this modeling strategy will become more obvious when a three dimensional modeling is undertaken.

To further increase the accuracy, consistency and stability of the numerical solutions, the concept of the so-called artificial balancing tensor diffusivity (BTD) is introduced in this study (Dukowica and Ramshaw, 1979). The objective of introducing the BTD model is to overcome the negative numerical damping caused by the advection dominated flows when the Gelerkin FEM is adopted (Gresho et al., 1984). As mentioned before, the variations of lake or reservoir storage and the corresponding fluctuation of free surface as well as the deposition/erosion of the lake or reservoir bottom, are easily taken care by the ALE moving boundary technique.

In the computational aspects of hydrodynamic processes of stratified lakes or reservoirs, in order to avoid the numerical instability caused by the internal gravity wave, the procedures used by Sun (1980) are adopted in the present study. Namely, the temperature and sediment concentrations are computed first, after that, velocity field and water depth or free surface are obtained. It is worthwhile to observe that the matrix equations for T, C, and u are all time-invariant as far as the matrix inversion is concerned. In addition, they all have the same property of symmetric matrix equation solver, except the calculation of water depth, H, which is nonlinear and needs Newton-Raphson iteration technique. All other variables are uncoupled and can be calculated in a similar way.

The final governing equations after time discretization are solved using Galerkin's weighted residual method of FEM. The resultant simultaneous algebraic equations are solved using available standard routines to obtain the variations of temperature, velocities, concentration, density and free surface, in space and time.

6.6 Model verification

The mathematical model developed has to be verified for accuracy so that it can be employed for hydrodynamic analysis of real lakes. The U.S. Army Corps of Engineers has carried out a number of experiments on the Generalized Reservoir Hydrodynamics flume, which is designed to test the behaviors of reservoir hydrodynamics for gravitational underflow. The reservoir data generated in this flume have been used to verify a number of models (Johnson, 1980 &

1981, King, 1982, and Karpik and Raithby, 1990). Hence the two-dimensional model developed in the present work is used to simulate the Generalized Reservoir Hydrodynamics flume experiments for the verification of the model.

The length of the flume is 24.39 m and it varies in depth from 0.3 m at its inflow end to 0.91 m at its outflow end. It has two different cross-sections; the upper section 1, which is 6.1 m long, has a horizontal bottom and linear variation in width from 0.3 m to 0.91 m; the lower section 2 has a constant width of 0.91 m, and a bottom dropping of 0.61 m over its 18.2 m length. Refer Figs. 3a & 3b for the general layout and Fig. 3c for the associated FEM mesh, which contains 612 linear triangular elements and 342 nodes in the longitudinal-vertical profile.

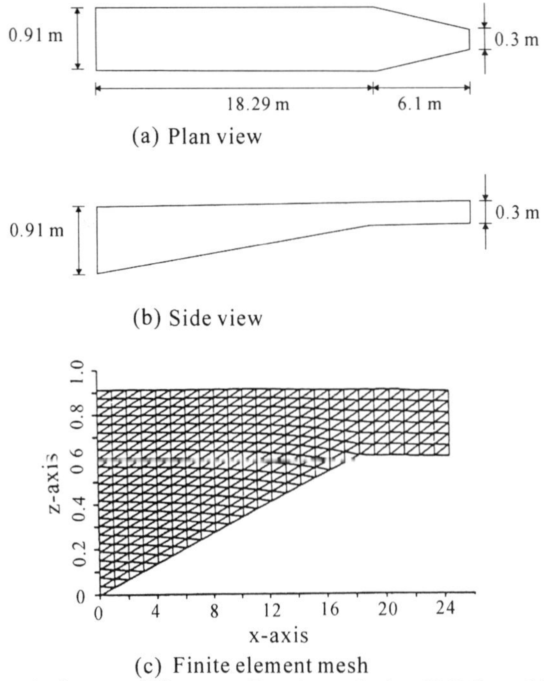

Fig. 3. Schematic diagram of Waterways Experiment Station GRH flume (a) plan view, (b) side view and (c) the finite element mesh.

In order to simulate the underflow type of density current due to the inflow of water at 16.67 C, the flume is filled with water at a uniform temperature of 21.44°C. The water is introduced under a baffle and allowed to flow into the system over the lower 0.15 m region. The outflow of water is removed from the outlet located on the dam 0.15 m from the bottom. The inflow rate of water is the same as the outflow at 0.00063 m^3/s.

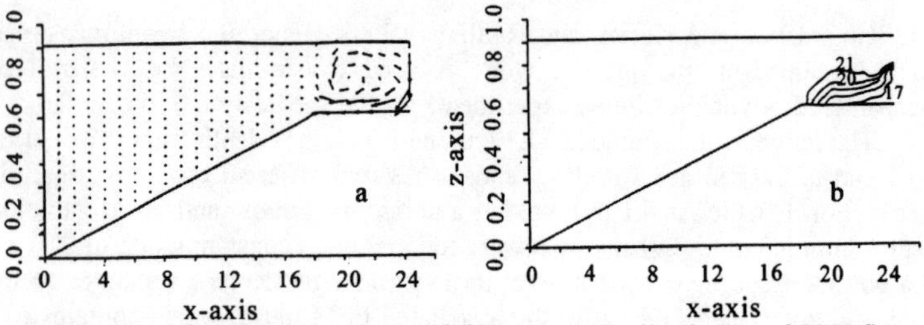

Fig. 4. Structures of (a) density current, and (b) temperature distribution of GRH flume at t=4 min.

Because the inflow temperature is lower than the ambient temperature, the underflow type of density current is expected from simulation. The velocity and temperature fields obtained by numerical simulations at time t=4 min. are shown in Figs. 4a & 4b, respectively. It is striking that the velocity and recirculation gyre in the inflow are established quickly, as compared with the other homogenous region where the velocity is very low. Thus the model is capable of predicting the establishment of a density underflow. As time goes by, the density current advances along the bottom with the recirculation gyre moving rapidly towards the dam.

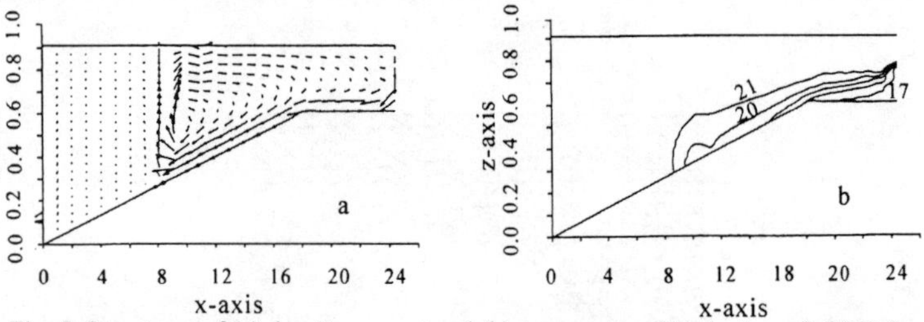

Fig. 5. Structures of (a) density current and (b) temperature distribution of GRH flume at t=14 min.

Figure 5 depicts the transient processes of the structures of density current at t=14 min. Because the vertical turbulent mixing is reduced by the buoyancy effect, the temperatures vary only in the region near the bottom. After the density current arrives at the dam at t=50 min, the flow field becomes more stable and the recirculation gyre occupies the entire region of the flume, as illustrated in Fig. 6.

Figure 7 shows the comparison between the numerically calculated and experimental results of the progress of the propagation speed of the density current front. It is interesting to note that the simulated and observed results are very consistent. The numerical results show much smoother variation than the laboratory measurements. However the general trend shows good agreement even for very small velocity in the order of 10^{-2} m/s.

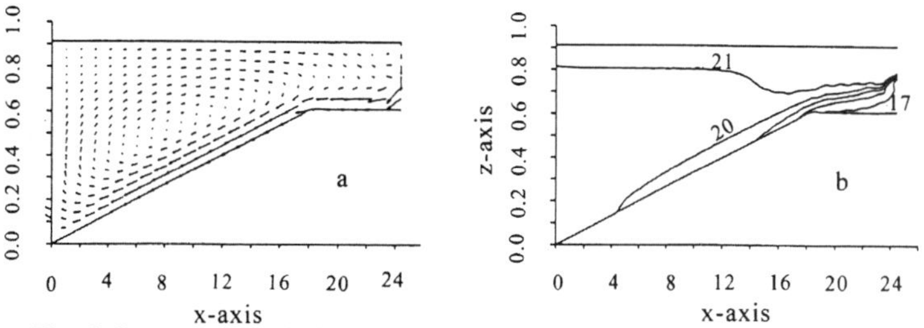

Fig. 6. Structures of (a) density current and (b) temperature distribution of GRH flume at t=50 min.

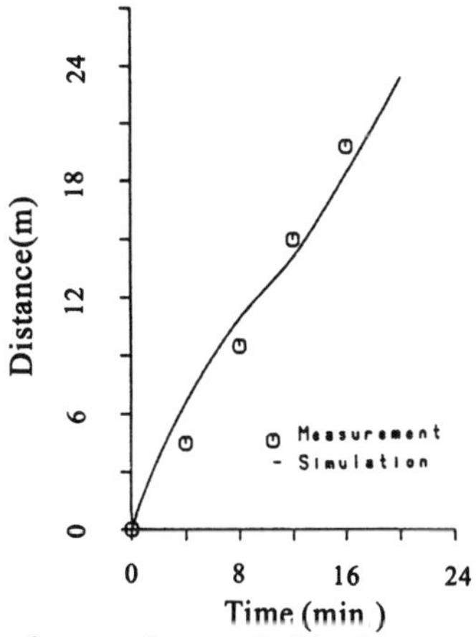

Fig. 7. Comparison of propagation speed of density current front of GRH flume.

6.7 Model application to Te-Chi reservoir

In this section we will present the application of the model to a real reservoir problem.

6.7.1 Description of the Te-Chi reservoir

The usage of measured field data is to test and verify the merit and credibility of the present model to the practical engineering applications. After having the model calibrated and verified with the laboratory data, now the model can be used for the simulations of hydrothermal characteristics of real reservoirs. The Te-Chi reservoir, located at central Taiwan is chosen as a pilot test site. The Te-Chi reservoir surrounded by high mountainous basins in an isle is a very long, narrow and deep reservoir, as shown in Fig. 8a. Due to these topographic characteristics, it is found to be an ideal reservoir suitable for a laterally averaged two-dimensional stratified lake model simulation. Its longitudinal length is about 13 km, with a major inflow coming from the upstream of the Ta-Chia River and nine other river tributes. These rivers have been considered as the minor lateral inflow sources. Figure 8b shows the longitudinal-vertical profile and its variation of lateral width. The bottom of the dam is located at 1268 m above the sea level and the maximum water depth is 140 m, with a total storage capacity of $2.5x10^8 m^3$. The intake of selective withdrawal for water utilities is located at a height of 75 m above the bottom of the dam. Numerical simulations for a total period of two years, from 1984 to 1985, are performed and compared with the measured field data.

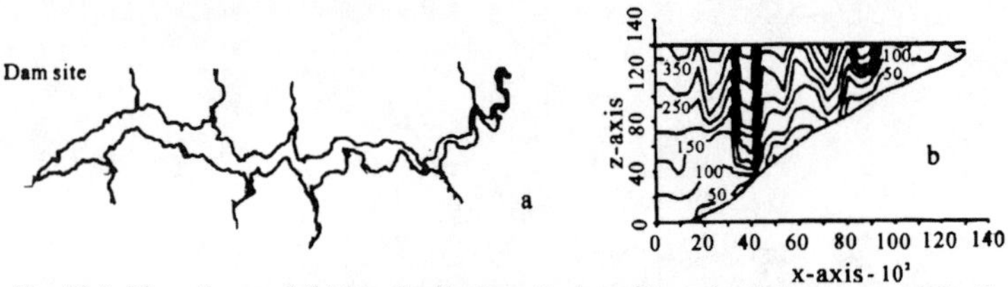

Fig. 8(a). Plan view and (b) longitudinal-vertical profile and width contour of the Te-Chi reservoir.

6.7.2 Hydrological input data

The hydrological input data required for the numerical simulation of the 1984 annual cycle of the reservoir are: the rate of major and minor lateral inflows, outflow rate, and the corresponding inflow temperatures and sediment concentrations. The annual inflow into the reservoir and outflow from the reservoir for the year 1984 are depicted in Figs. 9a and 9b respectively. In dry seasons, that is, in autumn and winter seasons, the inflow rate is very small. On the other hand in spring and summer seasons due to the storms or typhoon rainfalls, the inflow rate is very high. The maximum inflow of $300 m^3/s$ occurs at 156th day of 1984 (January 1 corresponds to day 1 of the year) when the reservoir was assaulted by a typhoon storm of nine days from 150th to 158th day. The watershed lies in the area of vast high mountains, which lie 3000 m above sea level. The surface temperature, inflow temperature, as well as intake temperature for the year 1984 are depicted in Fig. 9c. It is interesting to observe that the inflow temperature is always lower than the ambient surface temperature of the reservoir, since the Te-Chi reservoir is
located on a high mountain area. Furthermore, by a regression analysis from the measured data, the sediment inflow concentration has been correlated to the inflow discharge by an exponential function, which will provide the boundary condition for the sediment concentration equation.

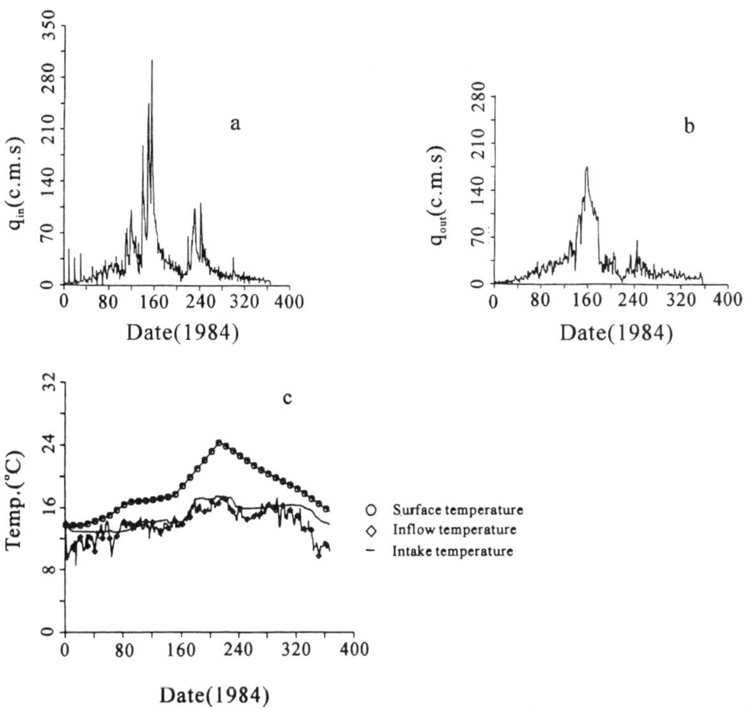

Fig. 9 (a). Inflow, (b) outflow and (c) surface temperature, inflow temperature and intake temperature in 1984 of the Te-Chi reservoir.

6.7.3 Meteorological input data

In addition to the hydrological data, the meteorological data such as the cloud cover, dry bulb temperature, dew point temperature, wind speed, wind direction and solar radiation are also required as input data. The heat flux on the water surface can be obtained by using a thermal equilibrium model from the meteorological data. This model is conceptually based on Newton's law of cooling, assuming convective heat transfer at the water surface. This law states that the convective heat transfer at a surface is the product of a heat transfer coefficient and the difference between the surface water temperature and the equilibrium temperature obtained from Eq. (13). The heat flux received by the Te-Chi reservoir for the year 1984 is calculated using the equilibrium temperature model. The heat flux thus calculated is used as the boundary condition at the water surface for the solution of the heat transport equation. Finally, the wind shear stress at the water surface, which is one of the boundary conditions for the momentum equation, is computed from the wind speed by an empirical formula given by Eq. (10).

6.7.4 Computational domain of the Te-Chi reservoir

The computational domain of the Te-Chi reservoir is discretized into 950 linear triangular elements with 522 nodes. The finite element mesh will move along with the variations of reservoir storage including the free surfaces, with the usage of the ALE method. The types of meshes used at the highest and lowest water levels are shown in Figs. 10a & 10b, respectively.

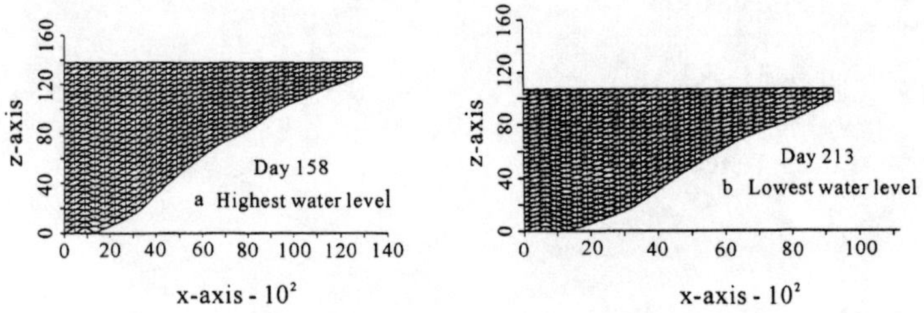

Fig. 10. Moving adaptive finite element grid system of (a) highest water level and (b) lowest water level in 1984 of the Te-Chi reservoir.

6.7.5 Results and discussion

Figure 11 shows the comparison between the variations of water levels calculated using the present numerical model and the measured values for the year 1984. The calculated water level is very consistent with the observed one, thus the mass conservation in this model is very well proved. The selective comparisons of the calculated and measured daily water temperature profiles at different cross sections for the year 1984 during the months of January, February, March, April, May and September are illustrated in Figs. 12 & 13. The matching between the calculated and the measured results is excellent except for the month of May, where some small deviations are observed. This discrepancy occurs both at the layers of epilimnion and thermocline and more at the layer of epilimnion.

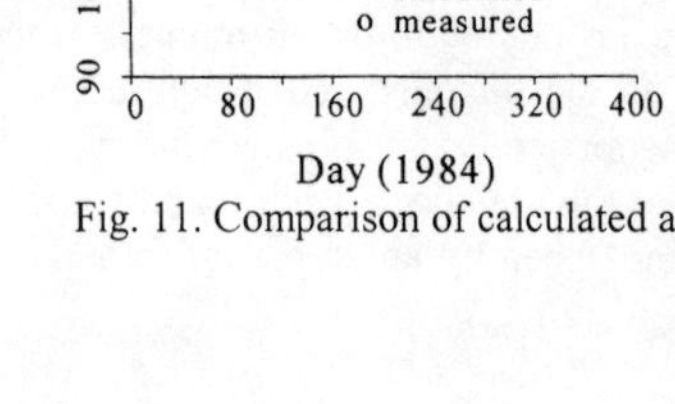

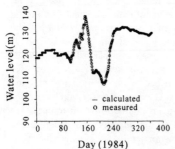

Fig. 11. Comparison of calculated and measured water level in 1984 of the Te-Chi reservoir.

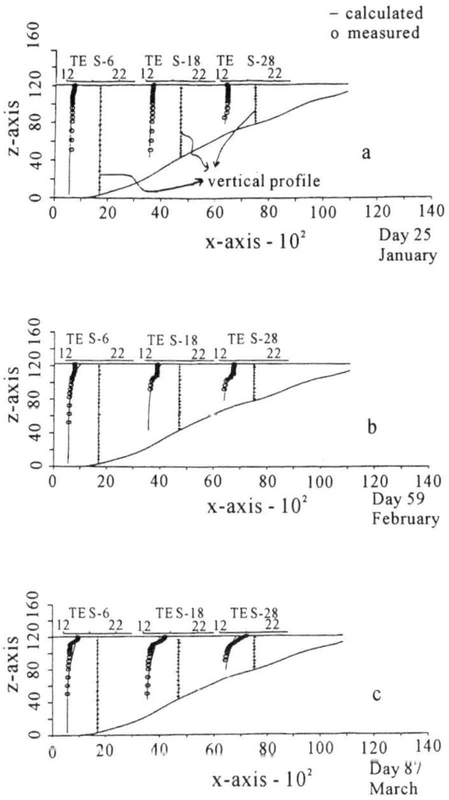

Fig. 12. Comparison of calculated and measured daily temperature profiles at a particular day during the months of (a) January, (b) February and (c) March in 1984 of the Te-Chi reservoir.

As can be observed from Fig. 9c, the inflow temperature of water into the reservoir is always lower than that of the reservoir for the year 1984. Hence an underflow type of density current is expected. Later, the layer of underflow is transformed into turbidity current, due to the presence of suspended sediments in the inflow. When the inflow rate is very small, the strength of turbidity current is very weak. However, while the inflow is very high, as occurring in the event of rainfall during storms, the turbidity current becomes very strong. The heavy rainfalls due to storms occurred from 150th day to 158th day for the year 1984. Figure 14 shows the density currents, temperature and sediment concentration distributions for the 150th day of the year. For the 150th day as depicted in Fig. 14b, the highest temperature is about 18°C located at the water surface and the

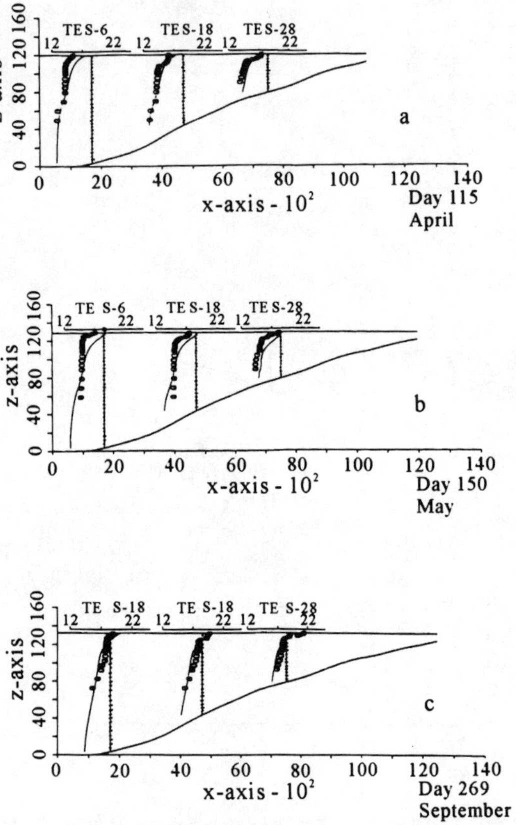

Fig. 13. Comparison of calculated and measured daily temperature profiles at a particular day during the months of (a) April, (b) May and (c) September in 1984 of the Te-Chi reservoir.

lowest one is about 12°C occurring near the bottom of the dam. The river inflow temperature is about 15° C which lies between the temperature of the epilimnion and the hypolimnion of the reservoir.

The sediment concentration of the river inflow is about 300 ppm. This results in a higher concentration range near the bed of the reservoir about 6 km away from the dam. The other sediment concentrations are lower than 50 ppm as shown in Fig. 14c. It is worth noting that the boundary condition of complete deposition at the reservoir bed has been used for the sediment transport equation. That is, for any sediment particle being carried into the reservoir, it will deposit at the bed with its settling velocity whenever it touches the bed and does not have any opportunity to be picked up again. Therefore, the sediment concentration decreases along the downstream direction. Due to the higher concentration and lower temperature of river inflows, the density of river inflows is higher than the density of upper layer of reservoir water body. Hence, buoyancy induced plunging phenomenon occurs after the river inflow comes into the reservoir and the underflow density current travels very fast downstream along the bed. Due to the sediment deposit at the bed, the concentration decreases along the downstream direction and the strength of underflow also decreases. Moreover, the selective withdrawal and the type of temperature distribution will affect the density current also. Therefore, the density current leaves the bed about 6 km away from the dam, and then the interflow occurs and advances horizontally to the intake of the selective withdrawal. Besides, due to the underflow and interflow of density currents, there are two circulation gyres generated, in which the upper circulation gyre is stronger than the lower one, as indicated in Fig. 14a.

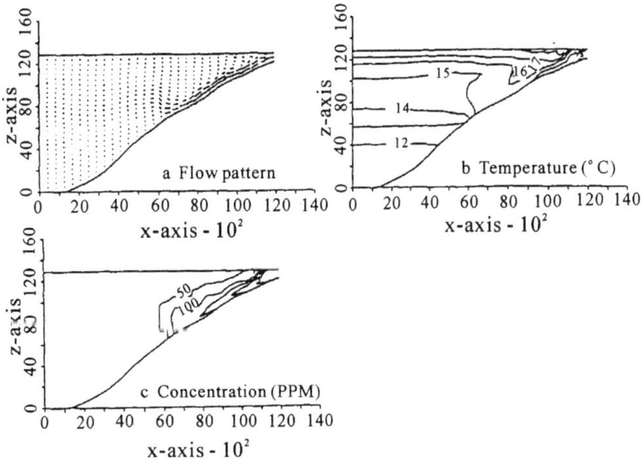

Fig. 14. Structures of (a) density current, (b) temperature, and (c) sediment concentration distributions on 150th day in 1984 of the Te-Chi reservoir.

7. PERSPECTIVE RESEARCH AREAS IN LAKE HYDRODYNAMICS

Real lakes or reservoirs do not satisfy the conditions of laterally averaged two-dimensional models. Hence three-dimensional lake or reservoir model is inevitable when we consider lakes or reservoirs with similar longitudinal and lateral scales, or large lakes or reservoirs, such as the Great Lakes in the northern America. In such models the Coriolis forces have to be included. Also a three-dimensional stratified lake or reservoir model (Simons, 1973) is generally desirable when water quality modeling is concerned.

7.1 Three-dimensional modeling

As far as the sophistication of the level of three-dimensional modeling is concerned, there are two types of models, the 3D Ekman hydrostatic model and the complete Navier-Stokes model. In the 3D Ekman hydrostatic model, the nonlinear inertial force terms as well as the horizontal diffusion terms are considered as secondary effects and are neglected in the model development (Young et al., 2000a). Boundary element method has also been used to reduce the dimension of the three-dimensional stratified flow modeling. For more details, the work of Young et al. (2000a) can be referred. Another alternative is the extension of the present model to the 3D stratified lake or reservoir model. Of course, the inclusion of Coriolis force in the horizontal momentum equation is necessary. In the mean time, one more horizontal momentum equation has to be added. Three-dimensional modeling is triggered by the need for more accurate predictions of water quality constituents and a better representation of the higher order hydrodynamic processes.

7.2 Improved version of turbulence closure models

In the complete Navier-Stokes 3D stratified lake or reservoir model, the primitive variable approach using the velocity and pressure as major unknowns, is commonly used. The semi-empirical formula or higher order parameter like $k-\varepsilon$ model (Spraggs and Street, 1975, and King, 1985b) can be used for the turbulence closure model. However, it is observed that due to the advancement of modeling techniques and computer capacity, more comprehensive turbulence closure models such as DNS or at least LES are becoming more popular and feasible nowadays.

7.3 Velocity-Vorticity formulation

The concept of using velocity and vorticity as primary variables or the so-called velocity-vorticity formulation has been proposed by the author as one of the methods to analyze lake hydrodynamics and some promising programs are underway. For further details the work of Young (1999) and Young et al. (2000b) can be referred. Both finite element and boundary element methods are employed to perform the three-dimensional homogeneous, as well as stratified flow modeling by the velocity-vorticity formulation.

Compared to the traditional 3D primitive variables, the velocity-vorticity formulation allows for an easier three-dimensional modeling. Initially, the Poisson type of equations are considered as the governing equations for the velocity field and once the vorticity field is known, the same equations can used for both homogeneous and stratified environments. The velocity field is governed by the following Poisson velocity vector equation, which is derived by taking the curl operator of the definition of the vorticity through the usage of the fluid incompressibility.

$\nabla^2\vec{u} = -\vec{\nabla} \times \vec{\omega}$
where $\vec{\omega} = \vec{\nabla} \times \vec{u}$ = vorticity.

The vorticity transport equations are obtained by taking the curl operator of the momentum equation of the Navier-Stokes equations and are given as below:

$$\frac{\partial\vec{\omega}}{\partial t} + [(\vec{u} - \bar{\vec{u}}) \cdot \vec{\nabla}]\vec{\omega} = \nu\nabla^2\vec{\omega} + \left[\left(\vec{\omega} + \Omega\,\vec{k}\right) \cdot \vec{\nabla}\right](\vec{u} - \bar{\vec{u}}) - \vec{\nabla} \times \left(\frac{\rho}{\rho_r} g\vec{k}\right)$$

where Ω = earth rotational speed. The vorticity transport equation provides a clear and thorough picture about the generation, advection, diffusion and dissipation of all the vorticity in the stratified environments. For example, the Coriolis force effect, $\Omega\,\vec{k}$ and the density stratification effect $\vec{\nabla} \times \left(\frac{\rho}{\rho_r} g\vec{k}\right)$ can be easily compared with the homogeneous and the non-rotating systems. Therefore, a better dynamic feature can be obtained from this approach. Finally, if the DNS model can be achieved as turbulence closure model, there is no need to provide those turbulent exchange coefficients and DNS will render all the necessary information for the stratified lakes or reservoirs. Even if we cannot afford a DNS model, a LES model may not be a bad substitute at the present stage. It is hoped that in the near future, the DNS model will become available, to be incorporated in a complete 3D stratified lake or reservoir model. With such models, the assumptions and approximations to be made will be reduced and a much better understanding of the hydrodynamics of 3D stratified lakes or reservoirs can be achieved, for both academic research and engineering applications.

Acknowledgements

The author would like to express his gratitude to the National Science Council of the Republic of China for their long-term supports of the research grants to investigate the hydrodynamics of stratified bodies of water. It is greatly appreciated. Thanks are also extended to Dr. Q. H. Lin for his excellent computer programming during this investigation and Dr. K. Murugesan for his contribution on major works of proof reading. It is the writer's privilege to be a visiting scholar during his sabbatical leave at the Department of Civil Engineering, the University of Wales, Swansea. Professor K. Morgan and Dr. O. Hassan, the Civil Engineering Department and the University provided very good academic environments to the author to complete this chapter. It is also acknowledged. Finally the author would like to thank our editors, in particular Professors H. Shen and A. Cheng for their patience and encouragement during this arduous study. Without their warmest supports and understanding, this chapter would never have been completed.

REFERENCES

Alavian, V .G. H., Jirka, R. A., Denton, M. C., Johnson, M.C., and Stefan, H.G. (1992) "Density currents entering lakes and reservoirs," *J. Hyd. Engg.*, ASCE, Vol. 118, pp. 1464-1489.

Bauer, S. W. (1978) "Three-dimensional simulation of time-dependent elevations and currents in a homogeneous lake of arbitrary shape using an irregular-grid finite-difference model," *Proc. of Symposium on Hydrodynamics of lakes,* pp. 51-63.

Brady, D., Graves, W. L., and Geyer, J. C. (1969) " Water heat exchange at power plant cooling lakes", *Report No. 5, Res. Proj. RP 49-5*, The Johns Hopkins University, Baltimore, MD, USA.

Buchak, E. M. and Edinger, J.E. (1982) User guide for LARM 2: A longitudinal-vertical time-varying hydrodynamic reservoir model, *Instruction Report E-82-3*, U.S. Army Corps of Engineers, Waterways Experiment Station, Vicksburg, Miss,USA.

Cheng, R. T. S. (1977) "Survey of numerical models for wind-driven lake circulation", *ASCE Fall Convention and Exhibit*, San Francisco, CA, USA.

Chung, S. W. and Gu, R. (1998) "Two-dimensional simulation of contaminant currents in stratified reservoir," *J. Hyd. Engg.,* ASCE, Vol. 124, pp. 704-711.

Csanady, G. T. (1975) "Hydrodynamics of large lakes," *Annual Review of Fluid Mechanics*, Vol. 7, pp. 357-385.

Donea, J., Fasolli-Stella, P. and Giuliani, S. (1977) "Lagrangian-Eulerian finite element technique for transient fluid structure interaction problems," *Trans. of the 4th Int. Conf. On Structural Mech. in Reactor Technology*, Paper B ½.

Donea, J., Giuliani, S. and Halleux, J. P. (1982) "An arbitrary Lagrangian-Eulerian finite element method for transient dynamic fluid-structure interaction," *Comp. Meth. in Appl. Mech. Eng.*, Vol. 33, pp. 689-729.

Dukowica, J. and Ramshaw, J. (1979) "Tensor Viscosity method for convection in numerical fluid dynamics," *J. Comp. Physics*, Vol. 32, pp. 71-91.

Edinger, J. R. and Buchak, E.M. (1979) "A Hydrodynamics two-dimensional reservoir model: Development and test application to Sutton Reservoir, Elk River, West Virginia," Prepared for U.S. Army Corps of Engineers, Ohio River Division, USA.

Gallagher, R.H., Liggett, J.A. and Chan, S.T.K. (1973) " Finite element shallow lake circulation analysis," *J. Hyd. Div.,* ASCE, Vol. 99, pp. 1083-1098.

Gallagher, R. H. and Chan, S. T. K. (1973) "Higher order finite element analysis of lake circulation," *Computers and fluids,* Vol. 1, pp. 119-132.

Gallagher, R. H. (1975) *Finite Element Analysis: Fundamentals*, Prentice-Hall, New York, USA.

Graf, W. H. and Mortimer, C. H. (Editors) (1979) *Lake Hydrodynamics*, Elsevier, Amsterdam, Netherlands.

Gray, D. D. and Girgini, A. (1976) "The validity of the Boussinesq approximation for liquids and gases," *J. Heat. Mass Transfer,* Vol. 9, pp. 545-551.

Gresho, P. M., Chan, S. T, Lee, R. L. and Upson, C. D. (1984) "A modified finite element method for solving the time-dependent, incompressible Navier-Stokes equations, Part I: Theory", *Int. J. Numer. Meth. Fluids*, Vol. 4, pp. 557-598.

Harleman, D. (1961) "Stratified Flows", *Handbook of Fluid Dynamics*, V. K. Streeter (Ed. in-chief), McGraw-Hill, London, England, UK, Sec. 26.1-26.18.

Herleman, D. R. F. (1982) "Hydrothermal analysis of lakes and reservoirs," J. *Hyd. Div.,* ASCE, Vol. 108, pp. 302-325.

Hansen, N. E. O. (1975) "Effect of wind stress on stratified deep lake", *J. Hyd. Div.,* ASCE, Vol. 101, pp. 1037-1051.

Hansen, N. E. O. (1978) "Effects of boundary layers on mixing in small lakes," *Proc. Of Symposium on Hydrodynamics of lakes,* pp. 341-356.

Harlows, F. H. and Welch, J.E. (1965) "Numerical calculation of time-dependent viscous incompressible flow of fluid with a free surface," *Phys. Fluids*, Vol. 12, pp. 2182-2189.

Hirt, C. W., Cook, J. L. and Bulter, T. (1970) "A Lagrangian method for caculating the dynamics of an incompressible fluid with free surface," *J. Comp. Phys.*, Vol. 5, pp. 103-124.

Hirt, C. W., Amsden, A. A. and Cook, J. L. (1974) "An arbitrary Lagrangian-Eulerian computing method for all flow speeds," *J. Comp. Phys.*, Vol. 14, pp. 227-253.

Hughes, T. J. R., Liu, W. K. and Zimmerman, T.K. (1981) "Lagrangian-Eulerian finite element formulation for incompressible viscous flows," *Comp. Meth. Appl. Mech. Eng.*, Vol. 29, pp. 329-349.

Hutchinson, G. E. (1957) *A Treatise on Limnology*, Vol. 1, *Geography, Physics and Chemistry,* Wiley, New York, USA.

Hutter, K. (Editor.), (1984), *Hydrodynamics of Lakes*, CZSM-Lecture, Springer Verlag, Vienna-New.York.

Hutter, K. (Editor), (1987), "Hydrodynamics modeling of lakes", in Chapter 22, Vol. 6, *Encyclopedia of Fluid Mechanics*, Golf Publishing Co., Houston, Texas, USA.

Imberger, J., Palterson, J., Hebbert, B. and Loh, I. (1978), "Dynamics of reservoirs of medium size," *J. Hyd. Div.,* ASCE, Vol. 104, pp. 725-743.

Imberger, J. and Patterson, J. C. (1981) "A dynamic reservoir simulation model-DYRESM: 5," *Proc. of Symposium on Transport models for inland and coastal waters*, Fischer, H. B. Editor, pp. 310-360.

Johnson, B. H. (1980) "A review of multidimensional reservoir hydrodynamic modeling," *Proc. of the Symp. on Surface Water Impoundments*, Stefan, H. F. Editor, ASCE, Vol. 1., pp. 497-507.

Johnson, B. H. (1981) "*A review of numerical reservoir hydrodynamic modeling*", U.S. Army Corps of Engineers, Waterways Experiment Station, Vicks bury, Miss, USA.

Kacimov, A. R. (2000) " Three-dimensional groundwater flow to a lake: an explicit analytical solution," *J. Hydrology*, Vol. 240, pp. 80-89.

Karpik, S. R. and Raithby, G.D. (1990) "Laterally averaged hydrodynamics model for reservoir predictions," *J. of Hydr. Eng.,* ASCE, Vol.116, pp. 783-798.

Kawahara, T. (1977) "A theoretical analysis of the thermally stratified layer in the lake",in *'Heat transfer and turbulent buoyant convection',* Spalding,D. B. & Afghan, N. Editors, Vol. 1, Hemisphere Pub. Co., pp. 93-103,

King, I. P, (1982) "A three dimensional finite element model for stratified flow," *Proc. of 4th Int. Symp. on Finite Elements in Fluids*, Tokyo, Japan, pp.513-520.

King, I. P. (1985a) "Finite element modeling of stratified flow in estuaries and reservoirs", *Int. J. Num. Meth. Fluids*, Vol. 5, pp. 943-955.

King, I. P. (1985b) "Strategies for finite element modeling of three dimensional hydrodynamic systems," *Adv. Water Resources*, Vol. 8, pp. 69-76.

Lee, H.Y. and Yu, W. S. (1997) "Experimental study of reservoir turbidity current," *J. Hyd. Engg.*, ASCE, Vol. 123, pp. 520-528.

Lin, Q. H. (1992) *Finite element analysis of hydrodynamic flows with moving boundaries*, thesis presented to the National Taiwan University, at Taipei, Taiwan, in partial fulfillment of the requirements for the degree of Doctor of Philosophy.

Lin, Q. H. and Young, D. L. (1995) "Finite element method for vertical 2-D Navier-Stokes equations with moving boundaries," *J. of Chinese Soil and Water Conservation,* Vol. 26, pp. 79-88.

Mamayer, O. I. (1957) "The influence of stratification on vertical turbulence mixing in the sea", *Izv. Geophy*, Ser. Vol. 1,pp. 870-875.

Markofsky, M. (1978) "Density induced transport processes in lakes and reservoirs," *Proc. of Symposium on Hydrodynamics of lakes*, pp. 99-109.

Meyer, Z. (1986) "Vertical circulation in density-stratified reservoirs," Chapter. 23, Vol. 2, *Encyclopedia of Fluid Mechanics*, Gulf Publishing Co., Houston, Texas, USA.

Mortimer, C. H. (1974) Lake Hydrodynamics, *Mitt. Internat. Verein Limnol.* Vol. 20, pp. 124 -197.

Mortimer, C. H. (1984) "Hydrodynamics of lakes", *CSIM lectures,* Hutter , K. Editor, pp. 287-319.

Munk, W. and Anderson, E. R. (1948) "Notes on a theory of the thermocline", *J. of Marine Research*, Vol. 7, pp. 276-295.

Noh, W. F. (1964) "CEL: A time-dependent, two-space-dimensional, coupled Eulerian-Lagrangian code," in *Methods in Computational Physics,* Vol. 3, Alder, B.J., Fernbach, S. and Rotenburg, M. Editors, Academic Press, New York, pp. 117-179.

Pedlosky, J. (1982) *Geophysical Fluid Dynamics*, Springer Verlag, Berlin-Heiderburg-New York.

Plate, E. J. and Wengefeld, P. (1978) "Exchange processes at the water surface" *Proc. of Symposium on Hydrodynamics of lakes*, pp. 114 -141.

Ramming, S. G. (1978) "The dynamics of shallow lakes subject to wind-an application to lake Neusiedl, Austria", *Proc. of Symposium of Hydrodynamics of lakes,* pp. 65-75.

Rimsha, V. A. and Rochenko, R. V. (1958) "*Investigations of heat loss from water surfaces in winter time* (in Russia), Summarized in Pally, Mecagno and Kennedy, water regime surfaces heat loss from heated stream, Tech Report No. 155, Inst. of Hydr. Res., University of Iowa, USA.

Rodi, W, (1980) *Turbulent Models and their Application in Hydraulics*, IAHR Publication, Delft, Netherlands.

Salmon, J. R. S., Liggett, J. A. and Gallagher, R. H. (1980) "Dispersion analysis in homogeneous lakes", *Int. J. Num. Methods. Engg.*, Vol. 15, pp. 1627-1642.

Segerlind, L. (1984) *Finite element analysis*, John Wiley, Singapore.

Simons, T. J. (1973) *Development of three-dimensional numerical methods of the Great Lakes*, Scientific Series No. 12, Canada Center for Inland Waters, Burlington, Ontario, Canada.

Simpson,J. E. (1995) *Gravity Currents*, John Wilcy, London, UK.

Spalding, D. B. and Svensson, U. (1977) "Erosion of thermocline' in *'Heat transfer and turbulent buoyant convection'*, Edited by D.B. Spalding & N. Afghan, Vol. 1, pp. 113-122, Hemisphere Pub. Co.

Spraggs, L. D. and Street, R. L. (1975) "*Three-dimensional simulation of thermally-influenced hydrodynamic flows,*" Report TR-190, Dept. of Civil Eng., Stanford University, Stanford, CA, USA.

Stoker, J. J., (1957), *Water Waves*, Interscience, New York, USA.

Street, R. L., Roberts, B. R. and Spraggs, L. D. (1977) "Simulation of thermally- influenced hydrodynamic flows," ASCE *Fall Convention and Exhibit*, San Francisco, CA., USA.

Sun, W. Y. (1980) "A forward-backward time integration scheme to treat internal gravity waves," *Monthly Weather Review*, Vol. 108, pp. 402-407.

Sundermann, J. (1978) "Numerical modeling of circulation in lakes," *Proc. of Symposium on Hydrodynamics of lakes,* pp. 30-67.

Thacker, W.C. (1977) " Irregular grid finite difference techniques: simulations of oscillations in shallow circular basins," *J. Physical Oceanography*, Vol. 7, pp. 284-292.

Turner, J. S. (1973) *Buoyancy Effects in Fluids*, Cambridge University Press, Cambridge, UK.

Verevochkin, Y. G. and Startsev, S. A. (2000) "Numerical simulation of convection and heat transfer in water absorbing solar radiation," *J. Fluid Mech.,* Vol. 421, pp. 293-305.

Winter, T. C. (1999) "Relation of streams, lakes and wetlands to groundwater flow systems", *Hydrogeology Journal*, Vol. 7, pp. 28-45.

Wittmiss, J. (1978) "Applications of a transient mathematical model to lake Kosen," *Proc. Of Symposium on Hydrodynamics of lakes*, pp. 31-40.

Yih, C. S. (1980) *Stratified Flows,* Academic Press, New York, London

Young, D. L. (1980) "Hydrothermal modeling by the finite element method", *Proc. of International Conf. On Water resources development*, Taipei, Taiwan, ROC, Vol. 1, pp. 283-300.

Young, D. L. (1999) "Nonlinear 3-D mathematical modeling of Pollutant Transport in Estuaries by the method of velocity-vorticity formulation," *Proc. of the 21st Nat. Coastal Eng. Conf.*, Hsin-Chu, Taiwan, pp. 224-231.

Young, D. L. Liggett, J. A. and Gallagher, R. H. (1976a) "Steady stratified circulation in a cavity," *J.Eng. Mech. Div.,* ASCE,Vol. 102, pp. 1-17.

Young, D. L., Liggett, J. A. and Gallagher, R. H. (1976b) "Unsteady stratified circulation in a cavity," *J. Eng. Mech.Div.,* ASCE,Vol. 102, pp. 1009-1023.

Young, D. L. and Liggett,J. A. (1977) " Transient finite element shallow lake circulation", *J. Hyd. Div.,*ASCE, Vol. 103, pp. 109-121.

Young, D. L. and Lin, O. H. (1993) "Free surface flow simulation by the arbitrary Lagrangian-Eulerian finite element method", in *Advances in Hydroscience and Engineering,*Wang, S.S.Y. Editor, Vol. 1, pp. 1148-1153.

Young, D. L. Her, B. C. and Eldho,T. I. (2000a) "Boundary integral modeling of three-dimensional pollutant transport in stratified estuaries," *J. of Engineering Mechanics,* ASCE ,Vol.126, pp. 1083-1092.

Young, D. L., Liu, Y. H. and Eldho, T. I. (2000b) "A combined BEM-FEM model for the velocity-vorticity formulation of Navier-Stokes equations in three-dimensions", *Int. J. of Eng. Ana. With Boundary Elements*, Vol. 24, pp. 307-316.

Zienckiewicz, O. C. and Morgan, K. (1983) *Finite Elements and Approximation,* John Wiley, New York, USA.

Zienckiewicz, O. C. (1977) *Finite element analysis,* John Wiley, New York, USA.

Chapter 11

WATER QUALITY MODELING IN RESERVOIRS

Jan-Tai Kuo[1] and Ming-Der Yang[2]

ABSTRACT

The purpose of this chapter is to review the progress of water quality modeling in reservoirs and its use for eutrophication management. Commonly used models including zero-dimensional (Vollenweider's type) models and multi-dimensional models such as CE-QUAL-W2 and WASP5 are introduced. Case studies on application of those models to reservoirs in Taiwan are demonstrated.

Traditional water quality sampling, which is time-consuming and expensive, can only provide point-basis water quality condition as initial data for water modeling. Remote sensing provides an efficient tool to acquire water quality data periodically for a large area. In this chapter, remotely sensed data (SPOT satellite imagery) were adopted in a water quality model for model calibration. Also, Geographic Information System (GIS) provides a useful platform to store water quality data and to present water quality variations. Water quality modeling, remote sensing, and GIS were integrated and applied on water quality simulation and forecasting for reservoirs in Taiwan.

1. INTRODUCTION

High concentration of phytoplankton in a reservoir is related to the eutrophication process and is mainly a result from point source and nonpoint source pollution due to human activities of discharging municipal, industrial and agricultural waste into the reservoir. An abundance of algae biomass is detrimental to the water environment and causes many problems such as diurnal oxygen variation, taste and odor in water supply, filter clogging in water treatment plants, and damage to water-contact sports and recreation. A water quality model can be used to explore the dynamics of the eutrophication process in an interactive aquatic ecosystem and to improve the engineering aspects of water

[1]Professor, Dept. of Civil Engineering, and Director, Hydrotech Research Institute, National Taiwan University, 1 Sec. 4, Roosevelt Rd., Taipei, Taiwan, kuoj@ccms.ntu.edu.tw
[2]Assistant Professor, Dept. of Civil Engineering, National Chung-Hsing University, 250 Kuo-Kuang Rd., Taichung 402, Taiwan, mdyang@dragon.nchu.edu.tw

quality prediction and management. Additional information introductions regarding the lake eutrophication can be found in Somlyódy and Straten (1986), Thomann and Mueller (1987), and Stefan (1994).

Distribution of concentration of constituents is a reservoir is influenced by hydrodynamic transport (advection and turbulent diffusion) and biochemical kinetics. The hydrodynamics is driven by many factors, such as winds, currents, topography, density distribution, inflow and outflow. The biochemical kinetics in composed of settling, interactive reactions with other constituents, light effect, and other physiochemical processes.

To assess the importance of hydrodynamic and biochemical processes on the distribution of phytoplankton, one has to examine the temporal or spatial scales of interest. For instance, if we wanted to study the diurnal change of phytoplankton biomass, the variation of solar radiation over a day could be a very important factor. If we were interested in the seasonal distribution of phytoplankton, then the seasonal cycle of temperature in the reservoir would be an important factor.

Di Toro et al. (1971) published a paper on modeling phytoplankton in the Sacramento-San Joaquin Delta and made a significant contribution to the engineering application of eutrophication studies. Prior to 1980, most water quality models for lakes and reservoirs considered the equations of mass conservation for constituents only. The inputs of flow regime and turbulent diffusion to a water quality model were based on observed data. Data of conservative substances (e.g. chloride) or heat/temperature distribution were used for model calibration. A modeler adjusted the assumed hydrodynamic regime until the agreement between the observed data and calculated results are favorable (Thomann et al., 1979).

Kuo and Thomann (1983) carried out modeling of three-dimensional, seasonal, phytoplankton in the embayments of Lakes by coupling a finite-element circulation model to a finite-segment water quality model. Today, as a result of advancement of technology and computers, coupling of hydrodynamic computation with water quality modeling becomes an easy task. For example, WASP5 (Ambrose et al., 1991) contains a hydrodynamic subprogram, DYNHYD5, which solves the one-dimensional continuity and momentum equations for a branching or a channel-junction (link-node) network. CE-QUAL-W2 (Cole and Buchak, 1995) is a two-dimensional, laterally averaged, hydrodynamic and water quality model which can simulate hydrodynamic transport, temperature and 21 constituents in a two-dimensional water body. Also, MIT research program made a great effort on formulating deterministic lake water quality models with specific attention to the interaction between hydrodynamics and water quality processes (Shanahan, 1991).

This chapter reviews the commonly used models for reservoir and lake water quality management. Those models range from simplified zero-dimensional (Vollenweider's type) models to two-and three-dimensional models. Case studies for Feitsui Reservoir using simplified models and a two-dimensional model, CE-QUAL-W2, are demonstrated. Readers can refer to many other good references for more information on lake and reservoir water quality modeling (Orlob, 1981 and 1983; Watanabe et al., 1983: Thomann and Mueller, 1987; Stefan, 1994; Chapra, 1997), reservoir limnology (Thornton et al., 1990), hydrodynamic modeling (Gray, 1986), decision support techniques (Henderson-Sellers and Davies, 1991), and general introduction (Baher, 1996). In addition, concepts and utilities of ecological models were introduced by Chen (1970 and 1975). Biological methods to control excessive aquatic growths were described by Karenkel (1980) and Yang (1998). A hydrophysical model and an ecological model were coupled as a lake eutrophication model by Shananhan and Harleman (1986).

Water quality simulation requires enough field data that are usually available only if a comprehensive water quality sampling program has been conducted in advance. On-site water quality assessment is expensive and time-consuming. Thus, the number of samples often could not be large enough to cover the entire water body that caused a difficulty to perform a highly frequent sampling program and to provide data set for model calibration and verification. The lack of field data becomes a barrier for making water quality forecasts more feasible in water quality management. Moreover, simulated results were usually displayed in a numerical format or X-Y plots without an association with geographic information. In this chapter, an approach of using the remote sensing technique and Geographic Information System (GIS) are introduced for resolving the problems encountered in the inputs and outputs of water quality modeling.

2. GOVERING EQUATIONS

The spatial and temporal variations of constituents in lakes and reservoirs are influenced by complex hydrodynamic transport and biochemical processes (kinetics). The hydrodynamic transport includes two components, i.e. advection and turbulent diffusion. Advective transport represents the organized movement of water, and turbulent diffusion is caused by the disorganized movement and variance of turbulence. The circulation of water (or the hydrodynamics) in lakes is driven by the combined influence of factors such as wind, lake topography, coriolis force, friction, and density distribution. The biochemical or physical kinetics of phytoplankton in lakes is composed of setting, interactive reactions with other constituents, light effect, and other physicochemical processes.

General Equations

The general hydrodynamic equations of lake circulation for the incompressible and turbulent flow can be represented as:

Conservation of Momentum:

$$\frac{\partial u}{\partial t}+u\frac{\partial u}{\partial x}+v\frac{\partial u}{\partial y}+w\frac{\partial u}{\partial z}-fv=-\frac{1}{\rho_0}\frac{\partial p}{\partial x}+\frac{\partial}{\partial x}\left(\eta_{xx}\frac{\partial u}{\partial x}\right)+\frac{\partial}{\partial y}\left(\eta_{xy}\frac{\partial u}{\partial y}\right)+\frac{\partial}{\partial z}\left(\eta_{xz}\frac{\partial u}{\partial z}\right)$$

$$\frac{\partial v}{\partial t}+u\frac{\partial v}{\partial x}+v\frac{\partial v}{\partial y}+w\frac{\partial v}{\partial z}+fu=-\frac{1}{\rho_0}\frac{\partial p}{\partial y}+\frac{\partial}{\partial x}(n_{yx}\frac{\partial v}{\partial x})+\frac{\partial}{\partial y}(n_{yy}\frac{\partial v}{\partial y})+\frac{\partial}{\partial z}(n_{yz}\frac{\partial v}{\partial z})$$

$$\frac{\partial p}{\partial z}=-\rho g \tag{1}$$

Continuity Equation:

$$\frac{\partial u}{\partial x}+\frac{\partial v}{\partial y}+\frac{\partial w}{\partial z}=0 \tag{2}$$

Heat Conservation:

$$\rho C_p\left(\frac{\partial T}{\partial t}+u\frac{\partial T}{\partial x}+v\frac{\partial T}{\partial x}+w\frac{\partial T}{\partial z}\right)=C_\rho\left[\frac{\partial}{\partial x}\left(D_x\frac{\partial T}{\partial x}\right)+\frac{\partial}{\partial y}\left(D_y\frac{\partial T}{\partial y}\right)+\frac{\partial}{\partial z}\left(D_z\frac{\partial T}{\partial z}\right)\right] \tag{3}$$

Equation of State:

$$\rho=\rho(T) \tag{4}$$

The symbols for variables are:

u,v,w = three-dimensional velocity components, [L/T]
f = the Coriolis parameter, [rad./T]
ρ_0 = ambient or reference density, [M/L^3]
ρ = density, [M/L^3]
p = pressure, [M/L^3/T^2]
g = acceleration of gravity, [L/T^2]
T = temperature, [Θ]
Cp = specific heat, [Btu/M/T]
$n_{xx}, n_{xy}, n_{xz}, n_{yx}, n_{yy}, n_{yz}$ = components of eddy viscosity, [L^2/T]
Dx, Dy, Dz = components of eddy diffusivity, [L^2/T]

The governing equations for constituent transport in lakes can be represented by the following set of equations:

Conservation of Mass for Constituents:

$$\frac{\partial C_k}{\partial t}+u\frac{\partial C_k}{\partial x}+v\frac{\partial C_k}{\partial y}+w\frac{\partial C_k}{\partial z}=\frac{\partial}{\partial x}\left(D_x\frac{\partial C_k}{\partial x}\right)+\frac{\partial}{\partial y}\left(D_y\frac{\partial C_k}{\partial y}\right)+\frac{\partial}{\partial z}\left(Dz\frac{\partial C_k}{\partial z}\right)$$
$$\pm S_k+W_k$$

$$k = 1, 2, \ldots, N \qquad (5)$$

where

C_k = the concentration of kth constituent, [M/L^3]
S_k = kinetic interaction among all C_k, [M/L^3/T]
W_k= input of constituent C_k, [M/L^3/T]
N = total number of constituents

The boundary conditions for the above equations include the waste loadings from tributaries in a closed basin or the specified concentration of constituents at the open boundaries of an embayment.

Unknown variables in Eqs. (1) through (5) are *u*, *v*, *w*, p, T, and C_k (k=1 , 2 ,……, N); other parameters such as f , η , and D have to be specified. Hence, (N+6) dependent variables are to be solved from a system of (N+6) linear and nonlinear simultaneous partial differential equations. Theoretically, these simultaneous equations can be solved by analytical or numerical methods if complete initial and boundary conditions are given. However, it would be very difficult to solve them analytically in three-dimensional framework unless the above governing equations are further simplified.

There will be at least two shortcomings if a numerical approach of solving these (N+6) simultaneous equations is adopted. Firstly, a large number of variables, equations, and boundary conditions are involved which would make it difficult to solve. This drawback is especially true in a large and complex water environment such as that of a large lake. Secondly, the time and length scales for the hydrodynamic variables (e.g., *u*, *v*, *w*, p, and ρ) can be significantly different from those for water quality variables (e.g., phytoplankton concentration); the time step and grid size used in the numerical methods may be different. The modeling of a highly interactive aquatic ecosystem in lakes or reservoirs, a large number of biochemical variables require considerable computation. A water quality model with a comparatively large grid size is usually accurate enough to represent the averaged spatially varying concentrations of the constituents. However, a smaller length scale is needed in hydrodynamic simulation due to the importance of spatial variation of physical properties (e.g., topography) on the accuracy of current prediction. As to the time scale in the time-variable modeling, the time step is also smaller in the hydrodynamic modeling than in water quality modeling since the response of current to weather condition is rather faster when compared to the response of biochemical variables. A smaller time scale is also required in the hydrodynamic model under the consideration of numerical stability. If the same scales are used for all variables, the cost in the numerical computation would be greatly increased. In practice, the order of magnitude of the temporal and spatial scales should be chosen according to the purpose of the models, their accuracy requirements, and use of different numerical methods.

3. PHYTOPLANKTON MODEL

Phytoplankton model is the core of most lake/reservoir water quality models. Eutrophication is a result of the over-discharge of nutrients from municipal, industrial, and agricultural sources. This problem occurs when the nutrient enrichment exceeds the self-purification capacity of lakes. How to accurately delineate and appropriately predict the impact of pollution on an ecosystem is the most important issue in dealing with water quality problems. Environmental engineers have made a great deal of progress in the development of water chemistry and mathematical modeling techniques. Government have also spent great efforts in monitoring, simulating, and controlling of eutrophication for several decades. There are a number of water quality models for different considerations. Various mathematical models have been developed and applied to streams, lakes, and estuaries（e.g. Lung, 1986; Thomann and Mueller, 1987; Kuo and Wu, 1991; Kuo et al., 1994; Cole, 1994）.

The mathematical model introduced here for phytoplankton biomass analysis employs the finite segment method (essentially, the finite-difference scheme), modified from the so-called "Delaware Estuary Model" (Thomann, 1963). A simulated water body consists of a number of segments of finite fluid volume; and each segment is assumed to be completely mixed. Mass balance equations can be formulated within each fluid volume. The governing equation is given as:

$$\underbrace{V_i \frac{dC_{ki}}{dt}}_{\text{change rate}} = \underbrace{\sum_j Q_{ji} C_{kj}}_{\text{mass transport due to advection}} + \underbrace{\sum_j E_{ji}\left(C_{kj} - C_{ki}\right)}_{\text{mass exchange due to dispersion}} \pm \underbrace{\sum S_k}_{\text{biochemical kinetics}} \pm \underbrace{\sum W_k}_{\text{sources/sinks}} \qquad (6)$$

$$k = 1, 2, 3, \ldots\ldots, N$$

in which

C_{ki} = concentration of a constituent k in segment i, [M/L^3]
C_{kj} = concentration of a constituent k in adjacent segment j, [M/L^3]
V_i = segment volume, [L^3]
Q_{ji} = net advective flow rate between segment i and j, [L^3/T]
E_{ji} = bulk rate of dispersive exchange between C_{kj} and C_{ki} for all segments j adjacent to segment i, [L^3/T]
S_k = kinetic interactions among the N variables, [M/T]
W_k = direct inputs or losses of the substance k, [M/T]
N = total number of constituents

The exchange of constituent by dispersion is assumed to be directly proportional to the concentration different between any two adjacent segments.

If N constituents are considered in a water body with a number of I segments, then a total of N×I linear and nonlinear differential equations are required to be solved. The numerical integration is essential for these equations and the solution gives the temporal variation of the constituent concentrations in each spatial segment.

There are eight subsystems involved in a complete water quality model, which are: (1) phytoplankton chlorophyll, (2) herbivorous zooplankton, (3) carnivorous zooplankton, (4) non-living organic nitrogen, (5) ammonia nitrogen, (6) nitrate nitrogen, (7) non-living organic phosphorus, and (8) available inorganic phosphorus. The variation of phytoplankton biomass and other constituents in a water body depends upon physical transport phenomena as well as biochemical reactions. Phytoplankton grows as a function of water temperature, incident solar radiation, and inorganic nutrients (especially inorganic nitrogen and phosphorus). Phytoplankton is preyed upon by herbivorus zooplankton, is decayed by endogenous respiration, and eventually settles to the bottom. Herbivorous zooplankton lives on phytoplankton and is eliminated by carnivorous zooplankton predation and endogeneous respiration. The carnivorous zooplankton, in turn, is preyed upon by the next higher trophic level. The respiration rates of phytoplankton and zooplankton grazing rate are temperature dependent. Parts of the organic nitrogen and phosphorus which result from phytoplankton and zooplankton endogeneous respiration and excretion recycle to ammonia nitrogen, nitrate nitrogen, and available phosphorus that turn into nutrients for phytoplankton growth.

The governing equations for the biochemical kinetics of phytoplankton chlorophyll-a are:

$$S_1 = V_i(G_p - D_p)C_{1i} - V_i \frac{W_{ij}}{H_i} C_{1i} + V_j \frac{W_{ji}}{H_j} C_{1j}$$

$$G_p = G_{\max}(T)\left[\frac{2.718f}{K_e H_i}\left(e^{-a1} - e^{-a0}\right)\right]\left(\frac{C_{5i}+C_{6i}}{k_{mn}+C_{5i}+C_{6i}}\right)\left(\frac{C_{8i}}{k_{mp}+C_{8i}}\right)$$

$$D_p = K_1(T) + C_g(T)C_{2i}$$

$$\alpha_1 = I_a / I_s . e^{-k_e H}$$

$$\alpha_0 = I_a / I_s \qquad (7)$$

where

S_1 = sources and sinks for chlorophyll-a, [M/T]

G_p = growth rate, [1/T]

D_p = death rate, [1/T]

C_{1i}, C_{1j} = concentration of chlorophyll-a in segment i and segment j, respectively, [M/L^3]

W_{ij}, W_{ji}= sinking velocities of the phytoplankton between segments i and j, [L/T]

H_i, H_j= depth of segment i and j, respectively, [L]
V_j = volume of the segment j, [L^3]
$G_{max}(T)$ = temperature-dependent maximum growth rate, [1/T]
f = photoperiod, [L_O]
K_e = extinction coefficient, [L]
I_a = mean daily incident solar radiation, [1y./T]
I_s = optimum light intensity, [1y./T];
C_{5i} = concentration of nitrogen in segment i, [M/L^3]
C_{6i} = concentration of nitrate in segment i, [M/L^3]
K_{mn} = half-saturation constant for total inorganic nitrogen, [M/L^3]
C_{8i} = concentration of orthophosphate in segment i, [M/L^3]
K_{mp} = half-saturation constant for inorganic phosphorus, [M/L^3]
$K_1(T)$ = temperature- dependent endogenous respiration rate constant, [1/T]
$C_g(T)$ = temperature-dependent grazing rate of zooplankton biomass, [L^3/T/M carbon]
C_{2i} = concentration of herbivorous zooplankton in segment i, [M/L^3]

4. ZERO-DIMENSIONAL AND VOLLENWEIDER'S TYPE MODELS

For the purpose of simplification, a lake can be simulated using zero-dimensional models under the assumption its water body is completely mixed horizontally and vertically. Zero-dimensional models calculate major physical and chemical interactions in lakes, and provide a gross variation of nutrient concentrations. Three types of simplified nutrient mass balance models are introduced as follows.

4.1 Steady-state model with a constant volume

The simplest mass balance model for calculating nutrient concentrations in a completely mixed lake with an assumption of constant fluid volume is a Vollenweider's type model (Vollenweider, 1968; 1975); which is described as (see Fig. 1):

$$V\frac{dC}{dt} = QC_{in} + B_s - KCV - QC \tag{8}$$

where

V = the volume of the lake (assumed to be constant herein), [L^3]
C = the average nutrient concentration in the lake, [M/L^3]
T = time, [T]
Q = the inflow rate, [L^3/T]
C_{in} = the nutrient concentration of inflow, [M/L^3]
K = the net sedimentation rate of the nutrient, [1/T]

B_s = sediment release, [M/T]

Under steady-state conditions (dC/dt=0), the solution of Eq. (8) is:

$$C = \frac{W}{Q + KV} \tag{9}$$

where $W=Q_{in}C_{in}+B_s$ is the nutrient loading. This steady-state model is suitable for estimating the long-term and averaged nutrient concentration in a lake, but is not applicable to the unsteady cases where there are significant time variations of hydraulics and water quality variables.

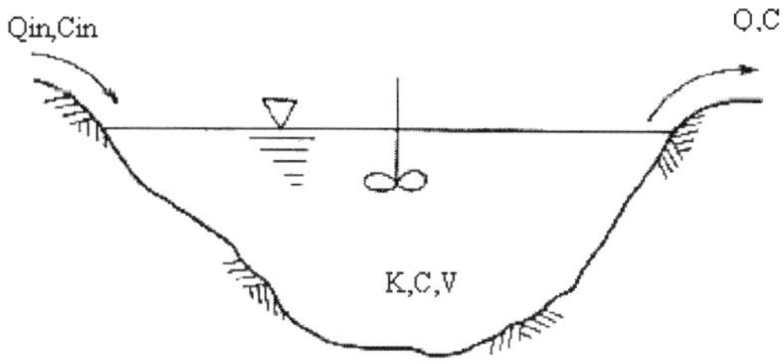

Fig.1 Notation for a completely mixed lake

4.2 Unsteady model with a constant fluid volume

The nutrient balance model of Eq. (8) can be solved analytically. The solution for an initial condition of $C=C_0$ at $t=0$ is:

$$C = \frac{W + B_s}{V(1/t_0 + K)}[1 - \exp^{-t(1/t_0+K)}] + C_0 \exp^{-t(1/t_0+K)} \tag{10}$$

where t_0 [1/T] is the lake detention time. Equation (10) can be used to calculate the temporal distribution of total phosphorus concentration in a lake subject to constant coefficients (e.g. lake volume, phosphorus loading rate, inflow, outflow, etc.) in the simulating period. Chapra (1977) used this model to calculate the distribution of total phosphorus in the Great Lakes.

4.3 Unsteady model with a time-variable volume

For lakes with significant difference between their inflow rate and outflow rate during the studying period, the lake volume will not be constant anymore. Thus, the preceding models cannot reflect the effect of variation of lake volume.

In reality, lake volume is subject to significant variation, therefore a lake model of time-variable volumes can improve the accuracy of the result. An unsteady-state model with time-variable volumes was developed which solves the following nutrient (phosphorus) mass balance equation by a power series method (Kuo and Wu, 1991):

$$\frac{d(CV)}{dt} = Q_{in}C_{in} + B_s - KCV - Q_{out}C \tag{11}$$

where V [L^3] is the lake volume and is a function of time, Q_{out} [L^3/T] is the outflow rate and is not necessarily equal to Q_{in} [L^3/T]. V can be expressed in the following form by continuity:

$$V=V_o(Q_{in}\text{-}Q_{out})t \tag{12}$$

where V_o is an initial lake volume at t=0. If Q_{in}, Q_{out}, C_{in}, B_s, and K are assumed to be constant during time interval Δt, the solution to Eq. (11) can be expressed as a power series:

$$C(t) = \sum_{N=0}^{\infty} c_n t^n \qquad \text{for} \quad 0 \leqq t \leqq \Delta t \tag{13}$$

Substituting Eqs. (12) and (13) into Eq. (11), the coefficients c_n for the power series can be expressed as:

$$c_1 = \frac{1}{V_0}[Q_{in}C_{in} + B_s - C_0(Q_{in} + KV_0)] \qquad for \quad n = 1 \tag{14}$$

$$c_n = \frac{1}{nV_0}\{-[KV_0 + nQ_{in} - (n-1)Q_{out}]c_{n-1} - K(Q_{in} - Q_{out})c_{n-2}\} \quad \text{for n} \geq 2 \tag{15}$$

where C_o is the initial condition. ΔT should be selected according to the condition of variation of hydraulics and water quality in the lake, and should not be much smaller than the detention time of the lake, due to the assumption of completely mixed lake. The selection of n depends upon t, Q_{in}, Q_{out} and V.

Although zero-dimensional models cannot predict the specific change of water quality at individual locations within lakes, they estimate the behavior and trend of such water bodies if the temporal scale of simulations is sufficiently long, such as months or years.

5. ONE-DIMENSIONAL WATER QUALITY MODELS

In reality, lakes are usually not completely mixed. Thus, one-dimensional models were developed to simulate water quality varying differently in the horizontal (X-) or vertical (Z-) direction. Recalling and rewriting Eq. (5), a general three-dimensional equation can give us a picture of a comprehensive water quality process:

$$\frac{\partial}{\partial t}(AC)=\frac{\partial}{\partial x}\left(AD_x\frac{\partial C}{\partial x}\right)+\frac{\partial}{\partial y}\left(AD_y\frac{\partial C}{\partial y}\right)+\frac{\partial}{\partial z}\left(AD_z\frac{\partial C}{\partial z}\right)$$
$$-\frac{\partial}{\partial x}(AuC)-\frac{\partial}{\partial y}(AvC)-\frac{\partial}{\partial z}(AwC)\pm AS\pm AW \qquad (16)$$

where

u,v,w = velocity of X-, Y-, Z- direction, [L/T]
A = section area, [L^2]
D_x, D_y, D_z, = turbulent diffusion coefficient, [L^2/T]
C = substance concentration, [M/ L^2]
S = sources or sinks of internal reactions, [M/ L^3/T]
W = external sources and sinks, [M/ L^3/T]

5.1 Horizontal one-dimensional water quality model

Neglecting the variation of substance concentrations over Y and Z directions in Eq. (16), the governing equation of a one-dimensional advection-dispersion mass transport model, such as QUAL2E, presents the interaction of each water quality constituent over space and time as (Brown and Barnwell, 1985):

$$\frac{\partial C}{\partial t}-\frac{\partial\left(A_x D_L\frac{\partial C}{\partial x}\right)}{A_x\partial x}\quad\frac{\partial\left(A_x\bar{u}C\right)}{A_x\partial x}+(rC+p)+\frac{W}{V} \qquad (17)$$

where,

C = the constituent concentration, [M/L^3]
A_X = cross-sectional area, [L^2]
D_L = longitudinal coefficient, [L^2/T]
$\bar{u}$ = mean velocity, [L/T]
r = growth/decay rate, [1/T]
p = internal sources and sinks, [M/T/L^3]
W = external sources or sinks, [M/T]
V = incremental volume, [L^3]

On the right-hand side of Eq. (17), the terms represent dispersion, advection, constituent changes, external sources and/or sinks, respectively. r represents a first order growth/decay rate, such as the growth rate and respiration rate for algal simulation. p represents internal sources/sinks, such as nutrient loss from algal growth and benthos sources for phosphorus and nitrogen simulation.

5.2 Vertical one-dimensional water quality model

In high-latitude area or somewhere with a high elevation, water quality of deep and wide lakes have a severe change in vertical direction rather than in horizontal direction. During a summer season, water in the surface layer is warmer than in lower layers that hinders the up-to-down transportation of water. This phenomenon, called thermal stratification, has an important effect on water quality variations. A vertical one-dimensional water quality model has to be used if the vertical variability is significant.

Considering a stratification phenomenon in reservoirs, a vertical one-dimensional water quality model focuses on the vertical variation and neglects longitudinal and lateral change. The water body is divided into several elements (or layers) in Z-direction, and complete mixing is assumed within each element. Thus, Eq. (16) can be rewritten as follows:

$$\underbrace{V\frac{\partial C}{\partial t}}_{\text{time variation}} = \underbrace{\Delta Z A D_z \frac{\partial^2 C}{\partial z^2}}_{\text{vertical dispersion}} + \underbrace{\Delta Z Q_z \frac{\partial C}{\partial z}}_{\text{circulatin}} + \underbrace{Q_{in} C_{in}}_{\text{inflow}} - \underbrace{Q_{out} C}_{\text{outflow}} \pm \underbrace{VS}_{\text{biochemica reaction}} \pm \underbrace{VW}_{\text{sources/sinks}} \tag{18}$$

where

ΔZ = thickness of layer, [L]
V = volume of layer, [L^3]
A= cross-sectional area, [L^2]
Dz = effective dispersion coefficient, [L^2/T]
Q_{jn} = inflow, [L^3/T]
Q_{out} = outflow, [L^3/T]
Q_z = advection coefficient, [L^3/T]
C_{jn} = concentration of inflow, [M/L^3]
S = sources or sinks of internal reactions, [M/ L^3/T]
W = external sources and sinks, [M/ L^3/T]

Applying numerical solution of finite difference, one can solve Eq. (18) in the form of the following equation (Corps of Engineering, 1978):

$$V_j\left(\frac{\partial C}{\partial t}\right)_j = C_{j-1}\left[Q_{uj} + \left(\frac{A_z D_z}{\Delta Z}\right)_j\right] - C_j\left[\left(\frac{A_z D_z}{\Delta Z}\right)_j + \left(\frac{A_z D_z}{\Delta Z}\right)_{j+1} + Q_{dj} + Q_{u(j+1)} + Q_{in} + \frac{\partial V_j}{\partial t}\right]$$

$$+ C_{j+1}\left[\left(A_z D_z\right)_{j+1} + Q_{d(j+1)}\right] + \sum Q_x C_x \pm V_j\left(\frac{dC}{dt}\right)_j \tag{19}$$

where

j = the jth layer

Q_u = upward flow, [L^3/T]
Q_d = downward flow, [L^3/T]
Q_{in} = inflow, [L^3/T]
Q_x = inflow from tributaries, [L^3/T]
C_x = concentrations of inflow from tributaries, [M/L^3]
A_z = cross section area, [L^2]
D_z = effective dispersion coefficient, [L^2/T]

For a series unit j-1, j, and j+1, the concentrations of C_{j-1}, C_j, and C_{j+1} could be rewritten as:

$$C_{j-1} = \alpha_{j-1,1} C_{j-2} - \alpha_{j-1,2} C_{j-1} + \alpha_{j-1,3} C_j + \beta_{j-1}$$

$$C_j = \alpha_{j,1} C_{j-1} - \alpha_{j,2} C_j + \alpha_{j,3} C_{j+1} + \beta_j \tag{20}$$

$$C_{j+1} = \alpha_{j+1,1} C_j - \alpha_{j+1,2} C_{j+1} + \alpha_{j+1,3} C_{j+2} + \beta_{j+1}$$

where

$$\alpha_{j,1} = Qu_j + (\frac{A_z D_z}{\Delta Z})_j$$

$$\alpha_{j,2} = \left(\frac{A_z D_z}{\Delta Z}\right)_j + \left(\frac{A_z D_z}{\Delta Z}\right)_{j+1} + Q_{dj} + Q_{u(j+1)} + Q_{in} + \frac{\partial V}{\partial t}$$

$$\alpha_{j,3} = \left(\frac{A_z D_z}{\Delta Z}\right)_{j+1} + Q_{d(j+1)} \tag{21}$$

$$\beta_j = Q_x . C_x + V_j \left(\frac{dC}{dt}\right)_j \qquad j = 1,\ 2, \ldots\ldots, \ n$$

Approximation method is adopted to solve the above first-order differential equations as:

$$C_{t+\Delta t} = C_t + \frac{\Delta t}{2}(\dot{C} + \dot{C}_{t+\Delta t}) = \frac{\Delta t}{2}\dot{C}_{t+\Delta t} + (\frac{\Delta t}{2}\dot{C} + C_t) \tag{22}$$

6. MULTI-DIMENSIONAL WATER QUALITY MODELS

To accurately characterize eutrophication, more sophisticated hydrodynamic and water quality models should be used to simulate the hydraulic and water quality variations. Multi-dimensional models simulating the water quality variation in both vertical and horizontal directions are the most powerful tool to model the complexity of real lakes. A two- dimensional model assumes that the water body can be represented by several layers of rectangular grids on a vertical Cartesian X-Z plane and the variations of water properties in the Y-axis are negligible (laterally averaged). The governing equations written for a control volume in the middle (kth) layer can be written as (Buchak and Edinger, 1982):

$$\frac{\partial}{\partial t}(uBh)+\frac{\partial}{\partial x}(u^2Bh)+(u_bw_bb)_{k-\frac{1}{2}}-(u_bw_bb)_{k+\frac{1}{2}}+\frac{1}{\rho}\frac{\partial}{\partial x}(PBh)-\frac{M_x}{\rho}\frac{\partial^2}{\partial x^2}(uBh)$$
$$+\frac{1}{\rho}(\tau_{xz}b)_{k+\frac{1}{2}}-\frac{1}{\rho}(\tau_{xz}b)_{k-\frac{1}{2}}=0 \quad (23)$$

$$\frac{\partial p}{\partial Z}=-\rho g \quad (24)$$

$$(w_bb)_{k-\frac{1}{2}}=(w_bb)_{k+\frac{1}{2}}-\frac{\partial}{\partial x}(uBh)-\frac{\partial Q}{\partial x} \quad (25)$$

$$\frac{\partial}{\partial t}(BhT)+\frac{\partial}{\partial x}(uBT)+(w_bbT)_{k+\frac{1}{2}}-(w_bbT)_{k-\frac{1}{2}}-D_{Hx}\frac{\partial^2}{\partial x^2}(BhT)$$
$$-\left[D_{Hz}\frac{\partial}{\partial z}(BT)\right]_{k+\frac{1}{2}}+\left[D_{Hz}\frac{\partial}{\partial z}(BT)\right]_{k-\frac{1}{2}}=\frac{H_nBh}{V} \quad (26)$$

$$\rho=999.841+5.29157*10^{-5}T^3-8.45123*10^{-3}T^2+6.59583*10^{-2}T \quad (27)$$

where

u = the layer-averaged and laterally averaged horizontal (X-direction) velocity, [L/T]
B = layer-averaged width, [L]
h = layer thickness, [L]
u_b = the laterally averaged horizontal (X-direction) velocity, [L/T]
w_b = the laterally averaged vertical (Z-direction) velocity, [L/T]
P = pressure, [M/L/T^2]
ρ = density, [M/L^3]
b = the width of the reservoir at any given depth, [L]
g = gravitational acceleration, [L/T^2]
Q = tributary inflow, [L^3/T]
T = water temperature, [Θ]
τ_{XZ} = shear stress, [M/L/T^2]
M_x = a constant horizontal momentum exchange coefficient, [M/L/T]
D_{Hx} = a constant horizontal heat exchange coefficient, [L^2/T]
D_{Hz} = a constant vertical heat exchange coefficient, [ΘL^2/T]
H_n = heat sources and sink term, [L^3/T]
V = volume of the reservoir, [L^3]

The subscripts k-1/2 and k+1/2 represent the interface between the kth layer and its upper layer (k-1) and lower layer (k+1).

Equation (23) is the momentum equation in the X-direction (the longitudinal direction of the reservoir), and Equation (24) is the reduced vertical momentum equation which is derived by assuming that the vertical acceleration of water is negligible while being compared with gravitational acceleration (the hydrostatics assumption). Equation (25) is the fluid continuity equation, Eq. (26) is the heat transport equation, and Eq. (27) is the equation of state.

One of the most versatile multi-dimensional water quality models in the public domain is the Water Quality Analysis Simulation Program (WASP) originally developed by researchers at Manhattan College and U.S. EPA Great Lake Research Laboratory (Ambrose et al., 1991). A new version of WASP is WASP5 that was initiated with funding from the EPA Large Lakes Research Station and the Great Lakes Program. WASP5 permits users to structure one-, two-, and three-dimensional models, and allows the specification of time-variable exchange coefficients, advection flows, waste loads, and boundary conditions.

7. APPLICATIONS AND CASES STUDIES

7.1 Introduction

Once the water quality hydrodinamic and biochemical problem is defined, the next step is to choose a right model to simulate the variation of the water body. There are several model factors that users need to consider whether a model is applicable, such as complexity, simulated constituents, dimensional level, applicable water types, and temporal scale. Table 1 lists several popular water quality models and their characteristics. Feitsui Reservoir and Techi Reservoir were used in the case study to demonstrate the water quality simulation using zero-, one-, and two-dimensional models. Both reservoirs are two of the biggest reservoirs in Taiwan (see Figs. 2 and 3). Vollenweider's model and CE-QUALW2 were applied to the Feitsui Reservoir, and QUAL2E was applied to the Techi Reservoir.

Feitsui Reservoir is located in northern Taiwan (about 30 km away from Taipei, see Fig. 3) and provides a long-range source of water supply for domestic and industrial use. The Feitsui Reservoir dam, which is located downstream of the Peishih Creek and 170 m above sea level, has a watershed of 303 km^2 and a detention time of about 129 days. The water quality of Feitsui Reservoir is under strict watch and control. There are 18 water quality sampling points in the reservoir area and tributaries. The water quality sampling and watershed water quality monitoring are to be done once a month. A zero-dimensional model (Vollenweider's type) and a two-dimensional model (CE-QUAL-2E) were applied to Feitsui Reservoir for water quality simulation and management.

Table 1 List of water quality models

Name	Complexity	Dimensional Level	Geographic Features	Developers
Vollenweider	low	0-D	lakes, reservoirs	—
HSPF	moderate	1-D	lakes, reservoirs, rivers	US EPA
QUAL2E	moderate	1-D	rivers	US EPA
CE-QUALW2	high	2-D	lakes, reservoirs, rivers	US Army Engineer Waterways Experiment Station
WQRRS	high	2-D	lakes, reservoirs, rivers	The Hydrologic Engineering Center, US Army Corps of Engineers
WASP	high	0-D, 1-D, 2-D	lakes, reservoirs	Manhattan College & U.S. EPA

Techi Reservoir is located in the upstream part of the Ta-Chia River approximately 1,400 m above sea level in central Taiwan. Techi Reservoir has a watershed area of 592 km^2. The Techi Dam is operated by the Taiwan Power Company for a single purpose of hydroelectric power generation. The studied reservoir extends 9.3 km upstream from the dam. A part of the Techi Reservoir watershed was developed as an agricultural area for economic purposes two decades ago. Large amount of nutrients from fertilizers and farm wastes enter the reservoir and stimulate an excessive algal growth. As a result, eutrophication becomes a big concern on the Techi Reservoir during the summer. Long-term water quality monitoring programs have been executed on Techi Reservoir for over a decade. Water samples are taken on a scheduled basis (bi-monthly). The sampling data include temperature, dissolved oxygen (DO), chlorophyll-a, turbidity, biological oxygen demand (BOD), total phosphorus, organic nitrogen, ammonia nitrogen, nitrate nitrogen, and pH. A stream numerical model QUAL2E was used to simulate water quality conditions of surface water for the Techi Reservoir and will be introduced later.

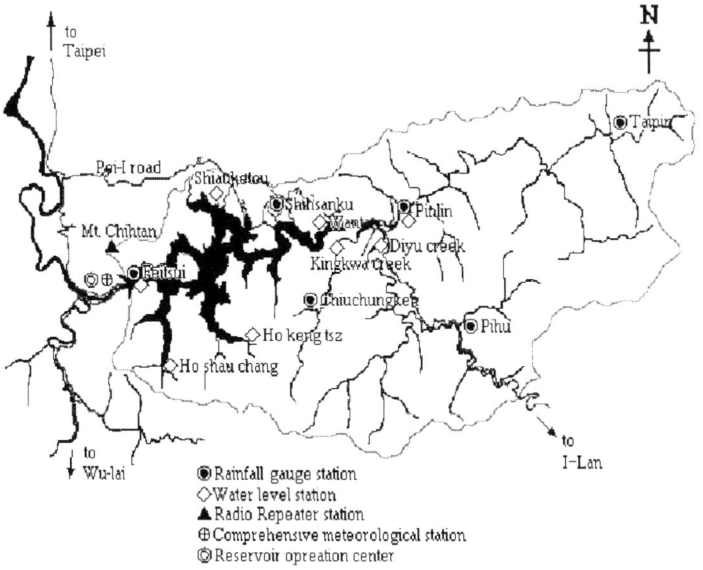

Fig.2 Location of Feitsui Reservoir and its sampling sites

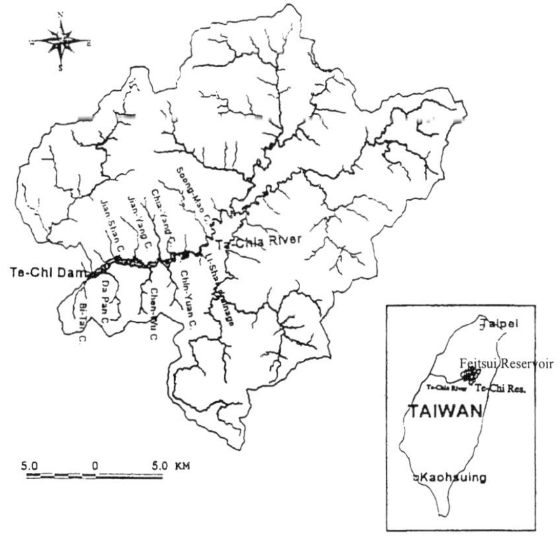

Fig. 3 The location and watershed of Techi Reservoir

7.2 Case Study of Feitsui Reservoir

7.2.1 Zero-Dimensional Water Quality Model

The population of algae increases through photosynthensis and the uptake of available nutrients in water. Phosphorus (P) and nitrogen (N) are the most important nutrients for algal growth. Whether the removal of phosphorus or nitrogen is the best way to control eutrophication depends on which is the controlling nutrient. In general, N/P ratios of 20 or greater reflect a phosphorus-limited system while N/P ratios of 5 or less reflect nitrogen limited system (Thomann and Mueller, 1987).

Figure 4, which shows the N/P ratios of Feitsui Reservoir from 1991 to 1997, reveals that phosphorus is the controlling nutrient (N/P>20). The minimums of $P/(P+K_{mP})$ and $N/(N+K_{mN})$, in which K_{mP} and K_{mN} are the affinity constants of nutrients N and P, decide whether phosphorus or nitrogen is the controlling factor for algal growth. In Fig. 5, phosphorus was shown as the controlling factor for Feitsui Reservoir, same as in Fig. 4. Thus, a simplified water quality model (a phosphorus limited system) was used to simulate Feitsui Reservoir and to estimate the reducing proportion of phosphorus to meet the criterion of oligotrophication.

Vollenweider model was applied to the simulation of phosphorus for Feitsui Reservoir. Figure 6 shows the simulation results of total phosphorus in 1997 compared with the measured data. The reservoir volume was set as a constant of 270×10^6 m^3 (yearly average) for the constant-volume case, while the reservoir volume may vary with time in the other cases. Other parameters were set the same values, including settling velocity of 10 m/year and the loss rate of total phosphorus of 0.287 each year. It can be observed that the time-variable-volume model gave a more accurate simulation for Feitsui Reservoir (which had a significant difference between inflows and outflows during the studying period) than the constant-volume model did. Furthermore, the results from repeatedly rerunning the model revealed that at least 40% reduction of the total phosphorus within inflow has to be cut so that the water quality of Feitsui Reservoir can be controlled and meet the criteria of oligotrophication (less than 8 μg/L).

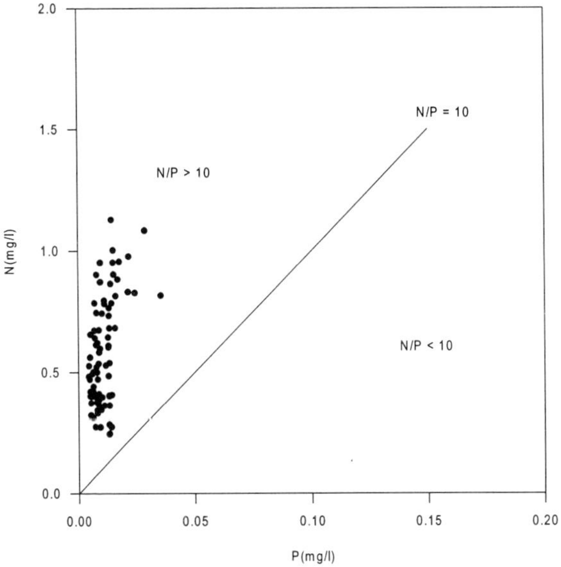

Fig. 4 N/P ratio of Feitsui Reservoir between 1991-1997

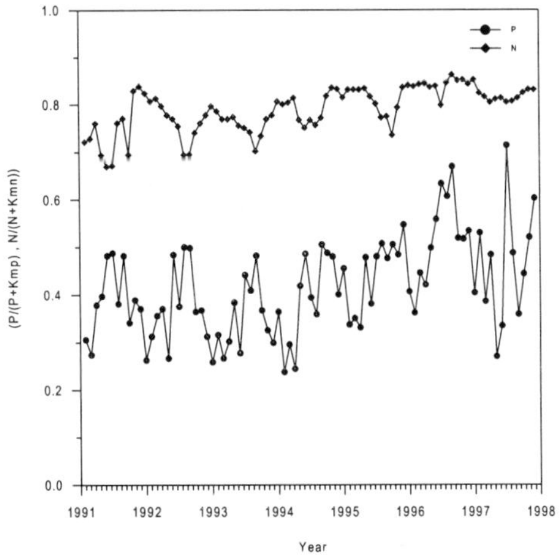

Fig.5 Controlling factor for algal growth of Feitsui Reservoir between 1991-1997

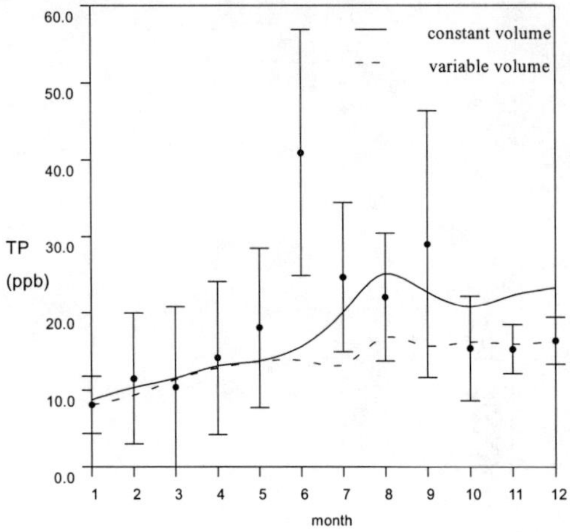

Fig. 6 Zero-dimensional simulated values and measured values of total phosphorus

7.2.2 Two-Dimensional Water Quality Model

To more accurately characterize Feitsui Reservoir, a more sophisticated water quality model CE-QUAL-W2 was used to simulate the hydraulic and water quality variations. CE-QUAL-W2 is a two-dimensional (vertical/longitudinal) hydrodynamic and water quality model that is applicable to a stratified lake like Feitsui Reservoir. However, CE-QUAL-W2 is a complex and time-consuming model, and needs many input files — a control file, a bathmetry file, a meteorological file, inflow files, inflow and boundary constituent files, inflow and boundary temperature files, outflow files, withdrawal files, and initial condition files. Figure 7 is a two-dimensional segmentation system of the Feitsui Reservoir.

Figure 8 shows the bi-monthly variation of water temperature simulation in 1996. Severe stratification occurred in May, July, and September. The stratified condition prevented water from being vertically transported from surface layers to bottom layers. The difference of water temperature between the upper layers and lower layers reached about 20℃, and the thermalcline penetrated from 10 m to 40 m deep below surface. Figures. 9(a) ~ 9(f) show the results from the water quality simulation of Feitsui Reservoir in 1986 using CE-QUAL-W2. During the summer (a typhoon and thunderstorm season) a great amount of organic matters flowed into the reservoir. As a significant amount of nutrients was discharged into water with increased temperature, phytoplankton had a better environment to grow. Thus, the peak of chlorophyll-a concentration appeared in late summer at the upper segment (around Wantan) and late fall (with a time lag of about a couple of months) at the downstream segment (the dam site). In general, the concentration of phosphorus and nitrogen decreases while the water flowed down to the dam due to the loss of settling and decay.

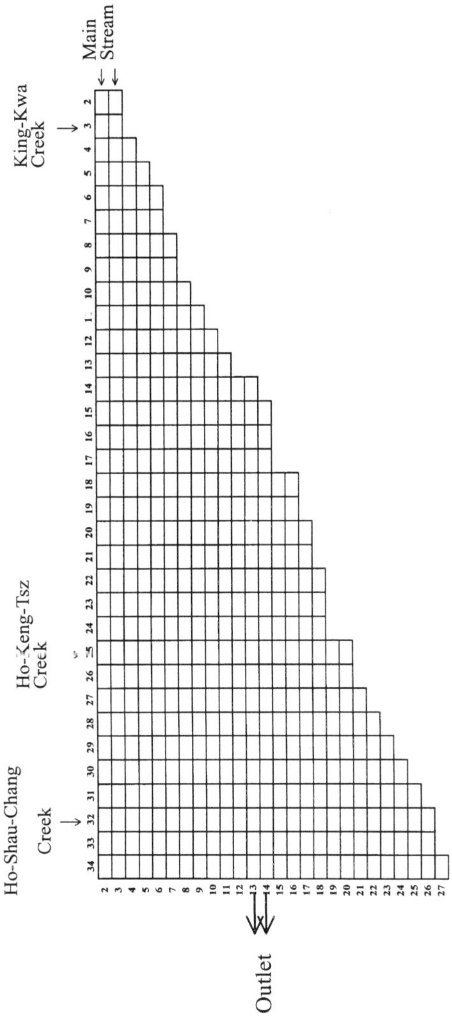

Fig. 7. Two-dimensional grid of Feitsui Reservoir

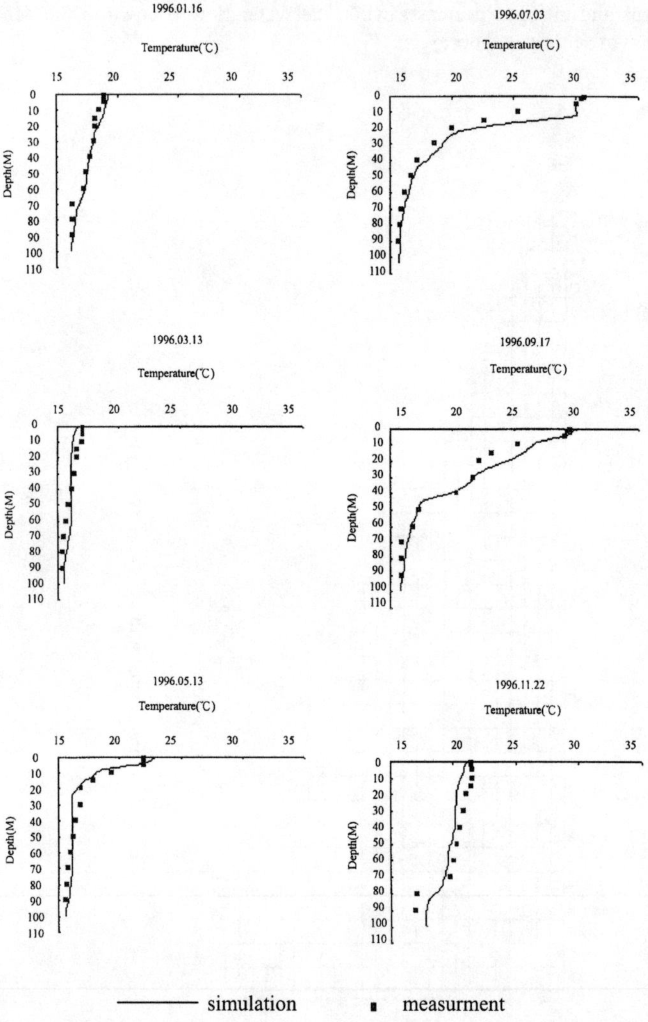

Fig. 8. Temperature simulation and field data at the dam site of Feitsui Reservoir in 1996

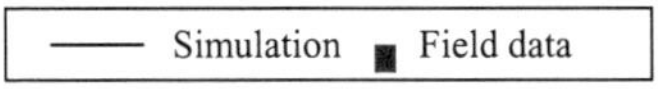

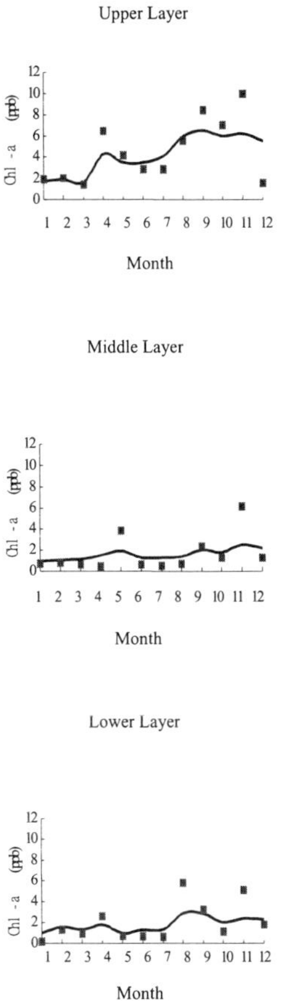

Fig. 9(a). Chlorophyll-a simulation and field data at Wantan of Feitsui Reservoir in 1996

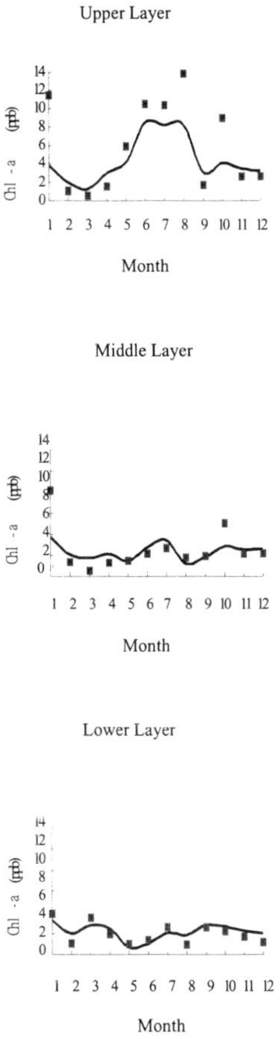

Fig. 9(b). Chlorophyll-a simulation and field data at the dam site of Feitsui Reservoir in 1996

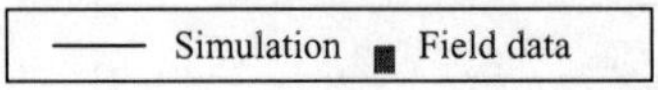

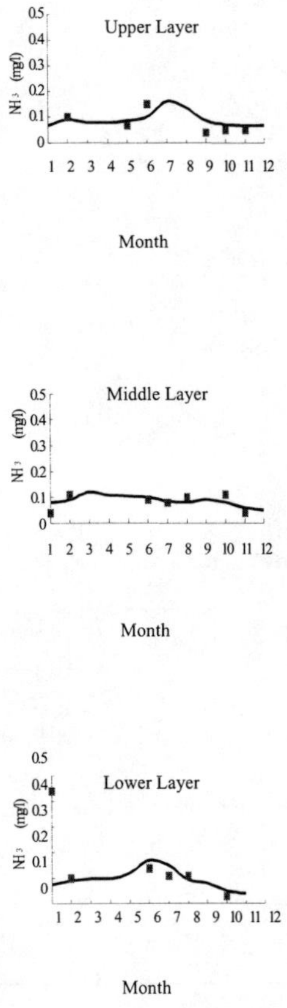

Fig. 9(c). NH_3 simulation and field data at the dam site of Feitsui Reservoir in 1996

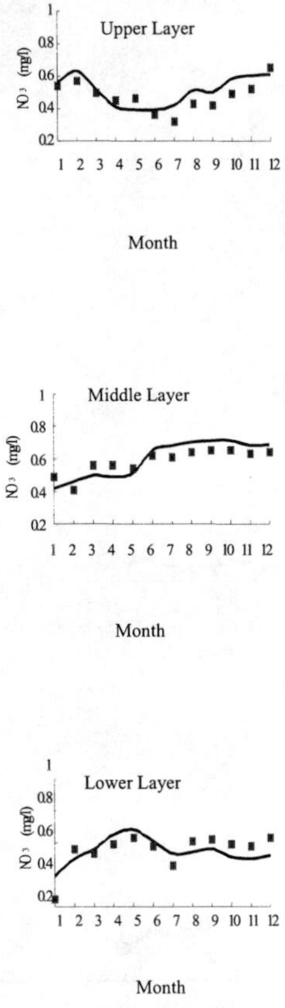

Fig. 9(d). NO_3 simulation and field data at the dam site of Feitsui Reservoir in 1996

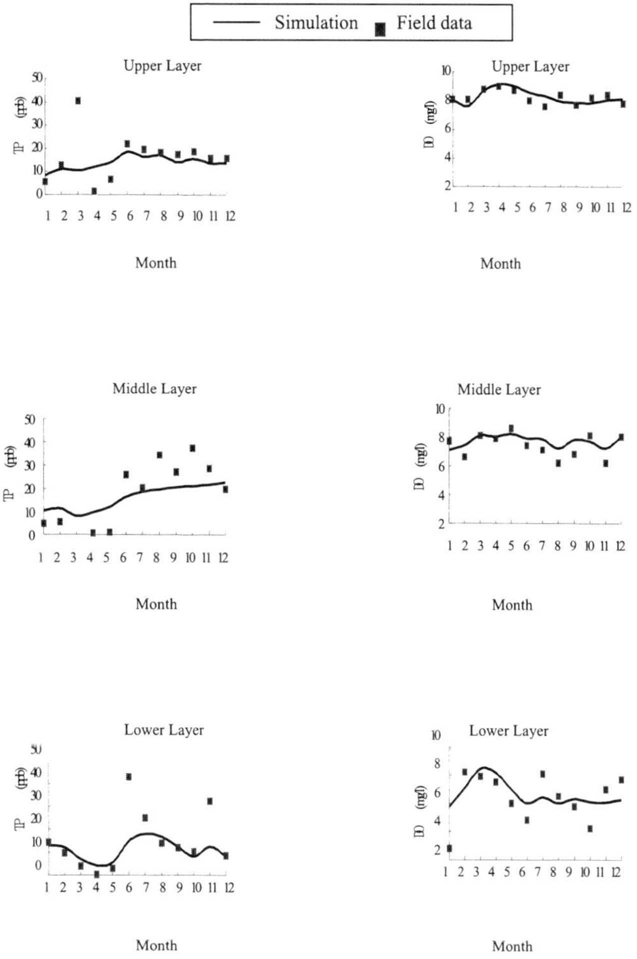

Fig. 9(e). TP simulation and field data at the dam site of Feitsui Reservoir in 1996

Fig. 9(f). DO simulation and field data at the dam site of Feitsui Reservoir in 1996

Finally, CE-QUAL-W2 was used to evaluate how much the controlling nutrient had to be reduced under Taiwan EPA's criteria. The simulation concluded that about 50% cut-off would keep Feitsui Reservoir at mesotrohpic or oligotrophic status all year that is similar to the evaluation derived from zero-dimensional water quality modeling.

8. SENSITIVITY ANALYSIS AND MODEL CALIBRATION

Model users usually are asked to input the model parameters before a simulation. In most cases, only a small portion of parameters is obtained from field investigation on the simulated water body. Most values of parameters have to be estimated according to published data for similar ecosystems in literature or in experiments. To input more proper values of parameters, model calibration is an essential step. However, it is difficult to check a bundle of parameters simultaneously. A sensitivity analysis is can determine what parameters play the most important roles and should be set first.

8.1 Sensitivity Analysis

Sensitivity analysis is a mechanistic procedure to analyze the response of a model to an imposed perturbation by changing the values of parameters. Most of model manuals provide reference tables listing the range of parameters. Based on the results of sensitivity analysis, the model parameters can be separated into several groups. Users can fix all other parameters except those parameters influencing the simulation results most. Two simple methods, parameter perturbation and first-order sensitivity analysis, are briefly described in this section. Monte Carlo, more complex with advance statistical technique, is another approach to sensitivity analysis (Chapra, 1997).

Parameter perturbation: By raising and lowering a certain percent of a designated parameter and fixing all other parameters, the corresponding variations of the water variables reflect the model sensitivity to this parameter. Recalling a mass-balance equation for a well-mixed lake and substituting K with $K\text{-}\Delta K$ and $K\text{-}\Delta K$ in Eq. (9), the variation of the nutrient concentration can be calculated as:

$$\Delta C = \frac{C(K+\Delta K) - C(K-\Delta K)}{2} \tag{28}$$

First-Order sensitivity analysis: This method employs the derivative of Eq. (9). By applying Taylor-series expansions to the variable and keeping the first-order term only, the varied variable can be calculated as:

$$C(K+\Delta K) = C(K) + \frac{\partial C(K)}{\partial K}\Delta K \tag{29}$$

$$C(K-\Delta K)=C(K)-\frac{\partial C(K)}{\partial K}\Delta K \tag{30}$$

ΔC can be solved for

$$\Delta C=\frac{C(K+\Delta K)-C(K-\Delta K)}{2}=\frac{\partial C(K)}{\partial K}\Delta K \tag{31}$$

Comparing ΔC resulting from different K's, users are allowed to classify the importance of model parameters, and subsequently execute model calibration.

8.2 Model Calibration

Model calibration is an essential process to test and tune a model by comparing the simulated results with field data. So far, there is no efficient systematic method for model calibration, especially for complex water quality models in which a great amount of variables and parameters are involved (Voinov, 1991). The simplest methods are the "parameter tuning" and "trial-and-error". By minimizing discrepancy between simulated results and field data (usually at least one full year's data), users can set up a water quality model with one set of rational parameters for a specific water body. However, this guesswork is time-consuming and considerably relies on users' experience with water quality simulation. The reasons that formal statistical methods are not commonly used in model calibration are the complexity of the models and the limitation of a large field data set (Henderson-Sellers and Davies, 1991). Moreover, field data are always not enough in both spatial and temporal considerations.

Recently, computerized approaches were tested to establish a more systematic calibration method. By optimizing the simulation or searching best-fit curves, modelers can obtain a better set of values of parameters. There are several different quantitative criteria to result in a best-fit, such as a least-squares fit technique.

8.3 Simplified Case Study

A nonlinear calibration model was developed by minimizing the average difference between observed and simulated values using a least squares method (Yang et al., 2000). The calibration model with a mathematical approach provides an alternative method to estimate biological parameters of algae besides *in situ* sampling and experiment. The selected study site of a model calibration is Techi Reservoir. Because a significant stratification occurred in Techi Reservoir during the summer, only the epilimnion, about a 10-m depth of water from the surface, was simulated. In this study, QUAL2E was run for a one-dimensional dynamic simulation.

QUAL2E, a horizontal one-dimensional water quality model designed by the U.S. Environmental Protection Agency (EPA), simulates the spatial and temporal variations of water quality variables, including DO, BOD, water temperature, algae reported as chlorophyll-a, organic phosphorus, and dissolved phosphorus. QUAL2E users are allowed to choose a steady state or dynamic simulation and subdivide the stream system into many different reaches (up to 50 reaches), depending on their hydraulic characteristics. In addition, each reach may consist of many elements (total maximum of 500 elements) of seven different types: the headwater element, the standard element, the element just upstream from a junction, the junction element, the last element in the system, the input element, and the withdraw element.

8.3.1 Development of a calibration model

For each water cell, a mass balance equation for the algae (Eq. (8)) can be rewritten as follows (Brown and Barnwell, 1985):

$$\frac{dC}{dt} = V(\mu - \rho - \frac{v_p}{D})C + QC_{in} - QC \tag{32}$$

where

t = time, [T]
C = the laterally-averaged algal concentration, [M/L^3]
C_{in} = the concentration in the incoming flow, [M/L^3]
V = cell volume, [L^3]
μ = the specific growth rate of algae, [1/T]
ρ = the death rate of algae, [M/L^3]
v_p = the settling rate of algae, [L/T]
Q = water flow, [L^3/T]
D = stream average depth, [L]

In general, water flow record can be obtained from hydrologic monitoring stations; cell area and volume are defined by the user; and settling rate of algae is set as a constant in this simulation.

The numerical solution of constituent concentrations (Eq. (17)) in QUAL2E is solved by the finite difference method. Consider chlorophyll concentration only in the calibration model, r is the sum of growth rate, respiration rate, and settling rate for algae, and p is zero because there are no internal sources/sinks for algal growth.

With a superscript n as a time step and a subscript i as a distance step, the derivative of a constituent concentration is estimated using finite differences. The entire equation can be written as:

$$\frac{C_i^{n+1}-C_i^n}{\Delta t}=\frac{(AD_L)_i C_{i+1}^{n+1}-(AD_L)_i C_i^{n+1}}{V_i \Delta X_i}-\frac{(AD_L)_{i-1} C_i^{n+1}-(AD_L)_{i-1} C_{i-1}^{n+1}}{V_i \Delta X_i}$$
$$-\frac{Q_i C_i^{n+1}-Q_{i-1} C_{i-1}^{n+1}}{V_i}+r_i C_i^{n+1}+p_i+\frac{s_i}{V_i} \quad (33)$$

Furthermore, Equation (33) can be rearranged as:

$$a_i C_{i-1}^{n+1}+b_i C_i^{n+1}+c_i C_{i+1}^{n+1}=Z_i \quad (34)$$

To reveal how the biological parameters affect the constituent concentration, a difference of b is introduced to Eq. (34) as follows:

$$a_i(C_{i-1}+\Delta C_{i-1})+(b_i+\Delta b_i)(C_i+\Delta C_i)+c_i(C_{i+1}+\Delta C_{i+1})=Z_i \quad (35)$$

where ΔC_i is the difference between the concentrations of chlorophyll-a at element i, and Δb_i is the difference between the biological parameters at element i. In Eq. (35), a_i, b_i, and c_i are constant, which are the same values in Eq. (34). Substituting Eq. (34) into Eq. (35) and assuming $\Delta b_i \Delta C_i$ small enough to be ignored, we get:

$$a_i \Delta C_{i-1}+b_i \Delta C_i+\Delta b_i C_i+c_i \Delta C_{i+1}=0 \quad (36)$$

In Eq. (36), all terms are known, except for ΔC_i and Δb_i. For a steady flow, Δb_i is only dependent upon the difference between the algal biological parameters (Δr_i). Since r includes the growth rate, respiration rate, and settling rate, which are assumed constant for all computational elements in the entire aquatic system, the difference between the biological parameters of algae can be written as:

$$\frac{\Delta b}{\Delta t}=-\Delta r=-\Delta\mu+\Delta\rho+\frac{\Delta v_s}{D} \quad (37)$$

By applying Taylor series, which leads to $\Delta f(x)=\partial f/\partial x\, dx$, Δr is a result of the sum of the first-order partial derivatives, and can be found by multiplying the parameter correction with respect to interested biological parameters, such as μ_{max}, K_L, and K_S.

A sensitivity analysis was executed during the first stage. By inputting a 100% increase on one parameter each time and checking algae concentration, we can observe its impact on algal growth. The effect of a given parameter on the algal growth rate is shown as a percentage in Table 2. The respiration rate and the algal maximum growth rate have the most significant influence on the algal growth among those model parameters; thus they are included in the first class of parameter calibration.

By giving an initial approximation of the maximum algal growth rate and respiration rate and comparing the simulated results with the satellite-derived data, which serves as a set of calibration data, the calibration model computes a set of correction values for the input parameters. A new set of parameters was calculated by adjusting the correction values on the old values, and was input into QUAL2E again. A new set of correction values for the model parameters was estimated by comparing the new simulated result with the calibration data. By minimizing the discrepancy between simulated and satellite-derived chlorophyll-a concentration, the most suitable parameters are identified.

Table 2 Sensitivity analysis of algal model parameters

Parameters	Notation	Values	Unit	Impact on Algae
Algal respiration rate	ρ	0.05 - 0.5	1/day	-100%
Algal settling velocity	V_p	0.1 - 0.3	m/day	-10%
Algal maximum growth rate	μ_{max}	1.0 - 3.0	1/day	100%
Phosphorus affinity constant	K_S	1.0 - 5.0*	μg/L	-6%
Light saturation constant	K_L	100 - 400*	ly/day	-5%

*Reference values were adopted from Thomann and Mueller's (1987); others were taken from Brown and Barnwell's (1985).

8.3.2 Model test

To evaluate the utility of the calibration model, an artificial test was formulated. One set of biological parameters (μ_{max} = 2.50 d^{-1} and 0.20 d^{-1}) was saved as the input data to QUAL2E to create an artificial calibration data set for chlorophyll-a. The test runs were carried out by preparing four sets of initial approximations for a combination of low and high algal maximum growth rates and respiration rates, including (μ_{max}, ρ) = (1.0, 0.3), (4.0, 0.3), (1.0, 0.1) and (4.0, 0.1). The parameter estimates were determined by minimizing the difference between the predicted values and the measured values. The results from the four test runs are shown in Fig. 10.

In Fig. 10, all iterations have a spiral trajectory, no matter if a low or high parameter was given as the initial approximation. In all four cases, the calibration model tracked the maximum growth rate and respiration rate back to the artificial values with errors within 0.5% of the original values. Evidently, this calibration model is suitable for the calibration of a set of previously presumptive biological parameters.

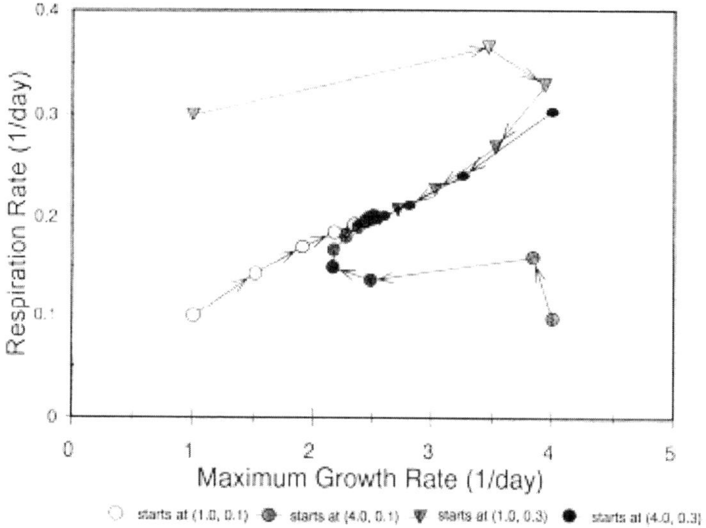

Fig. 10. Result of parameter calibration from test runs

8.3.3 Model experiments on Techi Reservoir

The calibration model was applied to Techi Reservoir and satellite-derived data was used. Chlorophyll-a concentration was calculated for each computational element by taking an average value for all pixels within a computational element. The calibration model was executed for 25 cases by giving various sets of initial approximation conditions. Starting from (μ_{max}, ρ) = (1.0, 0.1), each test run of an increment of maximum growth rate of 0.5 d^{-1} or respiration rate of 0.05 d^{-1} was added to the initial approximation. The results of the iteration of this calibration are shown in Fig. 11. Drawing an imaginary 45^0 line in Fig. 11 (from lower left corner to upper right corner), we found that the initial approximation points at the lower right have a trajectory ending on this 45^0 line, whereas the initial approximation points in the upper left of this figure have divergent results. Those iterations, starting with a respiration rate of less than 0.1 d^{-1}, result in negative values for μ_{max} and ρ, therefore, are not physically realistic. Those reasonable best-fit values of the maximum growth rate and respiration rate from the convergent points are marked by asterisks in Fig. 11 and can be expressed in terms of a linear equation through a statistical regression as:

$$\rho = 0.0576\,\mu_{max} + 0.134, \qquad R^2 = 0.965,\ n = 14 \qquad (38)$$

The regression equation presents a set of parameter values that yield a best fit for chlorophyll-a distribution in Techi Reservoir. All best fits satisfied an ending criterion of the minimum averaged difference between the QUAL2E-derived and SPOT-derived chlorophyll-a concentrations. The calibration model generated a series of values from the convergent approximations, which are similar to the results found by Shastry et al. (1973).

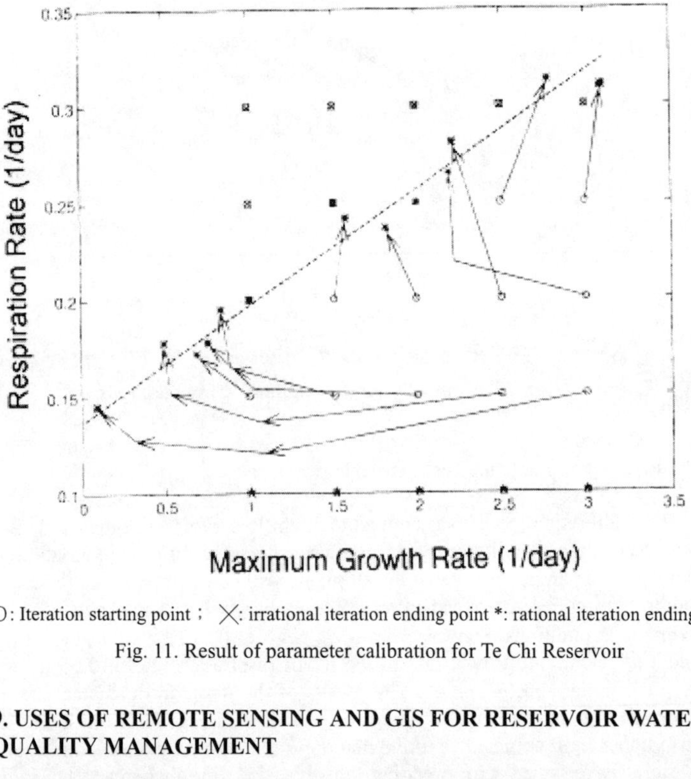

O: Iteration starting point ; X: irrational iteration ending point *: rational iteration ending

Fig. 11. Result of parameter calibration for Te Chi Reservoir

9. USES OF REMOTE SENSING AND GIS FOR RESERVOIR WATER QUALITY MANAGEMENT

Water quality simulation requires initial conditions that are usually available only if a comprehensive water quality sampling program has been conducted. Conventional approaches to water quality assessment require *in situ* sampling and expensive time-consuming laboratory work. Due to these two limitations, the number of water samples is often not large enough to cover the entire water body. Also, it is difficult to run a highly frequent overall sampling program and to provide data set for model calibration and verification. This becomes an obstacle for making water quality forecasts more feasible in water quality management. Moreover, simulated results are usually displayed in a numerical format or X-Y figures without an association with geographic information. Besides developing proper mathematical models, other problems in the input and output of water quality modeling remain to be solved.

9.1 Remote Sensing Techniques in Water Quality Monitoring

Remote sensing technique is an alternative to water quality monitoring over a range of temporal and spatial scales. A number of studies have shown that applications of remote sensing can meet the demand of acquiring many water quality samples over a large spatial area and also provide temporal view of water quality. A semi-automatic data acquisition and handling system was developed by Scarpace et al. (1979) to use multi-date Landsat image for assessing the trophic status of inland lakes in Wisconsin. Images from satellite Landsat and aircraft remote sensing systems were used in the assessment of water quality, such as temperature, chlorophyll-a, turbidity, and total suspended solids for lakes and reservoirs (Lillesand et al., 1983; Ritchie et al., 1991), for estuary water bodies (Verdin, 1985; Harding et al., 1995), and for tropical coastal areas (Ruiz-Azuara, 1995). SPOT satellite image was also proved to successfully assess water quality conditions (Lathrop and Lillesand, 1989; Yang and Yang, 1999). Most previous studies focused on the discovery of the relationship between remote sensing data and *in situ* measurements.

Water quality is a reflection of the controlling bio- and geo-chemical processes and shows a great variability over a range of temporal and spatial scales (Lathrop and Lillesand, 1989). As long as a water quality parameter has a spectral signature within the spectrum window, remote detection is feasible. There are several major concerns with remote sensing for monitoring water quality. The most important considerations are the spectral range and the optimum band width, which are crucial for detecting water quality characteristics. Water quality is considered to be associated with the concentration of three constituents: phytoplankton, dissolved organic substances, and inorganic suspended solids.

Clear water strongly absorbs energy in wavelengths greater than 0.75 μm, which is the near infrared range. In a eutrophic water body the chlorophyll in the algae largely increase the reflectance of a water body in the green and near-infrared spectral regions. The higher the chlorophyll concentration in the water body, the stronger the reflected energy in both wavelength ranges. Moreover, a eutrophic lake contains a high concentration of inorganic materials, such as phosphorus and nitrogen, that increase the water turbidity. Therefore, the spectral curve will be different from the clear water curve and will be a result of the combination of clear water and soil.

Through a linear or non-linear regression model, an empirical calibration of the satellite data with concurrent surface reference information is undertaken to correlate the digital data with water quality variables. Because of its straightforward interpretation, the empirical approach was widely adopted to determine of water quality variables using remotely sensed data (e.g. Khorram, 1981; Whitlock and Kuo, 1982; Bhargava, 1983; Lillesand et al., 1983; Bukata et al., 1985; Lathrop and Lillesand, 1986 and 1989; Sathyendranath and Trevor,

1989; Ritchie and Cooper, 1991; Rundquist et al., 1996).

After an examination of regression equations for the correlation between water quality variables and the digital data, the results show that the band ratio exponential model is the best-fit regression model. The examination results were evaluated from two points of view, statistical significance and theoretical explanation. Final regression models were suggested for chlorophyll-a (*CHLA*), Secchi depth (*SD*), and phosphorus (PO4) for Techi Reservoir as:

$$lnCHLA = 9.373 + 10.080\, ln(XS3/XS2) \quad (R^2 = 0.951, P = 0.005) \tag{39}$$

$$lnSD = -3.32 + 4.39\, ln(XS1/XS2) \quad (R^2 = 0.953, P = 0.004) \tag{40}$$

$$lnPO4 = 10.851 - 10.015\, ln(XS1/XS2) \quad (R^2 = 0.827, P = 0.032) \tag{41}$$

Statistical parameters were presented as indicators of the significance of the regression model. The square of the correlation coefficient (R^2) represents the level of relationship between the two sets of data. The SPOT- derived water quality data shows the finest spatial resolution (20m) of water quality conditions in Techi Reservoir compared to any format ever seen before. The SPOT-derived thematic map shows a high chlorophyll-a concentration (above 250 μg/L) in the upper stream reaches and with a reduced concentration (less than 50 μg/L) in the downstream reaches. This was due to a large amount of nutrients being released from the watershed with an intense farming activity coupled with a strong settlement effect of particulate matters while moving downstream. The band ratio of red/green (XS2/XS1) shows a great sensitivity to changes in turbidity, as represented by Secchi depth in this study. Comparing the thematic maps of chlorophyll-a and Secchi depth, both satellite-derived maps reveal consistent information that the water quality conditions of Techi Reservoir improved when water flowed from upstream down to the Techi Dam.

9.2 Geographic Information System in Water Quality Modeling

To move the application of water quality modeling toward a more practical level, an important task is to incorporate numerical water quality modeling with a friendly interface for input and output. Geographic information system (GIS) is one of the most efficient tools that can improve the storage and display of water quality information. Traditional numerical output or X-Y figures can be promoted into a two-dimensional colorful animation through GIS operation. The visualizing technique is helpful for most people to "see" the variation of water quality conditions.

Basically, GIS is a convenient tool to handle a variety of data sets, to provide an effective assessment of environmental controls, and to derive an analysis in a decision-making process. In short, GIS is an integrated system incorporating data, equipment, procedures, and users, and involves a more complete understanding of information. GIS provides techniques for analyzing landscape information and displaying spatial information in maps, tables, or graphics.

Better Assessment Science Integrating Point and Nonpoint Sources (BASINS), which was developed by the U.S. EPA's Office of Water, is a multipurpose environmental analysis system for use by regional, state, and local agencies in performing watershed- and water-quality-based studies (Lahlou et al., 1999). It integrates several key environmental data sets with improved analysis techniques to be applicable in various stages of environmental management planning and decision making

Through the use of GIS, BASINS has the flexibility to display and integrate a wide range of information (e.g., land use, point source discharges, water supply withdrawals) at a scale chosen by the user. In BASINS physiographic data, monitoring data, associated assessment tools, and simulation models are integrated in a customized GIS environment through a dynamic link. The results of the simulation models can also be displayed visually and can be used to perform further analysis and interpretation. User-friendly interface design makes water quality analysis easier.

BASINS was developed to promote better assessment and integration of point and nonpoint sources in watersheds and water quality management. The modeling tools include several main models: QUAL2E (a water quality and eutrophication stream model), TOXIROUTE (a model for routing pollutants through a stream system), and HSPF (a model for estimating instream concentrations resulting from loadings from point and nonpoint sources).

9.3 Integration of Remote Sensing, GIS, and Water Quality Modeling

The purpose of the following case study is to integrate water quality modeling with remote sensing and GIS techniques for use in practical water quality simulation. A short-term water quality forecasting system was developed by using SPOT satellite data, an image processing and GIS package — ERDAS Imagine, and a water quality model QUAL2E. Techi Reservoir was used as the study site for demonstrating the water quality monitoring and simulation system developed. The objective of the project is to make a short-term prediction of water quality and to present the variation temporally and spatially on a geo-referenced map by means of dynamic animation images.

A "liked-model" approach was chosen for the water quality forecasting system because of the complexity of water quality modeling. This complexity of mathematical calculation could not be easily operated within the current GIS software. In water quality models, a water body is subdivided into many finite segments or cells. All variables are represented as averaged values within a cell, much like a pixel of digital image. Chlorophyll-a is a representative variable for algal growth that is proportional to the concentration of phytoplankton biomass. For each water cell, a mass balance equation for the chlorophyll-a was calculated for the changes of inflow, outflow, growth, death, and settling. A stream numerical model QUAL2E was used to simulate water quality conditions and linked with ERDAS Imagine. Matlab (Matrix Laboratory) was used as a bridge to transfer remote sensing data from ERDAS Imagine to QUAL2E for water quality simulation, and then to transfer the simulated results from QUAL2E back to ERDAS Imagine for display. Figure 12 shows the flow chart describing the structure of the water quality forecasting system.

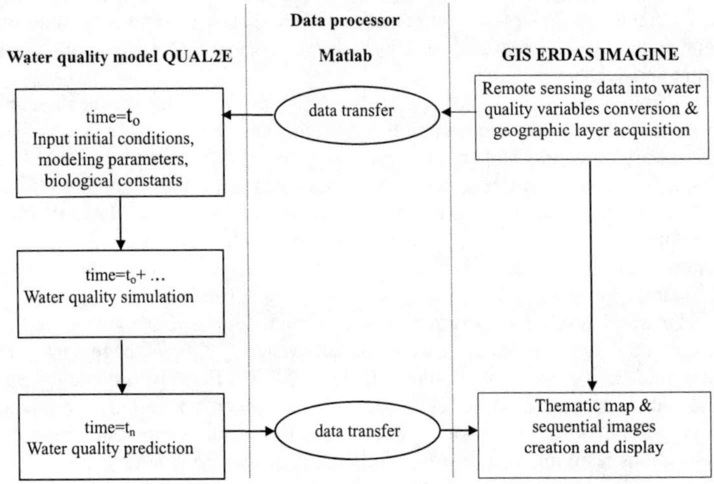

Fig. 12. Sketch of water quality forecasting system

Techi Reservoir was divided into 22 reaches, which were selected because of hydraulic characteristics. Each reach consists of small homogenous elements that are assigned to one of seven different element types — headwater elements, standard elements, elements just upstream from a junction, junction elements, last element in system, input elements, and withdraw elements. Once the initial conditions and environmental parameters are developed, QUAL2E calculates the transportation and interactions of pollutants and consequently predicts the changes in algae, nutrient, and temperature for a given time step. Among the variety of initial water quality variables, chlorophyll and nutrient concentrations are the most important concern for study and can be derived from the SPOT images.

In the study, QUAL2E was run for an one-dimensional dynamic simulation that needs meteorological, hydrological, hydraulic and biological data inputs. The meteorological and hydraulic data from Techi Reservoir, including temperature, wind, atmospheric pressure, and precipitation, were acquired from the Taiwan Power Company who is the operator of Techi dam, and used as inputs to QUAL2E for the simulation.

A significant stratification occurs in Techi Reservoir during the summer, thus only the upper level of water about 10-m down the surface was used in the one-dimensional dynamic simulations. The water body was divided into 93 computational elements along the stream flow direction (see Fig. 13). Lateral averaged water quality variables are computed for each element. All 93 elements belong to one of the 22 reaches. Each element has a length of about 100 m (1 pixel is 20 m on the SPOT image) and a cross-stream width which is equal to the down-stream river width and ranges from 80 m to 600 m. Each element was identified as a specific type, depending on its position and function in the model. The upstream section of the Ta-Chia River located near the Soong-Mao Creek was treated as a headwater source element because of its major contribution of flow. The rest of the seven tributaries are treated as point sources to input elements. The attributes of each reach, such as cross width, surface area, and constants in hydraulic functions were tabulated in ERDAS Imagine.

A six-day water quality forecast was made from the simulation by giving a steady pollution source with a chlorophyll concentration of 340 μg/L and phosphorus of 35 μg/L, which was the water quality condition at the upper stream location (near Soong-Mao Creek) and derived from the SPOT image taken on August 31, 1994. These simulations for six days were converted from numerical form into image form (geo-registered thematic layers) and displayed in ERDAS Imagine. The images were smoothed by a 7×7 low pass filter to remove the sharp difference between adjacent elements for simulated results.

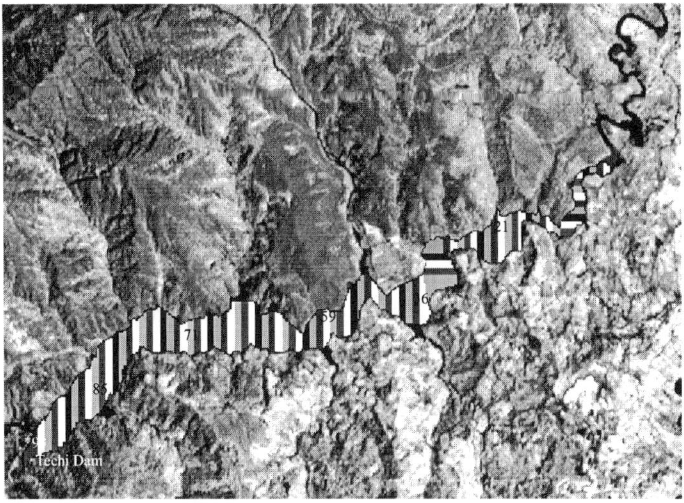

Fig. 13 Computational element of the Techi Reservoir defined in the water quality model QUAL2E.

Artificial images of chlorophyll, Secchi depth, and phosphorus were generated. Each single water quality variable was presented varying levels of concentration with corresponding colors (graytone in this book) on a thematic map. Figures 14 and 15 show the algae variation during the six days following August 31, 1994. Comparing these sequential images, one can observe that the high chlorophyll-a concentration area extended to the downstream reaches under steady pollution sources. By overlying water quality thematic layers to the background image, the distribution of water quality variation was clearly shown with a geographic emphasis.

Because of a steady pollutant input at the headwater and point sources, the chlorophyll-a concentration did not have significant change in the upper reaches, but significantly increased in the middle reaches. For example, the computational elements around Chin-Yuan Creek showed an almost two-fold increase of chlorophyll-a concentration (form 50 μg/L to 100 μg/L) within one week. The sequential images produced from the forecasting system can also be displayed dynamically as an animated forecast product. This visualization technique helps users to easily visualize the progression of water quality.

10. SUMMARY

Water quality models have been recognized as an efficient tool for water quality management. However, a mathematical model can only partially represent a part of the reality by making simplifying assumptions. It is an important task to understand the characteristics of models and choose a proper model for a specific application. The limitation and uncertainty of models are the other issues that users have to pay attention to. A mathematical mode is an abstraction of a complex water body and an aquatic ecosystem in real life. The set of quantitative equations can only represent a part of the complete bio-chemical process. Recently, increasing attention has been drawn to other problems besides models themselves, such as sensitivity, model calibration, input data acquisition, and output post process.

There are two major advantages of integrating remote sensing with water quality modeling. First, both systems store data in a compatible format (raster-basis). Pixels on digital images are similar to computational elements in water quality modeling. Generally, the size of the computational elements used in water quality models is larger than the spatial resolution of remote sensing images. Therefore, one can resample finer pixels to coarser elements, e.g. resampling 20 m to 100 m in the previous study. Secondly, remote sensing images displayed in GIS helps recognize the geographic variation of water quality. Since water quality is sensitive to geographic characteristics and the satellite data are registered on a coordinate system, it becomes easier to interpret the spatial variation of water quality and to identify potential pollution sources. Moreover, digital remote sensing data are suitable for use in a numerical model.

To overcome the difficulties of field sampling and to maintain a periodic real-time forecasting capability, remote sensing was shown as an efficient tool in the study. By integrating GIS with the modeling system, the simulated results were presented in two-dimensional thematic maps and in dynamic sequential images. The visualizing technique makes possible an easier understanding of water quality conditions. This water quality forecasting system will be more feasible and powerful as the earth observation satellite techniques become more sophisticated — finer resolutions, shorter data transmission time, and a lower image acquisition cost in the coming years.

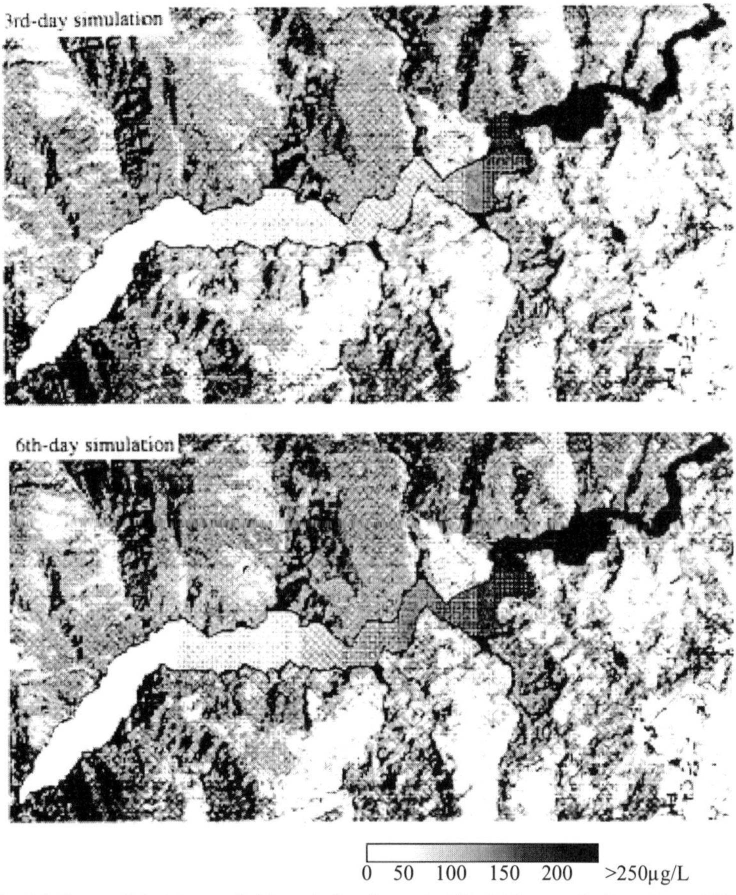

Fig. 14. Sequential pictures of chlorophyll-a forecast at Techi Reservoir since August 31, 1994

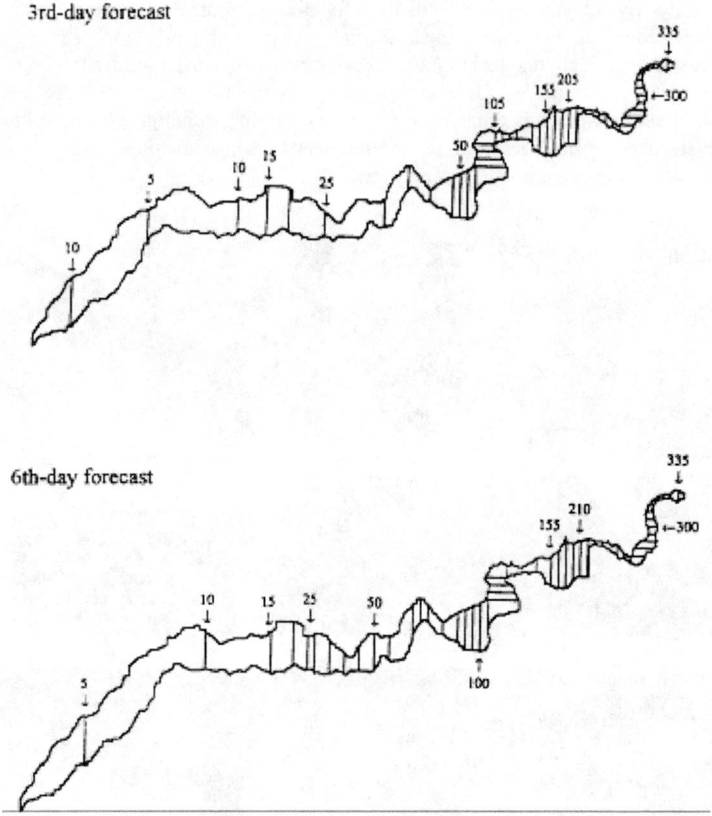

Fig. 15. Contour map (5 μg/L intervals) showing the forecasts of chlorophyll-a concentration in Fig. 14.

REFERENCES

Ambrose, R.B., T.A. Wool, J.L. Martin, J.P. Connolly, and R.W. Schanz (1991). *WASP5.x, A Hydrodynamic and Water Quality Model — Model theory, User's manual, and Programmer's Guide*. Environmental Research Laboratory Office of Research and Development, U.S. EPA. Athens, Georgia. 320p.

Baker, L.A. (1996). Chapter 9: Lakes and Reservoirs, in *Water Resources Handbook* (ed. by Mays). McGraw-Hill, New York.

Bhargava, D.S. (1983). Very low altitude remote sensing of the water quality of rivers, *Photogrammetric Engineering & Remote Sensing*, **49**(6):805-809.

Brown, L.C. and T.O. Barnwell (1985). *Computer Program Documentation for the Enhanced Stream Water Quality Model QUAL2E*. Environmental Research Laboratory, Office of Research and Development, U.S. Environmental Protection Agency. Athens, VA. 141p.

Buchak, E.M., J.E. Edinger (1982). *User's Guide for LARM2: a Longitudinal-vertical, Time-Varying hydrodynamic Reservoir Model*. Instruction Report, U.S. army Corps of Engineer Waterways Experiment Station, Vicksburg, Miss.

Bukata, R.P., J.H. Jerome, J.E. Bruton, and S. C. Jain (1985). *Application of Direct Measurements of Optical Parameters to the Estimation of Lake Water Quality Indicators*. Scientific Series No. 140, Inland Water Directorate, Canada Centre for Inland Waters, Burlington, ONT. 35p.

Chapra, S.C. (1997). *Surface Water-Quality Modeling*. The McGraw-Hill Companies, Inc. New York, 844 p.

Chen, C.W. and G.T. Orlob (1975). *Ecological Simulation for Aquatic Envioronments in Systems Analysis and Simulation in Ecology*, Vol. III. (ed. by Patton) . Academic Press, New York. 475 pp.

Chen, C.W. (1970). Concepts and utilities of ecological model. *J. Sanitary Engrg. Div.*, ASCE, **96** (SA5) :1085-1097.

Cole, T.M. (1994). *CE-QUAL-W2, Version 2.0. Water Operations Technical Support*. Waterways Experiment Station, US Army Corps of Engineers, **94**(1): 7p.

Cole, T.M. and E. M. Buchak (1995). *CE-QUAL-W2: A Two-dimensional, Laterally Averaged, Hydrodynamic and Water Quality Model*. US Army Engineer Waterways Experiment Station, Vicksburg, MS.

Corps of Engineers, Hydrologic Engineering center (1978). *Water quality for River-Reservoir systems, computer program description*.

DiToro, D.M., D.J. O'Connor, and R.V. Thomann (1971). A dynamic Model of the phytoplankton population in the Sacramento-San Joaquin Delta. *Advances in Chemistry*, **106**:131-180.

Donigian, A.S., J.C. Imhoff, B.R. Bicknell, and J.L. Kittle (1984). *Application Guide for Hydrological simulation program FORTRAN (HSPF)* EPA-60013-84-065, U.S. EPA, Athens, GA.

Gray, W.G. (editor) (1986). *Physics-based Modeling of Lakes, Reservoirs, and Impoundments*. ASCE, New York.

Harding, L.W., E.C. Itsweire, and W.E. Esalas (1995). Algorithm development for recovering chlorophyll concentrations in the Chesapeake Bay using aircraft remote sensing, 1989-91. *Photogrammetric Engineering and Remote Sensing*, **61**(2):177-185.

Henderson-Sellers, B. and R.I. Davies (1991). Chapter 5: Model Validation and Sensitivity: Case Studies in the Global Context in *Water Quality Modeling Volume IV: Decision Support Techniques for Lakes and Reservoirs* (ed. by Henderson-Sellers and French). CRC Press, Inc. Florida. 310 p.

Khorram, S. (1981). Use of ocean color scanner data in water quality mapping, *Photogrammetric Engineering & Remote Sensing*, **47**(5):667-676.

Krenkel, P.A. and V. Novotny (1980). *Water Quality Management*. Academic Press, New York. 475 p.

Kuo, J.T. and J.H. Wu (1991). A nutrient model for a lake with time-variable volumes. *Wat. Sci. Tech*, **24**(6):133-139.

Kuo, J.T., and R. V. Thomann (1983) "Phytoplankton modeling in the embayments of lakes". *Journal of Environmental Engineering*, ASCE, **109**(6),1311-1332.

Kuo, J.T., J.H. Wu, and W.S. Chu (1994). Water quality simulation of Te-Chi Reservoir using two-dimensional models. *Wat. Sci. Tech*, **30**(2):63-72.

Lahlou, M., and L. Shoemaker, S. Choudhury, R. Elmer, A. Hu, H. Manguerra, A. Parker (1999). *Better Assessment Science Integrating Point and Nonpoint Sources (BASINS) Version 2.0 User's Manual*. Office of Water United /states Environmental Protection Agency, Washington, DC.

Lathrop, R.G. and T.M. Lillesand (1989). Monitoring water quality and river plume transport in Green Bay, Lake Michigan with SPOT-1 imagery. *Photogrammetric Engineering and Remote Sensing*, **55**(3):349-354.

Lathrop, R.G. and T.M. Lillesand (1986). Use of Thematic Mapper data to assess water quality in Green Bay and central Lake Michigan, *Photogrammetric Engineering & Remote Sensing*, **52**(5):671-680.

Lillesand, T.M., W.L. Johnson, R.L. Deuell, O.M. Lindstrom, and D.E. Meisner (1983). Use of Landsat data to predict the trophic state of Minnesota lakes. *Photogrammetric Engineering and Remote Sensing*, **49**(2):219-229.

Lung, W.S. (1986). Assessing phosphorus control in the James River basin. *Journal of Environmental Engineering*, **112**(1):44-60.

Orlob, G.T. (1981). Chapter 8: Models for stratified Impoundments in *Models for Water Quality Management* (ed. by Biswas) . McGraw-Hill Inc, USA.

Orlob, G.T. (1983). Chapter 7: One-dimensional Models for Simulation of Water Quality in *Lakes and Reservoirs in Mathematical Modeling of Water Quality : Streams, Lakes, and Reservoirs* (International series on applied systems analysis: 12) (ed. by Orlob) . John Wiley & Sons, New York.

Poiani, K.A. and B.L. Bedford (1995). GIS-based nonpoint source pollution modeling: Considerations for wetlands, *Journal of Soil and Water Conservation*, **49**(6):613-619.

Ritchie, J.C. and C.M. Cooper (1991). An algorithm for estimation surface suspended sediment concentrations with Landsat MSS digital data. *Water Resources Bulletin*, **27**(3):373-379.

Ruiz-Azuara, P. (1995). Multitemporal analysis of "simultaneous" Landsat imagery (MSS and TM) for monitoring primary production in a small tropical coastal lagoon. *Photogrammetric Engineering and Remote Sensing*, **61**(2):877-198.

Rundquist, D.C., L. Han, J.F. Schalles, and J.S. Peake (1996). Remote measurement of algal chlorophyll in surface waters: The case for the first derivative of reflectance near 690 nm, *Photogrammetric Engineering and Remote Sensing*, **62**(2):195-200.

Sathyendranath, S. and T. Platt (1989). Remote sensing of ocean chlorophyll: consequence of nonuniform pigment profile, *Applied Optics*, **28**(3):490-495.

Scarpace, F.L., K.W. Holmquist, and L.T. Fisher (1979). Landsat analysis of lake quality. *Photogrammetric Engineering and Remote Sensing*, **45**(5):623-633.

Shanahan, P. and D.R.F. Harleman (1986). Chapter 10: Lake Eutrophication Model: A Coupled Hydrophysical-Ecological Model in *Modeling and Managing Shallow Lake Eutrophication With Application to Lake Balaton* (ed. by Somlyody and Straten) . Springer-Verlag, New York.

Shanahan, P., R.A. Luettich, Jr., and D.R.F. Harleman (1991). Chapter 3: Water Quality Modeling: Application to Lake and Reservoirs: Case Study of Lake Balaton, Hungary in *Water Quality Modeling Volume IV : Decision Support Techniques for Lakes and Reservoirs* (ed. by Henderson-Sellers and French) .CRC Press, Inc. Florida. 310 pp.

Shastry, J.S., L.T. Fan, and L.E. Erickson (1973). Nonlinear parameter estimation in water quality modeling. *J. Envir. Engrg. Div.*, ASCE, **99**(3):315-331.

Somlyódy, L. and G. van Straten (1986). *Modeling and Managing Shallow Lake Eutrophication With Application to Llake Balaton*. Springer-Verlag, New York.

Stefan, H.G. (1994). Chapter 2: Lakes and Reservoir Eutrophication : Prediction and Protection in *Water Quality and Its Control* (ed. by Hino) , Rotterdam, Netherlands.

Thomann, R.V. and J.A. Mueller (1987). *Principles of Surface Water Quality Modeling and Control*. Harper and Row, Publishers, Inc., New York. 644 p.

Thomann, R.V. (1963). Mathematical model for dieeolved oxygen. *Journal of the Sanitary Engineering Division*, ASCE, 89 (SA5) .

Thomann, R.V., R.P. Winfield and J.J. Segna (1979). *Verification Analysis of Lake Ontario and Rochester Embayment Three Dimensional Eutrophication Models*. EPA-600/3-79-094.

Thornton, K.W., B. L. Kimmel, and F. E. Payne (editors) (1990). *Reservoir Limnology : Ecological Perspectives*. John Wiley and Sons, Inc., New York.

Verdin, J.P. (1985). Monitoring water quality conditions in a large Western Reservoir whit Landsat imagery. *Photogrammetric Engineering and Remote Sensing*, **51**(3):343-353.

Voinov, A.A. (1991). Chapter 4: Lake Plesheyeveo – A Case Study in *Water Quality Modeling Volume IV: Decision Support Techniques for Lakes and Reservoirs* (ed. by Henderson-Sellers and French). CRC Press, Inc. Florida. 310 p.

Vollenweider, R.A. (1968). *The scientific basis of lake and stream eutrophication with particular reference to phosphorus and nitrogen as eutrophication factors*. Tech. Rep. OECD, DAS/CSI/68.27, Paris.

Vollenweider, R.A. (1975). Input-output models with special reference to the phosphorus loading concept in limnology. Schweiz. *J. Hydrol.*, **37**:53-84.

Watanabe, M., D.R.F. Harleman, and O.F. Vasilier (1983). Chapter 8: Two-and Three-dimensional Mathematical Models for Lakes and Reservoirs in *Mathematical Modeling of Water Quality: Streams, Lakes and Reservoirs* (International series on applied systems analysis) (ed. by Orlob) . John Wiley & Sons, New York.

Whitlock, Charles H. and Chin Y., Kuo (1982). Criteria for the use of regression analysis for remote sensing of sediment and pollutants. *Remote Sensing of Environment*, **12**:151-168.

Yang, M.D., and R. Sykes (1998). Trophic-dynamic modeking in a shallow eutrophic system ecosystem. *Ecological Modelling*, **105**:129-139

Yang, M.D., C. Merry, and R. Sykes (1999). Integration of water quality modeling, remote sensing, and GIS. *Journal of AWRA*, **35**(2):253-263.

Yang, M. D. and Y. F. Yang (1999). Eutrouphic Status Assessment Using Remote Sensing Data. *The EOS/SPIE Symposium on Remote Sensing*. Italy. 56-65p.

Yang, M.D., R. Sykes, and C. Merry (2000). Estimation of algal biological parameters using water quality modeling and SPOT satellite data. *Ecological Modelling*, **125**:1-13.

Chapter 12

LINEAR SYSTEMS APPROACH TO RIVER WATER QUALITY ANALYSIS

Clark C.K. Liu[1] and Jenny Jing Neill[2]

ABSTRACT

The water quality of a polluted river can be improved either by increasing the river's self-purification ability or by reducing the amount of waste loading the river receives. Traditional physically based water quality models do not evaluate separately a receiving river's purification ability and the effect of waste loading; thus, these models are not ideal analytical tools for water quality management. This chapter introduces and discusses an alternative river water quality modeling approach based on the linear systems theory. In a linear systems model, a receiving river's self-purification ability is represented completely by the model's impulse response function, whereas the amount of waste loading the river receives is represented by the model's input function. These two functions can be evaluated separately. Further, a simple convolution integration of these two functions gives the system output. Usually, the system output is the water quality condition of the receiving river. The linear systems model's usefulness as a water quality management tool is demonstrated in this chapter by applying it to studies of dissolved oxygen variations (1) in a steady-state river system that receives both point-source and nonpoint-source waste loading and (2) in a time-variable river system that receives point-source waste loading in the form of periodic function.

1. INTRODUCTION

Every natural river has an ability to assimilate wastes; this is referred to as the river's self-purification ability. While indiscriminate waste discharges cannot and should not be tolerated, complete prohibition of all waste discharges, without taking into consideration the receiving river's self-purification ability, is economically infeasible and environmentally unjustified. Consequently, effluent limitations must be established based on a detailed waste-assimilative-capacity

[1]Professor, Department of Civil Engineering, University of Hawaii at Manoa, Honolulu, HI 96822 USA, liu@wiliki.eng.hawaii.edu.
[2]Engineer, M&E Pacific, Inc., 1001 Bishop Street, Honolulu, HI 96813 USA, jenny_li@air-water.com.

analysis of the receiving river. Because a variety of hydrologic/hydraulic and biochemical reaction factors influence the waste-assimilative capacity of a receiving river, its characteristic value is difficult to establish. As a result, mathematical models have become essential analytical tools for river water quality analysis.

When organic wastes are discharged into a river, one of the noticeable signs of water quality degradation is the depression of the river's dissolved oxygen (DO) content. Therefore, DO content is an important water quality management index. The Federal Water Pollution Control Act Amendment of 1972 and later amendments, which became known as the Clean Water Act, require that all municipal wastewater effluent receive at least secondary treatment. Primary and secondary wastewater treatments are mainly required to reduce oxygen-demanding waste materials, which are measured by biochemical oxygen demand (BOD) (Liu, 1986). For receiving river system modeling, therefore, BOD is often the system input parameter and DO the system output parameter (Fig. 1).

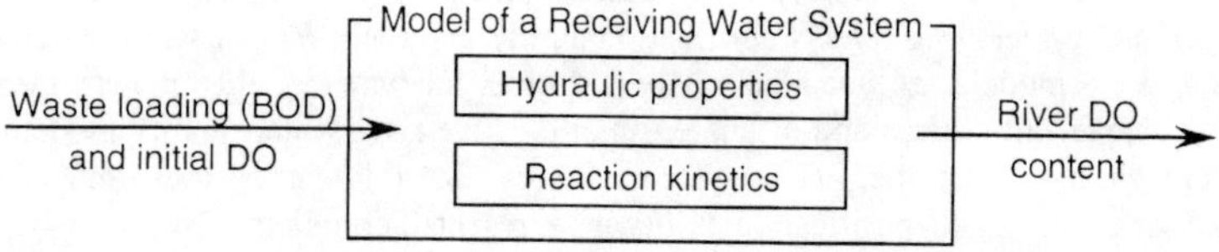

Fig. 1 Physically based river water quality model

Traditional governing equations for BOD-DO variation in a river system are based on mass conservation of the overall BOD and DO in the water, Fick's law of diffusion, and a number of other empirical formulas representing relevant reaction kinetics. They are referred to as physically based models because natural processes of transport and transformation that take place in a receiving river system are represented explicitly by parameters such as the dispersion coefficient, reaeration coefficient, and BOD deoxygenation coefficient (Fig. 1).

In this chapter, an alternative approach to river BOD-DO modeling is presented by using a linear systems approach (Fig. 2). Based on a linear systems approach, the DO content in the receiving river below a discharge point can be easily calculated by convolution integration of the loading function and the impulse response function, or Green's function. This alternative systems approach provides an efficient computation tool; more important, it allows separate evaluation of the self-purification ability of a river and its waste loading.

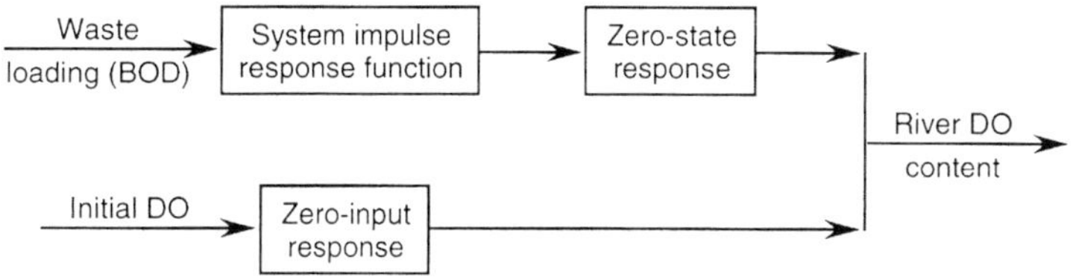

Fig. 2 Linear systems approach to river water quality modeling

2. TRADITIONAL PHYSICALLY BASED RIVER BOD-DO MODELING

2.1 Model Formulation

For a finite water element with density ρ and concentration C for a chemical of interest, the mass change of the chemical inside the element due to movement of the water element (Fig. 3) is equal to

$$\frac{D}{Dt}(C\rho dV)=\frac{\partial}{\partial t}(C\rho dV)+\frac{\partial}{\partial x}(C\rho dV)\frac{\partial x}{\partial t}+\frac{\partial}{\partial y}(C\rho dV)$$

$$+w\frac{\partial}{\partial z}(C\rho dV)\frac{\partial z}{\partial t}+r\rho dV. \qquad (1)$$

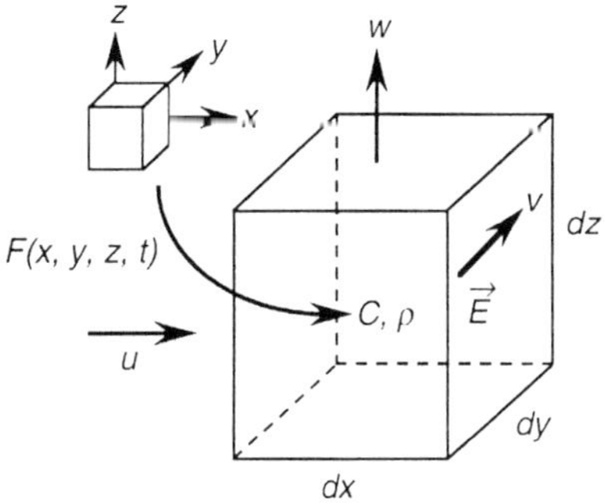

Fig. 3 A finite element of water in a receiving water

All symbols used in this chapter are defined in the Nomenclature section.

Assuming water density is a constant and velocities in the x, y, and z directions are u, v, and w, respectively, then by definition, $u=\partial x/\partial t$, $v=\partial y/\partial t$, and $w=\partial z/\partial t$, Eq. (1) becomes

$$\frac{D}{Dt}(CdV) = \frac{\partial}{\partial t}(CdV) + u\frac{\partial}{\partial x}(CdV) + v\frac{\partial}{\partial y}(CdV) + w\frac{\partial}{\partial z}(CdV) + rdV. \tag{2}$$

For a river, if we assume that the flow velocities in the y and z directions are zero and water density is constant, then Eq. (2) can be simplified to

$$\frac{D}{Dt}(CdV) = \frac{\partial}{\partial t}(CdV) + u\frac{\partial}{\partial x}(CdV) + rdV. \tag{3}$$

If the dispersion effect is considered, by Fick's law, the mass change due to dispersion can be represented as

$$\text{Dispersive flux} = \left\{\begin{array}{l} \left[-E_x\frac{\partial}{\partial x}(C\rho dydz) + \frac{\partial}{\partial x}\left(-E_x\frac{\partial}{\partial x}(C\rho dydz)\right)dx\right] \\ +\left[-E_y\frac{\partial}{\partial y}(C\rho dxdz) + \frac{\partial}{\partial y}\left(-E_y\frac{\partial}{\partial y}(C\rho dxdz)\right)dy\right] \\ +\left[-Ez\frac{\partial}{\partial z}(C\rho dxdy) + \frac{\partial}{\partial z}\left(-Ez\frac{\partial}{\partial z}(C\rho dxdy)\right)dz\right] \end{array}\right\} - \left\{-E_x\frac{\partial}{\partial x}(C\rho dydz) - E_y\frac{\partial}{\partial y}(C\rho dxdz) - Ez\frac{\partial}{\partial z}(C\rho dxdy)\right\} \tag{4}$$

with the assumption that the dispersions are along the principal directions. If we assume uniformity along the y and z directions, then Eq. (4) is simplified to

$$\text{Dispersive flux} = -\frac{\partial}{\partial x}\left[E_x\frac{\partial}{\partial x}(C\rho dydz)\right]dx. \tag{5}$$

Thus the total mass change inside the finite water element is the sum of Eq. (3) and Eq. (5), that is

$$\frac{D}{Dt}(CdV) = \frac{\partial}{\partial t}(CdV) + u\frac{\partial}{\partial x}(CdV) - \frac{\partial}{\partial x}\left[E_x\frac{\partial}{\partial x}(Cdydz)\right]dx + rdV. \tag{6}$$

This total mass change, based on the principle of mass conservation, should be equal to the influx of the chemical into the element, that is

$$\frac{\partial}{\partial t}(CdV) + u\frac{\partial}{\partial x}(CdV) - \frac{\partial}{\partial x}\left[E_x\frac{\partial}{\partial x}(Cdydz)\right]dx + rdV = f(x,t)dV. \tag{7}$$

Since $dV = dxdydz$ for the finite water element, the above equation is equal to

$$\frac{\partial C}{\partial t} + u\frac{\partial C}{\partial x} - E_x\frac{\partial^2 C}{\partial x^2} + r = f(x, t). \tag{8}$$

For river BOD-DO modeling, usually only organic material decay and reaeration are considered and both are taken as first-order reaction kinetics. Thus the governing equation for BOD, L, is

$$\frac{\partial L}{\partial t} - E_1\frac{\partial^2 L}{\partial x^2} + u\frac{\partial L}{\partial x} + k_d L = f_L \tag{9}$$

and the governing equation for the DO deficit, D, is

$$\frac{\partial D}{\partial t} - E_2\frac{\partial^2 D}{\partial x^2} + u\frac{\partial D}{\partial x} + k_2 D = k_2 L + f_D. \tag{10}$$

Analytical solutions to these equations with general initial and boundary conditions are not available. This is mainly because of the complication of their right-side terms, which physically represent inflow loading along the river. Several different kinds of methods are applicable for solving Eq. (9) and Eq. (10) with certain given boundary and initial conditions; these include the perturbation (Li, 1972) and transformation methods (Bennett, 1971).

Many river water quality models, such as the popular U.S. Environmental Protection Agency QUAL2E model (Brown and Barnwell, 1987), are solved by numerical approximations. One major problem of a numerical solution is numerical dispersion, which artificially changes the dispersive transport of BOD and DO.

2.2 Relevant Reaction Kinetics

As organic materials in wastewater effluent (BOD) decompose in the receiving river, DO decreases via deoxygenation. Through reaeration, oxygen enters the river from the atmosphere to replenish the loss of DO. Deoxygenation and reaeration are the two most important reaction kinetics of river BOD-DO modeling.

2.2.1 BOD Deoxygenation

The BOD of a water sample is the quantity of oxygen used by microorganisms in the aerobic stabilization of organic wastes. BOD is a function of time; its level increases with time as more organic wastes become stabilized. BOD reaches its maximum level when all organic wastes in the sample have been stabilized. This maximum level is referred to as ultimate BOD, or BOD_u, which is a constant for any water sample.

BOD decomposition is commonly assumed to be a first-order decay process.

$$\frac{dL}{dt} = -k_1 L \tag{11}$$

The laboratory BOD deoxygenation rate constant, k_1, can be determined by plotting the experimental BOD series on log-normal paper. In the natural environment, hydraulic and other physiochemical conditions are somewhat different from those in a laboratory; therefore, a laboratory-determined rate constant might not be directly applicable to field conditions. An empirical equation was developed that allows a modification of the laboratory-determined deoxygenation rate constant by taking river hydraulic properties into consideration (Tetra Tech, 1978).

More reliable BOD deoxygenation rate constants can be obtained based on data collected by intensive water quality surveys in the receiving river (Liu, 1986).

2.2.2 Reaeration

Atmospheric reaeration is a natural process of oxygen transfer from the atmosphere to a water body with a dissolved oxygen deficit. Atmospheric reaeration is also a first-order reaction.

$$\frac{dD}{dt} = -k_2 D \tag{12}$$

where D is the concentration of DO deficit. It is defined as the difference between the saturation level of DO at the river water temperature (C_s) and the actual in-stream DO.

In practice, reaeration rate constants are estimated by predictive equations such as the O'Connor and Dobbins equation (O'Connor and Dobbins, 1956), the Churchill equation (Churchill et al., 1962), and the Tsivoglou equation (Tsivoglou and Wallace, 1972). With many predictive equations available, a practicing engineer doing river water quality analyses is often perplexed by the fact that these equations give a wide range of predicted k_2 for a particular receiving river. In two of the review papers concerning this subject, the authors concluded that none of these equations permits predictions of reaeration rate constants that are sufficiently accurate for establishing effluent limitations (Wilson and McCleod, 1974; Rathbun, 1977).

Techniques are available to measure the reaeration rate constant in the field by using radioactive Krypton-85 as a tracer (Tsivoglou et al., 1968). A modified gas tracer method that uses a hydrocarbon gas, such as ethylene and propane, as the tracer was later developed by the U.S. Geological Survey (Rathbun et al., 1975). This method was successfully applied to the Canadaiqua

Outlet in central New York as part of a regular intensive river water quality survey (Liu and Fok, 1983).

2.2.3 Other Reaction Kinetics

In the early studies of river BOD-DO by Streeter and Phelps (1925), the only reactions considered were deoxygenation due to carbonaceous materials and reoxygenation due to atmospheric reaeration. More detailed studies that include other reactions such as nitrification and photosynthesis/respiration should be conducted.

In the presence of nitrifying bacteria, organic nitrogen and ammonia nitrogen in a river can be oxidized sequentially to nitrites and then to nitrates. This process is called nitrification. Because nitrification often lags several days behind the start of carbonaceous oxygen demand, it is referred to as second stage BOD (Viessman and Hammer, 1998). In the past, many wastewater effluents received only primary treatment or no treatment at all before being discharged into a receiving river. Under those conditions, the greatest DO depletion or "sag point" was due to carbonaceous BOD (CBOD) alone. With a mandatory minimum treatment imposed by the Federal Water Pollution Control Amendment of 1972 and subsequent amendments, essentially all wastewater effluents have since received secondary or advanced level treatment. For more accurate river waste-assimilative-capacity analysis, nitrogenous BOD (NBOD) as well as CBOD should be included.

Photosynthesis is the biological synthesis of organic compounds by chlorophyll-bearing plants in the presence of solar energy. A by-product of this process is oxygen. On the other hand, oxygen is consumed by living organisms during respiration. Both photosynthesis and respiration affect the DO content of a biologically active stream. A natural water body with significant biological oxygen production and consumption can be analyzed using a time-varying model (O'Connor and DiToro, 1970).

The formulation of a practical river water quality model is mainly based on the assumption that the river is in a steady state. The steady-state assumption is acceptable in engineering applications because a water quality analysis for a receiving river is normally conducted under low-flow conditions (i.e., when river flow consists mainly of a constant base flow). A method that separates the time-varying effects of photosynthesis and respiration from observed DO was developed (Liu, 1982). This method modifies the observed DO profile and allows the rational application of steady-state models to a biologically active river.

2.3 Model Validation

Before a river water quality model can be used in a predictive analysis, its validity must be established in terms of model calibration and validation (Thomann, 1980). In a physically based river water quality model, transport processes are represented by conceptual kinetic formulations. Rate constants for coefficients involved in these formulations must be determined by the interpretation of experimental data collected in the laboratory or in the field (Tetra Tech, 1978). Because of the complexity of natural river systems, developing a range of coefficient values for any kinetic coefficient is often more appropriate than attempting to define a single value. Once the estimated range of values of kinetic coefficients is defined, the values are plugged into the model, which is then solved to obtain predicted values of receiving river water quality, such as the DO content. Model calibration is completed when the calculated values are reasonably close to the observed ones.

After successful calibration, the model is subjected to tests using field data from an independent intensive water quality survey conducted on the same river. This is necessary to verify the model's predictive capability. Model calibration and verification often become rather difficult and confusing, especially when a sophisticated physically based model is used. Because a sophisticated model consists of many kinetic coefficients, any one of many combinations of coefficient values may lead to successful calibration, even though some transport kinetics are not truly represented. The uncertainties inherent in physically based river water quality modeling can be largely alleviated by following an alternative modeling approach based on the linear systems theory.

3. GREEN'S FUNCTION AND THE LINEAR SYSTEMS THEORY

3.1 Green's Function and Dirac Delta Function

An ordinary differential equation can be expressed in general form as

$$\boldsymbol{M}u(x) = f(x) \tag{13}$$

where $\boldsymbol{M}$ is an ordinary differential operator, $u(x)$ is an unknown function of x, and $f(x)$ is a known function of x. One method of finding the solution for this differential equation is to find the operator inverse to $\boldsymbol{M}$, namely, operator $\boldsymbol{M}^{-1}$, such that the product $\boldsymbol{M}\boldsymbol{M}^{-1}$ is the identity operator. Thus, multiplying the equation above by $\boldsymbol{M}^{-1}$ will result in

$$u(x) = \boldsymbol{M}^{-1} f(x) \ , \tag{14}$$

which is the solution to Eq. (13).

It should be expected that the inverse operator $\boldsymbol{M}^{-1}$ to an ordinary differential operator is an integral operator. Assume this operator has a kernel $g(x-\omega)$, so that

$$\boldsymbol{M}^{-1} u(x) = \int_{\Omega} g(\omega) u(x-\omega) d\omega. \tag{15}$$

where Ω is the integration domain and ω is an integration variable. The kernel of the integral operator, g, is called Green's function of the ordinary differential operator $\boldsymbol{M}$. Substituting this inverse operator back into Eq. (14), the solution to Eq. (13) becomes

$$u(x) = \boldsymbol{M}^{-1} f(x) = \int_{\Omega} g(\omega) f(x - \omega) d\omega \tag{16}$$

or, after changing the integration variable

$$u(x) = \boldsymbol{M}^{-1} f(x) = \int_{\Omega} g(x - \omega) f(\omega) d\omega. \tag{17}$$

With further manipulation, we have

$$\begin{aligned} u(x) &= Iu(x) = \boldsymbol{M}\boldsymbol{M}^{-1} u(x) = \boldsymbol{M} \int_{\Omega} g(x - \omega) u(\omega) d\omega \\ &= \int_{\Omega} \boldsymbol{M} g(x - \omega) u(\omega) d\omega \end{aligned} \tag{18}$$

where $I = \boldsymbol{M}\boldsymbol{M}^{-1}$ is the identity operator.

According to one of the properties of the Dirac delta function, for any continuous function, $\xi(x)$ (Sokolnikoff and Redheffer, 1966)

$$\xi(x) = \int_{\Omega} \delta(\omega - x) \xi(\omega) d\omega. \tag{19}$$

Since $u(x)$ is a set of continuous equations in the domain Ω, by comparing Eq. (18) with Eq. (19), it can be seen that Green's function, $g(x - \omega)$, for a given ordinary differential equation (Eq. (13)) is the solution to the equation:

$$\boldsymbol{M} g(x - \omega) = \delta(\omega - x) \tag{20}$$

or simply

$$\boldsymbol{M} g(x) = \delta(x). \tag{21}$$

Comparing Eq. (13) with Eq. (21), we find that the unknown function, $u(x)$, of an ordinary differential equation is Green's function when the known function, $f(x)$, takes the form of the Dirac delta function.

This method can be extended to partial differential equations. For a linear partial differential equation in n-dimensional space, we define a partial differential operator, $\boldsymbol{M}$, and express the equation in the form

$$\boldsymbol{M} u(x_1, x_2, \ldots, x_n) = f(x_1, x_2, \ldots, x_n). \tag{22}$$

Assume that the inverse of this operator is in a form such that

$$M^{-1}u(x_1, \ldots, x_n) = \iint_{\Omega} \cdots \int g(x_1 - \omega_1, \ldots, x_n - \omega_n) \\ u(\omega_1, \ldots, \omega_n) d\omega_1 \cdots d\omega_n \tag{23}$$

and so

$$u(x_1, \ldots, x_n) = M^{-1} \iint_{\Omega} \cdots \int g(x_1 - \omega_1, \ldots, x_n - \omega_n) \\ u(\omega_1, \ldots, \omega_n) d\omega_1 \cdots d\omega_n \\ = \iint_{\Omega} \cdots \int M^{-1} g(x_1 - \omega_1, \ldots, x_n - \omega_n) \\ u(\omega_1, \ldots, \omega_n) d\omega_1 \cdots d\omega_n . \tag{24}$$

It can also be concluded that Green's function for a linear partial differential equation should be the solution to the equation with a Dirac delta function as its input (Friedman, 1990)

$$Mg(x_1, \ldots, x_n) = \delta(x_1, \ldots, x_n) = \delta(x_1) \ldots \delta(x_n). \tag{25}$$

Thus the solution to Eq. (22) is

$$u(x_1, \ldots, x_n) = M^{-1} \iint_{\Omega} \cdots \int g(x_1 - \omega_1, \ldots, x_n - \omega_n) \\ f(\omega_1, \ldots, \omega_n) d\omega_1 \cdots d\omega_n . \tag{26}$$

By changing the integration variable, it becomes

$$u(x_1, \ldots, x_n) = M^{-1} \iint_{\Omega} \cdots \int g(\omega_1, \ldots, \omega_n) \\ f(x_1 - \omega_1, \ldots, x_n - \omega_n) d\omega_1 \cdots d\omega_n . \tag{27}$$

3.2 Linear Systems Modeling Approach

Green's function, which transforms a partial differential equation to an integral equation, is similar to kernel functions associated with the systems modeling approach. The systems modeling approach was introduced by Volterra (1959) for use in simulating electromagnetic and elastic processes. Volterra (1959) showed that any hereditary process, in which the current state of the system depends on the stated variables and their entire previous history, could be represented by a power series:

$$y(t)=x(t)+\int_{-\infty}^{t} h_1(t,s)x(s)ds+\int_{-\infty}^{t}\int_{-\infty}^{t} h_2(t,s_1,s_s)x(s_1)x(s_2)ds_1ds_2+\cdots$$

$$+\int_{-\infty}^{t}\cdots\int_{-\infty}^{t} h_n(t,s_1,s_s,\ldots,s_n)\mathop{\pi}_{i=1}^{n} x(s_i)ds_i \tag{28}$$

where y and x are the system output and input, respectively. The first term on the right side of the equation, $x(t)$, denotes the transient state of the system. The terms h_1, h_2, ..., h_n are the kernel functions of the multiple integral expansion; their forms determine the system behavior.

According to the terminology used by Volterra (1959), Eq. (28) is a Volterra integral equation of the second kind. A nonlinear analysis of the watershed response to rainfall was conducted using a truncated Volterra series that consists of only the first three terms on the left side of Eq. (16) (Liu and Brutsaert, 1978).

The linear systems model of a dynamic system can be formulated by taking only the first two terms on the left side of Eq. (28):

$$y(t)=x(t)+\int_{-\infty}^{t} h_1(t,s)x(s)ds. \tag{29}$$

For a process to be time-invariant and to obey the principle of dissipation of hereditary action, the kernel function in Eq. (29) [$h_1(t, s)$] must be in the form of $h_1(t-s)$ (Volterra, 1959). Further, if the transient state of this linear system is zero, then it is the same as Eq. (17). In other words, the kernel function of a linear system based on the Volterra integral series is Green's function of this system. This kernel function is also called the impulse response function of a linear system (Liu and Brutsaert, 1978).

The linear systems modeling technique has been applied to surface hydrology and is called the unit hydrograph method (Dooge, 1959; Nash, 1959; Amorocho, 1963; Liu and Brutsaert, 1978). More recently, it has been used in the simulation of chemical transport in soils and groundwater (Jury, 1982; Liu, 1988; Liu et al., 1994).

The impulse response function of a linear systems model describes the overall effects of the system's hydraulic and reaction mechanisms. Thus, the success of systems modeling depends largely on how accurately and efficiently the system impulse response function is evaluated. Generally, it can be determined by one of three methods: (1) system identification, (2) system parameterization, and (3) system physical parameterization.

In the system identification method, the determination of the impulse response function is based on given input and output data; no a priori information on the system behavior is assumed. It is sometimes referred to as the "black-box" method. One noted example is the unit hydrograph method for which the instantaneous unit hydrograph is evaluated based on given effective rainfall (input) and direct runoff (output) (Dooge, 1973). As the functional series of the Volterra model is essentially a smoothing operation, small errors of input and output data may result in large deviations of the calculated impulse response function (Distefeno, 1974). To overcome this problem, it is sometimes necessary to introduce system optimization techniques (Liu and Brutsaert, 1978) or to conduct frequency-domain analysis (Bras and Rodriguez-Iturbe, 1985).

When some a priori information of system behavior is available, the impulse response function can be determined by adjusting parameter values of a specific distribution function. This is the system parameterization, or the "grey-box" method. Jury (1982) suggested the use of a log-normal distribution to express the impulse response function, which he calls the transfer function, of a system model for solute transport in field soils. By so doing, the impulse response function is determined if two parameters in a specific log-normal distribution, i.e., mean and standard deviation, are known. More recently, the gamma distribution function was used as the general form of the impulse response function of soil transport systems (Liu et al., 1994).

In the system physical parameterization method, the impulse response function of a system model is expressed as a function of the physical parameters that describe relevant transport and reaction mechanisms. Therefore, this method requires the formulation of a corresponding physically based model. The impulse response function is essentially the solution of a governing equation of the physically based model with a Dirac delta function input (Sposito et al., 1986).

In this chapter, an example of linear systems modeling for river water quality analysis, based only on the physical parameterization method, is presented.

4. LINEAR SYSTEMS MODELING OF A RIVER BOD-DO SYSTEM

DO content is usually an important water quality management index. When organic wastes are discharged into a river, one of the noticeable signs of water quality degradation is the depression of the river's DO content. Since the DO level is a major concern for maintaining the ecological well-being of a receiving river, we will derive a river linear systems model to study the DO variation with various kinds of BOD input/loading (Fig. 4).

Mathematically, a linear systems model of a river's BOD-DO variation takes the form of

$$D(\theta,t) = D_s(\theta,t) + \int_{\theta_0}^{\theta}\int_{t_0}^{t} X(\zeta,\tau)h(\theta-\zeta,t-\tau)d\zeta d\tau \tag{30}$$

where θ is the travel time of a transport system defined as x/u; x is the travel distance; u is the river velocity; t is the observation time; θ_0 is the start of travel time; t_0 is the start of observation time; $D(\theta, t)$ is the state or the output of the system at time t; $D_s(\theta, t)$ is the zero-input response of the system, or the transient state of the system without the impact of input; $X(\zeta, t)$ is the input to the system; and $h(\theta, t)$ is the impulse-input response function of the system, which is the impulse-input response function of DO. The integral part of the model is basically an application of Green's function, as discussed above.

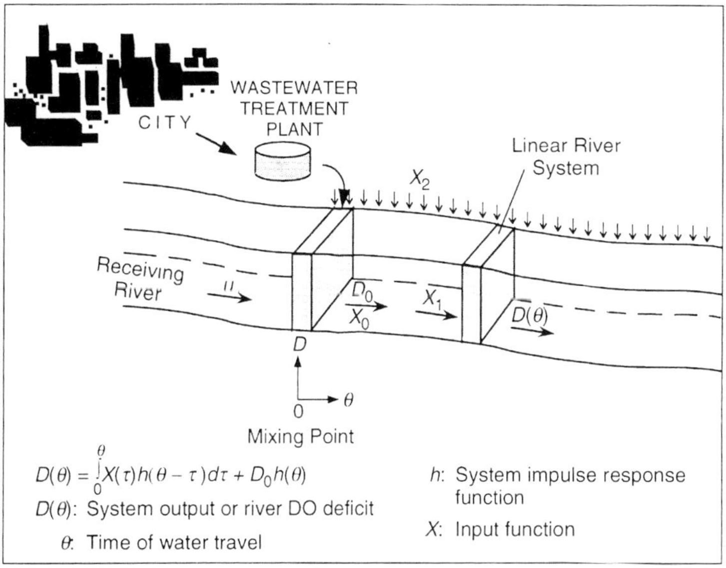

Fig. 4 Schematic of a river system model

4.1 Response Function and Input Functions

The river system's impulse-input response function, according to the physical parameterization method, is the solution of the governing equation (Eq. (10)) with a Dirac delta function input, that is

$$\frac{\partial D}{\partial t} - E_2 \frac{\partial^2 D}{\partial x^2} + u\frac{\partial D}{\partial x} + k_2 D = k_d \delta(x, t). \tag{31}$$

The derivations of the response function and the input functions are discussed below.

4.2 Steady-State Nondispersive Rivers

For a nontidal river, water quality modeling is normally conducted under a steady-state, low-flow condition when the dispersion effect is not important, and thus the classical Streeter–Phelps equation is applicable

$$u\frac{dD}{dx} + k_2 D = X(x) \tag{32}$$

Rewritten in terms of travel time, it becomes

$$\frac{dD}{d\theta} + k_2 D = X(\theta). \tag{33}$$

The left side of Eq. (33) describes the system behavior, whereas the right side gives the system loading or input. For a steady-state condition, the linear system is only a function of travel time. The linear systems model (Fig. 4) then takes the form

$$D(\theta) = D_s(\theta) + \int_{\theta_0}^{\theta} X(\zeta) h(\theta - \zeta) d\zeta. \tag{34}$$

The river system is defined as an elemental river volume that moves downstream. At the mixing point, or $\theta = 0$, this elemental volume has a DO deficit of D_0 and receives a point-source waste loading of L_0.

The impulse response function, $h(\theta - \zeta)$, of the river system is independent of the observation time. It is the solution of Eq. (33) with a Dirac delta function input, that is, $X(\theta) = \delta(\theta)$, and readily shown as

$$h(\theta - \zeta) = k_d \exp[-k_2(\theta - \zeta)]. \tag{35}$$

It is a function of the river velocity, deoxygenation, and reaeration coefficient only.

For the BOD-DO system of a river, the overall input $X(\theta)$ usually is the summation of three kinds of loading. These are point-source waste loading, nonpoint-source waste loading, and distributed organic waste loading due to benthic materials on the river bottom, namely, $X_1(\theta)$, $X_2(\theta)$, and $X_3(\theta)$, respectively.

As a linear system moving downstream, point-source waste input is subject to first-order decomposition with a constant rate, k_d. The input function due to point-source waste loading of BOD at $\theta = \theta_0$ to this system is equal to

$$X_1(\theta) = L_0 \exp[-k_d(\theta - \theta_0)]. \tag{36}$$

The input function due to nonpoint-source waste loading of BOD to the system with an inflow rate, $L_d(\varphi)$ (e.g., from refuse dump leakage along a river), is

$$X_2(\zeta) = \int_{\theta_0}^{\zeta} L_0(\zeta)\exp\left[-k_d(\zeta - \varphi)\right]d\varphi. \tag{37}$$

The input function due to distributed organic waste loading of benthic materials fixed on the river bottom can be expressed simply as

$$X_3(\theta) = S_b(\theta - \theta_0). \tag{38}$$

The zero-input response of a river system can be expressed as

$$D_s(\theta) = D_0 \exp(-k_2\theta). \tag{39}$$

Overall, the DO variation along a river due to BOD loading and benthic materials at the river bottom is described by the linear systems model:

$$\begin{aligned} D(\theta) &= D_s(\theta) + \int_{\theta_0}^{\theta} X(\zeta)h(\theta - \zeta)d\zeta \\ &= D_s(\theta) + \int_{\theta_0}^{\theta} X_1(\zeta)h(\theta - \zeta)d\zeta + \int_{\theta_0}^{\theta} X_2(\zeta)h(\theta - \zeta)d\zeta + \int_{\theta_0}^{\theta} X_3(\zeta)h(\theta - \zeta)d\zeta \\ &= D_0 \exp(-k_2\theta) + \int_{\theta_0}^{\theta} \left[L_0 \exp(-k_d\zeta)\right]k_d \exp\left[-k_2(\theta - \zeta)\right]d\zeta \\ &\quad + \int_{\theta_0}^{\theta} \left\{\int_{\theta_0}^{\zeta_1} L_0(\zeta_2)\exp\left[-k_d(\zeta_1 - \zeta_2)\right]d\zeta_2 \right\}k_d \exp\left[-k_2(\theta - \zeta_1)\right]d\zeta_1 \\ &\quad + \int_{\theta_0}^{\theta} S_b(\zeta)k_d \exp\left[-k_2(\theta - \zeta)\right]d\zeta. \end{aligned} \tag{40}$$

For the first line of Eq. (40), the first term on the right side, $D_s(\theta)$, gives the zero-input response of a linear steady-state river system. The second term on the right side, $\int_{\theta_0}^{\theta} X(\zeta)h(\theta - \zeta)d\zeta$, gives the zero-state response of the system (see Fig. 2). Note that the zero-state response of the system is produced by the three types of waste loading identified above.

4.3 Time-Variable Dispersive Rivers

Chemical transport in a tidal river or estuary, where the longitudinal gradient of the chemical concentration is the main concern, is subject to two hydraulic processes: (1) freshwater flow and (2) tidal reversals. For such a time-variable system, dispersion becomes an important transport mechanism. Under such conditions, Eq. (10) is used to derive the corresponding impulse-input response function for DO (see Appendix for derivation):

$$h(\theta-\zeta, t-\tau)=\frac{1}{\sqrt{4\pi E(t-\tau)}} \exp\left\{-\frac{u^2\left[(\theta-\zeta)-(t-\tau)\right]^2}{4E(t-\tau)}-k_2(t-\tau)\right\}. \quad (41)$$

The impulse-input response function varies with both travel time (θ) and observation time (t). Fig. 5 shows a plot of the impulse-input response function at different observation times.

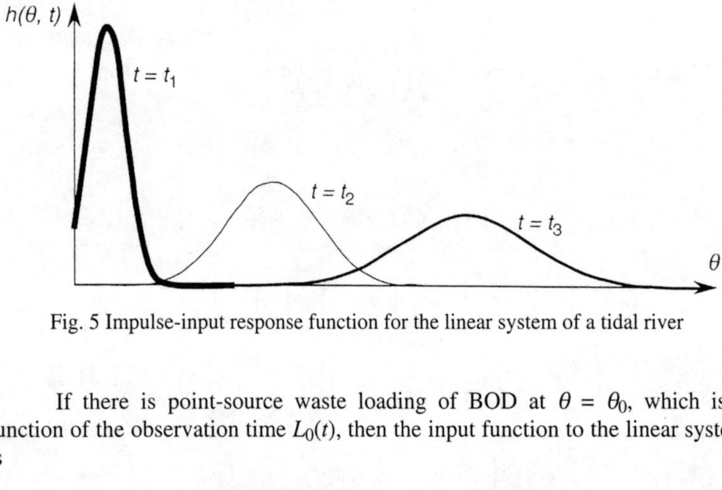

Fig. 5 Impulse-input response function for the linear system of a tidal river

If there is point-source waste loading of BOD at $\theta = \theta_0$, which is a function of the observation time $L_0(t)$, then the input function to the linear system is

$$X_1(\theta, t)=\int_{t_0}^{t} L_0(\tau)\frac{u(\theta-\theta_0)}{\sqrt{4\pi E(t-\tau)^3}} \exp\left\{-u^2\frac{\left[(\theta-\theta_0)-(t-\tau)\right]^2}{4E(t-\tau)}-k_d(t-\tau)\right\}d\tau. \quad (42)$$

If there is a spatially distributed waste loading of BOD, then the input function to the linear system is

$$X_2(\theta, t) = \int_{t_0}^{t} \int_{\theta_0}^{\theta} \frac{L_0(\zeta, \tau)}{\sqrt{4\pi E(t-\tau)}} \exp\left\{ -\frac{u^2\left[(\theta-\zeta)-(t-\tau)\right]^2}{4E(t-\tau)} - k_d(t-\tau) \right\} d\zeta d\tau. \quad (43)$$

The input function due to benthic materials on the river bottom remains unchanged as

$$X_3(\theta) = S_b(\theta - \theta_0). \quad (44)$$

The zero-input response of a tidal river system can be expressed as

$$D_s(\theta, t) = D_0(\theta)\exp(-k_2 t). \quad (45)$$

5. APPLICATION

5.1 BOD-DO Variations in a Steady-State Nondispersive River

A small, free-flowing river reach, as shown in Fig. 4, receives both point- and nonpoint-sources of waste loading; point-source loading is from the effluent of a wastewater treatment plant and nonpoint-source loading is from the leakage of refuse dumps along the river bank. River flow under the design low-flow condition, after mixing with the treatment effluent, is 5 m^3/s. The BOD concentration of the river water at the mixing point is 20 g/m^3. Nonpoint-source pollution produces an additional loading of 15 kg/m^3 BOD for each kilometer of river length. The river water temperature at the mixing point is 24°C, which corresponds to DO content at a saturation of 8.4 g/m^3. The river DO content at the mixing point is 5.9 g/m^3. The river velocity is 0.002 m/s, or 0.173 km/d. Based on data from an intensive river water quality survey, the BOD deoxygenation coefficient and reaeration coefficient were estimated at 0.23 per day and 0.46 per day, respectively.

Fig. 6 shows the impulse response function of a linear system which moves down the river (as shown in Fig. 4). Fig. 6 also shows the input functions due to point-source wastewater effluent from the treatment plant and nonpoint-source leakage from refuse dumps. Note that the results in Fig. 6 are plotted with respect to travel time (θ). Travel time is essentially an indication of the downstream location of the linear river system moving at the river velocity; e.g., a travel time of 2 days indicates a river location of 2 × 0.173 km, or 0.346 km, below the mixing point.

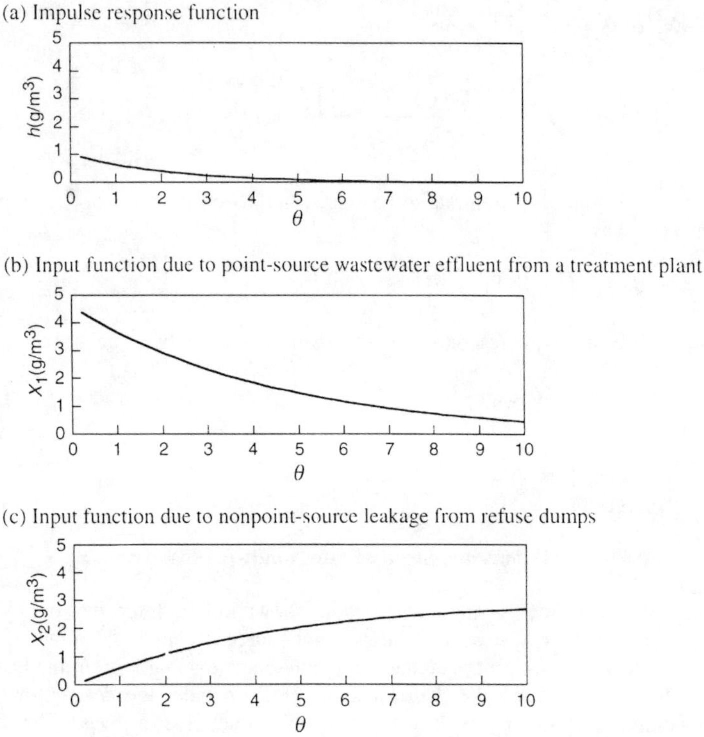

Fig. 6 Impulse response function and input functions of a linear river system

DO deficits in a receiving river are produced by point-source and nonpoint-source waste loadings and by the initial DO deficit. Variations of the river DO deficit were calculated using Eq. (34), and the results are shown in Fig. 7. Also shown in Fig. 7 is the overall DO variation along the river, which is calculated by superimposing all types of contributions to river DO deficit. The critical DO in this river system, produced by the given waste loading and initial DO deficit, is about 0.9 g/m^3 and occurs at a travel time of about 5.5 days or at 9.5 km below the mixing point.

The principal objective of water quality management planning is to find the most cost-effective way of improving the water quality of a polluted river. One management alternative involves the enhancement of the receiving river's self-purification ability by low-flow augmentation or by in-stream aeration. In-stream aeration increases the river's reaeration capacity by mechanical means (ReVelle and ReVelle, 1988). Fig. 8 shows that by increasing the reaeration coefficient

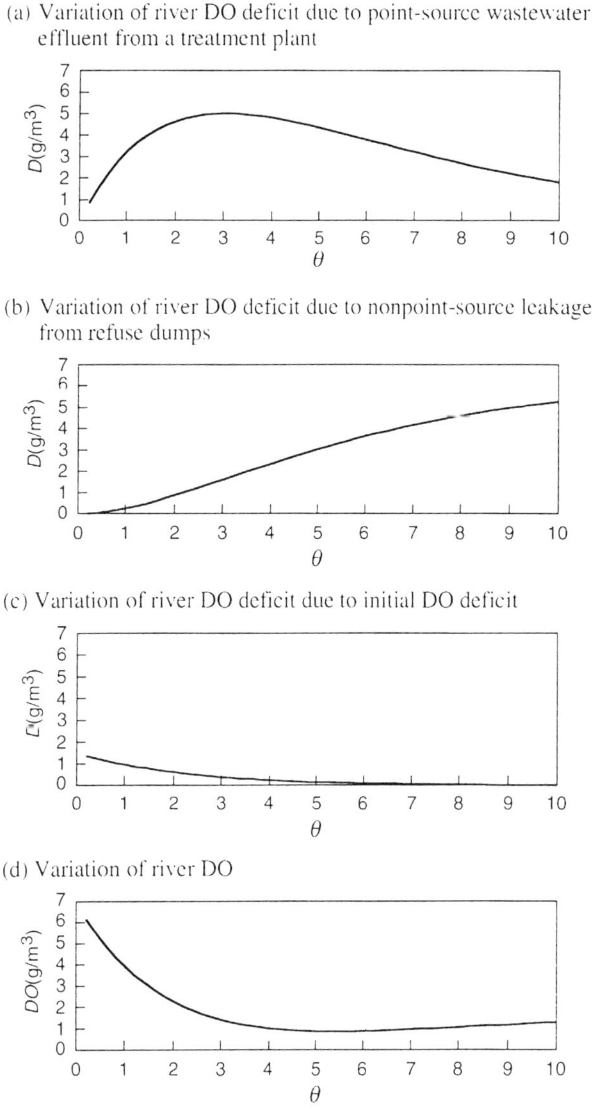

Fig. 7 River DO variations calculated by a linear systems model

from 0.46 per day to 0.92 per day, the minimum river DO content will rise from 0.9 g/m^3 to 4.2 g/m^3. Apparently, in-stream aeration is an effective water quality management alternative. Fig. 8 also shows the improvement of river DO content either by elimination of all leakages from refuse dumps along the river or by a 50% reduction of the point-source wastewater loading. Elimination of refuse dump leakages would result in a much more significant improvement of river DO content.

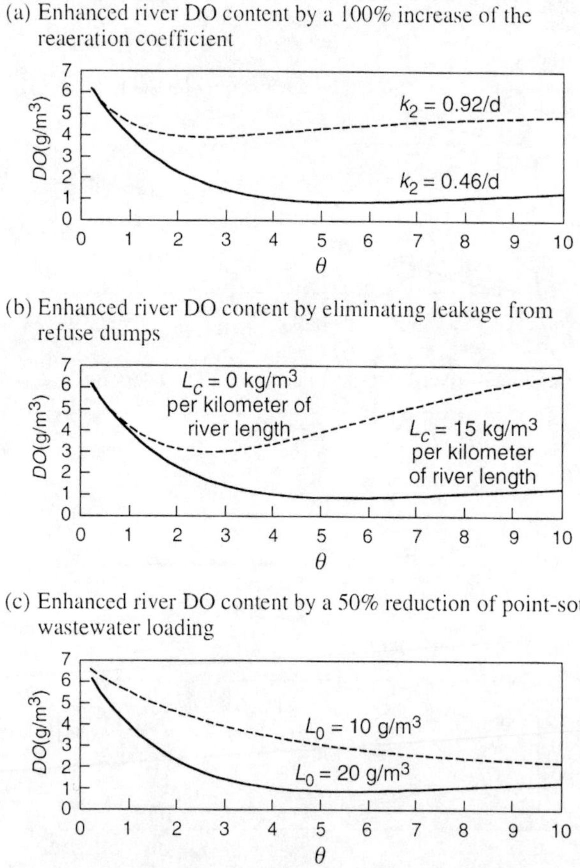

Fig. 8 Evaluation of water quality management alternatives

5.2 BOD-DO Variations in a Time-Variable Dispersive River

The linear systems model formulated in Section 4.3 is used to study BOD-DO variation in a river with a fluctuating BOD point-source that takes the following form:

$$L_0(t) = \bar{L}_0\left[1 + n\left(\sin\frac{2\pi t}{T}\right)\right] \quad \text{at } x = 0 \tag{46}$$

where $\bar{L}_0$ is the mean value of BOD at the discharge point, or $\theta = 0$. This problem was investigated earlier with a perturbation method (Li, 1972). Input conditions used in this problem are summarized in Table 1.

Table 1. Summary of input data for the linear systems model of a time-variable dispersive river

Parameter	Value
Dispersion coefficient, E	20.0 m^2/s
Flow velocity, u	0.2 m/s
Period, T	86,400 s
Amplitude, n	0.5
Deoxygenation coefficient, k_d	0.26/d
Reaeration coefficient, k_2	0.52/d

By following the linear systems modeling approach, the BOD variation along the river becomes

$$L(\theta, t) = \int_0^t L_w(t-\tau)\frac{u\theta}{\sqrt{4\pi E(t-\tau)^3}} \exp\left\{-u^2\frac{[\theta-(t-\tau)]^2}{4E(t-\tau)} - k_d(t-\tau)\right\}d\tau . \tag{47}$$

If the river is initially saturated with oxygen or the initial DO deficit is zero, and if there is no distributed BOD input and no benthic BOD demand, D_s, X_2 and X_3 are all zero. Then the DO variation is

$$D(x,t)=\int_0^t\int_0^\theta L(\zeta,\tau)\frac{k_d}{\sqrt{4\pi E_2(t-\tau)}}$$
$$\exp\left\{-\frac{\left[(\theta-\zeta)-u(t-\tau)\right]^2}{4E_2(t-\tau)}-k_2(t-\tau)\right\}d\zeta d\tau. \qquad (48)$$

A FORTRAN computer program was written to carry out the computations of Eq. (47) and Eq. (48). The details of the program documentation are included in a University of Hawaii Water Resources Research Center project report (Liu and Neill, 2000).

As shown in Figs. 9 and 10, the linear systems modeling approach provides very satisfying results for BOD-DO variations in a uniform tidal river, compared with the results of the physically based model as solved by the perturbation method. The general forms of Eq. (47) and Eq. (48) indicate that the

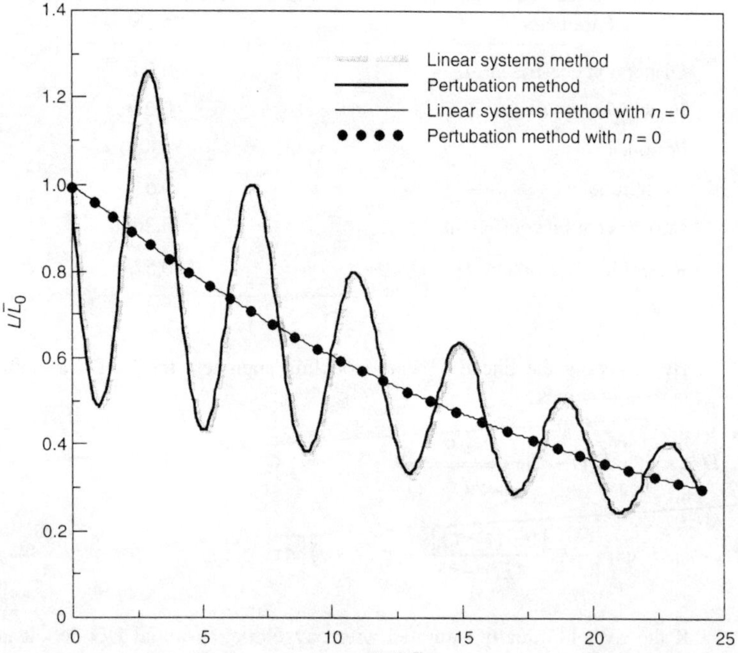

Fig. 9 BOD variation in a uniform time-variable river simulated using (1) the linear systems model and (2) the physically based model solved by the perturbation method

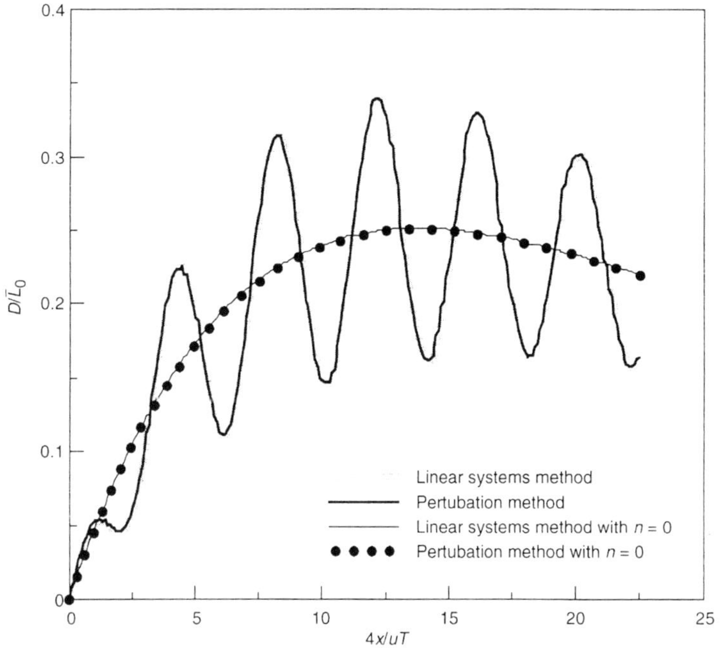

Fig. 10 DO variation in a uniform time-variable river simulated using (1) the linear systems model and (2) the physically based model solved by the perturbation method

linear systems modeling approach is much more flexible for handling different forms of the loading function than the traditional physically based modeling approach. For example, no analytical solution is available for a physically based tidal river model with distributed loadings, but a solution can be readily obtained using the linear systems modeling approach.

The dispersion effect on the distribution of BOD and DO in a river system was studied broadly in the past by other researchers (Li, 1972; Thomann, 1973). In this chapter, we demonstrate the use of the linear systems modeling approach to simulate the dispersion influence on the distribution of BOD and DO for the same periodical source, $L_W(t)$ (Eq. (46)). This is done by comparing the simulation results of Eq. (47) and Eq. (48) with those of Eq. (49) and Eq. (50).

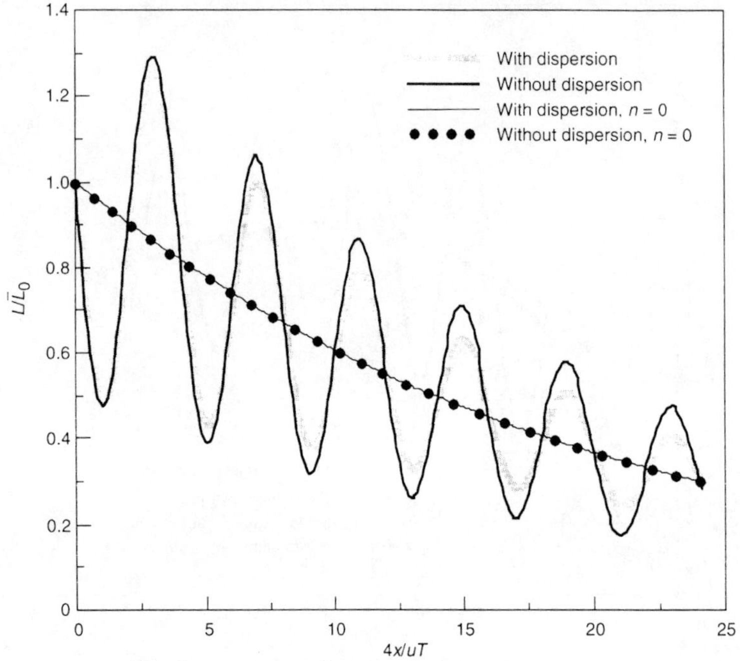

Fig. 11 The effect of longitudinal dispersion on BOD distribution in a time-variable river

$$L(\theta,t) = L_0(t-\theta)\exp(-k_d\theta) \tag{49}$$

$$D(\theta,t) = \int_{\theta_0}^{\theta} L[\zeta, t-(\theta-\zeta)]k_d \exp[-k_2(\theta-\zeta)]d\zeta . \tag{50}$$

The simulation results of river water quality with and without dispersion are plotted in Figs. 11 and 12 for comparison.

Figs. 11 and 12 illustrate the importance of the joint action of the source fluctuation and longitudinal dispersion on the distributions of BOD and DO, respectively. If the input source is not subject to periodic fluctuation (i.e., if $n = 0$ for Eq. (46)), the distributions are very similar to each other. However, when the source is subject to periodic fluctuation, there is a difference between the simulated results, especially as the system moves farther downstream. Similar findings were derived earlier by other researchers by following a physically based modeling approach (Li, 1972; Thomann, 1973).

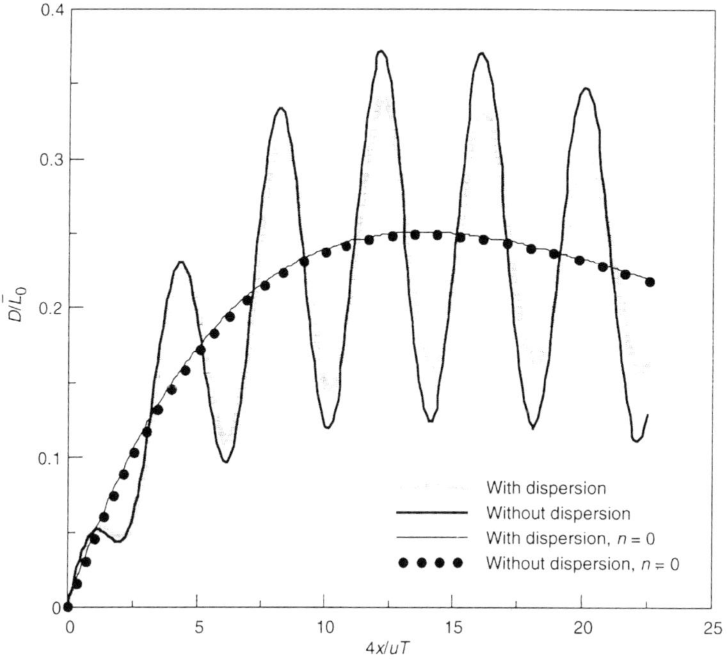

Fig. 12 The effect of longitudinal dispersion of DO distribution in a time-variable river

6. CONCLUDING REMARKS

BOD-DO variation in a receiving river has been investigated by following traditional physically based water quality models. A simple physically based river BOD-DO model was derived by Streeter and Phelps (1925) based on their Ohio River water pollution study. They treated a receiving river system as an ideal steady-state plug flow reactor. In so doing, the hydraulics of the system is represented entirely by travel time ($\theta = x/u$, where x is the downstream river distance and u is the river velocity) and the reaction kinetics of the system is represented by two coupled reactions of BOD deoxygenation and atmospheric reaeration.

The Streeter–Phelps model has been used by sanitary/environmental engineers for many years to evaluate the waste-assimilative capacity of a receiving river; it provides a basis on which the treatment requirement for wastewater discharged into the river can be determined. However, there has been demand for more sophisticated modeling tools since the Federal Water Pollution Control Act Amendment of 1972 (PL 92-500) was enacted by the U.S. Congress.

Because this legislation mandates that municipal and industrial wastewater dischargers build secondary or advanced treatment facilities (most of which involve construction costs in the millions or tens of millions of dollars), a slight change of effluent limitations would cause a huge difference in the construction cost. Therefore, a more precise evaluation of the receiving river's waste-assimilative capacity becomes necessary. In response to this new situation, more detailed receiving river models in terms of modified forms of the Streeter–Phelps equation have been developed.

The BOD-DO variation in a time-variable river system with a dispersive effect is traditionally solved by numerical methods. The application of popular numerical techniques such as finite difference and finite elements produces the problem of numerical dispersion and causes major uncertainties of modeling results (Fischer et al., 1979). Methods are available to control the problem of numerical dispersion, but complete elimination of the problem is not possible.

This paper presents an alternative river water quality modeling approach based on the linear systems theory. The system impulse response function of a linear system can be evaluated by using any of three methods: (1) system identification, (2) system parameterization, and (3) system physical parameterization. In this study, the method of physical parameterization is used. Using this method, the system impulse response function of a receiving river can be determined by solving the governing equation of the corresponding physically based model with a Dirac delta function input. This modeling approach is an extension of Green's function techniques that are applied to the solution of a linear differential equation.

This alternative modeling approach is demonstrated by applying it to two cases of river water quality analysis. In the first case, a linear systems model is developed for a steady-state, nondispersive river that receives both point- and nonpoint-source waste loadings. Results show that this model is an effective management tool for formulating and evaluating management alternatives.
In the second case, a linear systems model is developed for a time-variable dispersive river that receives point-source loading in the form of a periodic function. By applying the linear systems modeling approach, a solution can be readily obtained by numerical integration. Since no large system of simultaneous equations is involved, the problem of numerical dispersion can be alleviated. In addition, there is no need for large computer memory and computation time for data loading and storing.

A conjunctive application of the methods of system parameterization and physical parameterization has been successfully used in the study of solute transport in upper soils and groundwater contamination (Liu, 1988; Liu et al., 1991). Further work is needed to explore the potential of this conjunctive application to river water quality analysis. Both the system parameterization and physical parameterization methods allow an independent determination of the impulse response function of a receiving river system. By comparing the results obtained using these two methods, the following can be achieved: (1) a useful river water quality model can be developed, and (2) individual transport kinetics can be investigated more clearly.

7. ACKNOWLEDGMENTS

This work is derived from research supported by U.S. federal grants for the following projects: Compatibility of Physically-based and Linear System Solute Transport Modeling Approaches (Agreement no. 14-08-0001-G1489, 1987 to 1991, U.S. Geological Survey) and System Parameterization and Modeling of Chemical Transport Through Upper Soils (Agreement no. 14-08-0001-G2015, 1991 to 1992, U.S. Geological Survey). The contents of this chapter do not necessarily reflect the views and policies of the U.S. Geological Survey. This is a contributed paper CP-2002-01 of the Water Resources Research Center at the University of Hawaii at Manoa, Hawaii.

8. NOMENCLATURE

C	Concentration of a chemical
D	Concentration of DO deficit
D_0	Initial concentration of DO deficit in a river
D_s	Zero-input response of DO linear system
E	Longitudinal dispersion coefficient
E_x, E_y, E_z	Fisk's diffusion coefficients in x, y, and z directions, respectively
$f(x)$	One-dimensional known function of x, or a loading function
$f(x_1, \ldots, x_n)$	n-dimensional known function of x
f_D	Loading input for DO deficit
f_L	Loading input for BOD
$g(x)$	One-dimensional Green's function
$g(x_1, \ldots, x_n)$	n-dimensional Green's function
h	Impulse-input response function for a linear systems model
k_1	Laboratory deoxygenation rate constant
k_2	Reaeration rate constant
k_d	In-stream deoxygenation rate constant
L	Concentration of BOD
L_0	Point-source BOD input
$L_c(\theta)$	Distributed input loading to BOD distribution for the steady-state case

$\overline{L}_0$	Mean value of point-source BOD input at a discharge point
$L_w(t)$	Periodic point-source input to BOD distribution
$\boldsymbol{M}$	Ordinary/partial linear differential operator
r	Reaction rate
S_0	Distributed loading of DO due to benthic materials at the river bottom
t	Observation time
T	Period
t_0	Start of observation time
$u(x_1, \ldots, x_n)$	n-dimensional unknown function of x
V	Volume
u, v, w	Velocities in x, y, and z directions, respectively
w	Represents L or D
W	Width of a river
X	Input function for a linear system
x, y, z	Coordinates
β	Equal to $L/\overline{L}_0$
$\delta(x)$	One-dimensional Dirac delta function
$\delta(x_1, \ldots, x_n)$	n-dimensional Dirac delta function
ε	Equal to $4\ E_1/(u^2T)$ for BOD or $4\ E_2/(u^2T)$ for DO deficit
ε_t	Transverse mixing coefficient
θ	Travel time, defined as x/u
θ_0	Start of the travel time
θ_m	Mixing distance
ρ	Density
τ	Integration variable along t direction
τ^*	$= 4t/T$
τ'	$= t - \tau$
ζ	Integration variable along θ direction
ζ^*	$= 4x/(uT)$
ζ'	$= x - \zeta$

REFERENCES

Amorocho, J. (1963). Measures of the linearity of hydrologic systems. *J. Geophys. Res.,* **68(8)**, pp. 2237-2249.

Bennett, J.P. (1971). Convolution approach to the solution for the dissolved oxygen balance equation in a stream. *Water Resour. Res.*, **7(3)**, pp. 580-590.

Bras, R.I. and Rodriguez-Iturbe, I. (1985). *Random Functions and Hydrology*, Addison-Wesley, Reading, Massachusetts.

Brown, L.C. and Barnwell, T.O., Jr. (1987). The enhanced stream quality models QUAL2E and QUAL2E-UNCAS: Documentation and user manual. U.S. Environmental Protection Agency, EPA-600/3-87-007.

Churchill, M.A., Elmore, H.L. and Buckingham, R.A. (1962). The prediction of stream reaeration rates. *J. Sanitary Engineering Division* (ASCE), **88(SA4)**, pp. 1-46.

Distefeno, N. (1974). *Nonlinear Processes in Engineering*, Academic Press, New York.

Dooge, J.C. (1959). A general theory of the unit hydrography. *J. Geophys. Res.*, **64(2)**, pp. 241-255.

Dooge, J.C. (1973). Linear theory of hydrologic system. Agricultural Research Service, Technical Bulletin No. 148, U.S. Department of Agriculture, Washington, D.C.

Fischer, H.B., List, E.J., Koh, R.C.Y., Imberger, J. and Brooks, N.H. (1979). *Mixing in Inland and Coastal Waters.* Academic Press, 24/28 Oval Road, London NW1 7DX.

Friedman, B. 1990. *Principles and Techniques of Applied Mathematics*. Wiley, New York.

Jury, W. (1982). Simulation of solute transport using a transfer function. *J. Water Resour. Res.*, **18(2)**, pp. 363-368.

Li, W.H. (1972). Effects of dispersion on DO-sag in uniform flow. *J. Sanitary Engineering Division* (ASCE), **98(SA1)**, pp. 169-182.

Liu, C.C.K. (1982). Filtering of dissolved oxygen data in stream water quality analysis. *Water Resour. Bull.*, **18(1)**, pp. 15-20.

Liu, C.C.K. (1986). Surface water quality analysis, Chapter 1 in *Handbook of Environmental Engineering*, pp. 1-59. Humana Press, New Jersey.

Liu, C.C.K. (1988). Solute transport modeling in heterogeneous soils: Conjunctive application of physically based and the system approaches. *J. Contam. Hydrol.*, **3**, pp. 97-111.

Liu, C.C.K. and Brutsaert, W.H. (1978). A nonlinear analysis of the relationship between rainfall and runoff for extreme floods. *J. Water Resour. Res.*, **14(1)**, pp. 75-83.

Liu, C.C.K. and Fok, Y. (1983). Stream waste assimilative analysis using reaeration coefficients measured by tracer techniques. *Water Resour. Bull.*, **19(3)**, pp. 439-445.

Liu, C.C.K., Feng, J.S. and Chen, W. (1991). System modeling approach for solute transport through soils. *Wat. Sci. & Tech.*, **24(6)**, pp. 67-72.

Liu, C.C.K., Lin, P. and Firman, A. (1994). Modeling chemical transport in topsoil with a gamma distribution function. *Wat. Sci. & Tech.*, **30(2)**, pp. 121-129.

Liu, C.C.K. and Neill, J.J. (2000). Linear systems approach to river water quality analysis: Documentation and user manual. Proj. Rep. PR-2000-05, Water Resources Research Center, University of Hawaii at Manoa, Honolulu.

Nash, J.E. (1959). Systematic determination of unit hydrograph parameters. *J. Geophys. Res.*, **64(1)**, pp. 111-115.

O'Connor, D.J. and Dobbins, W.E. (1956). The mechanism of reaeration in natural streams. *J. Sanitary Engineering Division* (*ASCE*), **82(SA6)**, pp. 1115-1130.

O'Connor, D.J. and DiToro, D.M. (1970). Photosynthesis and oxygen balance in streams of reaeration in natural streams. *J. Sanitary Engineering Division* (*ASCE*), **96(SA2)**, pp. 547-571.

Rathbun, R.E. (1977). Reaeration coefficients of stream—State of the art, *J. Hydraulics Division* (*ASCE*), **103(HY4)**, pp. 409-424.

Rathbun, R.E., Shultz, D.J. and Stephens, D.W. (1975). Preliminary experiments with a modified tracer technique for measuring stream reaeration coefficients. U.S. Geological Survey Open-File Report No. 75-256, Bay St. Louis, Mississippi.

ReVelle, P. and ReVelle, C. (1988). *The Environmental Issues and Choices for Society*, 3rd ed. Jones and Bartlett Publishers, Boston.

Sokolnikoff, I.S. and Redheffer, R.M. (1966). *Mathematics of Physics and Modern Engineering*. McGraw-Book Book Co.

Sposito, G.I., White, R.E., Darrah, P.R. and Jury, W.A. (1986). A transfer function model of solute transport through soil, 3. The convection-dispersion equation. *Water Resour. Res.*, **22(2)**, pp. 255-262.

Streeter, H.W. and Phelps, E.B. (1925). Study of the pollution and natural purification of the Ohio River, Bulletin No. 146, U.S. Public Health Service, Washington, DC.

Tetra Tech, Inc. (1978). Rates, constants, and kinetics formulation in surface water quality modeling. EPA-600/3-78-105, U.S. Environmental Protection Agency, Athens, Georgia.

Thomann, R.V. (1973). Effect of longitudinal dispersion on dynamic water quality response of streams and rivers. *Water Resour. Res.*, **9(2)**, pp. 355-366.

Thomann, R.V. (1980). Measures of verification, in "Workshop on Verification of Water Quality Models," pp. 37-61. EPA-600/9-80-016, Athens, Georgia.

Thomann, R.V. and Mueller, J.A. (1987). *Principles of Surface Water Quality Modeling and Control*. Harper & Row, New York.

Tsivoglou, E.C., Cohen, J.B., Shearer, S.D. and Godsil, P.J. (1968). Tracer measurements of stream reaeration, II. Field studies. *J. Water Pollution Control Federation*, **40(2)**, pp. 285-305.

Tsivoglou, E.C., and Wallace, J.R. (1972). Characterization of stream reaeration capacity. EPA-R3-72-012, U.S. Environmental Protection Agency.
Viessman, W., Jr., and Hammer, M.J. (1998). *Water supply and pollution control*, Sixth ed. Addison-Wesley, Menlo Park, California.
Volterra, V. (1959). *Theory of Functionals and of Integral and Integro-differential Equations.* Dover, New York.
Wilson G.T. and McCleod, N. (1974). A critical appraisal of empirical equations and models for the prediction of the coefficient of reaeration of deoxygenated water. *Water Research*, **8**, pp. 341-366.

APPENDIX. MATHEMATICAL DERIVATION OF IMPULSE RESPONSE FUNCTION AND INPUT FUNCTIONS

Since the BOD and DO equations, or Eq. (9) and Eq. (10), are very similar, they can be represented by a generalized equation form as follows:

$$\frac{\partial w}{\partial t} - E\frac{\partial^2 w}{\partial x^2} + u\frac{\partial w}{\partial x} + Bw = f(x, t) \tag{A1}$$

where w represents either L or D. For BOD, the loading function f stands for any sort of organic material whose decomposition needs oxygen. For DO, it is any kind of loading which causes the variation of DO deficit in a river, including the loading of BOD, which is equal to $K_d\, L(x, t)$. E is the dispersion coefficient. For $L(x, t)$, $B = k_d$; for $D(x, t)$, $B = k_2$.

I. Impulse Response Function

To find the impulse response function of the system, we set the right side of the loading term $f(x, t) = \delta(x - \zeta, t - \tau)$. Then Eq. (A1) becomes

$$\frac{\partial w}{\partial t} - E\frac{\partial^2 w}{\partial x^2} + u\frac{\partial w}{\partial x} + Bw = \delta(x - \zeta, t - \tau). \tag{A2}$$

Let $x' = x - \zeta, t' = t - \tau$, then Eq. (A2) becomes

$$\frac{\partial w}{\partial t'} - E\frac{\partial^2 w}{\partial x'^2} + u\frac{\partial w}{\partial x'} + Bw = \delta(x', t'). \tag{A3}$$

Applying Laplace transformation with respect to t', Eq. (A3) becomes

$$s\hat{w} - w(x', t')\big|_{t'=0} - E\frac{\partial^2 \hat{w}}{\partial x'^2} + u\frac{\partial \hat{w}}{\partial x'} + B\hat{w} = \delta(x'). \tag{A4}$$

Note that $w(x', t')\big|_{t'=0} = 0$. Applying the Fourier transformation to Eq. (A4), we have

$$\left[-E(iw)^2 + u(iw) + (B+s)\right]F_{\hat{w}} = 1. \tag{A5}$$

The solution to Eq. (A5) is

$$F\hat{w} = \frac{C}{\sqrt{A^*}}\left[\frac{-1}{iw - \dfrac{u-\sqrt{A^*}}{2E}} - \frac{-1}{iw - \dfrac{u+\sqrt{A^*}}{2E}}\right] \tag{A6}$$

where $A^* = u^2 + 4E(B+s)$. The inverse Fourier transformation of Eq. (A6) is

$$\hat{w} = \frac{C}{\sqrt{A^*}}\exp\left(\frac{u-\sqrt{A^*}}{2E}x'\right) \qquad \text{for } x' > 0 \tag{A7}$$

and

$$\hat{w} = \frac{C}{\sqrt{A^*}}\exp\left(\frac{u+\sqrt{A^*}}{2E}x'\right) \qquad \text{for } x' < 0. \tag{A8}$$

The respective solutions to Eq. (A7) and Eq. (A8) with the inverse Laplace transformation is

$$w(x', t') = \frac{1}{\sqrt{4\pi Et'}}\exp\left[-\frac{(x'-ut')^2}{4Et'} - Bt'\right]. \tag{A9}$$

and

$$w(x, t) = \frac{1}{\sqrt{4\pi E(t-\tau)}}\exp\left\{-\frac{[(x-\zeta)-u(t-\tau)]^2}{4E(t-\tau)} - B(t-\tau)\right\}. \tag{A10}$$

The impulse-response function as a function of travel time and observation time is

$$w(\theta, t) = h(\theta, t) = \frac{1}{\sqrt{4\pi E(t-\tau)}}$$

$$\exp\left\{-\frac{u^2[(\theta-\zeta)-(t-\tau)]^2}{4E(t-\tau)} - B(t-\tau)\right\}. \tag{A11}$$

Thus the response function of BOD is

$$L(\theta, t) = h_{BOD}(\theta, t) = \frac{1}{\sqrt{4\pi E(t-\tau)}} \exp\left\{ -\frac{u^2\left[(\theta-\zeta)-(t-\tau)\right]^2}{4E(t-\tau)} - k_d(t-\tau) \right\} \tag{A12}$$

and the response function of DO is

$$D(\theta, t) = h_{DO}(\theta, t) = \frac{1}{\sqrt{4\pi E(t-\tau)}} \exp\left\{ -\frac{u^2\left[(\theta-\zeta)-(t-\tau)\right]^2}{4E(t-\tau)} - k_2(t-\tau) \right\}. \tag{A13}$$

In particular, if the BOD distribution is the only loading for DO, i.e., if $f(\theta, t) = k_d L(\theta, t)$, then the impulse response function for DO is

$$D(\theta, t) = h_{DO}(\theta, t) = \frac{k_d}{\sqrt{4\pi E(t-\tau)}} \exp\left\{ -\frac{u^2\left[(\theta-\zeta)-(t-\tau)\right]^2}{4E(t-\tau)} - k_2(t-\tau) \right\}. \tag{A14}$$

II. Input (Loading) Functions

A. Point-Source Input

Assuming that the initial value is zero everywhere and assuming that a single point-source loading at $x = x_0$ can be treated as a boundary condition, that is, $f(x, t) = 0$, then Eq. (A1) is reduced to

$$\frac{\partial w}{\partial t} - E\frac{\partial^2 w}{\partial x^2} + u\frac{\partial w}{\partial x} + Bw = 0. \tag{A15}$$

Performing Laplace transformation with respect to t, the equation becomes

$$-E\frac{\partial^2 \hat{w}}{\partial x^2} + u\frac{\partial \hat{w}}{\partial x} + (s+B)\hat{w} = 0. \tag{A16}$$

The solution to this equation is

$$\hat{w} = C_1 \exp\left(\frac{u + \sqrt{A^{**}}}{2E} x \right) + C_2 \exp\left(\frac{u - \sqrt{A^{**}}}{2E} x \right) \tag{A17}$$

where $A^{**} = u^2 + 4E(s + B)$, and C_1 and C_2 are arbitrary constants. Physically, w and therefore $\hat{w}$ are bound at $x \rightarrow +\infty$, so C_1 has to be zero. Then Eq. (A17) is reduced to

$$\hat{w} = C_2 \exp\left(\frac{u - \sqrt{A^{**}}}{2E} x \right) . \tag{A18}$$

Assuming that the boundary condition is the single point-source loading, or $w(x, t)\big|_{x = x_0} = f(t)$, then

$$\hat{w} = \hat{f}(s) \exp\left[\frac{u - \sqrt{A^{**}}}{2E} (x - x_0) \right] \tag{A19}$$

where $\hat{f}(s)$ is the Laplace transformation of $f(t)$. Performing the inverse Laplace transformation of the above equation results in

$$w(x, t) = \int_{t_0}^{t} f(\tau) \frac{x - x_0}{\sqrt{4\pi E (t - \tau)^3}} \exp\left\{ -\frac{\left[(x - x_0) - u(t - \tau) \right]^2}{4E(t - \tau)} - B(t - \tau) \right\} d\tau \tag{20}$$

or

$$w(\theta, t) = \int_{t_0}^{t} f(\tau) \frac{u(\theta - \theta_0)}{\sqrt{4\pi E (t - \tau)^3}} \exp\left\{ -\frac{u^2 \left[(\theta - \theta_0) - (t - \tau) \right]^2}{4E(t - \tau)} - B(t - \tau) \right\} d\tau . \tag{A21}$$

So, for BOD variation in a river receiving a single point-source BOD loading, the equation is

$$L(\theta, t) = \int_{t_0}^{t} L_0(\tau) \frac{u(\theta - \theta_0)}{\sqrt{4\pi E(t-\tau)^3}} \exp\left\{ -\frac{u^2\left[(\theta - \theta_0) - (t-\tau)\right]^2}{4E(t-\tau)} - k_d(t-\tau) \right\} d\tau. \tag{A22}$$

and for DO variation in a river receiving a single point-source BOD loading, the equation is

$$D(\theta, t) = \int_{t_0}^{t} D_0(\tau) \frac{u(\theta - \theta_0)}{\sqrt{4\pi E(t-\tau)^3}} \exp\left\{ -\frac{u^2\left[(\theta - \theta_0) - (t-\tau)\right]^2}{4E(t-\tau)} - k_2(t-\tau) \right\} d\tau. \tag{A23}$$

In particular, if BOD point-source loading is a constant, L_0, then BOD distribution is equal to

$$L(x, t) = L_0 \int_{t_0}^{t} \frac{x - x_0}{\sqrt{4\pi E(t-\tau)^3}} \exp\left\{ -\frac{\left[(x - x_0) - u(t-\tau)\right]^2}{4E(t-\tau)} - k_d(t-\tau) \right\} d\tau$$

$$= L_0 \left\{ \begin{array}{l} \exp\left[-\frac{x - x_0}{2E}\left(\sqrt{u^2 + 4Ek_d} - u\right)\right] \\ \operatorname{erfc}\left[\frac{(x - x_0) - (t - t_0)\sqrt{u^2 + 4Ek_d}}{\sqrt{4E(t - t_0)}} \right] \\ + \exp\left[\frac{x - x_0}{2E}\left(\sqrt{u^2 + 4Ek_d} + u\right)\right] \\ \operatorname{erfc}\left[\frac{(x - x_0) + (t - t_0)\sqrt{u^2 + 4Ek_d}}{\sqrt{4E(t - t_0)}} \right] \end{array} \right\} \tag{A24}$$

With this constant point-source BOD loading, DO distribution is equal to

$$D(x, t) = k_d L_0 \int_{t_0}^{t} \frac{x - x_0}{\sqrt{4\pi E(t-\tau)^3}} \exp\left\{ -\frac{\left[(x - x_0) - u(t-\tau)\right]^2}{4E(t-\tau)} - k_2(t-\tau) \right\} d\tau$$

$$= k_d L_0 \left\{ \begin{array}{l} \exp\left[-\frac{x - x_0}{2E}\left(\sqrt{u^2 + 4Ek_2} - u\right)\right] \\ \operatorname{erfc}\left[\frac{(x - x_0) - (t - t_0)\sqrt{u^2 + 4Ek_2}}{\sqrt{4E(t - t_0)}} \right] \\ + \exp\left[\frac{x - x_0}{2E}\left(\sqrt{u^2 + 4Ek_2} + u\right)\right] \\ \operatorname{erfc}\left[\frac{(x - x_0) + (t - t_0)\sqrt{u^2 + 4Ek_2}}{\sqrt{4E(t - t_0)}} \right] \end{array} \right\}. \quad \text{(A25)}$$

B. Distributed Loading

If there is distributed loading along the river, i.e., if $f(\theta) \neq 0$, then the result, based on the linear systems theory, should be

$$\begin{aligned} w(\theta, t) &= \int_{t_0}^{t} \int_{\theta_0}^{\theta} f(\zeta, \tau) h(\theta - \zeta, t - \tau) \\ &= \int_{t_0}^{t} \int_{\theta_0}^{\theta} f(\zeta, \tau) \frac{1}{\sqrt{4\pi E(t - \tau)}} \\ &\quad \exp\left\{ -\frac{u^2 [(\theta - \zeta) - (t - \tau)]^2}{4E(t - \tau)} - B(t - \tau) \right\} d\zeta d\tau . \end{aligned} \quad \text{(A26)}$$

So for BOD, a distributed loading of $f_d(\theta, t)$ results in

$$\begin{aligned} L(\theta, t) &= \int_{t_0}^{t} \int_{\theta_0}^{\theta} f_d(\zeta, \tau) h_{BOD}(\theta - \zeta, t - \tau) \\ &= \int_{t_0}^{t} \int_{\theta_0}^{\theta} f_d(\zeta, \tau) \frac{1}{\sqrt{4\pi E(t - \tau)}} \\ &\quad \exp\left\{ -\frac{u^2 [(\theta - \zeta) - (t - \tau)]^2}{4E(t - \tau)} - k_d(t - \tau) \right\} d\zeta d\tau \end{aligned} \quad \text{(27)}$$

and for DO, a distributed loading of $f_d(\theta, t)$ results in

$$D(\theta, t) = \int_{t_0}^{t} \int_{\theta_0}^{\theta} f_d(\zeta, \tau) h_{DO}(\theta - \zeta, t - \tau)$$

$$= \int_{t_0}^{t}\int_{\theta_0}^{\theta} f_d(\zeta,\tau)\frac{1}{\sqrt{4\pi E(t-\tau)}}$$

$$\exp\left\{-\frac{u^2\left[(\theta-\zeta)-(t-\tau)\right]^2}{4E(t-\tau)}-k_2(t-\tau)\right\}d\zeta d\tau. \tag{A28}$$

In particular, if $f_d(\theta, t) = k_d L(\theta, t)$ only, then the DO deficit is

$$D(\theta,t) = \int_{t_0}^{t}\int_{\theta_0}^{\theta} L(\zeta,\tau)h_{DO}(\theta-\zeta,t-\tau)$$

$$= \int_{t_0}^{t}\int_{\theta_0}^{\theta} L(\zeta,\tau)\frac{k_d}{\sqrt{4\pi E(t-\tau)}}$$

$$\exp\left\{-\frac{u^2\left[(\theta-\zeta)-(t-\tau)\right]^2}{4E(t-\tau)}-k_2(t-\tau)\right\}d\zeta d\tau. \tag{A29}$$

In particular, if the loading is constant and uniform, then for BOD, Eq. (A27) becomes

$$L(\theta,t) = \int_{t_0}^{t}\int_{\theta_0}^{\theta} L_d h_{BOD}(\theta-\zeta,t-\tau)$$

$$= \frac{2L_d}{u}\int_{t_0}^{t}\exp\left[-k_d(t-\tau)\right]\mathrm{erf}\left[\frac{u(\theta-\theta_0)-(t-\tau)}{\sqrt{4E(t-\tau)}}\right]d\tau \tag{30}$$

and for DO, Eq. (A28) becomes

$$D(\theta,t) = \int_{t_0}^{t}\int_{\theta_0}^{\theta} D_d h_{DO}(\theta-\zeta,t-\tau)$$

$$= \frac{2D_d}{u}\int_{t_0}^{t}\exp\left[-k_2(t-\tau)\right]\mathrm{erfc}\left[\frac{u(\theta-\theta_0)-(t-\tau)}{\sqrt{4E(t-\tau)}}\right]d\tau \tag{31}$$

when D_d represents constant and uniform DO loading.

Index